International Conference on Science and Engineering in Mathematics, Chemistry and Physics

(ScieTech 2014)

Journal of Physics: Conference Series Volume 495

Jakarta, Indonesia
13-14 January 2014

Editors:

Ford Lumban Gaol

Benfano Soewito
P. N. Gajjar

ISBN: 978-1-63266-720-5
ISSN: 1742-6588

Printed from e-media with permission by:

Curran Associates, Inc.
57 Morehouse Lane
Red Hook, NY 12571

Some format issues inherent in the e-media version may also appear in this print version.

Printed by Curran Associates, Inc. (2014)

For permission requests, please contact the Institute of Physics
at the address below.

Institute of Physics
Dirac House, Temple Back
Bristol BS1 6BE UK

Phone: 44 1 17 929 7481
Fax: 44 1 17 920 0979

techtracking@iop.org

Additional copies of this publication are available from:

Curran Associates, Inc.
57 Morehouse Lane
Red Hook, NY 12571 USA
Phone: 845-758-0400
Fax: 845-758-2634
Email: curran@proceedings.com
Web: www.proceedings.com

TABLE OF CONTENTS

Journal of Physics: Conference Series **495** (2014) 011001 doi:10.1088/1742-6596/495/1/011001

2014 International Conference on Science & Engineering in Mathematics, Chemistry and Physics (ScieTech 2014)

Media Hotel, Jakarta, Indonesia, 13 - 14 January 2014

Preface

2014 International Conference on Science Engineering in Mathematics, Chemistry and Physics (ScieTech 2014), was held at the Media Hotel, Jakarta, Indonesia, on 13–14 January 2014.

The ScieTech 2014 conference is aimed to bring together researchers, engineers and scientists in the domain of interest from around the world. ScieTech 2014 is placed on promoting interaction between the theoretical, experimental, and applied communities, so that a high level exchange is achieved in new and emerging areas within Mathematics, Chemistry and Physics.

We would like to express our sincere gratitude to all in the Technical Program Committee who have reviewed the papers and developed a very interesting Conference Program as well as the invited and plenary speakers.

This year, we received 187 papers and after rigorous review, 50 papers were accepted.

The participants come from 16 countries. There are 5 (Five) Paralell Sessions and Four Keynote Speakers.

It is an honour to present this volume of *Journal of Physics: Conference Series (JPCS)* and we deeply thank the authors for their enthusiastic and high-grade contributions. Finally, we would like to thank the conference chairmen, the members of the steering committee, the organizing committee, the organizing secretariat and the financial support from the conference sponsors that allowed the success of ScieTech 2014.

The Editors of the Scietech 2014 Proceedings:

Dr. Ford Lumban Gaol
Dr. Benfano Soewito
Dr. P.N. Gajjar

2014 International Conference on Science & Engineering in Mathematics, Chemistry and Physics (ScieTech 2014)

Media Hotel, Jakarta, Indonesia, 13 - 14 January 2014

Keynote Speaker

Keynote title:
Palladacycle Catalyzed Asymmetric P-H Addition Reactions

Speaker:
Professor Pak-Hing LEUNG
Nanyang Technological University

Abstract:
Chiral aza- and phospha-palladacycles are found to function as highly efficient catalysts for the asymmetric P-H addition reaction. The versatile protocol allows for the asymmetric hydrophosphination of olefenic C=C bonds thus yielding a family of tertiary C*-phosphines as well as C*P*-phosphines and diphosphines, depending on the nucleophile employed. The addition of two equivalents of HPPh2 to symmetrical bifunctionalized alkynes leading to generation of two new C* centers is also achievable. The air-sensitive nucleophiles and the unsaturated substrates containing unprotected functionalities such as aldeyhde, keto, ester, cyano and alcohol can be utilized directly under this mild and facile reaction conditions. The methodology is equally efficient when applied to the generation of P-N ligand systems via hydrophosphination of unsaturated pyridyl-based substrates as well as systems with C=N moieties. This synthetic strategy therefore offers to be a versatile approach for the generation of a wide range of chiral tertiary phosphine ligands with potential applications in further catalytic processes.

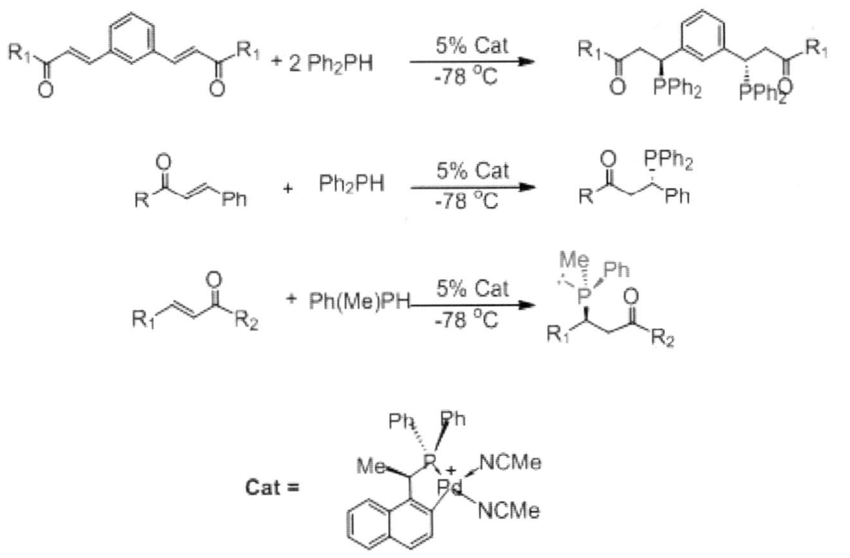

About the Speaker:
Professor Pak-Hing LEUNG (was born in Hong Kong (1956). Tertiary Education: Polytechnic of North London (BSc Hons 1982, FYP supervisor: Peter A Tasker), Australian National University (PhD 1986, supervisor: S. Bruce Wild), University of Toronto University of Chicago (posdoc fellow, 1987-1989, supervisor: Brice Bosnich). He was Professor and Deputy Head of the Department of Chemistry at the National University of Singapore till 2005 after which he was invited to be the founding Head of the Chemistry Division of the new School of Physical and Mathematical Sciences (SPMS) at the Nanyang Technological University, Singapore. He has served as Vice-Dean at SPMS and Head of the Division of Chemistry and Biological Chemistry. So far, he has received a University Researcher Award (NUS, 1998), 14 Teaching Awards (since 1993, both from NUS and NTU). He has published more than 140 peer-reviewed articles in the area of asymmetric synthesis, stereochemistry of phosphorus and organometallic compounds

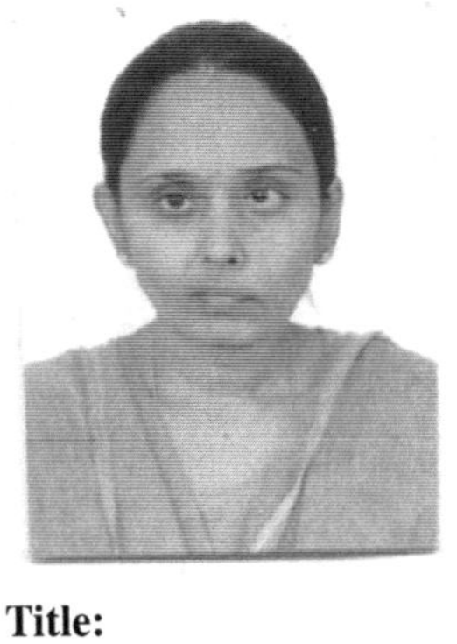

Title:
Mathematical Models of HIV/AIDS and Opportunist Diseases.

Speaker:
Prof. (Dr.) Nita H. Shah
Professor in Mathematics Department of Mathematics,

Gujarat University
Ahmedabad, Gujarat, India

Abstract:
Mathematical models help in understanding the population dynamics of HIV/AIDS. An epidemic is defined as an outbreak of a disease that infects a substantial portion of the community. In twentieth century, HIV/AIDS has created a havoc killing millions globally. Once infected, infective virus remains silent for 8 - 10 years, without any symptoms of HIV infection. This period is known as latent period. During this latent period, person's immune system and CD4+ cells get severely damaged by the virus resulting loss of fighting immunity against diseases. At this stage, person is diagnosed To have AIDS. Because of low immunity, patient catches malaria, tuberculosis opportunist diseases very easily. In this talk, aim is to develop transfer flow for the diseases in different compartment; viz. susceptible, exposed, infective etc. through differential equations. The basic reproduction number will be discussed. Using control theory, disease free equilibrium is derived which is locally asymptotically stable. Simulation study is discussed to predict the dynamics of HIV/AIDS with opportunist diseases.

About the Speaker:
Dr. Nita H. Shah is Professor of Mathematics in Department of Mathematics, University School of Sciences, Gujarat University, India. Her research interest is in Operations Research and in particular, Modelling real life problems. She hold PhD from Gujarat Universty and Post Dodctoral Fellow from University of New Bruncwick, Canada. She ahas published more than 5 books in the area of Mathematics.

Title:
Higher grading affine Toda solitons from quantum algebras.

Speaker:
Dr. Alexander Zuevsky
Senior Researcher, Czech Academy of Science, Prague, Czech Republic

Abstract:
We construct special types of quantum soliton solutions using quantum Lie algebras associated to affine Lie algebras

About the Speaker:
Dr Alexander hold PhD from Moscow institute of physics and technology in the area of Mathematical Physics. His research interest are Representation theory:, Automorphic Forms, Mathematical Physics. He is Senior Researcher, Czech Academy of Science, Prague, Czech Republic and member International Association of Mathematical Physics. He has Research Fellowships, Max-Planck Institut für Mathematik,Bonn and

Journal of Physics: Conference Series **495** (2014) 011003 doi:10.1088/1742-6596/495/1/011003

2014 International Conference on Science & Engineering in Mathematics, Chemistry and Physics (ScieTech 2014)

Media Hotel, Jakarta, Indonesia,
13 - 14 January 2014

2014

2014

Journal of Physics: Conference Series **495** (2014) 011003 doi:10.1088/1742-6596/495/1/011003

Journal of Physics: Conference Series **495** (2014) 011003 doi:10.1088/1742-6596/495/1/011003

Journal of Physics: Conference Series **495** (2014) 011003 doi:10.1088/1742-6596/495/1/011003

The Media Hotel, Jakarta, Indonesia,
13 - 14 January 2014

Journal of Physics: Conference Series **495** (2014) 011004 doi:10.1088/1742-6596/495/1/011004

Peer review statement

All papers published in this volume of *Journal of Physics: Conference Series* have been peer reviewed through processes administered by the proceedings Editors. Reviews were conducted by expert referees to the professional and scientific standards expected of a proceedings journal published by IOP Publishing.

Journal of Physics: Conference Series **495** (2014) 012001 doi:10.1088/1742-6596/495/1/012001

New View of Relativity Theory

Luiz Cesar Martini

Electrical and Computing Engineering Faculty
University of Campinas - Brazil

martini@decom.fee.unicamp.br

Abstract. This article results from *Introducing the Dimensional Continuous Space-Time Theory* that was published in reference 1. The Dimensional Continuous Space-Time Theory shows a series of facts relative to matter, energy, space and concludes that empty space is inelastic, absolutely stationary, motionless, perpetual, without possibility of deformation neither can it be destroyed or created. A elementary cell of empty space or a certain amount of empty space can be occupied by any quantity of energy or matter without any alteration or deformation. As a consequence of these properties and being a integral part of the theory, the principles of Relativity Theory must be changed to become simple and intuitive.

1. Introduction

The famous Lorentz transformations relating the coordinates of space and time that constitute the bases of relativity theory previse the space and time dilatation [3]. The time dilatation in fact exists but the space dilatation is not contemplated in this New View of the Relativity Theory.

The Lorentz transformations lead to a space contraction in front of a mobile object with v_0 speed and a space dilatation behind this same object. In the New View of Relativity Theory, the dilatation space effect does not exist. In the New View of the Relativity Theory it is not possible that the distances have some contraction or dilatation with the changes of the referential.
Also according to the New View of Relativity Theory there is no distinction between inertial and accelerated referentials.

In this New View of the Relativity Theory the empty space is the absolute referential which the origin point would be fixed in the center of the Big-Bang. However the existence of an absolute referential does not interfere in the Relativity Principle [4][5].

1.1. Postulates in the New View of the Relativity Theory

1.1.1. Postulates:

Generalization of the Einstein Postulate:
The Light Speed in the empty space is almost the same to the inertial and accelerated referentials.
Invariance space postulate:
The space is undeformable and always the same. The route in every referential is always the same.

Absolute Referential Postulate:
Inside the Universe there is an absolute referential. This referential is in all parts of the universe and is fixed inside the empty space which is totally stationary, invariant, perpetual and undeformable without the possibility of movement of its absolutely stationary position.

Any empty space point can be fixed as an absolute referential. A stationary origin of the absolute referential can certainly be found in the Big-Bang center. The Light speed and the length of the space set the time relations in every referentials.

2. Primary relation between the absolute referential and a referential moving with v_0 speed relative to the absolute referential

Considering the simplest case of a mobile object moving at v_0 velocity inside the empty space.

Fixing two referentials, the first one is absolute in the empty space and the other one is fixed at v_0 speed relative to the empty space.

To establish the time relations between these two referentials it is necessary to get a moving point at the speed of light. As this point is moving at the speed of light using the Einstein Relativity Principle its velocity is the same in all referentials and the point will be able to set the time relations between the two referencials.

To get a better comprehension suppose a point in the empty space. From this point two mobile objects will travel, the first one at v_0 speed and the second one at the speed of light c .

In the absolute referential the mobile object at the speed of light runs a length s given by $s = ct$; t is the time into the absolute referential. In this same absolute referential this point will be distant from the referential fixed at the v_0 speed by a distance given by $s_1 = ct - v_0 t$.

In the New View of the Relativity Theory, using the postulate of the invariance space into the referential fixed in v_0 speed the distance is the same as the absolute referential $s_1 = ct_0 = [c - v]t$. In this case t_0 is the time that the point at the light speed will be taking to move the distance s_1 in the referential fixed at the v_0 speed.

So the relation between t_0 and t will be given by $t_0 = \left(1 - \frac{v_0}{c}\right)t$.

3. Time instantaneity, Lorentz Factor and the independence of the absolute referential.

Considering two referentials, the first one at v_0 speed and the other one at $-v_0$ speed; both relative to the absolute referential.

To establish a time relation between the two referentials it must be considered a point moving at light speed during an interval time t into the absolute referential. Considering t_{0_1} the time spent by the object at light speed to run the s_1 length in the referential fixed at v_0, and considering t_{0_2} the time spent by the object at light speed to run the s_2 length in the referential fixed at $-v_0$,

$$s_1 = ct_{0_1} \tag{1}$$

$$s_2 = ct_{0_2} \tag{2}$$

Considering t the time spent in the absolute referential,

$$s_1 = ct - v_o t \tag{3}$$

$$s_2 = ct + v_0 t \tag{4}$$

Than

$$t_{0_1} = \left(1 - \frac{v_0}{c}\right) t \tag{5}$$

and

$$t_{0_2} = \left(1 + \frac{v_0}{c}\right) t \tag{6}$$

in the absolute referential there is time instantaneity but there is no instantaneity between t_{0_1} and t_{0_2}.

This situation occurs because s_1 is different than s_2, so t_{0_1} is different than t_{0_2}.

To make $t_{0_1} = t_{0_2}$, s_1 must be equal s_2. In this case the times in the absolute referential will be different from each other.

Making that $t_{0_1} = t_{0_2} = t_0$, than

$$t_0 = \left(1 - \frac{v_0}{c}\right) t_1 \tag{7}$$

and

$$t_0 = \left(1 + \frac{v_0}{c}\right) t_2 \tag{8}$$

in this case it is possible to obtain instantaneity between t_{0_1} and t_{0_2}, but not between t_1 and t_2. To obtain time instantaneity in every referentials, multiplies the equations:

$$t_{0_1} t_{0_2} = \left(1 - \left(\frac{v_0}{c}\right)^2\right) t^2 \tag{9}$$

and after

$$t_0^2 = \left(1 - \left(\frac{v_0}{c}\right)^2\right) t_1 t_2 \tag{10}$$

Getting the relation:

$$t_{0_1} t_{0_2} t_1 t_2 = t_0^2 t^2 \tag{11}$$

concluding:

$$t_0^2 = \left(1 - \left(\frac{v_0}{c}\right)^2\right) t^2 \tag{12}$$

The above equation must be the relation between time t_0 with v_0 module speed referential and time t in the absolute referential.

The must be no confusion between t_{0_1} or t_{0_2} with t_0 and t_1 or t_2 with t.

$t_{0_1}, t_{0_2}, t_1, t_2$ are times spent by an object at light speed to run certain distances according to its own referentials.

On the other hand t and t_0 are the times in its own referentials. The explanation of this relation between t and t_0 is justified in the Dimensional Continuous Space-Time Theory because the time is transverse to the mobile movement and it is caused by the relation between time and space.

3.1. The Relativity Principle is satisfied and it is not necessary to consider the absolute referential.
The absolute referential is only necessary to find the time of the Universe and not the matter time relatively to its own referential. The Universe time is the empty space time and it is the one that runs faster than times relative to the others referentials.

Considering one referential fixed at v_0 speed and other one at v speed both relative to absolute referential.
On the base of the referentials there will be

$$t_0 = \sqrt{\left(1 - \left(\frac{v_0}{c}\right)^2\right)} \cdot t \tag{13}$$

and

$$t_v = \sqrt{\left(1 - \left(\frac{v}{c}\right)^2\right)} \cdot t \tag{14}$$

Considering the referential fixed on the v_0 speed. The question is how to determinate the v' speed of the mobile object at v speed relative to the referential fixed at v_0.

To solve this question it is necessary to find the lengths

$$s_1 = vt - v_0 t \tag{15}$$

$$s_2 = vt + v_0 t \tag{16}$$

multiplying the two equations:

$$s_1 s_2 = \left(v^2 - v_0^2\right) t^2 \tag{17}$$

doing $s_1 s_2 = s^2$ and $s = v' t_0$.

The length will be:

$$s = \sqrt{\left(v^2 - v_0^2\right)} \cdot t \tag{18}$$

So v must satisfy

$$s = v' t_0 = v' \sqrt{1 - \left(\frac{v_0}{c}\right)^2} \cdot t \tag{19}$$

so

$$v' = \frac{\sqrt{v^2 - v_0^2}}{\sqrt{1 - \left(\frac{v_0}{c}\right)^2}} \tag{20}$$

Therefore the two equations below must be satisfied

$$tv = \frac{\sqrt{1 - \left(\frac{v}{c}\right)^2}}{\sqrt{1 - \left(\frac{v_0}{c}\right)^2}} \cdot t_0 \tag{21}$$

and

$$t_v = \sqrt{1 - \left(\frac{v'}{c}\right)^2} \cdot t_0 \tag{22}$$

From the last equation it easy to show the equality between equation 21 and that

$$t_v = \sqrt{1 - \frac{\left(\frac{v^2 - v_0^2}{c}\right)^2}{1 - \left(\frac{v_0}{c}\right)}} \cdot t_0 \tag{23}$$

4. Solution of the electromagnetism laws in all inertial referentials

In the 13th chapter of the reference 2 is presented a equivalence solution of the electrostatic and electromagnetic forces between inertial referentials using the Special Relativity Theory.

In the Feynman Book [2] there is an q elementary moving charge running to the right with v_0 speed in parallel with a conducting wire with negative charges moving to the right however with v speed. So the conventional electric current charges is moving with $-v$ speed to left and the distance from the q charge to the wire is r .

In the Feynman book shown that there is no electrostatic force in the fixed referential on the stationary wire, but there is a magnetic force in action on the q charge

$$F = \left(\frac{1}{4\pi\varepsilon_0 c^2}\right)\frac{2qA\rho_-}{r} v v_0 \tag{24}$$

where $\mu_0 = \varepsilon_0 c^2$.

Feynman analyzed the particular case when $v_0 = v$ obtaining

$$F = \left(\frac{q}{2\pi\varepsilon_0}\right)\left(\frac{A\rho_-}{r}\right)\frac{v^2}{c^2} \tag{25}$$

It is known that $F = q v_0 B$, where B is

$$B = \left(\frac{1}{4\pi\varepsilon_0 c^2}\right)\frac{2Av}{r}\rho_- \tag{26}$$

In the referential fixed in the stationary wire there is no electrostatic field because the negative moving charges inside the wire are neutralized by the positive charges of the atomic structure of the wire.

For the referential fixed on the q charge there Will be no magnetic field because in this referential the charge speed is $v' = 0$ so $F = q v' B = 0$.

On the other hand, using the relativity theory the Feynman book shows that there will appear the electrostatic force given by

$$F_0 = F' = \frac{1}{\sqrt{1 - \frac{v^2}{c^2}}} \cdot F \tag{27}$$

where F is the magnetic force when the referential is the stationary wire.
The New View of the Relativity Theory will show that

$$F_0 = F \tag{28}$$

The Lorentz Factor does not appear in the above equation. It Will be demonstrated that above result between forces is also valid when v_0 is different than v .

These results have great importance because the New View of the Relativity Theory changes some ideas about the Relativity Theory e.g. the length of a way is not elastic and it remains the same independently of the referential, only the elapsed time changes with the referential.

The relation between the t time fixed on the stationary wire and the time t_0 fixed on the v_0 speed of the moving charge q is controlled by the light speed and the length way.

4.1. *Developing computation using de new view of Relativity Theory.*
Analyzing the same situation in the Feynman book using now de New View of the Relativity Theory. The t time and the s_q length course traveled by the q charge are related by $s_q = v_0 t$.

The t_0 time fixed on the moving q charge can be found using a point traveling at c light speed and the s length way.

In the fixed wire referential the s length way between the point at light speed from the point at v_0 speed is $s = ct - v_0 t$.

The same space s in the q charge referential is $s = ct_0$ and it will be found that $t_0 = \left(1 - \frac{v_0}{c}\right)t$

On the other hand there is no point traveling at c speed.
There are only positive and negative charges in the wire, and the charge q outside the wire, traveling in parallel with the wire.
The common point in both referentials is the q charge.
It will be necessary to fix a point at c velocity above the q charge. So the t' time in the stationary wire referential to a point traveling at light speed on the q charge has a length s :

$$s = ct' \tag{29}$$

the t time in the same stationary referential wire is corresponding to the q charge at v_0 speed is

$$s = v_0 t \tag{30}$$

than

$$t' = \frac{v_0}{c} t \tag{31}$$

Now considering the charges with v speed inside the wire and the same wire referential. The distance between a point at c light speed and a point fixed in a charge inside the wire at v speed is $s' = ct' - vt'$.
The same length in the fixed v_0 referential is $s' = ct_0$ than

$$t_0 = \left(1 - \frac{v}{c}\right)t' = \left(1 - \frac{v}{c}\right)\left(\frac{v_0}{c}\right)t \tag{32}$$

Now considering the positive charges inside the wire and at $-v_0$ speed relative to the q charge moving with c speed.

So the space length will be $s_1 = ct' = c\left(\frac{v_0}{c}\right)t = v_0 t$.

To the positive charges relative to the q charge referential it is done $s_1 = ct_{0_1}$. Then

$s_1 = ct' = c\left(\frac{v_0}{c}\right)t$ then $t_{0_1} = \left(\frac{v_0}{c}\right)t$, t_{0_1} is the elapsed time of the moving positive charges relative to the q charge.

Now all it is prepared to compute the calculations that results on the force F_0 .

In two referentials the positive charges and negative charges will be conserved.

To the stationary wire referential the current of negative charges is $i_e = \frac{dq_e}{dt} = (\rho_-)A\frac{dl}{dt}$ and the current of positive charges is $i_p = 0$.

Into the fixed q charge referential there will be $i_{e_0} = \frac{dq_{e_0}}{dt_0}$.

As the area and the length are the same in both referentials there will be

$$i_{e_0} = (\rho'_-)\left[\frac{1}{\left(1-\frac{v}{c}\right)\left(\frac{v_0}{c}\right)}\right]A\frac{dl}{dt} \tag{33}$$

The charges and its variations must be preserved because compressing or expanding the time more or less charges goes through the cross section but the variation charges remains the same. $\frac{dq}{dt} = \frac{dq'}{dt'}$.

It means that incremental charges must be different dq_{e_0} is different than dq_e

$$dq_0 = \left(1-\frac{v}{c}\right)\left(\frac{v_0}{c}\right)dq \tag{34}$$

So the negative density of the charges in the v_0 referential is

$$\rho'_- = \left(1-\frac{v}{c}\right)\left(\frac{v_0}{c}\right)\rho_- \tag{35}$$

The positive charges relative to the wire referential will be resting and its time variation is zero. Relative to the fixed referential on the q charge, the positives charges are moving. This is a convection current.

In this case will be getting

$$i_{p_0} = \frac{dq_{p_0}}{dt_{0_1}} = \frac{d(A\cdot l\cdot\rho'_+)}{dt_{0_1}} \tag{36}$$

where

$$dt_{0_1} = \frac{v_0}{c}dt \tag{37}$$

as A and l are the same

$$i_{p_0} = \rho'_+\left(\frac{1}{v_0/c}\right)A\cdot\frac{dl}{dt} \tag{38}$$

Considering the convection current equal to ip current $\rho'_+ = \rho_-\left(\frac{v_0}{c}\right)$

The charges contribution on the fixed q referential will be $dq_0 = dq_{p_0} + dq_{e_0}$ that results

$$dq_0 = -(\rho_-)\left(\frac{v\,v_0}{c^2}\right) = (\rho_+)\left(\frac{v\,v_0}{c^2}\right) \tag{39}$$

The electrostatic force F_0 that exist in the fixed q referential now can be computed by the same procedure used in this chapter

$$F_0 = 2kq(\rho_+)A\left(\frac{1}{r}\right)\left(\frac{v\,v_0}{c^2}\right) \tag{40}$$

where $k = \dfrac{1}{4\pi\varepsilon_0}$.

That is

$$F_0 = F \tag{41}$$

If the q charge is negative the force will be an attraction force. If the negative charges inside the wire goes to the left it can be shown that the force is a repulsion force.

5. Accelerated referential

Considering the same case in the Feynman book, but the charge q will be with a_0 constant acceleration: $v_0 = a_0 t$ running to the right.

The distance between the charge and the conduction wire is r.

The negatives charges inside the wire has v speed to the right.

In the fixed wire referential the magnetic force will be

$$F = \left(\frac{k}{c^2}\right)2Aq(\rho_-)\left(\frac{v\,a_0\,t}{r}\right) \tag{42}$$

considering A is the cross section area of the wire.

The space length will be $s = \dfrac{a_0}{2}t^2$, to the fixed referential at v_0 speed, it will be fixed again the point at c speed on the moving q charge. So, to the fixed q charge referential the same considered point with c speed $ct' = s$, considering that t' is the time fixed on the stationary wire referential $t' = \dfrac{a_0}{2c}t^2$ then $ct_0 = ct' - vt'$.

Considering that t_0 is the time on the referential fixed in the $v_0 = a_0 \cdot t$ speed of the q charge

$$ct_0 = c\left(\frac{a_0}{2c}\right)t^2 - v\left(\frac{a_0}{2c}\right)t^2 \tag{43}$$

so

$$t_0 = \left(\frac{a_0}{2c}\right)t^2 - \left(\frac{v\,a_0}{2c^2}\right)t^2 \tag{44}$$

The electric current will be $i = \dfrac{dq_e}{dt}$ but

$$dt_0 = \left(\frac{a_0}{2c}\right)dt - \left(\frac{a_0 v}{c^2}\right)dt \tag{45}$$

So the densities positive and negative of the charges will be

$$\rho'_+ = \left(\frac{a_0\,t}{c}\right)\rho_+ \tag{46}$$

and

$$\rho'_{-} = \left[\left(\frac{a_0 t}{c}\right) - \left(\frac{v a_0 t}{c^2}\right)\right]\rho_{-} \tag{47}$$

then

$$\rho' = \rho'_{+} + \rho'_{-} \tag{48}$$

that results

$$\rho' = \left(\frac{v a_0 t}{c^2}\right)\rho_{+} \tag{49}$$

so it will get that

$$F_0 = k\left(\frac{2 A q (\rho_{-})}{r}\right)\left(\frac{v a_0 t}{c^2}\right) = F \tag{50}$$

6. Conclusion

Based in the Dimensional Continuous Space-Time Theory that shows the empty space is invariant, this necessary New View of The Relativity Theory is presented.

In this article there were many conclusions:

It was shown that the New View of The Relativity Theory the time elasticity is conserved but the empty space elasticity is not conserved resulting the invalidation of the Lorentz transformations.

Based in the length conservation and the light speed conservation it was demonstrated that the electromagnetism laws are conserved in all referentials including the accelerated ones.

It was shown that the electrostatic force F_0 in the q charge referential has the same intensity of the magnetic force in the stationary wire referential and is possible to find F_0 when v is different from v_0.

Finally it was demonstrated that the New View of The Relativity Theory is valid for inertial and accelerated referentials.

7. References

[1] Martini L., 2013, Introducing the Dimensional Continuous Space-Time Theory, *Journal of Physics: Conference Series*, Volume **423**, conference 1, 2013,Published online: 10 April, 2013, http://iopscience.iop.org/1742-6596/423/1

[2] Feyman R., 1977, *The Feynman Lectures on Physics*, Addison-Wesley, Vol II, Capt 13.

[3] Rainich G.Y., 1925, Electrodynamics in the General Relativity Theory, Transactions of the american Mathematical Society, vol. 27, No. 1, pp 106-136.

[4] Stewart J., 1993, *Advanced General Relativity*, Cambridge University Press.

[5] Einstein A, 1916, *Relativity: The Special and General Theory*, Methuen & Co Ltd, Translated: Lawson R., Transcription: Basgen B., Einstein Reference Archive, 2002.

Journal of Physics: Conference Series **495** (2014) 012002 doi:10.1088/1742-6596/495/1/012002

The Space-Time Model According to Dimensional Continuous Space-Time Theory

Luiz Cesar Martini

Electrical and Computing Engineering Faculty
University of Campinas - Brazil

martini@decom.fee.unicamp.br

Abstract. This article results from the Dimensional Continuous Space-Time Theory for which the introductory theoretician was presented in [1]. A theoretical model of the Continuous Space-Time is presented. The wave equation of time into absolutely stationary empty space referential will be described in detail. The complex time, that is the time fixed on the infinite phase time speed referential, is deduced from the New View of Relativity Theory that is being submitted simultaneously with this article in this congress. Finally considering the inseparable Space-Time is presented the duality equation wave-particle.

1. Introduction

In the Dimensional Continuous Space-Time Theory, time is not the fourth dimension but it is an integrant part of the space and it is a dynamic energy of the space.

So the time is the dynamic energy of the empty space responsible to provide movement energy for everything that has movement into the Universe.

Therefore the movement of everything is only possible thanks to the Energy provided by Space-Time. For example, everything that is above the absolute zero temperature is in a vibratory movement, our brain is thinking because has electrical movement, so, without time energy there is no electric movement, no vibratory movement and no temperature above zero Kelvin.

Space is the physical environment of the movement and time is the fuel of the movement. Space and time are inseparable. Time is dynamic and space is stationary. The dynamic time energy and the static energy of the empty space have the same intensity because each elementary axis of the potential empty space energy is occupied by a dynamic elementary time energy axis.

So, the time has elasticity and movement but the empty space does not have elasticity and mobility.

In [1] was deduced that the elementary axis of potential energy inside the matter and energy structure has a component that depends on the time given by

$$\frac{\alpha}{2}\cos\left(\omega_1 t_s + \frac{\pi}{4}\right) + \frac{\alpha}{2} \tag{1}$$

were α is a mathematic dimensional length given by

$$\alpha = \frac{1}{\pi} \tag{2}$$

and t_s is a subjective time that is increasing, differently from the real time t that oscillates like a pendulum between $-\frac{\pi}{c} \leq t \leq \frac{\pi}{c}$, w_l is the angular frequency of the time. In [1] was concluded that in the complete stationary state $\omega_1 = 1$.

However in this article it will be proved that mater in movement with v speed the angular frequency ω_1 is equal to 1 over the Lorentz Factor.

$$\omega_1 = \sqrt{1-\left(\frac{v}{c}\right)^2}$$

In [1] was deduced the primitive equation of the Continuous Space-Time given by

$$E(D) = \alpha e^{2\alpha_0 (D+\alpha)} e^{jt_s} e^{\alpha_0 f(t_s)} \tag{3}$$

So, in the reference [1] was presented the primitive equation of the continuous Space-Time in its derivative form that in the ref.[1] is the equation 3.23 and is presented below as the equations of the continuous space-time before the Big-Bang in the derivative static primitive form:

$$E(D) = \omega_1 \alpha e^{2\alpha_0 (D+\alpha)} e^{j\omega_1 t_s} e^{\alpha_0 f'(t_{ss}, t_{sr})} \tag{4}$$

where

$f'\{t_{ss}, t_{sr}\}$ is one dynamic derivation of a stat condition in the time function $f(t_s)$ and

$$f'(t_{ss}, t_{sr}) = -\frac{\alpha}{2} + \frac{\alpha}{2} = 0$$

where

$\omega_1 = 1$ for the static condition

$\omega_1 = \sqrt{1-\left(\frac{v}{c}\right)^2}$ for the general case with v speed.

the space constant is α_0 given by

$$\alpha_0 = \frac{1}{2}\ln(2) \tag{5}$$

the time function is $f(t_s)$ which t_s is a subjective time that increase indefinitely than

$$f(t_s) = -\frac{\alpha}{2}\cos\left(\omega_1 t_s + \frac{\pi}{4}\right) \tag{6}$$

the equations of the continuous space-time before the Big-Bang in the integrative static primitive form is.

$$E(D) = \frac{1}{\omega_1} \alpha e^{2\alpha_0 (D+\alpha)} e^{j\omega_1 t_s} e^{\alpha_0 f'\{t_{ss}, t_{sr}\}} \tag{7}$$

The function time $f(t_s)$ belongs to elementary cell of potential energy that produces mass, charge end energy and $f(t_s)$ is synchronized with time equation that is responsible by the stability of matter and energy.

2. Modelling the Space-Time

The primitive equation of the continuous Space-Time can be separated in two parts that are totally independents from each other

$$E(D) = \alpha TS \tag{8}$$

Where T is the time equation of the elementary cells given by

$$T = e^{\alpha_0 (D+\alpha)} e^{jt_s} \tag{9}$$

and S is the Space equation of the elementary cells given by

$$S = e^{\alpha_0 (D+\alpha)} e^{\alpha_0 f(t_s)} \tag{10}$$

to the elementary cells of potential energy D is equal to D_0 given by

$$D = D_0 = 56 \tag{11}$$

Both the time and space equations have infinite dimensional axis of zero width and a specific defined length. Their elementary cells of continuous space-time are cubic shaped.

3. Deduction of the angular frequency ω₁ to the matter in movement.

To the stationary matter in the empty space the time function of the matter and energy is

$$f(t_s) = -\frac{\alpha}{2}\cos\left(\omega_1 t_s + \frac{\pi}{4}\right), \text{ where } \omega_1 = 1$$

This time function receive energy from the time wave equation of the absolutely stationary empty space given by

$$t(t_s) = \alpha \cos(\omega_1 t_s) + \alpha, \text{ where } \omega_1 = 1 \tag{12}$$

If the matter is in movement at v speed, ω_1 will change.

The ω_1 angular frequency for the matter in movement is deduced from the New View of the Relativity Theory which space is not elastic.

Consequently the length λ_* of the time function to the stationary matter is the same to the matter in movement.

To the stationary matter

$$\lambda_* = 2\pi \frac{c}{\omega_1} = 2\pi c \tag{13}$$

λ_* keep on the same to the referential fixed on the matter in movement because in the New View of the Relativity Theory the space is not elastic.

Considering the fixed time t in the absolutely stationary empty space and t' the time in the referential fixed on the matter with v speed.

The relationship between t and t' is.

$$t' = t\sqrt{1-\left(\frac{v}{c}\right)^2} \tag{14}$$

If is done

$$\lambda_* = ct' \tag{15}$$

In the stationary referential the wave length is

$$\lambda_1 = ct \tag{16}$$

so

$$\lambda_* = ct' = ct\sqrt{1-\left(\frac{v}{c}\right)^2}$$

$$\lambda_1 = \frac{\lambda_*}{\sqrt{1-\left(\frac{v}{c}\right)^2}}$$

than

$$\omega_1 = 2\pi\frac{c}{\lambda_1}$$

$$\omega_1 = \sqrt{1-\left(\frac{v}{c}\right)^2} \tag{17}$$

So when $v = 0$ it has $\omega_1 = 1$

When $v = c$ it has $\omega_1 = 0$

In this case the time no more transfer movement energy to the mater, the time stops, the matter do not present relative movement and the temperature goes to zero Kelvin. The matter do not became older.

4. The complex time fixed in the infinite phase speed

Considering t_s the subjective time fixed in the referential stationary space and t'_s the subjective time fixed on the infinite phase speed of time.

According the New View of the Relativity Theory, given a t_1 time fixed in the stationary empty space relatively to s_1 course given by

$$s_1 = v_f t_1 \tag{18}$$

Now is necessary to find a point with c velocity to use the Light Speed Invariance Postulate, so

$$s_1 = v_f\, t_1 = c\, t_s \tag{19}$$

or

$$t_1 = \frac{c}{v_f} t_s \tag{20}$$

The relation between t'_s and t_1 is

$$t'_s = \sqrt{1-\left(\frac{v_f}{c}\right)^2}\; t_1 \tag{21}$$

As v_f is infinite

$$t'_s = j\frac{v_f}{c} t_1$$

it is found

$$t_s' = jt_s \tag{22}$$

Than the time T_s' fixed on the infinity phase speed of time has the same module as the time t_s in the absolute stationary empty space, but is complex.
So the time equation that is fixed on the infinite phase speed of time is

$$T = e^{\alpha_0(D+\alpha)} e^{jt_s} \tag{23}$$

Of which the number of dimensional axis of the time is $D = \infty$

the space constant is $\alpha_0 = \frac{1}{2}\ln(2)$ and the dimensional mathematic length is $\alpha = \frac{1}{\pi}$

The time equation is responsible to maintain stability to the matter and energy and provides movement to them.

5. The frequency spectrum of the time

The time at stationary empty space in an infinitesimal volume is represented by wave equation

$$t(t_s) = \alpha\cos(t_s) + \alpha \tag{24}$$

This wave equation is the same for all infinitesimal volumetric points that constitute the empty space. However the time only exist at the present moment, so the wave equation is rigorously describe by

$$t(t_s) = \left[\alpha\cos(t_s) + \alpha\right] i\left(t_s - t_p\right) \tag{25}$$

where t_p is the present time

$$\begin{aligned} i(t_s - t_p) &= 1 \text{ to } t_s = t_p \\ i(t_s - t_p) &= 0 \text{ to } t_s \neq t_p \end{aligned} \tag{26}$$

The present time phase is $\alpha\cos(t_p) + \alpha$, and the frequency spectrum of $t(t_s)$ is $X(f) = \alpha\cos(t_p) + \alpha$

The frequency spectrum is infinite, all frequencies with the same amplitude $\alpha\cos(t_p) + \alpha$ that depends on only the present time t_p.

To demonstrate the above affirmation substitute

$$i(t_s - t_p) = \lim_{\varepsilon \to 0}\left[\operatorname{ret}\left(\frac{t_s - t_p}{\varepsilon}\right)\right] \tag{27}$$

where

$$\operatorname{ret}\left(\frac{t_s - t_p}{\varepsilon}\right) = \begin{cases} 1; & \text{to } -\frac{\varepsilon}{2} < t_s - t_p < \frac{\varepsilon}{2} \\ 0; & \text{to others values} \end{cases} \tag{28}$$

ε is the volumetric infinitesimal dominium of the present time
So

$$t(t_s) = \left[\alpha\cos\left(t_s\right) + \alpha\right] \lim_{\varepsilon \to 0}\left[\operatorname{ret}\left(\frac{t_s - t_p}{\varepsilon}\right)\right]$$

The above wave equation is representing an infinitesimal space volumetric point.

On the other hand the time has infinite dimensional energy axis in a limited volume. This non specified limited volume has 2^D dimensional axis of zero width, where D is infinite.

Considering the wave equation in a unitary volumetric elementary cell. This cell has $1/\varepsilon$ volumetric infinitesimal points of wave equations.

The volumetric space incremental ε is the same ε incremental time because space and time constitute the same unit.

So in an elementary cell has

$$t(t_s) = \left[\alpha \cos\left(t_p\right) + \alpha\right] \lim_{\varepsilon \to 0} \left[\frac{1}{\varepsilon} \text{ret}\left(\frac{t_s - t_p}{\varepsilon}\right)\right]$$

Where $\lim_{\varepsilon \to 0} \left[\frac{1}{\varepsilon} \text{ret}\left(\frac{t_s - t_p}{\varepsilon}\right)\right]$ is the unitary impulse function at the present time t_p, $\delta(t_s - t_p)$ produce a frequency spectrum $G(f) = e^{-j2\pi f t_p}$

But as there is only the present time, with present time is a new time origin and, in this case there is no delay time to consider so

$$t(t_s) = \left[\alpha \cos(t_p) + \alpha\right] \delta(t_s - t_p) \tag{29}$$

will produce the frequency spectrum for the volumetric cell of time

$$X(f) = \left[\alpha \cos(t_p) + \alpha\right] \tag{30}$$

Where the time delay is inconsiderate because there is only present time t_p, so each instant t_p is a new origin of time.

It means that the time has infinite frequency components with same amplitude but this amplitude change with the present time t_p.

6. The inseparable Space-Time

The infinite speed phase of time has a complex time equation T and it is simultaneously present in all space of the Universe.

$$T = e^{\alpha_0 (D+\alpha)} e^{j t_s} \tag{31}$$

The empty space otherwise is stationary and reflex of the time and has a complex equation of antispace given by

$$S_t = e^{\alpha_0 (D+\alpha)} e^{-j t_s} \tag{32}$$

So the Space-Time equation is the combination of this two equations.
So T is to be replaced by

$$T_0 = T + S_t \tag{33}$$

$$T_0 = e^{\alpha_0 (D+\alpha)} \left[e^{j t_s} + e^{-j t_s}\right]$$

In this case the result would be

$$T_0 = e^{\alpha_0 (D+\alpha)} 2\cos(t_s)$$

On the other hand t_s is the subjective time that increase indefinitely. The real time t oscillates as a pendulum between $-\frac{\pi}{c} \le t \le \frac{\pi}{c}$

With the above constitution it can be proved that

$$e^{j\omega_1 t_s} \text{ is replaced by } \cos(\omega_1 c t_s) \cdot i(\omega_1 c t_s - n\pi) + j\sin(\omega_1 c t_s) \tag{34}$$

$$e^{-j\omega_1 t_s} \text{ is replaced by } \cos(\omega_1 c t_s) \cdot i(\omega_1 c t_s - n\pi) - j\sin(\omega_1 c t_s) \tag{35}$$

where n is a integer number

And the results should be

$T_0 = e^{\alpha_0(D+\alpha)} 2\cos(\omega_1 c t_s)\cdot i(\omega_1 c t_s - n\pi) + 0j$

where

$i(\omega_1 c t_s - n\pi) = 1$ to $\omega_1 c t_s = n\pi$ and zero other wise

Other condition is that the empty space is not elastic. This condition will change the result above because the elastic equation $e^{-j\omega_1 t_s}$ has his components in cosine and sine as grounded referential and this results in fact that

$$T_0 = \cos(\omega_1 c t_s) i(\omega_1 c t_s - n\pi) + \sin(\omega_1 c t_s) \tag{36}$$

So the resultant particle mass from the fundamental equation of the continuous space time [1] is

$$m = \left(\frac{1}{\omega_1}\right) m_0 \left[\cos(\omega_1 c t_s) i(\omega_1 c t_s - n\pi) + \sin(\omega_1 c t_s)\right] \tag{37}$$

where m_0 is resting mass of the particle.

In the above equation appears the duality-wave-particle.

Some results of this theory:

$N = 7$ Number of atomics Shells

$RT1 = \frac{1}{2\pi\sqrt{2}} - \frac{1}{4\pi\sqrt{2}}$ Partial reduction of the time function

$RT2 = \frac{1}{2\pi\sqrt{2}}$ Complete dimensional reduction of the time function

$M(0) = \frac{1}{2}\left(e^{\alpha_0 \frac{\alpha}{2}} + e^{-\alpha_0 \frac{\alpha}{2}}\right)$ Dimension alfactor of correction

$Z(0) = \frac{1}{2}\left(M(0) + \frac{1}{M(0)}\right)$ Dimensional factor correction

where the factor $M(0)$ is associated to the space and the $\frac{1}{M(0)}$ factor is associated to the anti-space.

The factor $Z(0)$ is associated to the conjugated actuation between space and anti space.

Light speed into the empty space

$$c = e^{\alpha_0 (D_0 + \frac{1}{\pi})} = 2{,}9974385630 \times 10^8 \,\text{m/s} \tag{38}$$

$$\text{PD} = -0{,}016$$

where PD is the percentage difference.

Static proton mass

$$m'_p = \frac{\alpha_0}{\pi}(-\alpha_0^{D_0})\alpha_0^{RT1} \frac{1}{M(0)} \text{kg} = 1{,}671795279 \times 10^{-27} \text{ kg} \tag{39}$$

PD = -0,049

Static neutron mass

$$m'_n = \frac{\alpha_0}{\pi}(-\alpha_0^{D_0})\alpha_0^{RT_1}Z(0)\text{kg} = 1,674341082\times 10^{-27}\text{kg} \tag{40}$$

$$\text{PD} = -0,034$$

Boltzmann constant [8]

$$k = \frac{1}{\alpha_0 \frac{\alpha_0}{\pi}\alpha_0^{2RT_1 - RT_2}} e^{-D_0} = 1,380299907\times 10^{-23}\,\text{J/K} \tag{41}$$

$$\text{PD} = -0,0262$$

Bohr Radius [8]

$$r_0 = \frac{1}{(m(8)-1)c} = \frac{1}{63c} = 5,295526678\times 10^{-11}\,\text{m} \tag{42}$$

$$\text{PD} = 0,070$$

where

$$m(8) = \sum_{l=0}^{7}(2l+1) = 64$$

$$m(8) - 1 = D_0 + N$$

Static electron mass

$$me = \left(\frac{\alpha}{\pi}\right)^2 \alpha_0^{D_0}\alpha_0^{5}\alpha_0^{RT2}\frac{1}{M(0)} \tag{43}$$

$$me = 9{,}127735158\text{E} - 31\,Kg$$

absolute percentage deviation = 0,201%

Eletric Charge

$$Q = K\, m_{pcc} \tag{44}$$

onde

$$m_{pcc} = \left(\frac{1}{\alpha_0}\right)^3 \alpha_0^{RT1} M(0)$$

$$K = (4\pi\varepsilon_0)G$$

and G is a the gravitational constant and ε_0 is the dielectric permittivity of vacuum.

Using the real value of ε_0 and G we obtain Q =1,601789478E-19 coulombs

absolute percent deviation = -0,024%

7. Conclusion

From the Dimensional Continuous Space-Time theory, the time and the space are inseparable. The Space-Time is tridimensional and the time does not is the fourth dimension. The time is a dynamic energy of the empty space and it is the fuel of the movement.

Everything that moves inside the Universe only have movement because receives energy from the time. The time is also responsible to keep stability for matter and energy.

Journal of Physics: Conference Series **495** (2014) 012002 doi:10.1088/1742-6596/495/1/012002

There are three kinds of time equation. One is the time wave equation of the empty space; the second is the time equation that gives stability to the matter, propagation energy and others energy forms.

The third is the time function that belongs to the structure of the matter and energy and receive energy from the time equation to keep the matter and energy structures.

The time of the empty space has infinite frequency spectrum and its phase speed is infinite.

The time of the empty space is the same inside the whole Universe.

The time only exist in the present time. From the action of the time equation on the matter and energy was found the duality-wave-particle.

In this theory the space does not have elasticity and the Space-Time is no more a mystery.

References

[1] Martini L., 2013, Introducing the Dimensional Continuous Space-Time Theory, *Journal of Physics: Conference Series*, Volume **423**, conference 1, 2013,Published online: 10 April, 2013, http://iopscience.iop.org/1742-6596/423/1

[2] Martini L., 2013, New View of the Relativity Theory, Submission in: *International Conference on Science & Engineering in Mathematics, Chemistry and Physics* (ScieTech 2014), Conference Date: 13 - 14 January 2014, Jakarta, Indonesia, http://scietech.org/

[3] Martini L., 2013, The New Big Bang Theory according to the Dimensional Continuous Space-Time Theory, Submission in: *International Conference on Science & Engineering in Mathematics, Chemistry and Physics* (ScieTech 2014) Conference Date: 13 - 14 January 2014, Jakarta, Indonesia, http://scietech.org/

[4] Feyman R., 1977, *The Feynman Lectures on Physics*, Addison-Wesley, Vol II, Capt 13.

[5] Rainich G.Y., 1925, Electrodynamics in the General Relativity Theory, Transactions of the american Mathematical Society, vol. 27, No. 1, pp 106-136.

[6] Stewart J., 1993, *Advanced General Relativity*, Cambridge University Press.

[7] Einstein A, 1916, *Relativity: The Special and General Theory*, Methuen & Co Ltd, Translated: Lawson R., Transcription: Basgen B., Einstein Reference Archive, 2002.

Journal of Physics: Conference Series **495** (2014) 012003 doi:10.1088/1742-6596/495/1/012003

The New Big Bang Theory according to Dimensional Continuous Space-Time Theory

Luiz Cesar Martini

Electrical and Computing Engineering Faculty
University of Campinas - Brazil

martini@decom.fee.unicamp.br

Abstract. This New View of the Big Bang Theory results from the Dimensional Continuous Space-Time Theory, for which the introduction was presented in [1]. This theory is based on the concept that the primitive Universe before the Big Bang was constituted only from elementary cells of potential energy disposed side by side. In the primitive Universe there were no particles, charges, movement and the Universe temperature was absolute zero Kelvin. The time was always present, even in the primitive Universe, time is the integral part of the empty space, it is the dynamic energy of space and it is responsible for the movement of matter and energy inside the Universe. The empty space is totally stationary; the primitive Universe was infinite and totally occupied by elementary cells of potential energy. In its event, the Big Bang started a production of matter, charges, energy liberation, dynamic movement, temperature increase and the conformation of galaxies respecting a specific formation law. This article presents the theoretical formation of the Galaxies starting from a basic equation of the Dimensional Continuous Space-time Theory.

1. Introduction

Every conclusion in this article is based on ref. 1. In the ref.1 the primitive space before the Big Bang was constituted only by elementary cells of potential energy. These cells did not have either mass or charge and it did not have particles in its composition. The elementary cells of potential energy were disposed side by side filling the entire primitive universe, which was totally stationary. The temperature of the Universe was the absolute zero Kelvin, see figure 1.

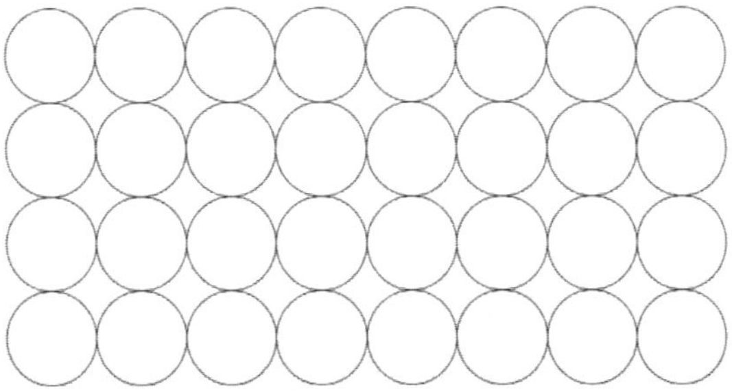

Figure 1. Model of the primitive Universe constituted only by elementary cells of potential energy

In At the exact instant of the Big Bang a unique elementary Cell of potential energy that was in static equilibrium lost its equilibrium and an enormous reduction of the cells size began and one particle between eight possible particles was produced, liberating an amount of energy, perturbing the neighbor cells that started to collapse and produce matter and more energy liberation. So the universe transformation and movement began.

The temperature increases and the universe transformation occurs in light speed, starting the Big Bang.

The enormous size reduction of the cells liberates a big empty space and the conformation of the Universe after the Big Bang has its beginning.

In the center of the Galaxies were created Black holes constituted of Proton clouds and turning around a mass of neutrons.

The Black holes create the gravity responsible by the Galaxy structural stability.

In the Dimensional Continuous Space-time Theory was found the fundamental equation of the continuous Space-Time that will be responsible for every mater and energy present inside the Universe after the Big Bang.

Then was defined the elementary embryonal equations of the space S1(s) and space S1(r) given by:

$$S_1(s) = S_1(r) = e^{\alpha_0(D+\alpha)} = \left(\sqrt{2}\right)^{(D+\alpha)} \tag{1}$$

where:
D is the dimension

α_0 is the space constant given by $\alpha_0 = \frac{1}{2}\ln(2)$ and α is the mathematic dimensional elementary length given by $\alpha = \frac{1}{\pi}$.

In the conformation of the structure Galaxies the *D* dimensional structure of one gravitational dimension axis with an amount of star systems is aligned with a number *n* of elementary axis of energy given by $n = 2^D$.

In this case *D* is a big number that incorporate many star systems. The size of this *D* dimensional axis is

$$S_1 = \left(\sqrt{2}\right)^{(D+\alpha)} \tag{2}$$

To constitute a *2D* dimensional axis from two *D* dimension axis, one set of *D* dimensional axis is placed orthogonally to the first *D* dimensional axis, producing

$$S_2 = \left(\sqrt{2}\right)^{(D+1+\alpha)} = \sqrt{2}\, S_1 \tag{3}$$

To produce a *3D* dimensional axis it is necessary a *2D* dimensional axis placed orthogonally to another *2D* dimensional axis, producing

$$S_3 = \left(\sqrt{2}\right)^{(D+3+\alpha)} = \sqrt{2}\, S_2 = 2S_1 \tag{4}$$

So to produce a *kD* dimensional axis it will be

$$S_k = \left(\sqrt{2}\right)^k S_1 \tag{5}$$

The S_4 structure is represented in the figure 2

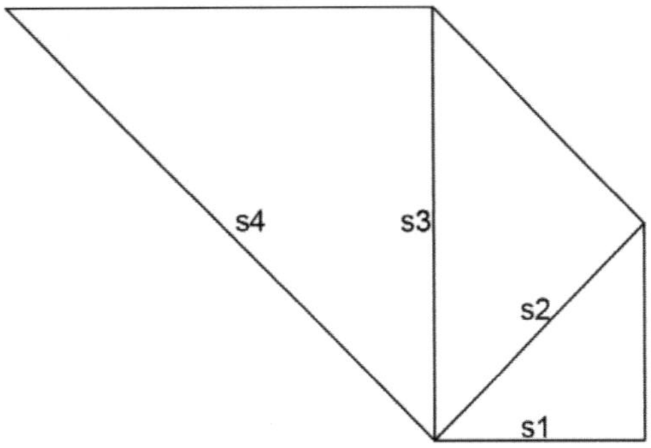

Figure 2.- structure of a gravitational *4D* dimensional axis

Where S_1, S_2, S_3, S_4 are the sustainer gravitational axis of the structure.
So a first possible Galaxy Structure is a *D* dimensional axis represented in the above figure

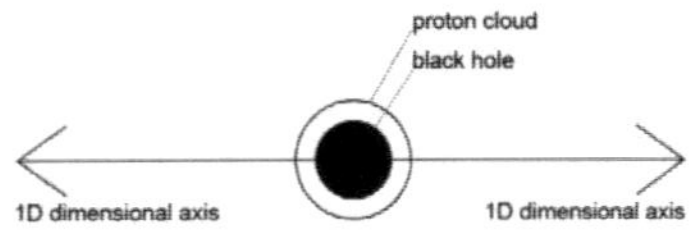

Figure3.- Possible Gravitational model of a bifilar *D* dimensional Galaxy structure

Another possible Galaxy Structure is the spiral galaxy based in a gravitational *kD* dimensional axis with one branch represented in figure 4

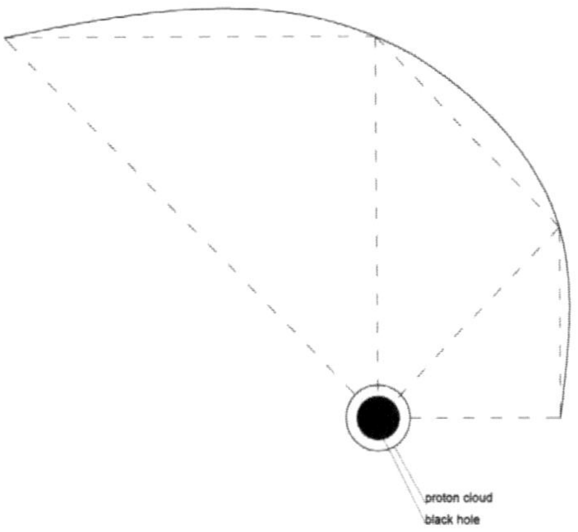

Figure 4.- Gravitational Model of a one branch 4D Dimensional Galaxy structure.

Another possibility is to have one structure with two branches with the equation

$$S = S_k + Z_k \tag{6}$$

where

$$Z_k = \left(\sqrt{2}\right)^k Z_k \tag{7}$$

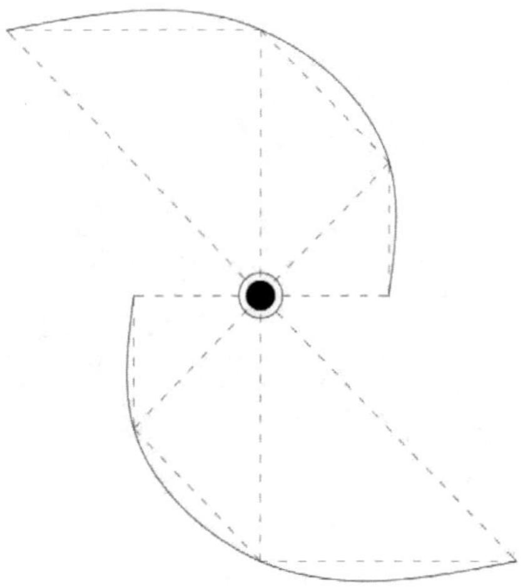

Figure 5.- Gravitational Model of a two branches *4D* dimensional galaxy structure

Another possibility is a four branches *kD* Dimensional Galaxy.
In this case it will be

$$S = S_k + Z_k + U_k + V_k \tag{8}$$

where

$$U_k = \left(\sqrt{2}\right)^k U_1 \tag{9}$$

$$V_k = \left(\sqrt{2}\right)^k V_1 \tag{10}$$

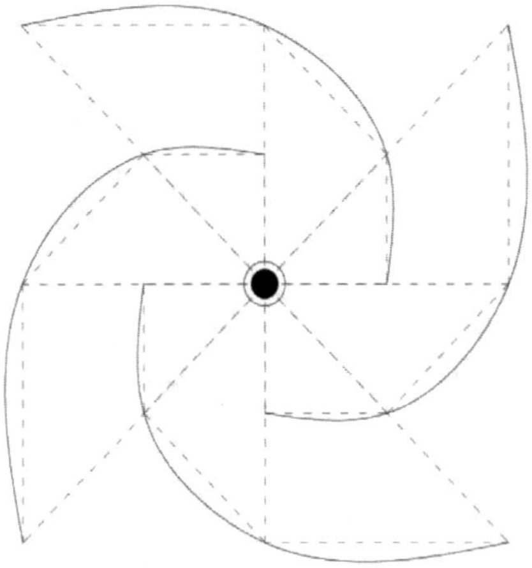

Figure 6.- Gravitational Model of a four branches *4D* Dimensional Galaxy structure

Journal of Physics: Conference Series **495** (2014) 012003 doi:10.1088/1742-6596/495/1/012003

Another possible structure is a circular array of D dimensional Galaxy given by

$$S = \sum_{n} S_{1,n} \tag{11}$$

where

$$S_{1,n} = \left(\sqrt{2}\right)^{(D+a)} \tag{12}$$

This is the Wagon Wheel Galaxy.

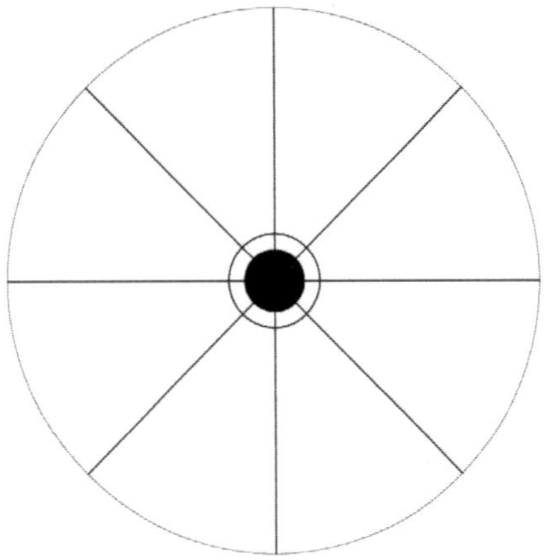

Figure 7.- Gravitational Model of a eight circular 1D dimensional Galaxy Array structure (wagon Wheel)

Another possibility is a Circular model galaxy that is formed by an array of *2D* dimensional axis given by

$$S = \sum_{n} S_{2,n} \tag{13}$$

where

$$S_{2,n} = 2S_{1,n} \tag{14}$$

In this case the circular conformation is obtained by a combination of each dimensional axis of the 2D dimensional axis

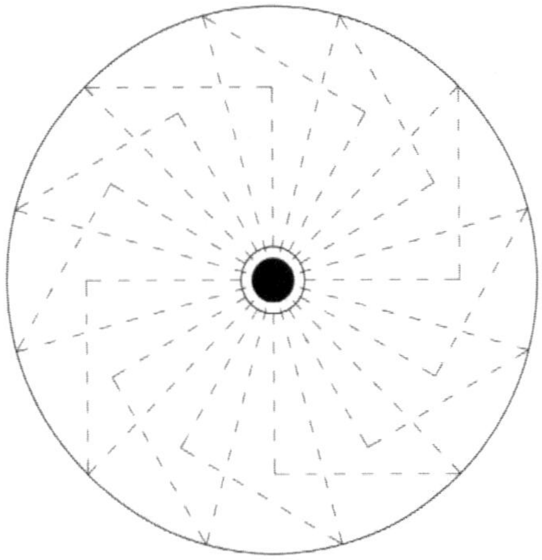

Figure8.- Gravitational Model of a circular Galaxy structure

Another possible Galaxy Structure is the elliptical Galaxy given by figure 9

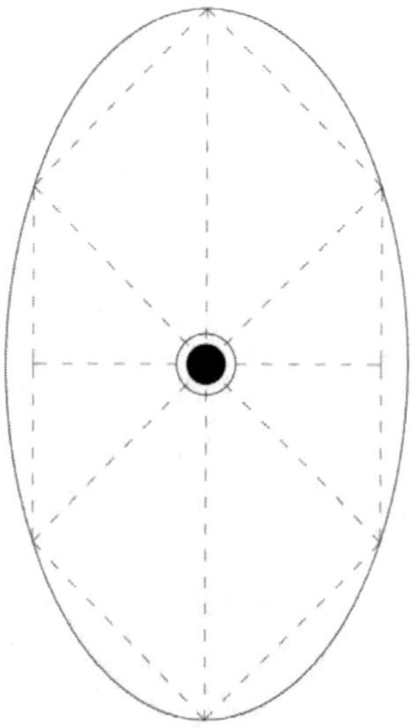

Figure 9.- Gravitational model of an elliptical Galaxy structure

2. Conclusion

This article presents a brief of a New View of the Big Bang Theory based in the Dimensional Continuous Space-Time Theory where some theoretical characteristics of the Universe and its empty space were presented. It means that exist one elliptical gravitational law that keep one rigid structure of the galaxy and no needs dark matter to justify the cohesion inside the galaxy.
One basic equation of the space conformation obtained from Ref.1 was employed in this article to deduce theoretical Galaxy structures. The existent Galaxies that do not have a law formation, probably because they result from collisions between Galaxies and the central Black Holes and others black holes randomly disposed are interfering in the formation law.

In this new Big Bang theory, the Big Bang started from a central point inside the Universe with light speed propagation in all directions. Far away from the central point some billions years, today the Big Bang is still occurring with new galaxies structures in continuous formation. In this reduced article it was not possible to present in details all concepts of the Dimensional Continuous Space-Time Theory.

On the other hand, the objective of this article was to present a introduction of a New View of the Big Bang Theory, which different and new theoretical considerations of the primitive Universe before the Big Bang and the resultant Universe after the Big Bang was presented.

3. References

[1] Martini L., 2013, Introducing the Dimensional Continuous Space-Time Theory, *Journal of Physics: Conference Series*, Volume **423**, conference 1, 2013,Published online: 10 April, 2013, http://iopscience.iop.org/1742-6596/423/1

[2] Linde A., Linde D., Mezhlumian A., ArthurPress.,1994 , From the big bang theory to the theory of a stationary universe, *American Physical Society's New Journal*, Volumen 49, page 1783-1826, Phys. Rev. D, Published: Feb, 1994, http://link.aps.org/doi/10.1103/PhysRevD.49.1783

[3] Ashtekar A., Pawlowski T., Singh P., 2006 , Quantum Nature of the Big Bang, *American Physical Society's New Journal*, Volumen **96**, Phys. Rev. D, Published: April, 2006, http:// link.aps.org/doi/10.1103/PhysRevLett.96.141301

Journal of Physics: Conference Series **495** (2014) 012004 doi:10.1088/1742-6596/495/1/012004

Linear free energy relationship and deuterium kinetic isotope effect observed on phospho and thiophosphoryl transfer reactions in some organophosphorous compounds

B J Lumbiny[1], **Z Hui**[2], **M A Islam**[1], **M A Quader**[3] **and M Rahman**[1]

[1] Department of Chemistry, University of Rajshahi, Rajshahi 6205, Bangladesh
[2]Department of Chemistry, Inha University, Incheon 402-751, South Korea
[3]Department of Chemistry, University of Dhaka, Dhaka 1000, Bangladesh

E-mail:blksjahan@yahoo.com

Abstract. Tetracoordinated organophosphorous compounds were synthesized, characterized and nucleophilic substitution reaction were investigated by varying substituents around phosphorous centre or in nucleophile considering its utility in biological and environmental system. The reactivity is expressed in terms of second-order rate constant, k_2 and measured conductometrically. Linear Free Energy Relationship (LFER) tools mainly Hammett (ρ), Brönsted (β) LFER coefficients and deuterium kinetic isotope effects (KIEs) being determined for the pyridinolysis of 4 - chlorophenyl 4 - methoxy phenyl chlorophosphate, **1** in acetonitrile at 5.0 °C. The experimental data's were compared with those of structurally similar organophosphorous compounds reported earlier in quest for the mechanistic information. Nice linear correlation being found for Hammett (logk_2 vs σ_X), having negative value of the ρ_X = –5.85 and Brönsted (logk_2 vs $pK_{a(X)}$) plots having large positive value for β_X = 1.18 for **1** can be interpreted as S_N2 process with greater extent of bond formation in transition state (TS) of **1**. The observed k_H/k_D values of **1** is 1.00 ± 0.05 and *net* KIE, 1.32 suggests the primary KIE and indicates frontside nucleophilic attack through the partial deprotonation of pyridine occurs by the hydrogen bonding in the rate-determining step.

1. Introduction

Tetracoordinated organophosphorous species play a key role in a wide range of organic and biological processes [1a]. Phosphate is a strong tribasic acid and is ionized at neutral pH in aqueous environments. So it has quite natural availability as biomolecules like phospholipids and very significantly form the monomeric unit of DNA/ RNA by linking up to ribose and heterocyclic nitrogenous base (*e.g.* adenosine, figure.1). Phosphate can link up to two groups as ester in nucleic acid and still remain ionized as shown in figure. 1. Phosphate esters and anhydrides, *e.g.* ATP, are thermodynamically unstable and kinetically stable, which is a hallmark of life [1b]. So Phosphoryl transfer reactions play a fundamental role in a wide range of biological processes including basic metabolism, energy transduction, gene expression, and cell signalling [1]. Besides Certain organophosphorous compounds are widely used as pesticides, neurotoxin, and other biologically active substances. So new findings on organophophorous compounds have been withdrawn much attention to the synthetic chemist as well as in drug designing.

Tetracoordinated organophosphorous compounds were synthesized, characterized and nucleophilic substitution reactions were investigated by varying substituents around phosphorous

[1] B J Lumbiny, Department of Chemistry, University of Rajshahi, Rajshahi 6205, Bangladesh.
[2] M A Islam, Department of Chemistry, University of Rajshahi, Rajshahi 6205, Bangladesh.

centre or in nucleophile. All of the classically used tools of physical organic chemistry mainly linear free energy relationship (LFER) and heavy atom (deuterium) kinetic isotope effect (KIEs) have been used in this study to interpret the (thio)phosphoryl group transfer reaction. LFER and KIEs methods are both powerful tools for the diagnosis of organic reaction mechanism, and relatively few reactions have been studied combining both methodologies.

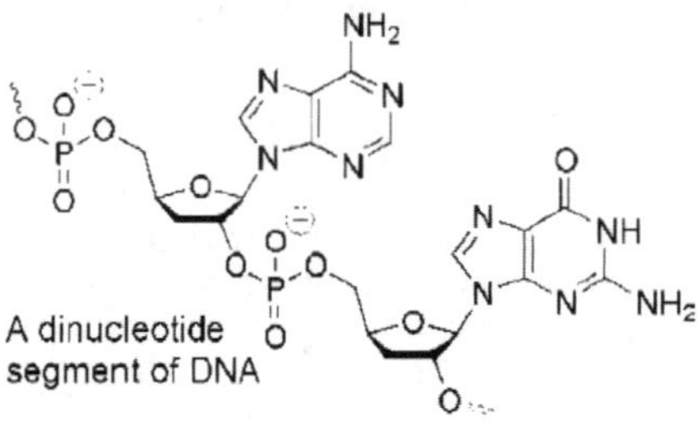

Figure 1. A dinucleotide segment where phosphate esters are linked up to two groups and still remain ionised [1b].

LFER is an empirical observation which can be derived when the shapes of the potential energy surfaces (PES) of a reaction are not substantially altered by varying the substituents [2]. LFER have been very useful in delineating the degree of nucleophilic participation and the extent of leaving group bond fission in the transition states (TS) of the reaction. The measurement of the KIEs has been found widespread use in mechanistic study of various reactions as KIEs are one of the few experimental probes for the TS of the rate limiting step of a reaction [3]. KIEs of atoms at different positions within a molecule can provide details of the TS structures. It is also very insightful and convenient tool for studies of reaction mechanisms. A library of new organophosphors compounds being synthesized and series of kinetics and mechanism of nucleophilic substitution reactions on phosphoryl group (P=O) and thiophosphoryl group (P=S) have been reported for long from the research group acknowledged [4-13]. The above studies are relied on physical organic chemistry methods; Hammett and Brönsted LFER, cross interaction constants (CICs), heavy atom kinetic isotope effects (KIE) which have provided unambiguous evidence for the interpretation of the mechanistic pathway.

It has been reported that nucleophilic substitution at a phosphoryl (P=O) or thiophosphoryl (P=S) centre generally proceeds either through stepwise mechanism with a trigonal bipyramidal pentacoordinate (TBP-5C) intermediate (upper route in figure 2) or an S_N2 mechanism with TBP-5C TS (lower route in figure 2), where the attacking and leaving groups occupy apical positions, *i.e.*, backside nucleophilic attack toward the leaving group [4-14].

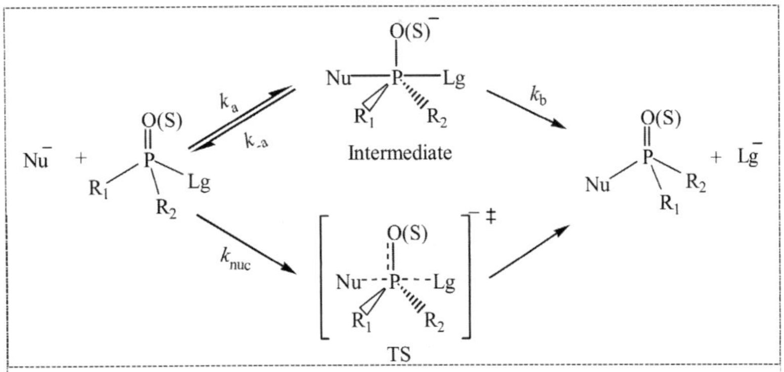

Figure 2. Schematic representation on nucloephilic substitution at P centre on phosphoryl (P=O) and thiophosphoryl (P=S) group transfer reaction.

Our very recent studies show that in case of concerted mechanism the nucleophile can approach towards reaction centre in two different ways. Front and back-side nucleophilic attack on the substrates is proposed mainly based on the deuterium kinetic isotope effects. A hydrogen-bonded, four-center type transition state is suggested for a frontside attack while the TBP-5C TS is suggested for a back-side attack as shown in figure 3 [5, 7, 12].

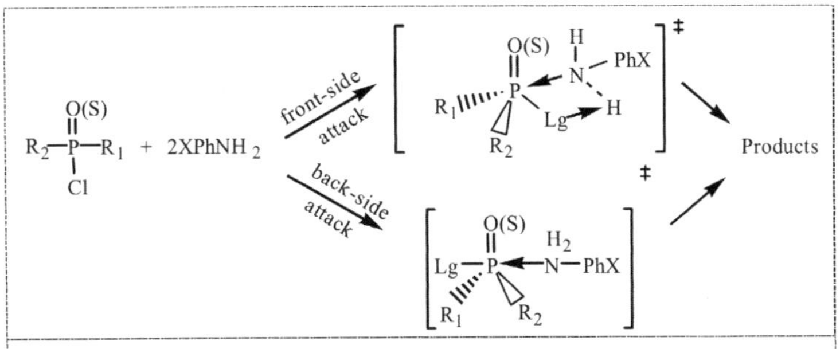

Figure 3. Schematic representation of nucleophilic attack on P centre in concerted mechanism.

The nucleophile attacks the backside toward the leaving group in most phosphoryl transfers, *i.e.*, the attacking and the leaving groups occupy apical positions [ap(Nu)-ap(Lg)] causing inversion of configuration. However, when the nucleophile attacks frontside toward the leaving group, *i.e.*, when the nucleophile and the leaving groups occupy apical and equatorial positions [ap(Nu)-eq(Lg)], or equatorial and apical positions [eq(Nu)-ap(Lg)], respectively, the configuration is retained [5, 12]. When backside and frontside nucleophilic attacks occur simultaneously, the relative importance of each reaction pathway leads to products with inversion or retention of the configuration, depending on the nucleophile, the leaving group, and the reaction conditions [5, 12]

This is also very strongly claimed from previous studies as the substituents around the P centre varies in leaving or nonleaving group, or substituents around nucleophilic centre, the reactivity, selectivity, stereochemistry and finally mechanistic pathway for S_N reaction varies a lot. Continuing our studies on phosphoryl transfer reactions the tetracoordinated organophosphorus compound 4-chlorophenyl 4-methoxy phenyl chlorophosphate [4-ClPhOP(=O)(4-MeOPhO)Cl], 1, being synthesized, characterised by spectrometric methods and element analysis and aminolyses reaction with substituted pyridines have been carried out at 5.0 °C in acetonitrile (figure 4). Recently the pyridinolysis of some organophosphorous compounds, structurally similar of those of 1, 4-chlorophenyl 4-methyl phenyl chlorophosphate [4-ClPhOP(=O)(4-MePhO)Cl], 2, 4-chlorophenyl phenyl chlorophosphate [4-ClPhOP(=O)(OPh)Cl], 3, in acetonitrile at 5.0 °C, and *O*-aryl phenyl phosphonochloridothioates [PhP(=S)(OPh-Y)Cl, 4, in acetonitrile at 35.0 °C were observed. The substituents around phosphorus centre (Y in 4) or in nucleophile (X with 2, 3, 4) being varied and compared with those of 1. The reported mechanism according to LFER is S_N2 without any change in mechanism for the system 2, 3 and 4 [4, 6, 13]. Comparing the reactivities, the linear free energy selectivity parameters and deuterium KIEs with the above comparable systems this study could be a further evidence to clarify aminolyses mechanism and stereochemistry to propose plausible TS structure.

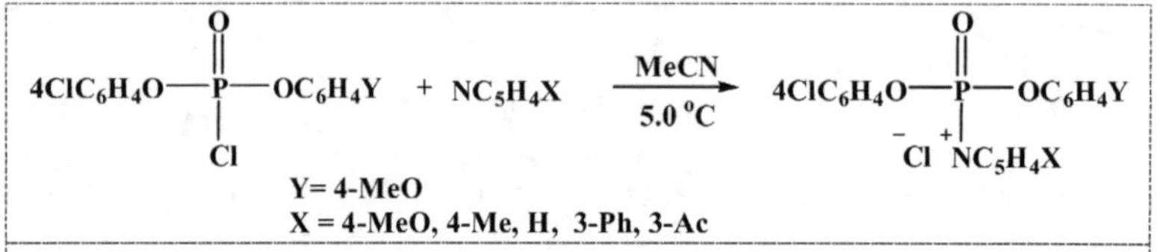

Figure 4. Reaction systems for the pyridinolysis of 1, in acetonitrile at 5.0 °C (present study)

2. Experimental

2.1. Materials
GR grade starting materials (4-Chlorophenyl dichlorophosphate, 4-methoxy phenol, triethylamine) substitued pyridines, deuterated pyridine (C_5D_5N; 99 atom% D), $CDCl_3$ (99.8 atom % D, 0.03% v/v TMS), from Aldrich, USA, were purchased and used. HPLC grade acetonitrile (less than 0.005% H_2O content) were used to prepare nucleophile solution to study.

2.2. Synthesis and characterization [4 -13]
Equimolar proportion of the starting materials were taken with methylene chloride and stirred in ice bath for 2 hrs. The compound 1 is obtained from the reaction mixture by column chromatography and characterized in the following way.

4-Chlorophenyl 4-methoxyphenyl chlorophosphate: Yellowish liquid; ^{1}H NMR (400 MHz, $CDCl_3$) δ 7.37 (d, J = 9.2 Hz, 2H), 7.26-7.19 (m, 4H), 6.90 (d, J = 9.2 Hz, 2H), 3.80 (s, 3H); ^{13}C NMR (100 MHz, $CDCl_3$) δ 157.81, 148.26, 143.21, 131.97, 130.10, 121.84, 121.32, 114.97, 55.65 (OCH_3); ^{31}P NMR (162 MHz, $CDCl_3$) δ 0.66 (s, 1P); IR (neat, cm^{-1}) 3003 (C-H, aromatic), 2951 (-CH_3 Asym) , 2844 (-CH_3 Sym), 1388 (P=O), 1239, 1160, (C-O-Ar) 1066, 990, 843 (P-O-Ar str); m/z, 332 (M^+); Anal. Calcd for $C_{13}H_{11}O_4PCl_2$: C, 46.87; H, 3.33; Found: C, 46.92; H, 3.58.

2.3. Kinetics [4-13]:
Reactions were carried out under pseudo first-order conditions at 5.0 °C in which amine concentrations were at least 20 times greater than the substrate concentration. Thus the pseudo first- order rate (k_{obsd}) was obtained experimentally by using Guggenheim equation [equation (1)] by nonlinear curve fitting method in ORIGIN program.

$$\lambda_t = \lambda_\infty - (\lambda_\infty - \lambda_0)e^{(-k_{obsd} \times t)} \quad (1)$$

[where, λ_0 = initial conductivity, λ_t = conductivity at any time, λ_∞ = conductivity at infinity]

This will ultimately produce second-order rate constant (k_2) from the slope of the plots (see table 1) of k_{obsd} vs [Nu], in equation (2) which gave very good linearity in all cases (regression > 0.997) reproducible to within ± 3%.

$$k_{obsd} = k_0 + k_2\,[Nu] \qquad (2)$$

Similarly deuterated pyridine were treated to obtain k_2, values for deuterium effect. In this study k_H is expressed as average of second order rate constants with pyridine and k_D indicates average of second order rate constant with deuterated pyridine. If $k_H \neq k_D$ a kinetic isotope effect (KIE) exists, expressed as the ratio, k_H/k_D [3]. Standard error being measured by the following equation (3)

$$\text{Standard error } \{= 1/k_D[(\Delta k_H)^2 + (k_H/k_D)^2 \times (\Delta k_D)^2]^{1/2}\} \qquad (3)$$

Another significant measure is net KIE which is determined by the relation below

$$(k_H/k_D)_{net} = (k_H/k_D)_{obsd}/(k_H/k_D)_{expd} \qquad (4a)$$

$$\log(k_D/k_H)_{expd} = \beta_X \times \Delta pK_a \qquad (4b)$$

Perrin and his coworkers reported that the basicities of β-deuterated analogs of benzylamine, *N*, *N*-dimethylaniline and methylamine increase roughly 0.02 pK_a units per deuterium and these effects are additive [15, 16]. For five deuterium atoms in *d*-5 pyridine this gives an expected ΔpK_a of approximately +0.1 unit.

2.4. Tools to interprete mechanism (theory)

In the early part of twentieth century the quantitative study of LFER was introduced by the Danish scientist Johannes Nicolaus Brönsted (1923) and Swedish scientist Louis Plack Hammett (1937).

2.4.1. Hammett Equation [2, 17, 18]

It treats the electronic effect of substituents on the rate or equilibria of organic reactions, can be expressed as follows equation 5(a) and 5(b)

$$\log(k_X/k_H) = \sigma\rho \qquad (5a)$$

$$\log(K_X/K_H) = \sigma\rho \qquad (5b)$$

k_X or K_X is the rate or equilibrium constant, respectively, for the given reaction of *m*- or *p*-XC_6H_4COOH, k_H or K_H refers to the reaction of C_6H_5COOH, i.e., X = H, σ is the substituent constant, ρ is the reaction constant or Hammett coefficients and obtained by plotting $\log k_2$ vs. σ_X.

2.4.1.1 Significance of Sign and Magnitude of ρ

The susceptibility of the reaction to substituents, (+ve) sign; a reaction favored by EWS and $\rho < 1$ indicates negative charge is built, (–ve) sign and $\rho > 1$ suggest reverse situation.

2.4.2. The Brönsted equation [19, 20]

The Gibbs free energy for proton dissociation is proportional to the activation energy (equation 6a, 6b) for the catalytic step. When the relationship is not linear, the chosen group of catalysts do not operate through the same reaction mechanism.

$$k_b = G_bK_b^{\beta} = G_b\,(K_W/K_a)^{\beta} = G_bK_a^{-\beta} \qquad (6a)$$

$$\text{or} \quad \log k_b = \beta pK_a + \text{Constant} \qquad (6b)$$

The Brönsted correlations of rate constants with nucleophile pK_a (β_X) is one measure of the degree of nucleophile bond formation in the rate determining TS. Reactions that have low values for proportionality constants (β_X) are considered to have a transition state closely resembling the reactant with little proton transfer, with a high value, resembles product .

2.4.3. Kinetic isotope Effects (KIEs)

Kinetic isotope Effects (KIEs) is a kinetic method that can in principal tells about bonding changes in the rate limiting step of a reaction as the rate of reaction varies when an atom is replaced by a different (usually heavier atom) [3b]. Isotopic substitution does not affect the PES of the molecule nor does it perturb the electronic energy levels. It is only those properties that are dependrnt upon atomic masses which are affected; for chemical purposes, the perturbation can be considered to be limited to vibrational frequencies [3c]. Vibrational energy will usually change during the course of a reaction or between reagent and transition state since some bonds are in the course of reaction being broken or made and their associated frequencies will be affected. Isotopic substitution should therefore affect reaction rates. The extent to which this occurs depends greatly upon the relative masses of the isotopes, and the mass of D doubles to that of H (greatest) so hydrogen isotope effects are the most thoroughly examined, the following discussion will refer to this case, rate constants being denoted, k_H, k_D, but the same principles apply to any pair of isotopes.

The following types of isotope effect are distinguished [3]

(a) primary kinetic isotope effect (PKIE): in which the bond to the isotopic atom is broken in the rate determining step, $k_H/k_D >> 1$;

(b) secondary kinetic isotope effect (SKIE), in which the bond to the isotopic atom (s) remains intact throughout the reaction, $k_H/k_D << 1$ or k_H/k_D, around, 1;

(c) solvent isotopic effects, which result from isotopic differences in the medium, e.g., if the solvent is changed from H_2O to D_2O, then $k_{(H2O)} / k_{(D2O)}$ is obtained as solvent isotope effect.

More precisely The net KIE of less than unity, $(k_H/k_D)_{net} < 1$, implies the secondary inverse KIE while the net KIE of greater than unity, $(k_H/k_D)_{net} > 1$, implies a primary KIE [21]. The primary KIE suggests that partial deprotonation of nucleophile occurs by hydrogen bonding in the rate-determining step retaining the configuration whereas The secondary inverse KIE is ascribed to the increment of out-of-plane bending vibrational frequencies of C−H(D) bonds in the TS because of steric congestion of the hydrogen (deuterium) atom in the C−H(D) moiety in the bond-making step and inversion of configuration is found.

4. Results and Discussion

The second-order rate constants ($k_2 \times 10^3/M^{-1}s^{-1}$) of the pyridinolysis [4-ClPhOP(=O)(4-MeOPhO)Cl], 1, [4-ClPhOP(=O)(4-MePhO)Cl], 2, [4-ClPhO)PhOP(=O)Cl], 3, with Y = H: (PhO)PhP(=S)Cl, 4, obey the following order of reactivity respectively (see table 1, k_2 is given upto three significant figures)

3 (31.6, P=O) > 2(26.2, P=O) > 1(24.0, P=O) > 4 (11.2, P=S)

At a glance, the reaction rates seem to be proportional to the positive charge on reaction center P. However, it is well known that P=O substrates are more reactive than P=S because of electronegativity difference between O and S, favoring O over S. Also the presence of electron donating substitent, 4-MeO in 1 retards the rate of reaction most than that of 2 and 3 follow the expected trend of nucleophilic substitution reaction..

The Hammett plots for substituent (X) variations in the nucleophile (log k_2 vs. σ_X, figure 5) and Bronsted plots (log k_2 vs. pK_a(X), figure 6) show linearity for 1, 2, 3, same as 4 suggesting no change in mechanism.

The reactivity selectivity parameters for 1, 2, 3, 4 are compiled in table 1 and compared. The obtained $\rho_X = -5.78$, and $\beta_X = 1.17$ for 2, $\rho_X = -5.66$, and $\beta_X = 1.14$ for 3, $\rho_X = -4.35$ to -4.75, and $\beta_X = 0.87$ to 0.95, of 4 are somewhat smaller than those in the present work, $\rho_X = -5.85$, and $\beta_X = 1.18$ in 1. The reaction mechanism for the pyridinolysis of 2, 3, and 4 follow S_N2 mechanism as reported earlier. The comparable ρ_X and β_X of 1 with those of 2, 3, 4 indicates concerted S_N2 mechanism for system 1, with a similar but later transition state structure (TS) i.e., a greater extent of bond-formation in (Figure 7).

Table 1. Second-Order Rate Constants, k_2 ($\times 10^2$ /M^{-1} s^{-1}) and Selectivity Parameters for the Pyridinolysis of [4-ClPhOP(=O)(4-MeOPhO)Cl], 1, [4-ClPhOP(=O)(4-MePhO)Cl], 2, (4–ClPhO)PhOP(=O)Cl. 3, (YPhO)PhP(=S)Cl, 4.

Y/X	4-MeO	4-Me	H	3-Ph	3-Ac	$-\rho_X$	β_X
σ	– 0.27	– 0.17	0.00	0.06	0.38	*Hammette coefficient*	*Bronsted coefficient*
pK_a (py)	6.47	6.00	5.17	4.87	3.26		
4-MeO, 1	101	20.7	2.40	0.656	0.016	5.85	1.18
4-Me, 2	111	23.5	2.62	0.709	0.020	5.78	1.17
H, 3	121	27.8	3.16	0.918	0.026	5.66	1.14
H, 4			1.12			4.35-4.75	0.87-0.95

The observed k_H/k_D values of **1** (1.00) is unity (see table 2) so there is a chance to be secondary kinetic isotope effect (SKIE). The *net* KIE of 1, 2, 3, 4, [$(k_H/k_D)_{net} = (k_H/k_D)_{obsd}/(k_H/k_D)_{expd}$] are 1.32, 1.31, 1.30, 1.35 respectively suggesting primary kinetic isotope effect (PKIE).

Table 2. Deuterium kinetic isotope effects (k_H/k_D) and *net KIE* for the Pyridinolysis of [4-ClPhOP(=O)(4-MeOPhO)Cl], 1, [4-ClPhOP(=O)(4-MePhO)Cl], 2, (4-ClPhO)PhOP(=O)Cl. 3, (YPhO)PhP(=S)Cl, 4.

Y	k_H ($\times 10^2$/ $M^{-1} s^{-1}$)	k_D ($\times 10^2$/ $M^{-1} s^{-1}$)	$(k_H/k_D)_{obsd}$	$(k_H/k_D)_{expd}$	*net KIE* $(k_H/k_D)_{obsd}$/ $(k_H/k_D)_{expd}$
4-MeO, 1	2.40 ± 0.07	2.40 ± 0.07	1.00 ± 0.05	0.762	1.32
4-Me, 2	2.62 ± 0.06	2.62 ± 0.06	1.04 ± 0.04	0.764	1.31
H, 3	3.16 ± 0.20	3.04 ± 0.14	1.04 ± 0.08	0.769	1.30
H, 4	1.12 ± 0.2	1.01 ± 0.1	1.11 ± 0.02	0.820	1.35

The primary KIE and indicates frontside nucleophilic attack through the partial deprotonation of pyridine occurs by hydrogen bonding in the rate-determining step.Taking into account the larger magnitude of *net* PKIE the extent of hydrogen bond formation in TS would be greater for 4 than for 1, 2 and 3.

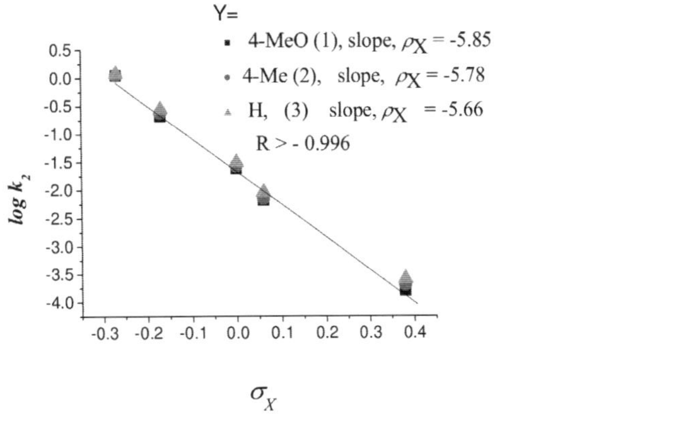

Figure 5. The Hammett plot (log k_2 vs. σ_X) for system 1, 2, 3

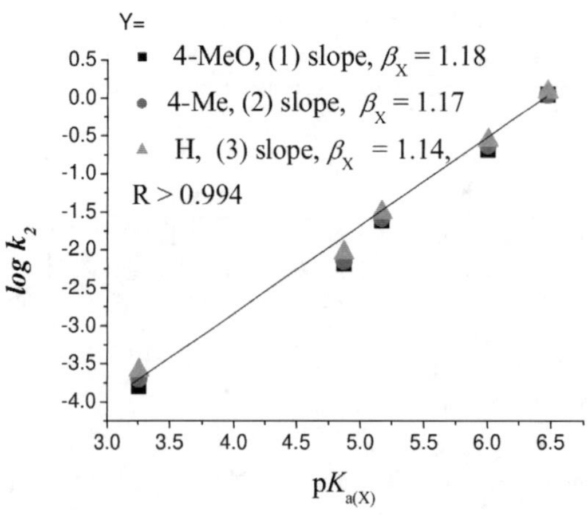

Figure 6. The Brönsted plot log *k* vs. p*K* (X) for the system 1, 2, 3

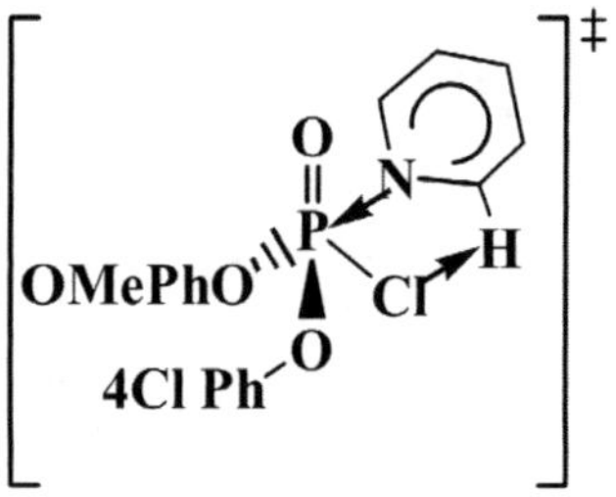

TS 1

Figure 7. Plausible TS structure for 1

4. Conclusion

Physical organic chemistry methods; linear free energy relationship (LFER) and most recently developed technique heavy atom kinetic isotope effects (KIEs) have been used in quest for the mechanistic information of the pyridinolysis of 4-chlorophenyl 4-methoxy phenyl chlorophosphate [4-ClPhOP(=O)(4-MeOPhO)Cl], 1, at 5.0 °C in acetonitrile (figure 4). The experimental data being compared with those of some structurally similar organophosphorus compounds, 4-chlorophenyl 4-methyl phenyl chlorophosphate [4-ClPhOP(=O)(4-MePhO)Cl], 2, 4-chlorophenyl phenyl chlorophosphate [4–ClPhOP(=O)(OPh)Cl], 3, at 5.0 °C, and *O*-aryl phenylphosphonochloridethioates [PhP(=S)(OPh-Y)Cl, 4, at 35.0 °C in acetonitrile. Reaction kinetics of these nucleophilic substituition reactions were measured conductometric method and reactivity was obtained as second order rate constant k_2. These k_2 values being used up to study LFER and KIEs and to interpret the mechanistic pathway.

Nice linear correlation being found for Hammett plot ($\log k_2$ vs σ_X, figure 5), having negative value of the $\rho_X = -5.85$ and Brönsted ($\log k_2$ vs $pK_{a(X)}$, figure 6) plots having large positive value for $\beta_X = 1.18$ for 1 can be interpreted as S_N2 process with greater extent of bond formation in transition state (TS) of 1 (figure 7). The observed k_H/k_D values of 1 is 1.00 ± 0.05 and *net* KIE, 1.32 suggests the primary KIE and indicates frontside nucleophilic attack through the partial deprotonation of pyridine occurs by the hydrogen bonding in the rate-determining step.

4. References

[1] (a) Zalatan J G, Catrina I, Mitchell R, Grzyska P K, O'Brien P J, Herschlag D and Hengge A C 2007 *J. Am.Chem. Soc.* **129** 9789 (b) Westheimer F H 1987 *Science. 235*, 1173.

[2] Williams A 2003 *Free Energy Relationship in Organic and Bio-Organic Chemistry* (UK: RS.C Press).

[3] (a) Olaf W, Houk K N, Black K A and Thomas B 1995 *J. Am.Chem. Soc.* **117** 8594 (b) Carrol F A 1998 *Perspectives on Structure and Mechanism in Organic Chemistry* (ITP) (c) Isaacs N S 1995 *Physical Organic Chemistry* (Longman: Harlow, 2nd ed) (d) Paneath P and O'Leary M H 1991 *J. Am.Chem. Soc.* **113** 1691

[4] Lumbiny B J 2012 *J. Asiat.Soc. Bangladesh. Sci.* **38** 17

[5] Adhikary K K, Lumbiny B J, Dey S and Lee H W 2011 *Bull. of Korean Chem. Soc.* **32** 2628

[6] Lumbiny B J, Adhikary K K, Lee B S and Lee H W 2008 *Bull. of Korean Chem. Soc.* **29** 1769

[7] Lumbiny B J and Lee H W 2008 *Bull. of Korean Chem. Soc.* **29** 2065

[8] Lee I, Kim C K, Li H G, Sohn C K, Kim C K, Lee H W and Lee B S 2000 *J. Am. Chem. Soc.* **122** 11162

[9] Adhikary K K, Lumbiny B J and Lee H W 2011 *Bull. of Korean Chem. Soc.* **32** 3743

[10] Guha A K, Lee H W and Lee I 2000 *J. Org. Chem.* **65** 12

[11] Adhikary K K, Lumbiny B J, Kim C K and Lee H W 2008 *Bull. of Korean Chem. Soc.* **29** 851

[12] Hoque M E U, Dey S, Guha A K, Kim C K, Lee B S and Lee H W 2007 *J. Org. Chem.* **72** 5493

[13] Lumbiny B J, Hui Z and Islam M A 2013 *APCBEE Procedia (Elsevier)* preprint w 035

[14] Hengge A C and Onyido I 2005 *Curr. Org. Chem.* **9** 61

[15] Perrin C I and Engler R E 1991 *J. Phys. Chem.* **95** 8431

[16] Perrin C I, Ohta B K and Kuperman J 2003 *J. Am. Chem. Soc.* **125** 15008

[17] Williams A 2000 *Concerted Organic and Bio-Organic Mechanisms* (CRC: Boca Raton)

[18] Hansch C, Leo A and Taft R W 1991 *Chem. Rev.* **91** 165

[19] Zuman P and Patel R C 1984 *Techniques In Organic Reactions Kinetics* (John Wiley and Sons, USA)

[20] Albert A and Serjeant E P 1984 *The Determination of Ionization Constants* (Chapman and Hall, New York, 3rd ed.)

[21] Melander L and Saunders W H 1980 *Reaction Rates of Isotopic Molecules* (Wiley-Interscience: New York)

Acknowledgement

A portion of experimental study was carried out in Chemical Kinetics Laboratory, Inha University, South Korea.

Development of CMOS Imager Block for Capsule Endoscope

S Shafie [1, 2], F A M Fodzi [1], L Q Tung[1], D X Lioe [1], I A Halin [1], W Z W Hasan [1], H Jaafar [1]

[1] Department of Electrical and Electronic Engineering, Faculty of Engineering, Universiti Putra Malaysia, 43300 UPM Serdang, Malaysia

[2] Institute of Advanced Technology, Universiti Putra Malaysia, 43300 UPM Serdang, Malaysia

E-mail: suhaidi@upm.edu.my

Abstract. This paper presents the development of imager block to be associated in a capsule endoscopy system. Since the capsule endoscope is used to diagnose gastrointestinal diseases, the imager block must be in small size which is comfortable for the patients to swallow. In this project, a small size 1.5V button battery is used as the power supply while the voltage supply requirements for other components such as microcontroller and CMOS image sensor are higher. Therefore, a voltage booster circuit is proposed to boost up the voltage supply from 1.5V to 3.3V. A low power microcontroller is used to generate control pulses for the CMOS image sensor and to convert the 8-bits parallel data output to serial data to be transmitted to the display panel. The results show that the voltage booster circuit was able to boost the voltage supply from 1.5V to 3.3V. The microcontroller precisely controls the CMOS image sensor to produce parallel data which is then serialized again by the microcontroller. The serial data is then successfully translated to 2fps image and displayed on computer.

1. Introduction

Nowadays endoscopy is essential in internal disease diagnosis. Endoscopy is the process of diagnosing the gastrointestinal tract of the patients, which is conducted by either inserting a tube through the mouth that is esophagogastroduodenoscopy or through rectum that is colonoscopy. Normally, a camera is mounted at the end of the endoscope and images are taken and sent to a display monitor. This is an effective way of diagnosis but the disadvantages are it causes pain and uncomfortable to the patients. Patients undergo gastroscopy may suffered from mild sore throat or feeling of distention from the insufflate air used during the procedure. Moreover, colonoscopy causes discomfort and pain to the patients as the tube needs to be pushed into the intestine. Apart from the discomfort, conventional endoscopes have its limitation, where it was not able to fully evaluate the disorder in the small intestine1. This is due to small intestine is very long and convoluted, therefore no available scopes are able to traverse the entire length of small intestine2, resulted in some small intestine diseases to be hardly diagnosed. Therefore, the capsule endoscopy is a solution for abovementioned problems. The evolution of endoscopy started around 10 years ago when Given Imaging produced the first ever capsule endoscope named M2A[3]. The design and development of CMOS imager block requires several important characteristics need to be taken into account, particularly the image quality such as resolution and frame rate, and the power consumption of the system as the system is a battery operated device. The power supply must be able to sustain for approximately 8 hours, for a full cycle of digestion until the capsule reaches the bowel[4]. Therefore, the selection of CMOS image sensor, microcontroller, LED and power management device is very important. The CMOS imager block can be connected to high data rates transmitter to transfer the image wirelessly to the receiver. The mentioned characteristics obscure the development of the capsule endoscope and it is really challenging in realizing the proposed system.

2. Imager Block Design and Components Selection

Designing the system aims at optimizing the power consumption and reduced the circuit board size by selection of electronics components to be associated in the CMOS imager block. The block diagram of the proposed CMOS imager block is shown in Figure 1.

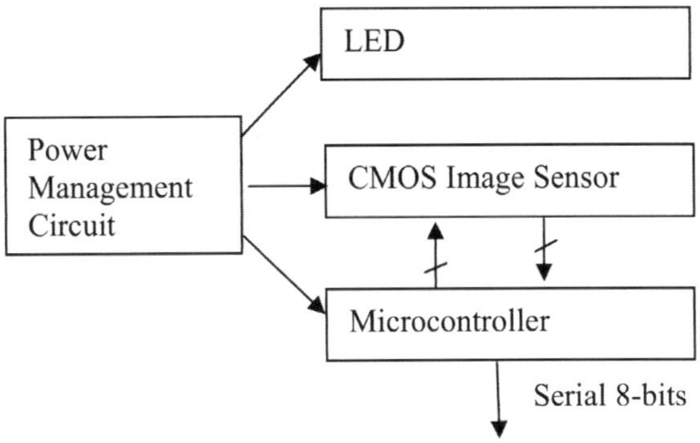

Figure 1. Block diagram of CMOS imager

2.1. Power Management Circuit

A voltage booster has been developed based on the Microchip MCP1654 IC to boost from the supply voltage from1.5V to 3.3V. This IC consumes only 19μA during operation and 1μA at no load. A diode is used in the circuit to prevent reverse bias which will damage the components. Then, regulators are used to supply necessary voltages to the CMOS Image sensor, LED and microcontroller. The CMOS image sensor requires 2.8V and 1.5V of voltage to function properly, while the microcontroller only requires single voltage of 2.8V. The LED can use any of them based on the circuit design. Figure 2 shows the circuit diagram of the proposed power management circuit and its PCB layout.

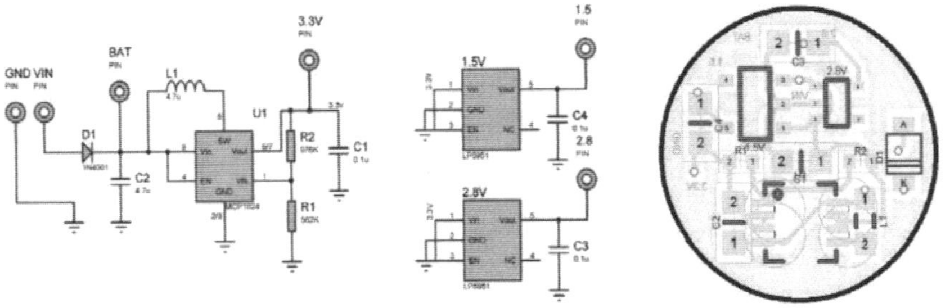

Figure 2. Power Management Circuit and its PCB Layout

2.2. CMOS image sensor

The CMOS image sensor with wide dynamic[6-8] range ability and with global shutter[9] is preferable for better image quality. However such type of image sensor is either too expensive or not available in the current market. Therefore, the low cost, small size Toshiba TCM8230MD CMOS image sensor has been chosen as the imager for this project. It provides VGA colour pixel resolution, able to capture up to 30fps, small size with the dimension of 6(W) × 6(D) × 4.5(H) and consumes low power. However in this design we are focusing of using the VGA image resolution with the frame rate of 2fps only. The CMOS image sensor can be controlled by I^2C bus with 2 wires in which one wire is used for clock line and another for the data line. The PIC 18F26K20 is used to produce the necessary clock and control pulses to operate the CMOS Image sensor. Figure 3 show the circuit diagram and PCB layout of the integrated CMOS image sensor, LED and microcontroller.

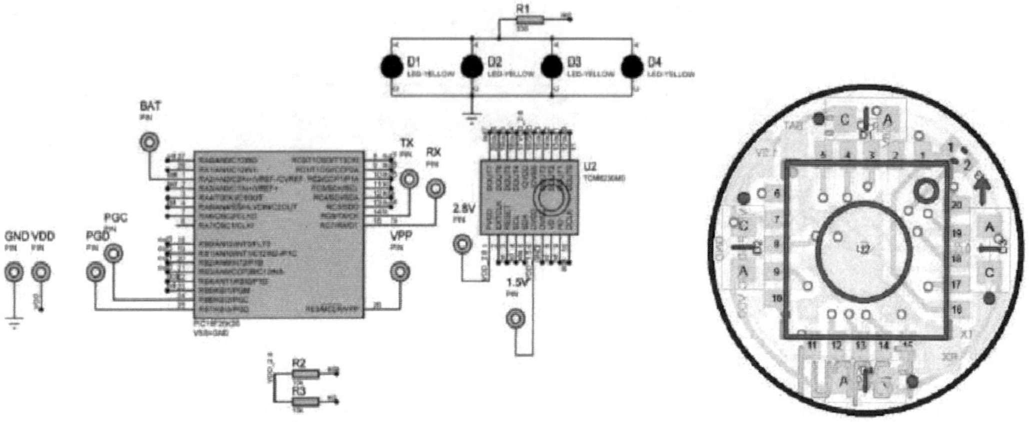

Figure 3. Circuit diagram and PCB layout of the integrated CMOS image sensor, LED and microcontroller.

2.3. Microcontroller as parallel to serial converter

The PIC 18F26K20 Microcontroller can act as parallel to serial converter in which the coding has been design as in Figure 4. The MSB and LSB are from DOUT7 to DOUT0, respectively. Table 1 shows the pin assignment for the microcontroller and CMOS image sensor.

Table 1. Pins for PIC18F26K20 and TCM8230MD

PIC 18F26K20	**TCM8230MD**
B0	DOUT3
B1	DOUT2
B2	DOUT1
B3	DOUT0
C0	DOUT7
C1	DOUT6
C2	DOUT5
C5	DOUT4

```
if(DOUT7==1)          get7=0b10000000;
else if(DOUT7==0)     get7=0b00000000;
if(DOUT6==1)          get6=0b01000000;
else if(DOUT6==0)     get6=0b00000000;
if(DOUT5==1)          get5=0b00100000;
else if(DOUT5==0)     get5=0b00000000;
if(DOUT4==1)          get4=0b00010000;
else if(DOUT4==0)     get4=0b00000000;
if(DOUT3==1)          get3=0b00001000;
else if(DOUT3==0)     get3=0b00000000;
if(DOUT2==1)          get2=0b00000100;
else if(DOUT2==0)     get2=0b00000000;
if(DOUT1==1)          get1=0b00000010;
else if(DOUT1==0)     get1=0b00000000;
if(DOUT0==1)          get0=0b00000001;
else if(DOUT0==0)     get0=0b00000000;

gettotal=get7+get6+get5+get4+get3+get2+get1+get0;
```

Figure 4. C language Coding to convert data from parallel to serial

3. Results and Discussion

The Imager block has successfully developed and measured. It consumes low power and small space area. The voltage booster in power management circuit was able to boost voltage supply from 1.5V to 3.3V as shown in Figure 5. From the figure, the input range can be vary from 0.6V to 3.3V, that means if the batteries voltage reduced to less than 1.5V, the components such as CMOS image sensor, LED and microcontroller can still operate correctly. The voltage regulator also regulates voltage supply to 2.8V and 1.5V as required by the components.

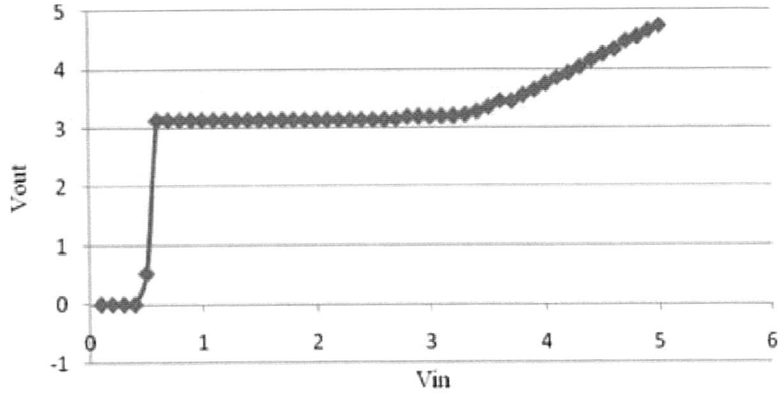

Figure 5. Measured output from the voltage booster circuit

The parallel data from CMOS image sensor output successfully converted to serial data by the microcontroller. The results are shown in Figure 6 where the measurement has been done using the logic analyser. In the figure, parallel data represented by DOUT7-DOUT0 while the serial data represented by R_X.

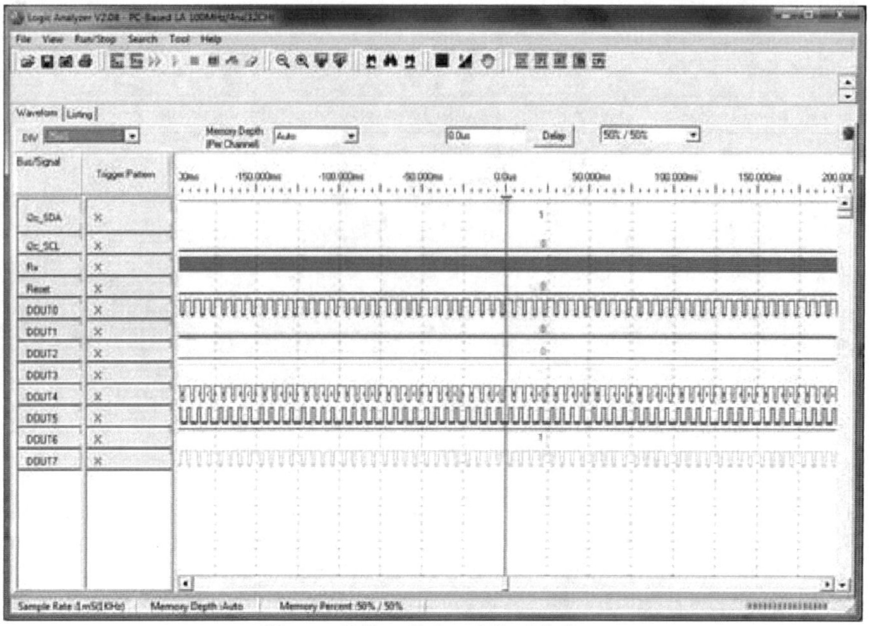

Figure 6. Parallel and Serial data

Figure 7 shows the matrix array data captured by developed imager block which has been transferred to the computer. This matrix array can easily be translated to the video image by using MATLAB graphical user interface. The value of each cell will decide the colour of reproduced image. In this project, RGB565 with 16 bits data string has been chosen.

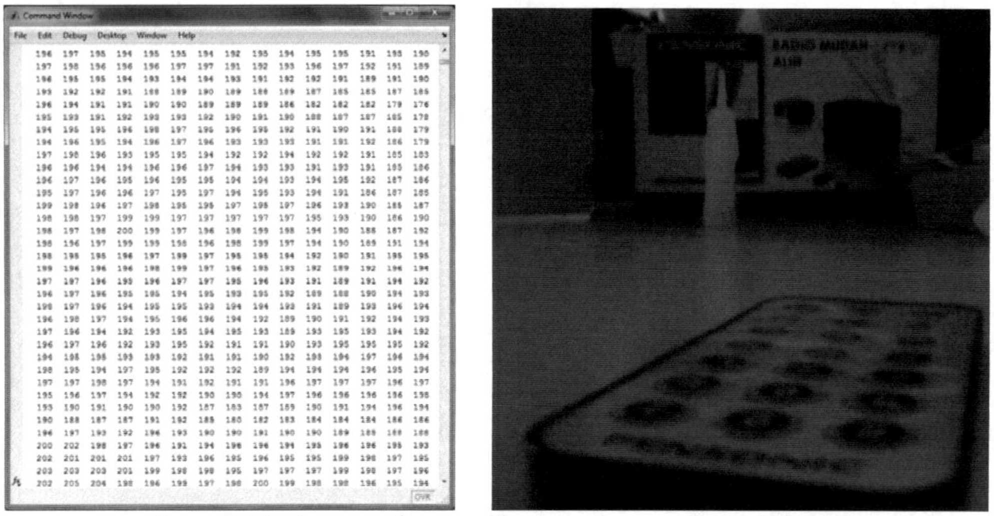

Figure 7. Matrix array of captured image and captured sample image

4. Conclusion

The imager block for capsule endoscope has successfully been developed and the functionality was verified. The power management circuit is able to boost up input voltage from 1.5V to 3.3V, as power supply requirement for components. Processor successfully generates control pulses for CMOS image sensor. Mounted 4 LED's produce relatively bright lighting for CMOS image sensor to capture sample image in dark environment, and finally parallel to serial converter was able to produce the desired outputs. The video image with 2 fps has been successfully formed from matrix array via developed MATLAB graphical user interface.

References

[1] Leighton, J., Sharma, V. et al., "Capsule Endoscopy Versus Push Enteroscopy for Evaluation of Obscure Gastrointestinal Bleeding with 1-Year Outcomes" 2006 Digestive Diseases and Sciences 51(5) pp. 891-899

[2] Song, B., Nam, D. et al., "Auto-focusing actuator and camera module including flexible diaphragm for mobile phone camera and wireless capsule endoscope." 2010 Microsystem Technologies 16(1) pp.149-159

[3] Given Imaging from (http://www.givenimaging.com)

[4] Lioe, D. X., Shafie, S., Ramiah, H., Sulaiman, N., Halin, I. A. "Low Power Transmitter for Wireless Capsule Endoscope", 2013 Journal of Physics: Conference Series

[5] Itoh, S., Kawahito, S., Terakawa, S., "A 2.6mW 2fps QVGA CMOS One-chip Wireless Camera with Digital Image Transmission Function for Capsule Endoscopes", 2006 IEEE International Symposium on Circuits and Systems

[6] Shafie, S. and Kawahito, S., "A Wide Dynamic Range CMOS Image Sensor with Dual Charge Storage in a Pixel and a Multiple Sampling Technique", 2008 SPIE IS&T, 6816(5) pp. 1-9

[7] Shafie, S., Kawahito, S. and Itoh, S., "A Dynamic Range Expansion Technique for CMOS Image Sensors with Dual Charge Storage in a Pixel and Multiple Sampling" 2008 Sensors, 8 pp. 1915 - 1926

[8] Shafie, S., Kawahito, S., Halin, I. A. and Hasan, W.Z.W., "Non-linearity in Wide Dynamic Range CMOS Image Sensors utilizing a Partial Charge Transfer Technique" 2009 Sensors, 9, pp.9452-9467

[9] Yasutomi, K., Itoh, S., Kawahito, S., "A two-stage charge transfer active pixel CMOS image sensor with low-noise global shuttering and a dual-shuttering mode"2011 Electron Devices, IEEE Transactions on 58(3) pp. 740-747

Acknowledgments

Authors would like to thanks the Ministry of Science, Technology and Innovation, Malaysia for supporting this work under eScience Fund.

Radial profile of moments of instantaneous axial velocity of a fully pulsed axisymmetric jet

Hariyo P. S. Pratomo[1,a,b,†], **Klaus Bremhorst**[a]

[a]Division of Mechanical Engineering, the University of Queensland
St. Lucia, Brisbane 4072, Queensland, Australia

[b]Department of Mechanical Engineering, Petra Christian University
Siwalankerto Street 142-144, Surabaya 60236, East Java, Indonesia

[†]Institute of Numerical Methods in Mechanical Engineering, TU Darmstadt
Dolivostraβe 15, Darmstadt 60236, Germany

E-mail: hariyo_p@peter.petra.ac.id

Abstract. Radial distribution of the first and higher moments of instantaneous axial velocity of a fully pulsed round jet were investigated along 80 diameters downstream from the jet exit. Response of the unsteady flow to the varying Reynolds and Strouhal numbers thus allowing a series of distinct pulses were sampled using a single wire hot-wire anemometer. It was found that the first moments, as compared to steady jet phenomena, do not maintain self-similarities over the flow domain but their shapes remain unchanged. At a lower Reynolds and Strouhal number, widening radial profiles of the first moment are obtained. Transverse profiles of the third and fourth turbulence moments explain the nature of random process in the mechanically excited jet. Their values were found to remarkably depart from the Gaussian values and to be significantly higher than those of the steady jet across the observed jet widths hence yielding more intense-highly irregular motions and more spreading jet. For all the tested cases, it was demonstrated that the direct effects of the controlled parameters on the flow field become stronger in the outer region than in the inner conterpart of the jet width.

Keywords: hot-wire anemometer, mass flow rate, pulsing frequency, unsteady jet

1. Introduction

Research interest into turbulence behavior in *free* shear flows has been a long tradition as attempts to understand their unique characteristics in various technical devices such as: mixing processes, sprays, and thrust augmentation. In principle, the turbulent free shear flows have simple flow domains which have no interaction with any solid surfaces and are triggered by mean velocity differences. Various techniques to generate the flows include jets, wakes, and mixing layers. In this paper, however, we restrict our attention to *round* jet which remains classical flows of relevance in the mixing processes, sprays, combustor systems, and other technical devices. This is because the velocity statistics of the

[1] To whom any correspondence should be addressed. Phone: +6231 298 3464. Fax: +6231 841 7658.
The first author is a research staff at Petra Christian University and currently is a research fellow at TU Darmstadt.

jet, described by the mean velocity and higher order turbulence moments are valuable in comprehending the nature of random process in the jet.

Much study on the turbulent round jet has been performed in the past from a relatively low to very high Reynolds number. As a central point in this present study, we signify a subtantial contribution of Dahm and Dimotakis [5] in which the unique characteristics of Reynolds number defined-flow structures were revealed; displaying qualitatively turbulence scales and jet width. *Interestingly*, from the planar images of their jet evolution [5] it was found that there is *no* dependence of the jet width on the varied Reynolds numbers but the finer turbulence scales caused by scale separation were immediately seen at a higher Reynolds number. A few years later, experimental studies by Panchapakesan and Lumley [7] and Hussein, Capp et. al [8] confirmed the self-determination of the jet width [5] demonstrating a linear spreading rate of the jet propagation and relatively similar empirical constants of the jet growth under a large range of the Reynolds number. From this point of view, it therefore is encouraging to optimize the functionality of the jet indicated by a more spreading jet by introducing an additional non-dimensional parameter to control the jet development. It is desired that the additional parameter would create stronger turbulence in which eventually would affect the jet width.

In this paper, velocity and higher order turbulence statistics of a *mechanically* exited jet namely fully pulsed jet was experimentally studied. The jet was defined by a large range of the Reynolds and Strouhal numbers based on its nozzle diameter. Physically, this unsteady jet produced a series of distinct pulses as a result of the modulation imposed. We assessed the first, third, and fourth moments of instantaneous axial velocity across the jet field along further downstreams with the use of a hot-wire anemometer. Comparisons of the higher order turbulence moments with those of a steady round jet [13] are also illustrated. Henceforth, this article is devoted to furthering an understanding on the nature of random process in the pulsed jet.

2. Research Method

The pulsed jet was generated from a circular nozzle of 12.77 mm diameter and pulsing valve, located in the Jet Laboratory, the Division of Mechanical Engineering, the University of Queensland. The apparatus is illustrated in Figure 1. This is the same facility which was used by Bremhorst and Harch [3], Bremhorst and Hollis [4] with the exception that the jet exit diameter was halved giving a 12.77 mm diameter [6, 12]. The nozzle consisted of a 50 mm extension with a honeycomb flow straightener, and mesh cone. A 200 mm disc was mounted on the nozzle to avoid directly entrainment of the fluid behind the nozzle, hence allowing the flow near the jet exit to be axisymmetric.

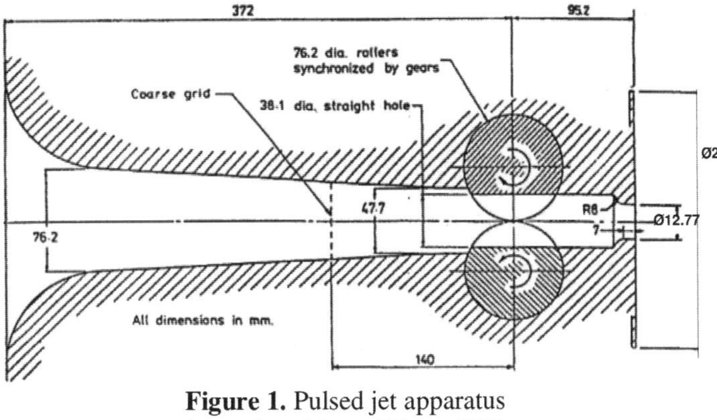

Figure 1. Pulsed jet apparatus
(reproduced from Winter [12])

Flow field of the jet was sampled by a hot-wire anemometer operated in a constant temperature mode. The hot-wire probe was a single normal wire which was made of Sigmond Cohn alloy 851 (79%Pt, 15%Rh, and 6%Ru) mounted on the prongs with a spot-welding technique. An overheat ratio of 1.3 was used to maintain the wire velocity sensitivity. The Kolmogorov length scale was estimated from the work of Gehrke [6] giving a value of 0.124 mm. A 10.16 μm diameter wire was used with the active length of 2 mm in order to give good spatial resolution. Hot-wire calibration gave a ±0.03% accuracy for the extended power-law equation improved by the look-up table method and 1.22% rms relative error of the fluctuating velocity component. Statistical accuracy evaluation found the sampling frequency, number of sample, and sampling time of 1000 Hz, 50,000 and 50 seconds; respectively to be chosen for accurate and repeatable measurements. It allowed a 0.26%, 0.45%, 1.10%, and 4.38% uncertainty of the estimated mean value, second, third, and fourth order turbulence moments; respectively.

Radial profile and propagation tests were initially to be performed to evaluate the symmetrical velocity distribution and propagation of the jet. For the preliminary tests [9], the jet was generated at the bulk exit velocity and pulsing frequency of 27.6 m/s and 10 Hz, respectively. It was shown that the jet is axisymmetric and follows a straight line trajectory [9]. Afterwards, the primary measurements were implemented at the bulk exit velocities of 13.7 and 34.4 m/s to obtain the radial profiles of mean axial velocity along 80 diameters downstream from the jet exit. The pulsing frequencies of 10 and 25 Hz were set to produce the pulsed jet.

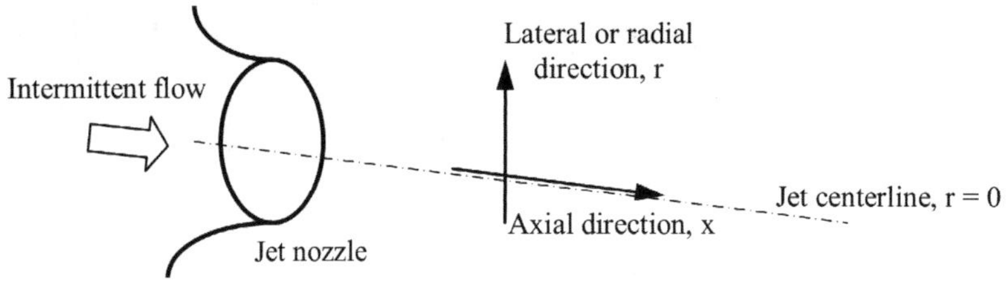

Figure 2. A schematic of the jet field

Flow field of the jet is illustrated in Figure 2. At a fixed point in the jet field, the instantaneous axial velocity of the jet, U_i (x, r) consists of the mean axial velocity, U (x, r) and the aggregate turbulence or fluctuating component, u (x, r). This was to be the case as the statistics does not depend on the time and circumferential coordinate, i.e the shapes of velócity distribution across the field in the horizontal and vertical planes were similar [10]. The Reynolds decomposition as in Equation 1 then holds for the evaluation of the mean axial velocity and fluctuating component. Since the flow was statistically stationary [10], estimation of the desired mean axial velocity can be obtained from a discrete equation over a finite time period as expressed in Equation 2 where N is number of samples.

$$U_i(x,r) = U(x,r) + u(x,r) \tag{1}$$

$$\hat{U}(x,r) = \frac{\sum_{i=1}^{n} U_i(x,r)}{N} \tag{2}$$

A Gaussian [4, 7] and exponential [3] equations, depending on the turbulent flow-sensors used (i.e laser Doppler and hot-wire anemometers), can be applied to represent the radial profile of the first moment of instantaneous axial velocity of the pulsed axisymmetric jet as in Equation 3 and 4; respectively. From their observation [4, 7], it was found that α was equal to 0.693 offering a good Gaussian curve fit for the radial profiles as in Equation 1.

$$\frac{U(x,r)}{U_0(x,r=0)} = e^{-\alpha\left(\frac{r}{r_{1/2.U}}\right)^2} \tag{3}$$

The exponential equation with $\alpha = 0.44$ [3] to provide a good fit of the radial profiles of the mean centerline axial velocity, is expressed as in

$$\frac{U(x,r)}{U_0(x,r=0)} = \left[1+\alpha\left(\frac{r}{r_{1/2.U}}\right)^2\right]^{-2}. \tag{4}$$

The third and fourth moment of turbulence can be determined with the aid of the Reynolds decomposition and the discrete equation. The third, $\overline{u^3}$ and fourth, $\overline{u^4}$ moments can be formulated as in

$$\overline{u(x,r)^3} = \overline{(U_i(x,r)-U(x,r))^3} \tag{5a}$$

or rewritten as

$$\overline{u(x,r)^3} = \frac{1}{(N-1)}\left[\sum_{i=1}^{N} U_i(x,r)^3 - NU(x,r)^3\right], \tag{5b}$$

and

$$\overline{u(x,r)^4} = \overline{(U_i(x,r)-U(x,r))^4} \tag{6a}$$

or simplified as

$$\overline{u(x,r)^4} = \frac{1}{(N-1)}\left[\sum_{i=1}^{N} U_i(x,r)^4 - NU(x,r)^4\right], \tag{6b}$$

respectively. Similar expressions to define the radial, V and azimuthal velocity components, W can also be used. Nevertheless, in this present study we only focus on the measurements of axial velocity component, U.

Other higher order turbulence moments which are dimensionless quantities, including: skewness and kurtosis or flatness factors can be determined with the help of equation 5a – 6b. The skewness, S_u and flatness, F_u of the fluctuating velocity, u are formulated, respectively, as in

$$S_u = \frac{\overline{u(x,r)^3}}{(\overline{u(x,r)^2})^{\frac{3}{2}}} \tag{7}$$

and

$$F_u = \frac{\overline{u(x,r)^4}}{(\overline{u(x,r)^2})^2} \tag{8}$$

These quantities are valuable in the case of turbulent flows to judge how far from Gaussian distribution a particular random variable of the flow might be. In the case of *strict* Gaussian distribution, the value of S_u and F_u are 0 and 3, respectively. Therefore, the closer a random variable with those two values, the more stationary or steady the random process is.

3. Research Results and Discussion

Some results of radial measurements are presented in Figures 3 – 4 displaying the normalized velocity signals at x/d = 20. The instantaneous axial velocity, U_i and time, t are normalized by the mean exit velocity, U_e and time for one complete cycle, t_p; respectively. For the pulsing frequency, f_p of 25 Hz, the value of t_p was 0.035 seconds. As can be seen in Figures 3 and 4, under a constant mean exit velocity the levels of U_i/U_e are relatively the same as the pulsing frequency is increased. However, from the measurements, an increase in the pulsing frequency resulted in the weakening axial aggregate turbulence components.

Appreciations for the radial profiles of the first moment under varying jet exit velocity and pulsing frequency over 80 diameters downstream are illustrated in Figures 5 – 6. As depicted in Figure 5, narrower radial profiles are produced as the Strouhal number is increased at the lower mass flow rate. The similar pattern was also demonstrated by Bremhorst and Harch [3] but with a higher mass flow rate. However, the narrower profiles as the consequence of increasing Strouhal number are relatively less evident at the higher jet exit velocity. Moreover, the narrower profile are also seen as the Reynolds number was increased at the lower Strouhal number (corresponding to the pulsing frequency of 10 Hz), as illustrated in Figure 6. Conversely, at the pulsing frequency of 25 Hz the narrower profiles as a result of the increasing mass flow rate are relatively less obvious. The present study was the *pioneering* work to establish the narrow radial profiles in response to the increasing Reynolds number as compared to the former experimental studies by Bremhorst and Harch [3], Bremhorst and Hollis [4], Gehrke [6], and Winter [12].

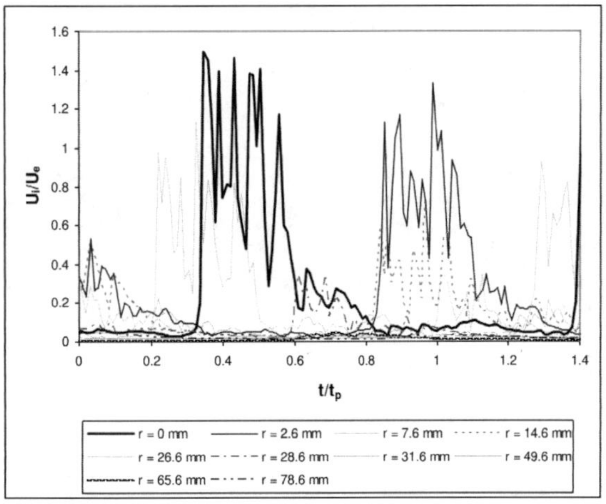

Figure 3. Velocity signals of radial measurement in horizontal direction at U_e = 34.4 m/s, f_p = 10 Hz, x/d = 20.

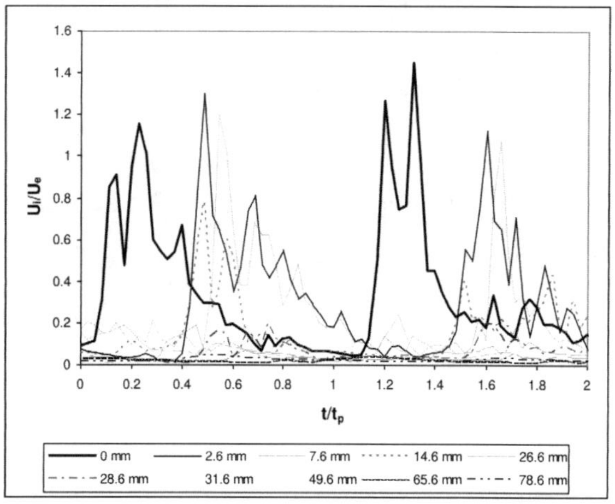

Figure 4. Velocity signals of radial measurement in horizontal direction at U_e = 34.4 m/s, f_p = 25 Hz, x/d = 20.

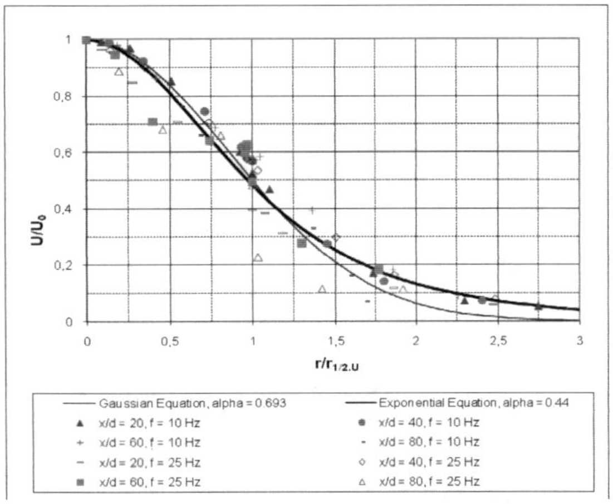

Figure 5. Radial profiles of the first moment of instantaneous axial velocity at U_e = 13.7 m/s.

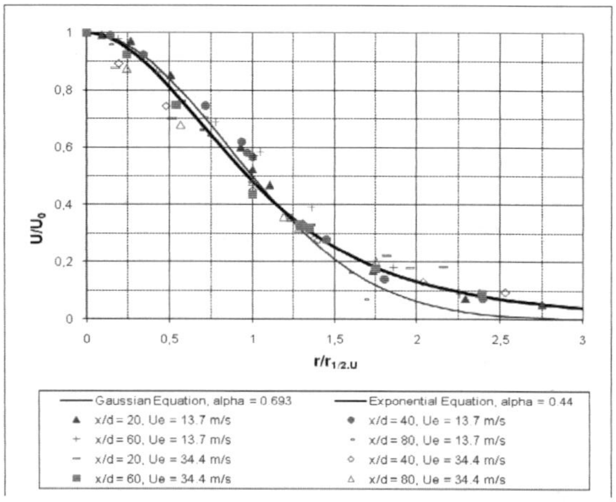

Figure 6. Radial profile of the first moment of instantaneous axial velocity at f_p = 10 Hz.

Comparison of the radial profile of the first moments between steady [2, 5, 7, 8] and pulsed jets are given in Figure 7. It is clearly seen that the significant changes in the radial profile are not the case for the steady jets investigated by Panchapakesan and Lumley [8], Hussein et. al [7], Bogey and

Bailly [2] under a very large range of the Reynolds numbers. From the red continuous line, all the laser Doppler anemometer (LDA) experimental [7, 8] and simulated [2] results of the jets over 140 diameters downstream ranging from a Reynolds number of 5000 up to 95,500 collapsed into a single curve, i.e a Gaussian curve (equation 3). This self similarity is an indication that there is no dependence of the flow structure of the steady jets on varying mass flow rate thus the steady jet width is not affected by controlling the Reynolds number [5]. The simulated results by Bogey and Bailly [2] were obtained by using Large Eddy Simulation (LES). Concerning the widening radial profiles of the pulsed jet at the lower Strouhal number, under different Reynolds numbers Tanaka [11] and Atassi, Charnay et. al [1] also demonstrated the similar evidences for their unsteady axisymmetric jets. In addition, it is seen that the present results depart from the Gaussian curve fit (Equation 3) thus agreeing with the exponential curve fit (Equation 4) but with different constants, for example as seen in Figure 8. Using a hot-wire sensor, similar trends on the radial profiles were also shown by Bremhorst and Harch [3] and Gehrke [6].

Transverse profiles of the skewness and flatness factors of the fluctuating components at 80 nozzle widths downstream which are in the pulsed dominated- and high turbulence steady jet regions [9] are shown in Figures 9 – 12. The results are plotted on the normalized positions in the radial direction associated with the jet width. As can be seen from the figures, both factors vary from the strict Gaussian values ($S_u = 0$ and $F_u = 3$) over the whole region. For all tested cases, the values of both factors have lower values in the locations close to the jet centerline and reach much larger values in the outer regions from the middle of jet widths ($\frac{r}{x} \approx 0.14$). Local maximum values ($S_{u,max}$ and $F_{u,max}$) are registered at the edges of the jet for all cases. That the nearer Gaussian values are found in the inner regions of the jet widths indicates that the influences of the Reynolds and Strouhal numbers on the turbulent statistics in the inner regions are weaker relative to those in the outer regions ($\frac{r}{x} \geq 0.14$). The physical phenomena in the outer regions are best explained by visual observations of the steady jet of Dahm and Dimotakis [5] at an instant of time as there are more intense-highly irregular interactions between the turbulent flow and the ambient fluid in the outer region. Moreover, in all cases, the further downstream where the pulsed dominated-region gradually collapsed [9] the more decreasing the values of both factors over the jet widths thus reaching relatively closer to the Gaussian values.

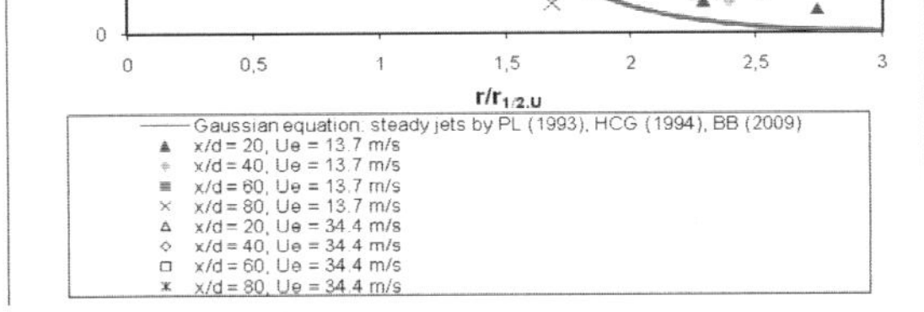

Figure 7. Comparison of the radial profiles between steady and pulsed jets.

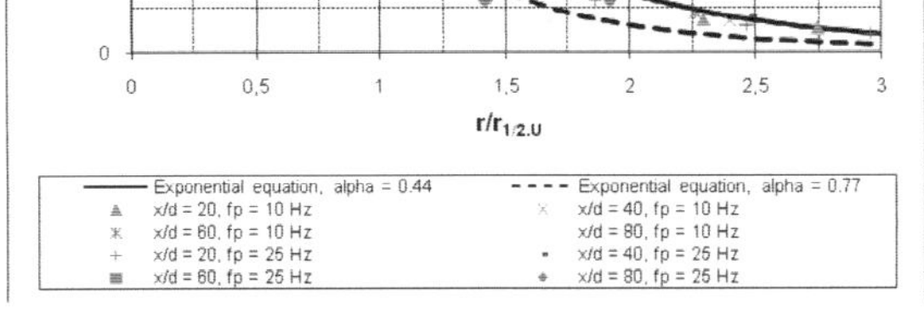

Figure 8. Exponential curve fits (Equation 4) of the first moment of pulsed jet.

Comparison of the skewness and flatness factors between the steady and pulsed jets are given in Figures 13 – 14. Both factors of the steady jet were experimentally studied by Wygnanski and Fiedler [13] at a position of 75 diameters from the jet nozzle at a high Reynolds number of 10^5 using a hot-wire sensor. Based on this benchmark, both factors of the pulsed jet at the position of x/d = 80 are plotted together with those earlier results [13] but we only focus on the highest Reynolds number used in which produced the narrower radial profiles relative to ones at the lower Reynolds number. It is important to note that the location of 80 diameters in the pulsed jet is supposedly the high turbulence steady jet region [9]. With this scheme, the comparison is sufficiently reasonable to judge the turbulence nature in the original steady and 'modulated steady' jet i.e. the unsteady one when already switched to a high turbulence steady jet.

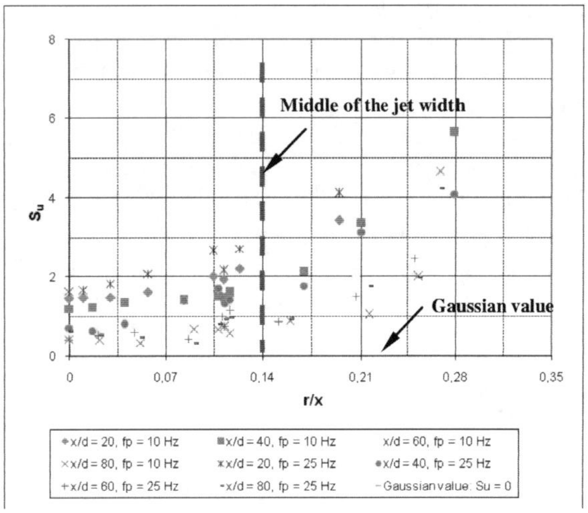

Figure 9. Skewness at U_e = 13.7 m/s.

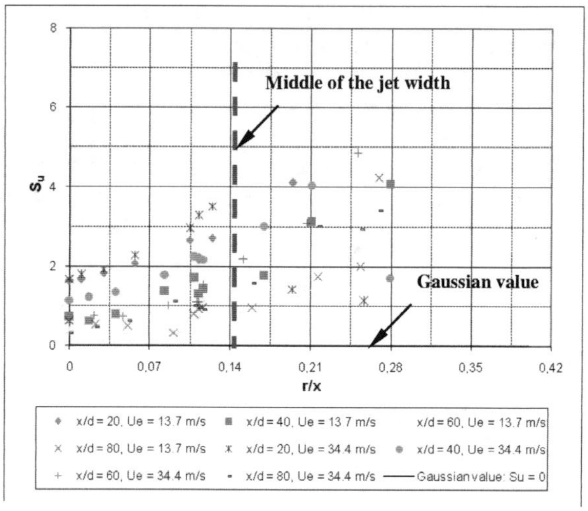

Figure 10. Skewness at f_p = 25 Hz.

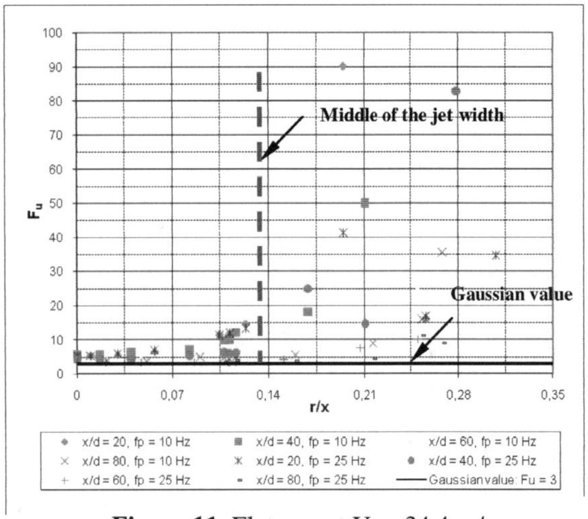

Figure 11. Flatness at U_e = 34.4 m/s.

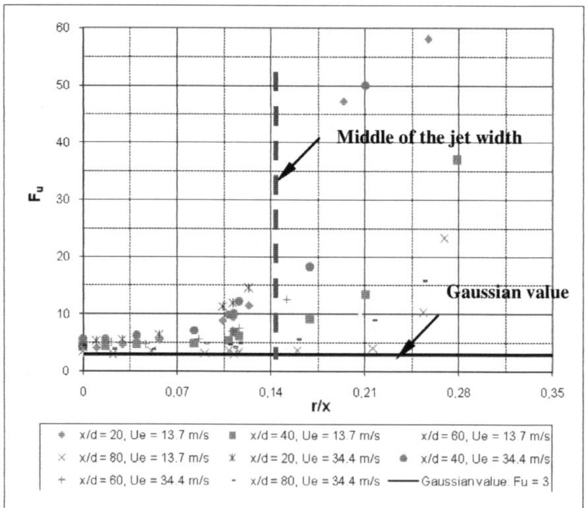

Figure 12. Flatness at f_p = 10 Hz.

From the figures, as the turbulent flow sensors were moved away from the jet centerlines, the value of both factors for the jets elucidate the aforesaid phenomena thus the departures from the strict

Gaussian values over the jet widths result. Concerning the turbulent statistics in the steady jet, the skewness in the inner region is relatively close with the Gaussian one ($S_u = 0$) for the locations very close to the jet centerline but that is not the case for the flatness factors as the closeness to the Gaussian value ($F_u = 3$) remains even up to the border of the inner region. Furthermore, comparing to the pulsed jet this is not the case in the inner region where it has the higher values than those of the steady couterpart. However, when comparing the values with those in the outer region of the pulsed jet, the influences of the Reynolds and Strouhal numbers on the turbulent statistics become significantly stronger in the outer region than in the inner region as previously explained. Such phenomena also occur for the steady jet but are only relevant for the influence of the Reynolds number as the jet is only defined by the parameter. As a *focal* point, it was found for the two distinct jets that the more intense-highly irregular motions occur in the outer region associated with the higher values of both factors than those of strict Gaussian values. The skewness and flatness factors of the pulsed jet were found to be substantially higher than those of the steady jet as a results of the modulations imposed. From Figure 13 and 14, it is also seen that the pulsed jet has a qualitatively wider flow field reaching the jet edge up to around $\frac{r}{x} \approx 0.28$ than that of the steady jet which reaches the jet edge of around $\frac{r}{x} \approx 0.22$.

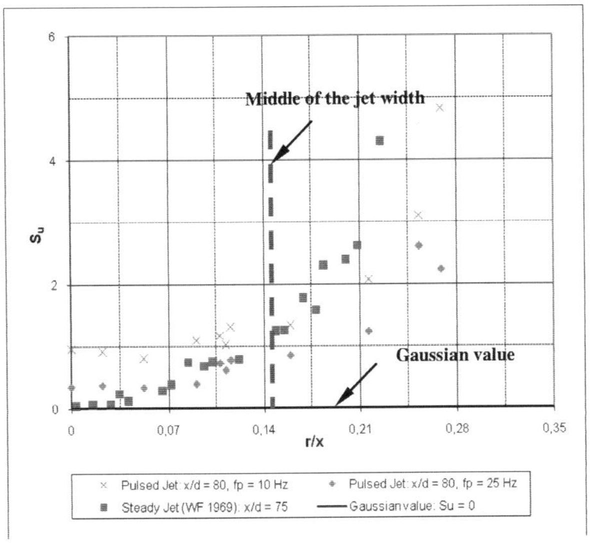

Figure 13. Comparison of the skewness between the steady and pulsed jets

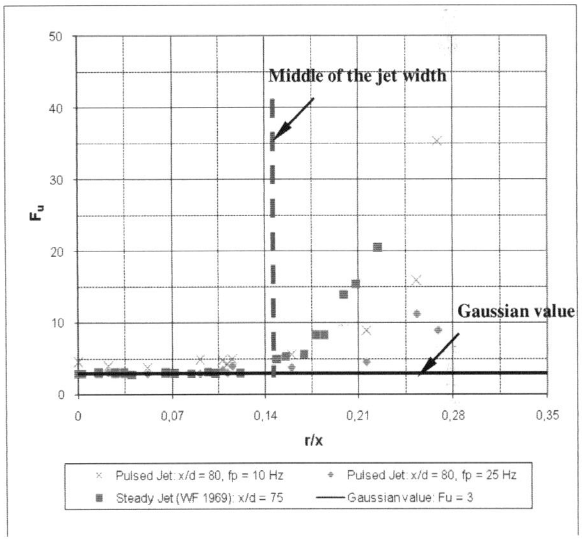

Figure 14. Comparison of the flatness between the steady and pulsed jets

4. Conclusion

Present results feature the responses of unsteady flow in a fully pulsed round jet to a range of the Reynolds and Strouhal numbers. The ensemble average technique allows evaluations of the first and higher moments of turbulence within a high degree of accuracy and proves a statistically stationarity of the intermittent flow [10]. Instead of the Reynolds number, imposing the additional parameter i.e the Strouhal number to define the pulsed jet certainly results in a significant change in the flow structure of the jet thus the self similarity in the first moment as is the case for a steady jet is no longer seen. Over the steady jet, key features of the pulsed jet include a wider jet width and more intense-highly irregular motions as a result of the mechanically excitations. However, a less fluctuating-pulsed jet is established at an increasing pulsing frequency. Therefore, the current findings offer the benefits for mixing enhancement in various industrial processes such as: sprays, combustor systems by using the fully pulsed axisymmetric jet as is the case in the earlier works [3, 4, 6, 12].

For the practical purposes, the present scheme also highlights the established knowledge on controlling the flow field of the fully pulsed jet. A less fluctuating-pulsed jet and narrower radial profile of the first moment are obtained at a constant mass flow rate with an amplified Strouhal number. Furthermore, at a constant Strouhal number as the mass flow rate is increased the augmentation of the fluctuating components at the jet centerline, u_0' and the narrower radial profiles of the first moment yield. For practical reasons, as an attempt to enhance the mixing process it therefore is desired to have a more fluctuating-pulsed jet and widening radial profile by setting a low mass flow rate and Strouhal number.

Transverse profiles of the third and fourth moments of the fluctuating components for the fully pulsed jet illustrate the nature of random process and were found to remarkably depart from the Gaussian values under the controlled parameters. The skewness and flatness factors of the mechanically excited jet were also found to be significantly higher than those of the steady jet over the jet widths thus leading to more intense-highly irregular interactions between the turbulent flow and the ambient fluid as well as more qualitatively spreading jet. Furthermore, stronger effects of the Reynolds and Strouhal numbers on the flow field take into account in the outer region than in the inner part of the jet width.

Concerning the complicated results of the radial profile of the first moment at the higher Reynolds number and Strouhal number, the present study can therefore be the *stimulus* for a further research in order to advance the knowledge on the turbulence nature in the jet. The crucial point here is to determine the *limit* whether the pulsed jet is still acceptable for the mixing enhancement compared to the steady jet. To have an efficient mixing, typically, one would avoid the narrower radial profiles produced at the higher jet exit velocity and pulsing frequency. As the radial profiles gradually become narrow under the increasing controlled parameters, the further research would therefore be useful to expand the upper limits of the current used parameters.

Acknowledgement
H. P. S. P. gratefully acknowledges the supervision of emeritus Professor Klaus Bremhorst, PhD, Australian Orders for his MPhil study at the University of Queensland as well as the research fellowship from the Technological Professional and Skill Development Project - Asian Development Bank (INO 1792) and Petra Christian University. It is expressly stated that all experiments were undertaken at Division of Mechanical Engineering, the University of Queensland. Further study and analysis of the final manuscript was undertaken by H. P. S. P. as a research fellow at TU Darmstadt under DIKTI Overseas Scholarship Programme. Finally, insightful comments from reviewers which have strengthened the manuscript and hot-wire repairs from Mr. George Dick are also gratefully acknowledged.

References
[1] Atassi, N., G. Charnay, J. Boree (1993). "Transient Behaviour of an Axisymmetric Turbulent Jet." Applied Scientific Research **51**: 137-142.
[2] Bogey, C. and Bailly C. (2009) "Turbulence and Energy Budget in a Self-preserving Round Jet: Direct Evaluation using Large Eddy Simulation". Journal of Fluid Mechanics **627**: 129-160.
[3] Bremhorst, K. and W. H. Harch (1979). "Near Field Velocity Measurements in a Fully Pulsed Subsonic Air Jet." Turbulent Shear Flows I, Springer Verlag, Berlin: 37-54.
[4] Bremhorst, K. and P. G. Hollis (1990). "Velocity Field of an Axisymmetric Pulsed, Subsonic Air Jet." AIAA Journal 28, no. 12: 2043-2049.
[5] Dahm, W. J. A. and P. E. Dimotakis (1990). "Mixing at Large Schmidt Number in the Self-Similar Far Field of Turbulent Jets." Journal of Fluid Mechanics 217: 299-330.

[6] Gehrke, P. J. (1997). The Turbulent Kinetic Energy Balance of a Fully Pulsed Axisymmetric Jet. PhD Thesis, Department of Mechanical Engineering. Brisbane, The University of Queensland, Australia: 378 pages.

[7] Hussein, J. H., S. P. Capp, William K. George (1994). "Velocity Measurements in a High-Reynolds-Number-Momentum-Conserving, Axisymmetric, Turbulent Jet." Journal of Fluid Mechanics **258**: 31-75.

[8] Panchapakesan, N. R. and J. L. Lumley (1993). "Turbulence Measurements in Axisymmetric Jets of Air and Helium, Part I: Air Jet." Journal of Fluid Mechanics **246**: 197-223.

[9] Pratomo, Hariyo P. S., Klaus Bremhorst (2012) "Experimental Study on the Propagation of a Pulsed Jet." Elsevier Procedia Engineering 50: 174-187.

[10] Pratomo, Hariyo P. S. (2006). Mean Velocity and Turbulence Measurements in a Pulsed Jet. MPhil Thesis, Division of Mechanical Engineering. Brisbane, The University of Queensland, Australia: 224 pages.

[11] Tanaka, Y. (1984). "On the Structure of Pulse Jet." Bulletin of JSME 27, no. 230: 1667-1674.

[12] Winter, A. R. (1991). A Laser Doppler Anemometer Investigation into Fully Pulsed Jet Flows with an Examination of Velocity Bias Error. PhD Thesis, Department of Mechanical Engineering. Brisbane, The University of Queensland, Australia.

[13] Wygnanski, I. and H. Fiedler (1969). "Some Measurements in the Self-Preserving Jet." Journal of Fluid Mechanics 38: 577-612.

The Influence of BTO-BHF Different Composition on its Ferroelectric Properties

M Hikam, D Suastiyanti, S R Adnan and B Soegijono

Department of Physics, Faculty of Mathematics and Sciences, Universitas Indonesia, Depok 16424 Indonesia.

E-mail: hikam@sci.ui.ac.id

Abstract. Barium Titanat and Barium Hexaferrite composite was synthesized by sol-gel method in different composition and holding sintering time. The crystallographic property was characterized by x-ray diffraction and analysed by GSAS program, the saturation and remanent electrical polarisation were measured using Sawyer Tower circuit. The best composition for multiferroic candidate is BTO:BHF 1:1 at 5 hours sintering time.

1. Introduction

As individual material, both Barium Titanat ($BaTiO_3$) and Barium Hexaferrite ($BaFe_{12}O_{19}$) are well known respectively as ferroelectric [1,2,3] and ferromagnetic [4,5] materials, however their combination is rarely looked at as multiferroic materials. Already known, these materials possess more than onetype of ferroic order such as ferromagnetic, ferroelectric, ferroelastic andferrotoroidics. This class of materials finds promisingapplications in nonvolatile memory, memory-cellcapacitor, electromagnetic-interference filter, sensorsand so forth [6,7].

In this paper we present our current work on synthesizing BTO - BHF and exploring the possibility of having them as multiferroic devices. We prepared BTO-BHF composant by sol-gel method in our laboratory at Physics Department, University of Indonesia. The specimen was then analyzed by x-ray diffraction and quantitatively by GSAS program. We further explored its electrical polarisation saturation and remanent to seek the possibility BTO-BHF composant as multiferroic materials.

2. Experiment Method

By using regular route, the BHF-BTO composite are synthesized by using sol gel technique[5,8]. We start to synthesized BHF and followed by BTO synthesized. The starting materials for BHF were iron nitrate $Fe(NO_3)_3.9H_2O$, barium nitrate $Ba(NO_3)_2$, citric acid ($C_6H_7O_8$) and ammonia solution (NH_4OH), all of with the analytical purity (99.0%) from Merck KGaA Chemicals, Damstadt, Germany. The appropriate amount of $Fe(NO_3)_3.9H_2O$ and $Ba(NO_3)_2$ in molar ratios of Ba^{2+}/Fe^{3+} of 1:12 and 1:11.5 were dissolved in a minimum amount of deionized water to get a clear solution. Citric acid was added into the prepared aqueous solution to chelate Ba^{2+} and Fe^{3+}, stirring for a few minutes until turning into the solution. Then ammonia solution was added to adjust the pH of 7 until the mixed solution getting into brown. The solution was evaporated to dryness by heating at 80-90°C on a hot plate with continuous stirring. As the water evaporated, the solution became viscous and finally it formed a very viscous brown gel then kept it in oven at 150°C for 5 hours. Most of the moisture in the viscous brown gel was evaporated and formed dried gel after heating at 450°C for 24 hours. After grinding in the mortar, the dried gel was sintered at 850°C for 10 hours respectively.

The starting materials for BTO were barium nitrate $Ba(NO_3)_2$, Titanium Oxide (TiO_2), citric acid ($C_6H_7O_8$), ammonia solution (NH_4NO_3), and nitric acid (HNO_3). The appropriate amount of $Ba(NO_3)_2$, TiO_2, were reacted by NH_4NO_3. Citric acid ($C_6H_7O_8$) were add into result compound with ratio $C_6H_7O_8$:NH_4NO_3 (1:1 and 2:1). The solution was evaporated to dryness by heating at 80-90°C on a hot plate with continuous stirring for 5 hours. As the water evaporated, the solution became viscous and finally formed a very viscous brown gel then kept it in oven at 100°C for 1 hours. Most of the moisture in the viscous brown gel was evaporated and formed dried gel after heating at 500°C for 12 hours. After grinding in the mortar, the dried gel was sintered at 700°C for 2 hours respectively.

Both result compound were mixed with weight fraction BTO:BHF (1:1, 1:2, 1:3) and heated at 925°C for 5, 10 and 15 hours in sintering process. XRD and General Structure Analysis System (GSAS)program were used for crystallographic properties andelectrical properties were characterized to know their ferroelectric properties [5,8,9].We have utilized Sawyer Tower circuit to measure these electrical properties. The circuit is conected to our home made instrument (Fig. 1) which consists of voltage attenuators, an adjustable HVPS, a DAQ card and a PC.

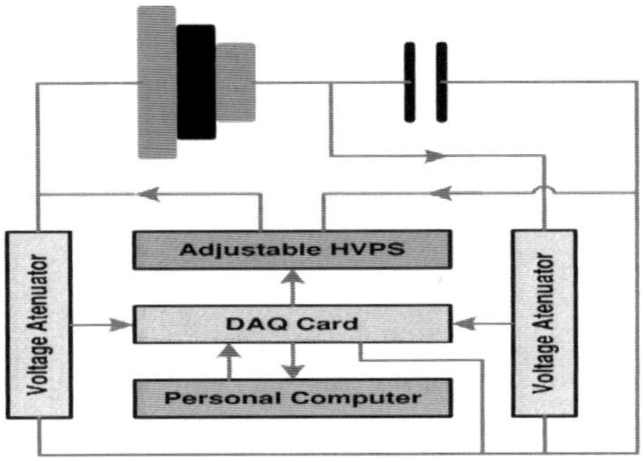

Figure 1. Electrical Hysterisis Loop Measurement

3. Result and Discussion

For three different composition of BTO:BHF and three different sintering temperature, we have examined the crystallographic property and the real composition by using x-ray diffraction (XRD). The results for 5 hours sintering at different weight fraction are shown in Fig 2 and the results for different sintering time at weight fraction of 1:1 are shown in Fig.3.

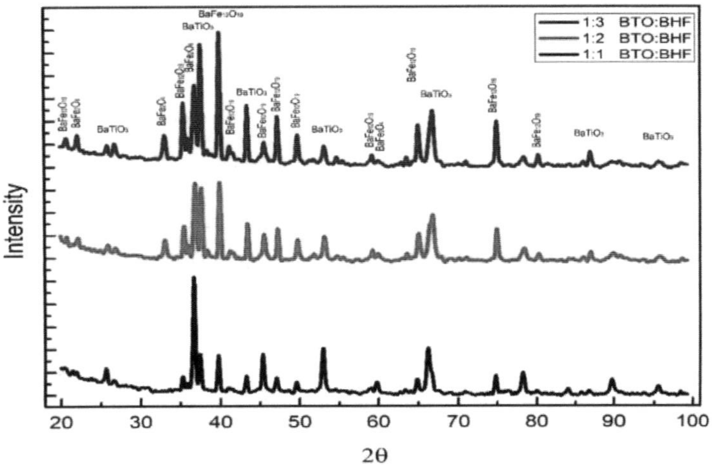

Figure 2. XRD pattern of BTO-BHF composite with variation composition for 5 hours

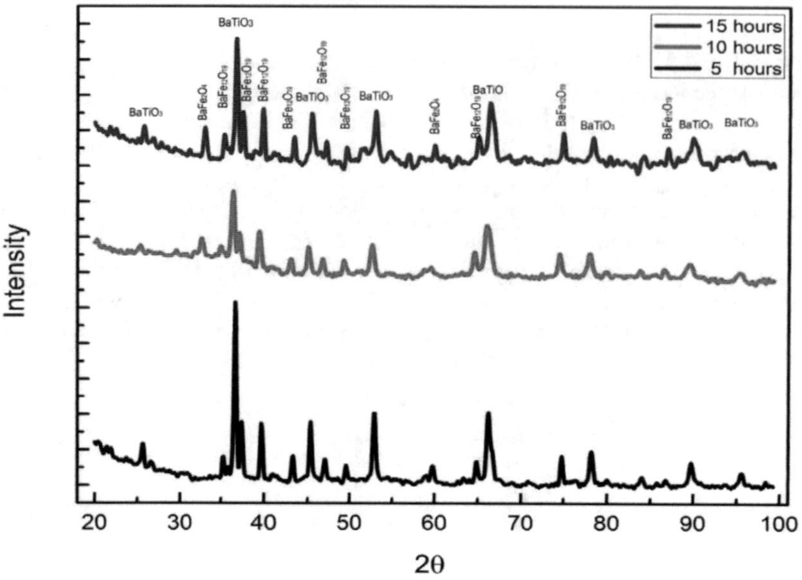

Figure3. XRD pattern of 1:1 composition with temperature variation of BTO-BHF composite

The ICDD database running on PCPDFWin is utilized to get the crystal structure, space group and lattice parameters. The search-match found the existence of $BaTiO_3$, $BaFe_{12}O_{19}$ as well as one satelite phase $BaFe_2O_4$. The peaks increase or decrease in accordance with the prescripted composition. There is no residual phase $BaFe_2O_4$ in weight fraction of 1:1 at 5 hours sintering. The XRD data is further examined deeply by refinement with HighScore Plus Software to know the exact composition, crystallographic parameters and crystallite size. However, what we concern the most is their composition. Figure 4 showsrefinement results for BTO:BHF with 1:1 intended composition respectivelly at 5, 10 and 15 hours sintering hold time. Figure 5 shows refinement results for BTO:BHF with 1:2 intended composition respectively at 5, 10 and 15 hours sintering hold time and Fig 6. Shows the similar case for BTO:BHF with 1:3.

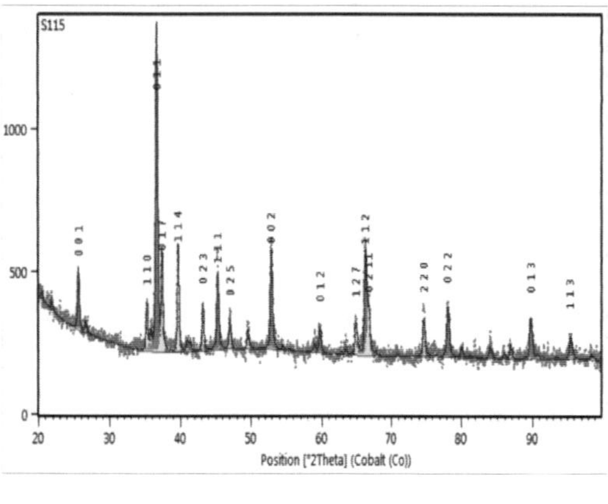

(a).

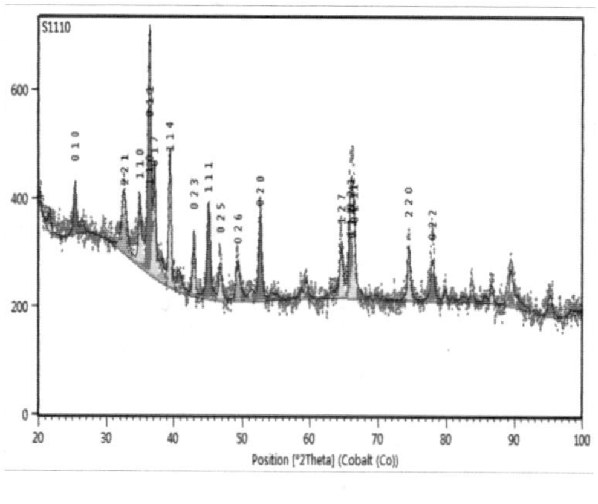

(b).

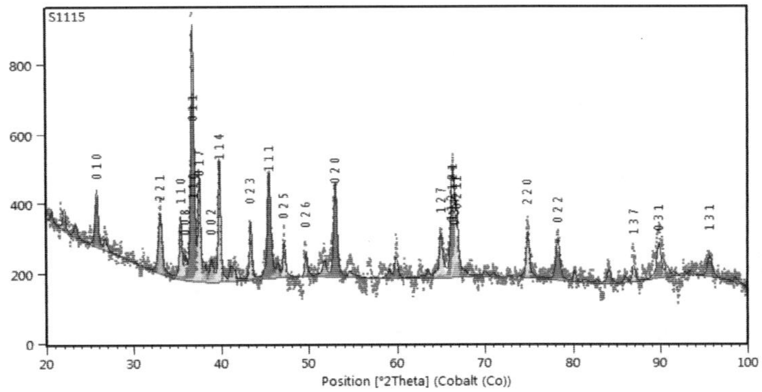

(c).

Figure 4. Refinement Results for BTO:BHF 1:1 (a) t = 5 hours (b) t=10 hours (c) t=15 hours

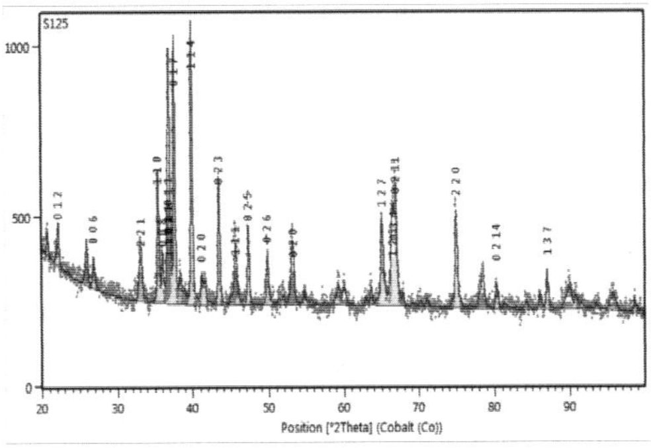

.(a).

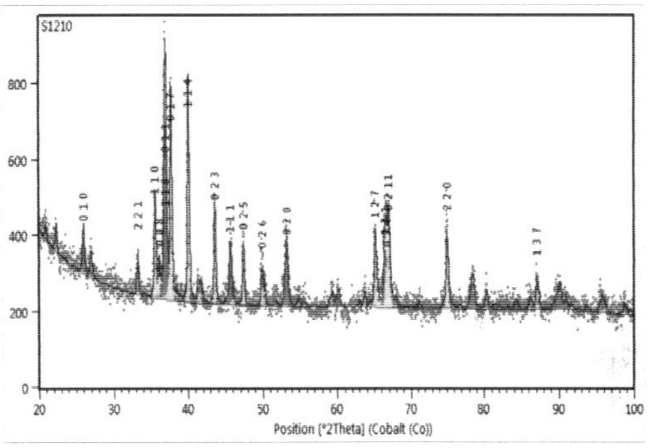

(b).

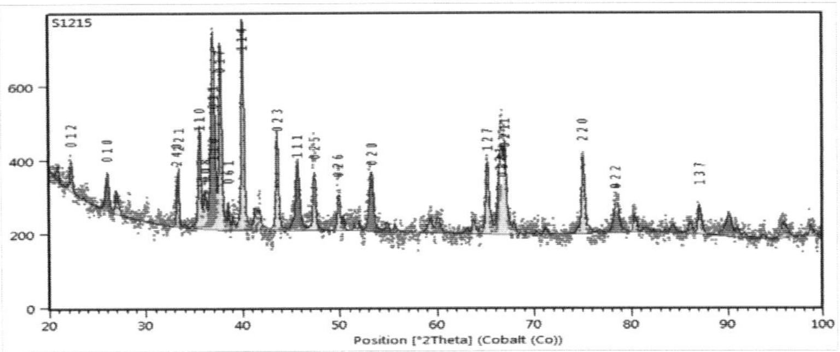

(c).

Figure 5. Refinement Results for BTO:BHF 1:2 (a) t = 5 hours (b) t=10 hours (c) t=15 hours

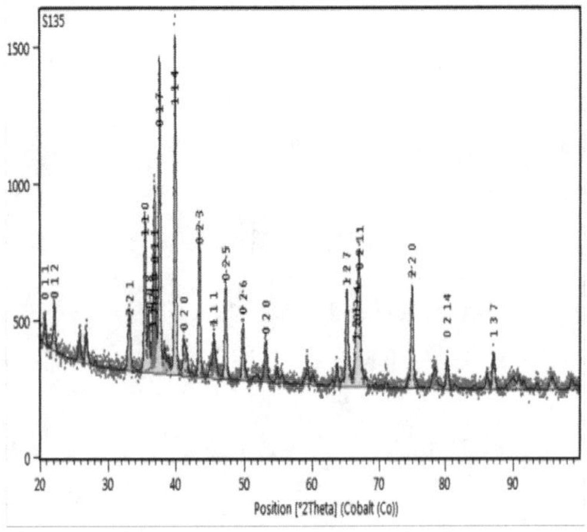

(a).

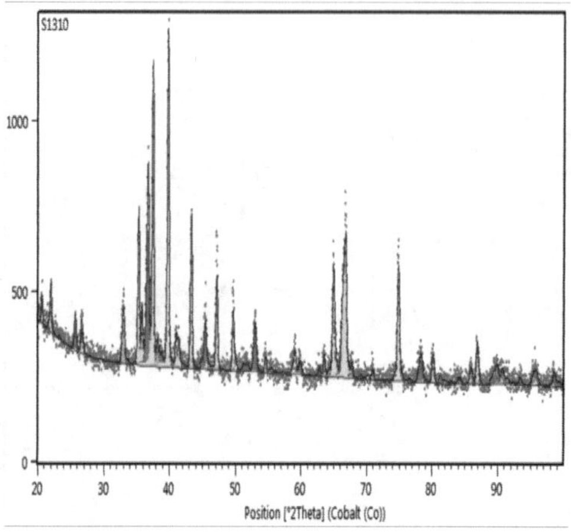

(b).

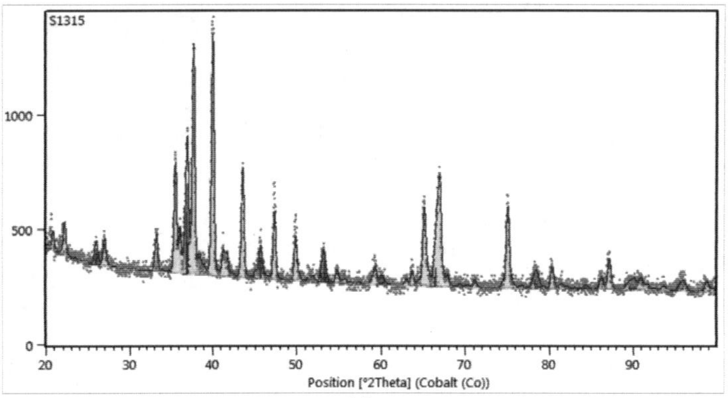

(c).

Figure 6. Refinement Results for BTO:BHF 1:3 (a) t = 5 hours (b) t=10 hours (c) t=15 hours

From the results of refinement from Fig. 4, 5 and 6, it could be seen that almost of all samples have residual phase $BaFe_2O_4$ in plane (221) except at weight fraction of 1:1 for 5 hours sintering time. The figures also show that longer sintering time and higher sintering temperature, the amount of $BaFe_2O_4$ phase getting increase as shown in Table 1.

Table 1. Composition (Wt%) of BTO-BHF Composite for several sintering temperature variaton

BTO:BHF	t : 5 hours			t : 10 hours			t : 15 hours		
	BTO	BHF	$BaFe_2O_4$	BTO	BHF	$BaFe_2O_4$	BTO	BHF	$BaFe_2O_4$
1:1	44.2%	55.8%	0%	27.5%	49.5%	7.02%	35.2%	48.6%	11.38%
1:2	25.4%	49.4%	5.25%	23.1%	42.1%	8.56%	22.8%	38.1%	16.24%
1:3	20.2%	70.5%	6.92%	23.5%	69.5%	9.16%	23.6%	65%	22.98%

If the sintering process getting longer and higher, it will hapen the decomposition of $BaFe_{12}O_{19}$ whereas Ba^{2+} ion with Fe_2O_3 will form the third phase $BaFe_2O_4$ in plane (221).We summarized the prescripted and real composition in Table 1. The table also shows the variation condition for several sintering times.

After we convinced the existence of BTO and BHF, we examine the electrical polarization saturation and remanent. The some results of electrical characterization are shown in Fig 7,8 and 9 as an electrical hysteresis loop.

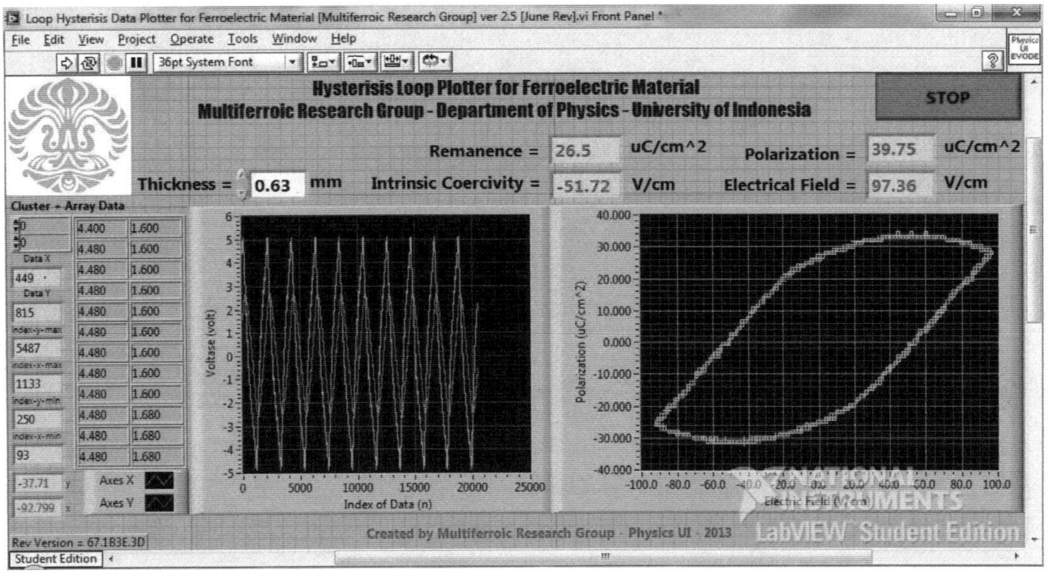

Figure 7. Hysteresis Loop of BTO:BHF 1:1 for 5 Hours Sintering Time

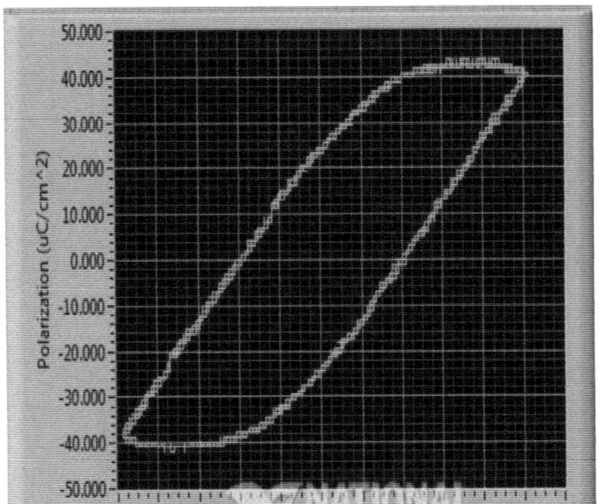

Figure 8. Hysteresis Loop of BTO:BHF 1:1 for 10 Hours Sintering Time

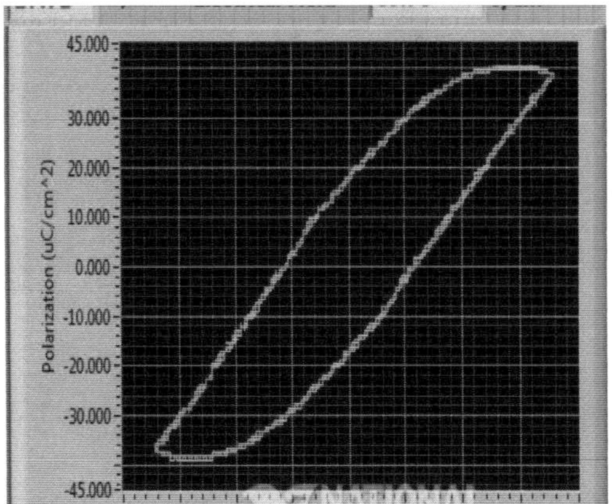

Figure 9. Hysteresis Loop of BTO:BHF 1:1 for 15 Hours Sintering Time

Clearly as indicated in the Fig. 7,8 and 9 that for all cases we found that all samples show a form of hysteresis loop with different value of electrical polarisation, coersive and remanent. The data could be converted in graph as shown in Fig 10, 11 and 12.

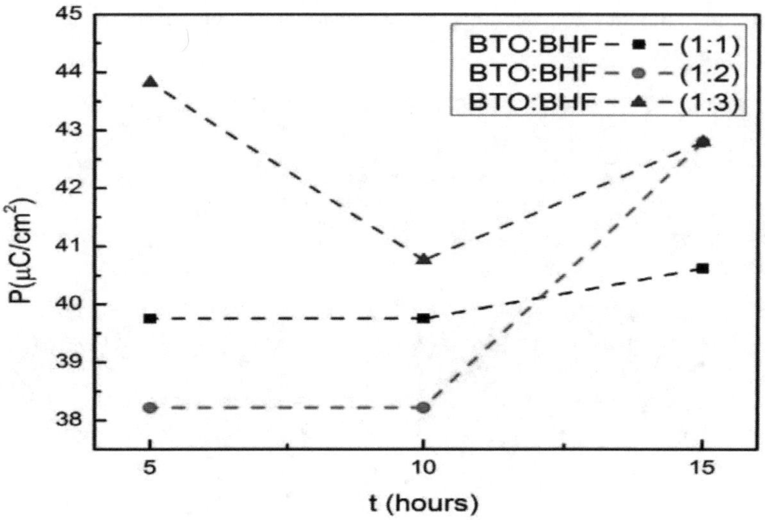

Figure 10. Electrical Saturation Polarisation for Several Sintering Process Duration

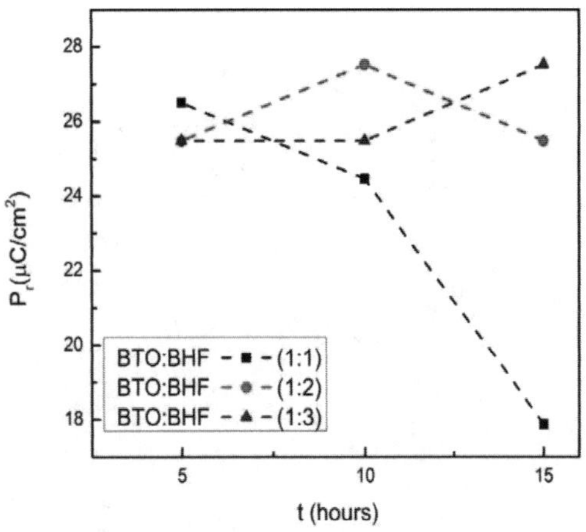

Figure 11. Electrical Saturation Polarisation for Several Sintering Process Duration

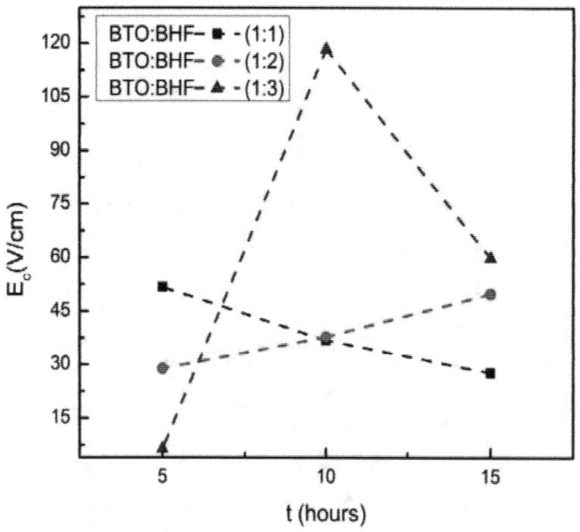

Figure 12. Coersive Electric Field of BTO-BHF Composite at different Composition

From Fig 10 it could be seen that the value of electrical saturation polarisation decreasesuntil 2 parts of weight fraction of BHF, but if the weight fraction of BHF exceeds from 2 parts (especially at ratio of 1:3) the value of electrical saturation polarisation increases. On theotherhand the Fig. 11 shows that the electrical remanent polarisation decreases as the BHF composition increases. We further examined the coersive field of the materials. Unfortunetely, the Fig. 12 indicates the strenght of the field is inconclusive if we compare with its composition, this is probably due to the unknown nature of the third phase, $BaFe_2O_4$.

4. Summary

We can conclude that BTO:BHF prepared by sol-gel method shows good crystalinity and the ferroelectric property is well observed. The composition 1:1 at 5 hours sintering time is the most stable and no residual phase, hence judging from its saturation and remanent electrical polarisation, it can be a good candidate for multiferroic devices.

5. Acknowledgment

This work was partially supported by SINas research grant 2013 under contract no. 26/SEK/INSINAS/PPK/I/2013, Ministry of Research and Technology, The Republic of Indonesia.

References

[1] Wahl R, Vogtenhuber D, and Kresse G2008*Phys. Rev.***B** 78104116
[2] Bell A J and Cross L E1984 *Ferroelectrics***59** 197
[3] Li YL, CrossLE, and Chenl LQ2005*J. Appl. Phys.***98** 064101
[4] Liu Y, Yin H S,Liu Y, Sun K, Li Y, Wang L, Xu W J 2011 *Adv. Mat. Research.***Vols 239-242**pp 3039-3042
[5] Suastiyanti D, Soegijono B, Hikam M 2013 *Adv. Mat. Research***Vol. 789** pp 87-92
[6] Spaldin N A, Cheong S W and Ramesh R 2010*Phy. Tod.***38**
[7] Fishman RS, Haraldsen JT, Furukawa N, and Miyahara S 2013*Phys. Rev.***B 87** 134416
[8] Suastiyanti D, Soegijono B, Hikam M 2013 *Adv. Mat. Research***Vol. 789** pp 118-123
[9] Andika R and Hikam M 2013*Adv. Mat. Research***Vol. 789** pp 101-104

Intrinsic Hysteresis Loops Calculation of BZT Thin Films

M Hikam and S R Adnan

Department of Physics, Faculty of Mathematics and Sciences, Universitas Indonesia, Depok 16424 Indonesia.

Email: hikam@sci.ui.ac.id and septian.rahmat@ui.ac.id

Abstract. The Landau Devonshire (LK) simulation is utilized to calculate the intrinsic hysteresis properties of Barium Zirconium Titanate (BZT) doped by Indium and Lanthanum. A Delphi program run on Windows platform is used to facilitate the calculation. The simulation is very useful to calculate and understand the Gibbs free energy and the relationship between spontaneous polarization and electric field.

Keywords.Landau Devonshire, BZT, Gibbs Energy, Hysteresis

1. Introduction

Barium Zirconium Titanate (BZT) is well known material utilized for ferroelectric devices such as FRAMs and sensor [1,2]. Therefore a theoretical consideration of the material behaviour is needed to predict its hysteresis characteristics, especially when the materials are altered by some dopants. Most of ferroelectric materials in practical applications are of first order transition, but the theoretical approach of this transition is much more complicated than the second order phase transition. Landau Devonshire (LD) is known as a static version of first order phase transition theory [3,4,6,7], it is used to understand ferroelectric fenomena such as phase transition and hysteresis properties. In this paper, the applicability of the LD equation for Barium Zirconium Titanate thin films Gibbs free energy and hysteresis loops is demonstrated. Different compositions of BZT doped by Lanthanum and Indium are compared and the ferroelectric bahaviour change is analyzed.

2. Simulation

The phenomenological investigation of static properties of Barium Zirconium Titanate (BZT) are studied with the aid of Landau Devonshire (LD) equation. The LD free energy density for materials may be written [3,4]:

$$\Delta G(P,E)=\frac{a(T-T_0)}{2\varepsilon_0}P^2-\frac{B}{4\varepsilon_0^2}P^4+\frac{C}{6\varepsilon_0^3}P^6-EP \qquad (1)$$

where P is polarization, E is electric field, T, T_0 are temperature of sample and a, B, C are constant. The permitivity ε_0 factor in Eq. 1 is to confirm that a, B, C are simple mechanical dimensions [3]. The term in P^6 in eq. 1 is maintained to make dimensionless variables, hence the equation can be rewritten as:

$$\Delta G(P,E)=\frac{1}{2}tP^2-\frac{1}{2}P^4+\frac{1}{6}P^6-EP \qquad (2)$$

where t is dimensionless variables from crystallographic data from sample. This constant (t) shows the phase of the materials and indicates its Gibbs free energy. One can draw the relation between the Gibbs energy as function of its polarization using the eq. 2 to get a rough idea of its behaviour. This energy is minimum, therefore:

$$\frac{\partial \Delta G(P,E)}{\partial P} = tP - 2P^3 + P^5 - E = 0 \tag{3}$$

Hence the energy can be written:

$$E = tP - 2P^3 + P^5 \tag{4}$$

The polarization of material can simply be written as the following:

$$p = \sum_{i=1}^{n} q_i r_i \tag{5}$$

where p is momen dipole, q_i is atomic charge and r_i is atomic radii. This equation was used to calculate momen dipole of BZT:

$$P_s = \frac{pxa}{V} \tag{6}$$

where P_s is spontaneous polarization and V is volume of crystal.

To obtain the spontaneous polarization, one should find the saddle point of the Gibbs free energy, hence the second derivative is zero. The eq. 3 becomes as the following.

$$\frac{\partial^2 \Delta G(P,E)}{\partial P^2} = t - 6P^2 + 5P^4 = 0$$

By solving this equation, one can obtain the spontaneous polarization as function of t values [3, 8].

$$P_s^2 = -\frac{3}{5}\left(1 + \left[1 - \frac{5t}{9}\right]^{1/2}\right) \tag{7}$$

Using spontaneous polarization of BZT calculated in the eq. 6, the value of t can be obtained. These t values are utilized to obtain the Gibbs energy from the LD theory. The similar scheme is also utilized to calculate the one of BZT doped by La and In. Here the calculation was on the assumption that both La and In are soft dopant where these ions have caused the raise of effective dipole moments. In this model, the interacting dopant replaces the atomic position of Ba on the crystal structure where the lattice parameters are shown in the table 1. This table also shows the complete result of dimensionless contants t for BZT and BZT doped by La and In.

Table 1. Lattice Parameters (Å) and dimensionless constant (t)

$BaZr_{0.1}Ti_{0.9}O_3$ Doped	a	c	V	t
0	3.967	3.988	62.759	2.536
1% La	3.807	3.849	55.784	3.209
2% La	3.999	4.018	64.255	2.419
3% La	3.904	3.92	59.745	2.798
4% La	3.979	4.00	63.329	2.490
1% In	3.974	4.052	63.991	2.439
2% In	3.982	4.034	63.964	2.441
3% In	3.979	4.051	64.137	2.428
4% In	4.028	4.09	66.359	2.268

The hysteresis loop curves depend on the temperature difference ($T - T_0$) or t in this case. On the extreme situation t has four conditions [4] which are t_{sc} supercooling, t_c critical temperature, t_{SH} super heating and t_2 upper temperature limit of ferroelectric phase induced by electrical field.

The calculation is performed by a home made Delphi program running on Windows platform. Delphi is an objected oriented programming language as well as an integrated development environment. We have used those facilities for calculation and created graphic based routines to make it easier for visual analysis. The flowchart of the calculation is shown in the fig. 1.

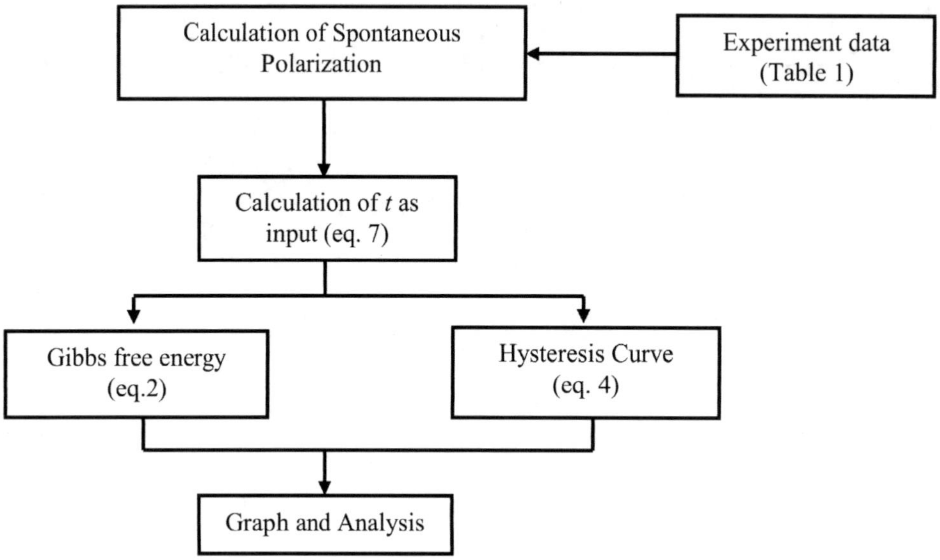

Figure 1. Simulation Method

3. Result and Discussion

Polarization switching process and electrical field influence of BZT and BZT doped by La and In (BLZT and BIZT) thin films can be illustrated by using the LD static model. The fig. 2 shows the Gibb free energy of BZT on varying the electrical field (e) at t: 2.536 x=0.1. It is clear that for positive polarization the minimum Gibbs free energy decreases on increasing the electrical field, on the contrary for negative polarization the minimum Gibbs free energy increases in the same situation. Hence, on increasing the electrical field the polarization of ferroelectric material reaches its maximum value.

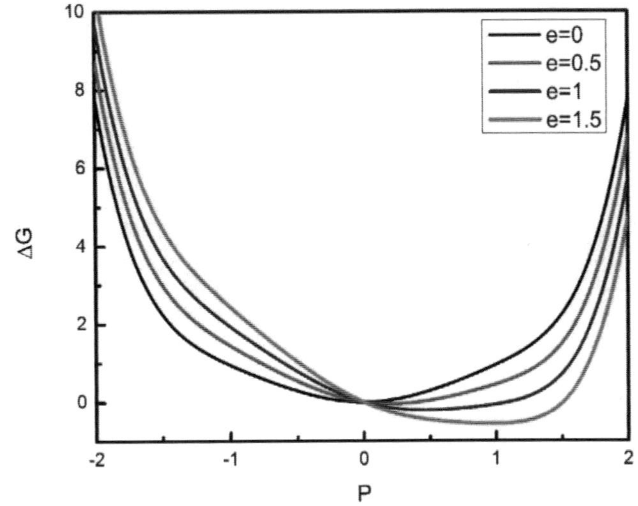

Figure 2. Gibbs Free energy of $BaZr_{0.1}Ti_{0.9}O_3$ at various electric field (e)

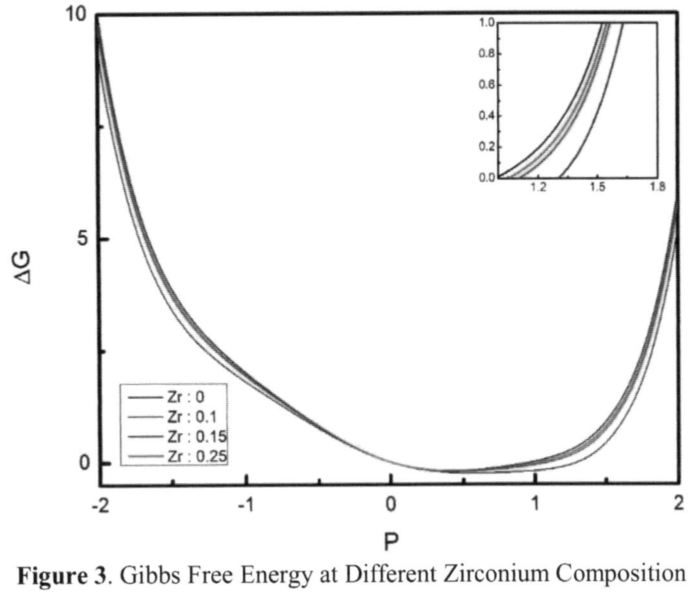

Figure 3. Gibbs Free Energy at Different Zirconium Composition

Fig. 3 shows the Gibbs free energy as function of polarization for different zirconium composition. The lowest energy occurs at x : 0.15 composition where Ti^{4+} ions replace the position of Zr^{4+} ions, here we can say that the Zr^{4+} ions are more stable compared with the Ti^{4+} ions, this finding is in accordance with the experiment results by other authors [1,2]. Nevertheless continously increasing the zirconium composition has caused the Gibbs free energy increases, this might be due to there is a limitation of the Zr^{4+} ions that can replace Ti^{4+} ions in BZT. Increasingly Zr^{4+} ions within BZT causes the degree of crystalinity of BZT decreases [2].

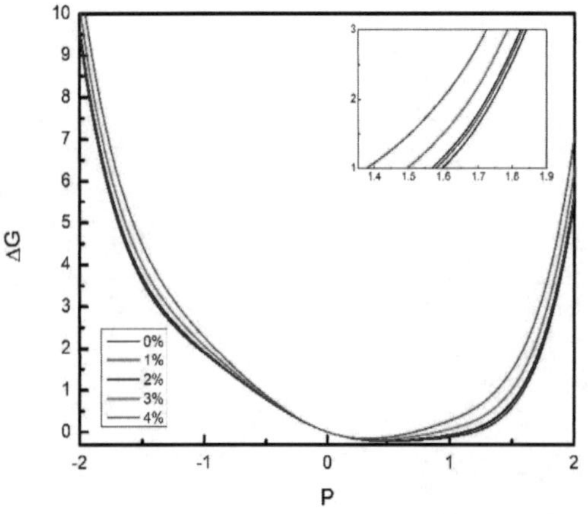

Figure 4. Gibbs Free Energy of BZT doped By Lanthanum.

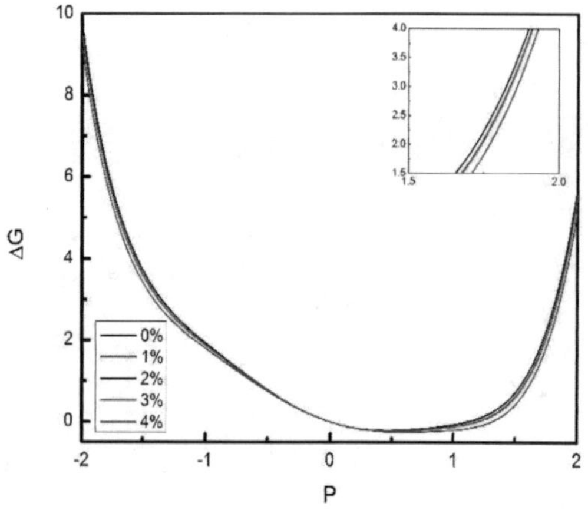

Figure 5. Gibbs Free Energy of BZT doped By Indium

The Gibbs free energy of BZT doped by La is shown in the fig. 4, it is clear that La doping onto BZT has caused the Gibbs free energy increases, the replacement of Ba^{2+} ions by La^{2+} ions has caused some damage of BZT crystal [5]. On the contrary, the situation is different for BZT doped by In, here the Gibbs free energy decreases on increasing the dopant composition. The lowest Gibbs free energy occurs at 4% Indium dopant composition.

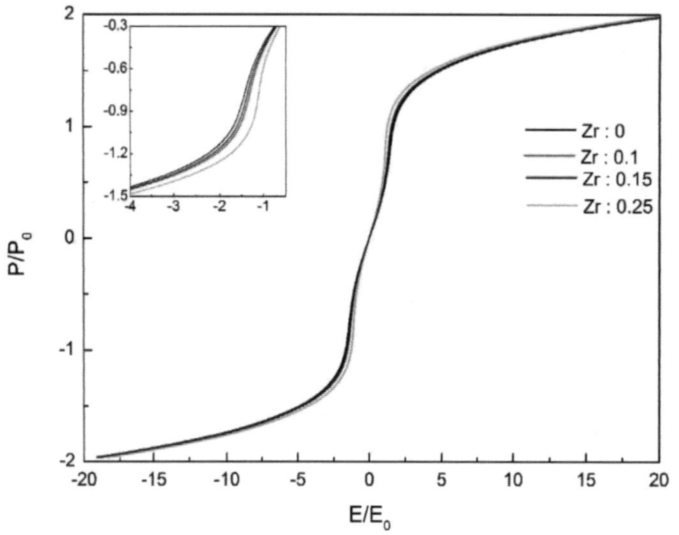

Figure 6. Hysteresis Loop of $BaZr_xTi_{1-x}O_3$

From the grouping of ferroelectric materials, it can be seen that the value of *t* related to the increasing electrical field could influence the stability of the phase of the material. For the values of *t* > 1, it shows the phase of material on the time of measurement is ferroelectric, based on estimation using the lattice parameter data of BZT. The calculation shows that there are slight displacement of the polarization in the fig. 6 on increasing the Zr composition. The highest polarization occurs at *x* : 0.25 zirconium composition.

When one looks closely both the fig. 2 and the fig. 6, it appears that the Gibbs free energy depends on the polarization and the electric field, meanwhile the fig. 6 itself shows that the polarization is closely related to the electric field and the hysteresis effect becomes slightly more prominent when one increases the concentration of the Zr composition. Therefore it is a clear indication that the Gibbs free energy increases on adding the Zr concentration. One should take this in a great consideration since the more Gibbs free energy of the specimen, the material becomes more unstable.

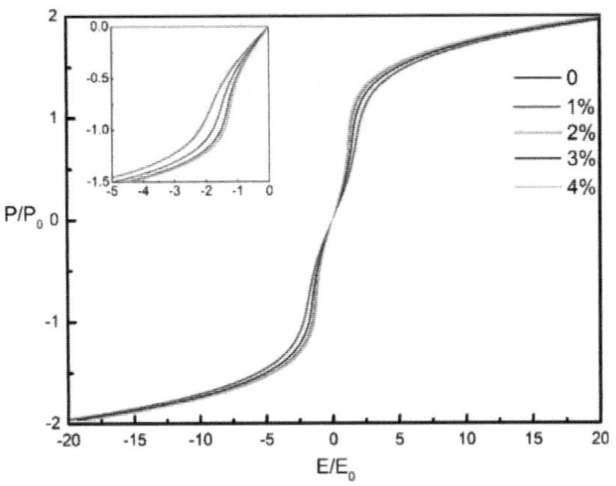

Figure 7. Hysteresis Loops of $BaZr_{0.1}Ti_{0.9}O_3$ Doped by Lanthanum at Different Composition

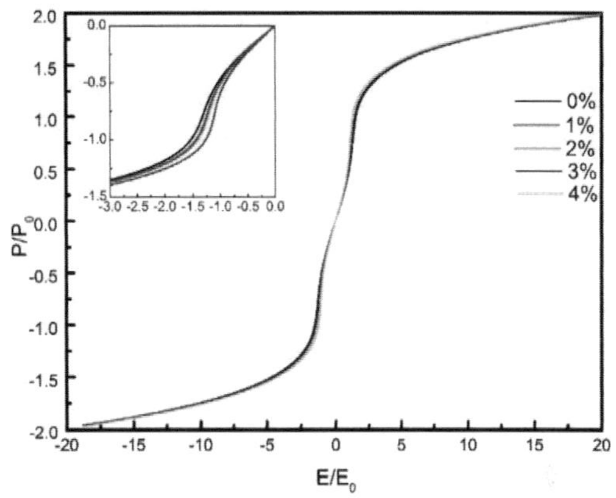

Figure 8. Hysteresis Loops of $BaZr_{0.1}Ti_{0.9}O_3$ Doped by Indium at Different Composition

For the BZT with Zr = 0.1, the material is further doped by Lanthanum and Indium. Compared with the original property of BZT in the simulation, a similar occurance also happens on the ferroelectric property of BZT doped by Lanthanum. The fig. 7 shows that the hysteresis loops shift on increasing the La dopant, whereas the fig. 8 shows that only a minor change arises for the case of In dopant.

In the future we will further examine the behaviour of BZT by more dopants using the calculation of dynamic model of Landau-Devonshire also known as Landau-Kalathikov (KL) model [3] to gain more understanding of its ferroelectric properties and to find the suitable dopant. Since the samples of BZT doped by In and La are already available, we would develop Sawyer Tower circuit to measure the electrical property of these materials and make a comparation with the calculated model.

4. Summary

From calculation using the Landau Devonshire (LD) model for BZT films, we found that lanthanum and Indium doped onto BZT have caused shifting both on the Gibbs free energy and the hysteresis curve. The lowest Gibbs free energy occurs at 4% Indium dopant composition, while the highest polarization occurs at x : 0.25 zirconium composition. The Simulation of the LD theory is proven workable to understand effect of dopant composition on Barium Zirconium Titanate.

5. Acknowledgment

This work was partially supported by SINas research grant 2013 under contract no. 26/SEK/INSINAS/PPK/I/2013, Ministry of Research and Technology, The Republic of Indonesia.

References

[1] Liu A, Xue J, Meng X, Sun J, Huang Z, Chu J 2008 *App. Surface Scie.* **254**5 660-5663
[2] Jiwei Z, Xi Y, Liangying Z, Bo S, Chen H 2004 *J. Crys. Growth* **262** 341-347
[3] Wang C 2010 *Theories and Methods of First Order Ferroelectric Phase Transitions, Ferroelectrics* (Rijeka: InTech)
[4] Tan E K, Osman J, Tilley D R 2001 *Sol. State Comm.* **117** 59-64
[5] Uchino K 2000 *Ferroelectric Devices* (New York: Marcel Dekker Inc)
[6] Devonshire A F 1954 *Phil. Mag.* **30** 10
[7] Song T K 2005 *J. Korean Phys. Soc.* **Vol.46** No.1 pp 5-9
[8] Yogaraksa T 2008 *Studi Pembalikan Polarisasi dan Model Histerisis pada Material Film Tipis Feroelektrik* $Ba_{0.5}Sr_{0.5}TiO_3$ *yang Didadah Mg, Mn, Cu dan Pb* (Jakarta: Thesis Universitas Indonesia)

Journal of Physics: Conference Series **495** (2014) 012009 doi:10.1088/1742-6596/495/1/012009

Temperature effects on diffusion coefficient for 6-gingerol and 6-shogaol in subcritical water extraction

Nor Ilia Anisa [1, 2], **Noor Azian** [1, 2,*], **Mohd Sharizan** [1, 2], **Yoshio Iwai**[3]
[1]Malaysia-Japan International Institute of Technology, Universiti Teknologi Malaysia, 54000 Kuala Lumpur, Malaysia
[2]Center of Lipids Engineering and Applied Research (CLEAR), Universiti Teknologi Malaysia, 54000 Kuala Lumpur, Malaysia
[3]Department of Chemical Systems and Engineering, Kyushu University, Fukuoka 812-8581, Japan

E-mail: azian.morad@gmail.com

Abstract. 6-gingerol and 6-shogaol are the main constituents as anti-inflammatory or bioactive compounds from zingiber *officinale* Roscoe. These bioactive compounds have been proven for inflammatory disease, antioxidatives and anticancer. The effect of temperature on diffusion coefficient for 6-gingerol and 6-shogaol were studied in subcritical water extraction. The diffusion coefficient was determined by Fick's second law. By neglecting external mass transfer and solid particle in spherical form, a linear portion of Ln (1-(Ct/Co)) versus time was plotted in determining the diffusion coefficient. 6-gingerol obtained the higher yield at 130°C with diffusion coefficient of $8.582 x 10^{-11}$ m^2/s whilst for 6-shogaol, the higher yield and diffusion coefficient at 170°C and 19.417×10^{-11} m^2/s.

1. Introduction

Ginger, the rhizome of Zingiber officinale is one of widely used species of the zingiber family (zingibereace) and is common used for various foods and beverages. Oleoresin (extract crude) has medicine values. The oleoresin of ginger contains the pungent and non-pungent constituents, also in addition to the volatile oil [1]. The major constituent of ginger is lipophilic rhizome extracts, have yielded potentially to active gingerols. The substances which are phenolic ketones are responsible for the pungent flavour of fresh ginger and known as; 4-, 6-, 8-, 10-, and 12-gingerol. During thermal processing or storage, the gingerols may be modified to a series of homologous compounds known as shogaols (8- and 10-shogaols). These gingerols and shogaols had been reported to possess the following activities; anti-inflammatory disease [2], antioxidatives [3] and anticancer [4].

Subcritical water extraction (SWE) is an alternative method on extraction the plant for pharmaceutical proposes since this method deals with the usage of water only. The SWE is operating at elevated temperature; above than boiling point to near critical point of water (100-374°C) at moderate pressure for maintain the water in liquid state. From Mustafa and Turner [5], the principle of green technology is to reduce the using of harsh organic solvent whilst encourage the use of novel extraction technique that are known to be more friendly. Indeed, SWE is direct method to recover the solutes without using the stage of clean-up which have advantages in time consuming and cost reduction [6].

Knowledge of mechanisms and kinetic study on extracting the solutes or compounds from natural plant matrix is crucial for optimizing the design and operating conditions. Indeed, the estimation of diffusion coefficient is important to determine the rate of mass transfer in extracting solutes from solid matrix [7]. Thus, in obtaining the optimum yield of extraction the bioactive compounds or solute from natural plant, the mechanisms of extraction should be well understood. In describing mechanism of extraction, the first step (1) is desorption of solute through the stagnant film layer into the solid plant matrix. This step may be dominated by the adhesive and cohesive forces between solid plant matrix, solvent and solute. The less bound solute is more easily dissolved into the solvent [8]. The second step (2) is the diffusivity of solvent into the solid plant matrix which this step involve the breaking of chemical bonds within the solid plant matrix. The third step (3) is dissolution of solutes into the bulk solvent due to concentration gradient and may be described by the partitioning equilibrium. The solubility of solutes in solvent will depends on the polarities of desired solutes and solvent. The last step (4) is elution or removal of the solutes from solid plant matrix through stagnant solvent [9, 10, and 11]. A mathematical model based on Fick's second law which introduced by Crank [12] describes the extraction process.

Thus, it is obvious that diffusivity of solute in a solvent plays a major role in describing mechanism of extraction. Diffusivity of solute also varies with process temperature. This study investigates the temperature effects on diffusivity for the ginger bioactives 6-gingerol and 6-shogaol.

2. Material and methods

2.1Material and methods

Dried and ground zingiber officinale Roscoe samples were obtained from local supplier. The mean particle size of 1.59 mm was obtained.

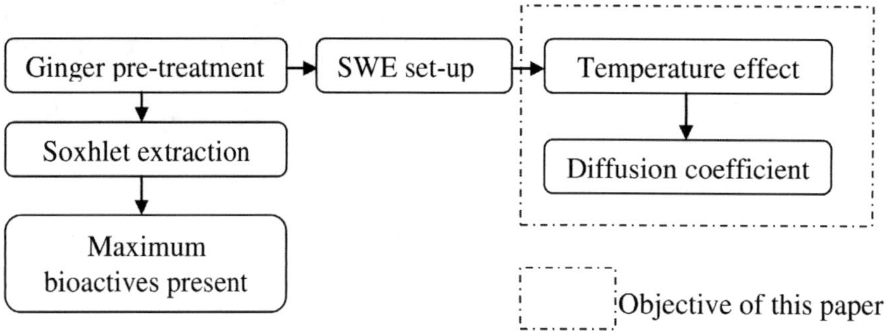

Figure 1. Overall research methodology

2.2Subcritical water extraction (SWE) CLEAR prototype

The SWE with 1L capacity was used in extraction process. Ratio of ginger to water was 1:10 with constant pressure at 3.5MPs and was operated in cumulative process within 60 minutes. The fresh solvent was introduced for each 10 minutes during the extraction process as stated by Sharizan [13].

2.3Soxhlet extraction

70g of dried and ground ginger was weighed and extracted with 280mL of ethanol for 8 hrs using soxhlet extraction apparatus. The extraction temperature was constant at the boiling point of ethanol (78.1°C) and monitored using infrared laser thermometer (AR300, China). The process is repeated for each prescribed time. Then, the extracted sample was purified in rotary evaporator (Buchi R-205, Switzerland) to remove the ethanol. The experiments were conducted in triplicate and standard deviations were calculated (<5%).

2.4 HPLC analysis

The HPLC (Waters 600-MS, USA) equipped with Photo iodide Array Detector (Waters, USA). The method of this analysis was adapted by Sharizan [13] as follows: mobile phases flow rate; 1.2 ml/min;

column temperature; 35-38oC; chromatographic run; 20 min; eluents: (A) acetonitrile and (B) water. The percentage of the solvent A to solvent B was increased gradually from 20 % to 50 %.

2.5 Diffusion Coefficient

From [12], the diffusion coefficient is described by Fick's second law;

$$\frac{\partial C}{\partial t} = D\left(\frac{\partial^2 C}{\partial x^2} + \frac{\partial^2 C}{\partial y^2} + \frac{\partial^2 C}{\partial z^2}\right), \text{ for multidirectional diffusion} \quad (1)$$

By assuming (i) the ginger matrix is uniformly in a spherical form, (ii) there is no interaction between the bioactives, (iii) the diffusivity of the extracted bioactives is constant, thus the initial and boundary conditions are obtained as follows:-

Initial condition $C_{(t=0)} = C_i$

Boundary conditions $C_{(r=r_b)} = 0$

$\partial C_{(r=r_c)} / \partial r = 0$

where r is radial distance coordinate from centre of spherical particle (m), r_b is radius of spherical particle and r_c is centre of spherical particle (r=0).

By considering the diffusivity is constant in one-direction, the equation is given by:-

$$\frac{\partial C}{\partial t} = \frac{D_e}{r^2}\frac{\partial}{\partial r}\left(r^2 \frac{\partial C}{\partial r}\right) \quad (2)$$

The content of solutes in solid varies with time and distance. The diffusion coefficient of solutes can be determined through the concentration changes of solutes in respect solvent with time:-

$$\frac{C_t}{C_o} = \frac{4r}{\pi}\sum_{n=1}^{\infty}\frac{1-(-1)^n}{n.\pi}\exp\left(-\frac{n^2\pi^2}{r^2}.D_e t\right) \quad (3)$$

where C_t and C_o are the concentration at time given (μg/g) and concentration of bioactives in solid (μg/g). *De* is diffusion coefficient (m^2/s) of solute in subcritical water and average radius of solid matrix (m).

For sufficiently prolonged tine, truncating it to the first term in the Eq. (3) can obtain [14]:-

$$1 - \frac{C_t}{C_o} = \frac{6}{\pi^2}\exp\left(-\frac{D_e \pi^2 t}{r^2}\right) \quad (4)$$

D_e could be determined from the slope (D_e= slope x r^2/π^2) of the linear portion of *Ln (1-(Ct/Co))* versus time.

3. Results and discussion

The effect of temperature on the extracted yield was determined for the temperature ranges of 110 to 170°C for both 6-gingerol and 6-shogaol as shown in Figure 2 (a) and (b), respectively. The maximum yield of 6-gingerol and 6-shogaol is determined through soxhlet extraction using ethanol as solvent are 8406.996 μg/g and 716.76 μg/g, respectively. 6-shogaol using SWE was found to be 690.401 μg/g at 170°C and this approaches the maximum yield of 716.76 μg/g or 96% yield as depicted in Figure 2 (b). This phenomena is due to the hydrolysis of 6-gingerol to 6-shogaol at higher temperature [15].

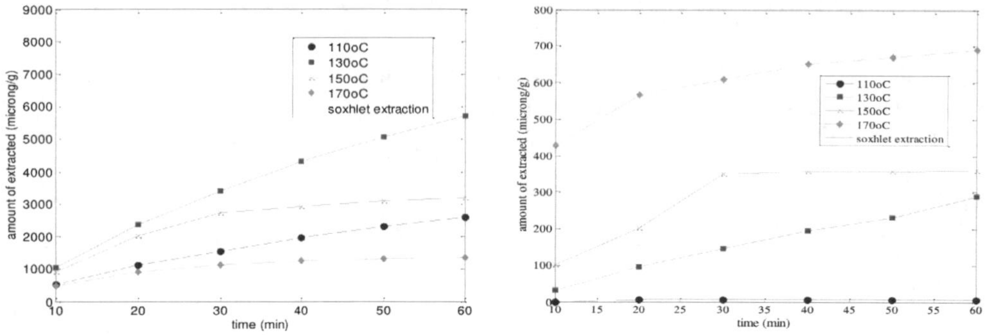

(a) (b)

Figure 2. The bioactive compounds extract a) 6-gingerol b) 6-shogaol

Referring to Figure 2 (a), the trends of extraction process at 150 and 170°C are classified by three stages; rapid, transition and constant extraction process. The rapid extraction stage occurred up to 20 minutes and in this stage it involves the 'free solutes' or 'easy accessible' solutes from solid matrix to the bulk solvent. The transition stage involves the depletion of the 'free solutes' from solid matrix from 20 to 30 minutes. The 6-gingerol reached the third stage (constant extraction process) within 30 minutes. While at lower temperature conditions, there is no indication of three stages of extraction process taking place. The extraction process occurs in two stages for the temperature of 110 and 130°C.

As referring to Figure 2 (b), at 110°C there was no significant of 6-shogaol observed. This is due to the higher of 6-gingerol present and the dehydration of 6-gingerol to 6-shogaol is slower at this temperature. Meanwhile at 150°C, trend of extraction process was analogous to the trend of 6-gingerol at 150 and 170°C. Thus, the extraction process also can be described by three stages; rapid, transition and constant. As mentioned for 6-gingerol, the stages are involving the 'tied solute' and 'free solutes' to be dissolved in bulk solvent. However, at 130°C, the extraction process of 6-shogaol was not high enough to observe the 'tied solutes'. This is might be the 'free solute' was dissolved in bulk solvent and the remaining solute was still in solid matrix. The highest yield which reached the maximum amount of 6-shogaol presents was observed at 170°C. From the figure, there were two stages of extraction process, but it would be changed to three stages at longer extraction time.

The stages of extraction process in Figure 2 are important to be investigated for optimisation and cost reduction of operating. Once the constant extraction process is achieved, the extraction process should be stopped.

A linear line is not appropriate when plotting *Ln (1-(Ct/Co))* versus time at higher temperature (150 and 170°C) as shown in Figure 3. This phenomenon can be explained from the remaining of solutes (6-gingerol and 6-shogaol) in the solid matrix (zingiber). At higher temperatures, water can diffuses more into the solutes, thus tends to dissolve the solute due to higher mass transfer rate and reduction of driving force.

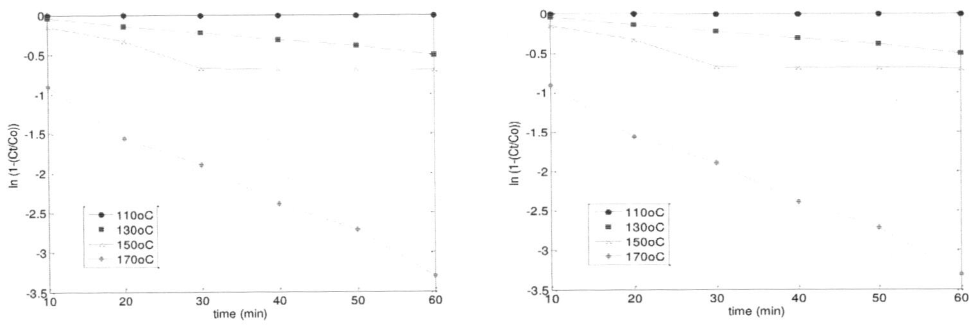

(a) (b)

Figure 3. Prediction of diffusion coefficient for (a) 6-gingerol and (b) 6-shogaol

Table 1 shows the calculated diffusion coefficient, D_e increased from $0.427x10^{-11}$ m^2/s to $19.417x10^{-11} m^2/s$ ascending with temperature. This is consistent with the findings by Ho et al. [17], who showed that the diffusion coefficient in extraction of lignans from flaxseed meal increased with temperature because of the enhancement in extraction rates and mass transfer of solutes from solid matrix. However, the diffusion coefficient of 6-gingerol did not obtained the similar trend. The diffusion coefficient for 6-gingerol is highest at 130°C was $2.903x10^{-11}$ m^2/s and drastically reduced to $0.940x10^{-11}$ m^2/s at 170°C. 6-shogaol is more lipophilic than 6-gingerol, thus at higher temperature the loss of a hydroxyl group from 6-gingerol to 6-shogaol is due to dehydration as mentioned earlier [15].

Table 1. Diffusion coefficient of 6-gingerol and 6-shogaol with respect to temperature

Temperature (°C)	**Diffusion coefficient, D_e x $10^{11}(m^2/s)$**	
	6-gingerol	**6-shogaol**
110	2.605	0.427
130	8.582	3.885
150	2.903	4.653
170	0.940	19.417

Karacabey et al. [18] have reported that calculated diffusion coefficients of trans-resveratrol were increased from 3.3 to 10.4 x 10^{-11} m^2/s with ascending temperature from 105 to 160°C. However as comparing to the 6-gingerol, the results did not indicated the same trend. This is because of 6-gingerol do not stable as the other compounds studied. The properties of bioactive compounds also play a role on the effectiveness of diffusivity. According to the LeBas equation, the molar volumes of 6-gingerol and 6-shogaol are 377.4 and 363.6 cc/mol, respectively. The diffusivity of solutes will be higher at lower molar volume pointing to 6-shogaol having a higher diffusivity compared to 6-gingerol [16].

4. Conclusion

For SWE, it was identified that the optimum temperature conditions were 130°C and 170°C for 6-gingerol and 6-shogaol, respectively. Diffusion coefficient is the highest at these temperatures; 8.582 x 10^{-11} m^2/s for 6-gingerol and 19.417 x 10^{-11} m^2/s for 6-shogaol. Thus, the yield of extraction also was correlated to the diffusion coefficient. Since the studies of diffusivity in subcritical water extraction is still rare, the diffusivity model ignored the external mass transfer resistance. In future, the comparison of external and non-external mass transfer resistance should be considered to improve the knowledge and findings in diffusivity of bioactives.

5. Acknowledgements
The financial support provided for this project by Malaysia-Japan International Institute of Technology is gratefully acknowledged.

References
[1] Azian M N *et al* 2010 *Journal of Food Engineering* **62** 359–364
[2] Lee S W *et al* 2011 *Food Chemistry* **128** 78–782
[3] Sang S *et al* 2009 *J Agric Food Chem.* **57** 10645–10650
[4] Harliansyah *et al* 2007 *Asian Journal of Biochemistry* **2** 421-426
[5] Mustafa A and Turner C 2011 *Analytica Chimica Acta* **708** 8-18
[6] Teo C C et al 2010 *Journal of Chromatography A* **1217** 2484–2494
[7] Varzakas T H *et al* 2005 *Enzyme and Microbial Technology* **37** 29-41
[8] Carr A G *et al* 2011 *Chemical Engineering Journal* **172** 1-17
[9] Anekpankul T *et al* 2007 *Separation and Purification Technology* **55** 343–349
[10] Waldebäck M *Pressurized Fluid Extraction. A sustainable technique with added values*. Uppsala, Acta Universitatis Upsaliensis, 2005.
[11] Islam M Z *et al* 2013 *Journal of Industrial and Engineering Chemistry* **19** 129–136
[12] Crank J *The mathematics of diffusion*, Oxford University Press, Oxford, 1975
[13] Sharizan M M S *Subcritical water extraction of 6-gingerol and 6-shogaol from Zingiber Officinale*. Msc Thesis, Universiti Teknologi Malaysia, 2012
[14] Cussler E L *Diffusion: Mass transfer in fluid systems*. Cambridge University Press, New York: USA; 1984
[15] Guerrero M S et al 2008 *Bioresource Technology* **99** 1311-1318
[16] Bhattarai S T et al 2001 *Journal of Pharmaceutical Sciences* **90** 1658-1664.
[17] Ho et al 2008 *Journal of Food Engineering* **89** 64-71
[18] Karacabey et al 2012 Food Bioprocess Technol **5** 359-371

Journal of Physics: Conference Series **495** (2014) 012010 doi:10.1088/1742-6596/495/1/012010

Applicability of nano zero valent iron (nZVI) in sono – Fenton process

M R Taha[1], A H Ibrahim[2], R C Amat[2] and A W Azhari[2]

[1] Professor, Department of Civil and Structural Engineering, Universiti Kebangsaan Malaysia (UKM), 43000, Malaysia.
[2] Lecturer, School of Environmental Engineering, Universiti Malaysia Perlis (UniMAP), 01000, Malaysia.

E-mail: haqiibrahim@gmail.com

Abstract. Fenton process is one of the advanced oxidation processes (AOPs) used to remove complex organic pollutants in wastewater. In this study, instead of iron sulfate ($FeSO_4$), nano zero valent iron (nZVI) was used as a major source of ferrous iron (Fe^{2+}). In order to enhance the process, ultrasound was utilized in this study. Results show that, with the aid of ultrasound, nZVI produced more Fe^{2+} compared to $FeSO_4$ at pH 2. Furthermore, combination of higher intensity and longer sonication time in Fenton process acceleratde the chemical oxygen demand (COD) removal from palm oil mill effluent (POME). Through the process, 80% of COD content was removed within 2 hours instead of 24 hours of silent degradation.

1. Introduction

In 2012, Malaysia was recorded as second largest world producer of palm oil. As a result of massive production of palm oil, huge amount of palm oil mill effluent (POME) was produced. As POME contains extremely high content of COD, several stages of treatment were required before it can be discharged into the environment. In most cases, anaerobic process was used to treat POME at the primary stage. However, oil millers as well as researchers are still looking for the best treatment method to be applied at the tertiary/polishing level.

One of the options to be consider for tertiary treatment is Fenton process. Study by Kellel [1] and Nieto [2] showed that Fenton process had successfully treated the olive oil mill effluent with similar characteristics to POME. In Fenton process, organic pollutants were attacked by strong oxidant called hydroxyl radical ($OH^\cdot$) and produced harmless products. $OH^\cdot$ was a result of the reaction between Fe^{2+} and hydrogen peroxide (H_2O_2). The reaction can be written as:

$$Fe^{2+} + H_2O_2 \rightarrow Fe^{3+} + OH^- + OH^\cdot \quad (1)$$

In conventional Fenton process, $FeSO_4$ was utilized as a major source of Fe^{2+} [3], [4]. However, to enhance the process, scholars had combined Fenton with other advanced oxidation processes (AOPs) methods such as photo-Fenton [5], electro-Fenton [6] and ozonation [7]. In addition, some scholars also used ultrasound [8] and nano materials [9] to improve the efficiency of Fenton process.

In this study, we utilized nano zero valent iron (nZVI) and ultrasound to enhance the Fenton process. This paper discuss the effect of pH, ultrasound intensity and its duration on Fe^{2+} production and the successful of sono-Fenton process to remove COD from diluted POME.

2. Method

2.1. Material/chemicals
Hydrogen peroxide (R&M Chemicals, 30%), Sulfuric acid (Merck, 95-97%), Sodium Hydroxide (Merck, M=40g/mol), Iron Sulfate (R&M Chemicals, U.K), nZVI particles (Nanofer Star by NANO IRON).

2.2. Ferrous ion (Fe^{2+}) production
A sono reactor was utilized in order to study the effect of pH, ultrasound intensity and its duration on Fe^{2+} production by nZVI. Using H_2SO_4 (0.1M), 150 ml of distilled water was adjusted to desired pH and 0.09 gram of nZVI was added into the solution. Design of Experiment 5.0 (DOE) software was used to build the experimental design in order to study the effect of pH, ultrasound intensity and it duration on Fe^{2+} production. The mixture was sonicated using ultrasonic probe (Sonic Ruptor 250, 20 kHz, OMNI International) at various intensity and duration as mentioned in table 1. Meanwhile, for comparison purposes, $FeSO_4$ was also used as a replacement for nZVI. The concentration of Fe^{2+} in the solution was determined through 1, 10-Phenanthroline Method [10] and measured using a HACH DR6000 UV- spectrophotometer.

Table 1: Independent variables in Central Composite Design (CCD) using Design of Experiment 5.0 software.

Factor	Units	Low Level (-1)	High Level (+1)
A - pH	pH	2	4
B - Intensity	%	10	50
C - Time	minute	3	10

2.3. Chemical Oxygen Demand (COD) removal test
Before the treatment process began, POME was diluted 25 times. COD content in diluted sample was analyzed by closed reflux method [10] and recorded as 1160 mg/L Using H_2SO_4 (0.5 M), diluted POME was adjusted to pH 2 and 4. 100 ml of diluted POME were then poured into the water-jacketed cylindrical glass sono reactor as pictured in figure 1.

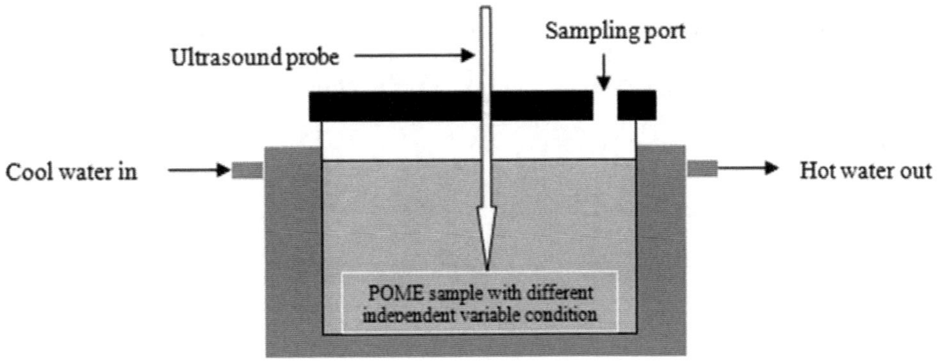

Figure. 1: Water-jacketed cylindrical glass sono reactor.

In every 100 ml diluted POME, 0.4 ml H_2O_2 and 0.06 g nZVI were added. The mixture was sonicated using an ultrasonic probe. The ultrasound intensity and duration setting was generated by DOE software. The independent variables and their levels for the experimental design were shown in table 2.

Table 2: Levels and independent variables for the experimental design.

Experiment	Sonication intensity (%)	Sonication time (min)	pH
1	20	5	2
2	40	5	2
3	20	15	2
4	40	15	2
5	20	5	4
6	40	5	4
7	20	15	4
8	40	15	4

Immediately after the sonication process, 15 ml of POME was taken out from the reactor for COD testing. The rest of the POME was then placed in an automatic shaker for the silent treatment process to continue. COD testing was done for every 1 hour, 2 hours and 24 hours of silent treatment.

3. Results and discussion

In traditional Fenton process, $FeSO_4$ was used as a major source for Fe^{2+}. Besides Fe^{2+} concentration, the oxidation efficiency of organic pollutants by Fenton process was also very much dependent on pH of the solution and H_2O_2 concentration [11]. Figure 2 shows the production of Fe^{2+} by $FeSO_4$ and nZVI in an aqueous solution at pH 2.

Due to shockwave and cavitation created by ultrasound irradiation, the nZVI particles were dispersed at high speed inside the POME sample. As a result of rapid collision of nZVI particles, the chemical reactivity was enhanced. Hence, with the support of ultrasound, nZVI produced more Fe^{2+} compared to $FeSO_4$. On the other hand, as $FeSO_4$ was easily dissociated into Fe^{2+} and SO_4^{2-}, ultrasound gives no impact on Fe^{2+} production by $FeSO_4$.

nZVI particle was formed by two unique layers. The outer layer was made up of iron oxides (i.e. FeO) while the inner part comprised of Fe^0 [12]. Iron oxides layer keeps Fe^0 from rapid oxidation, hence prevent the production of Fe^{2+}. Figure 3 shows the Fe^{2+} production by nZVI at different weight and pH of aqueous solution. At pH 3 and 4, very little Fe^{2+} was produced. This was due to the presence of iron oxides layer which protect the core from any oxidation process.

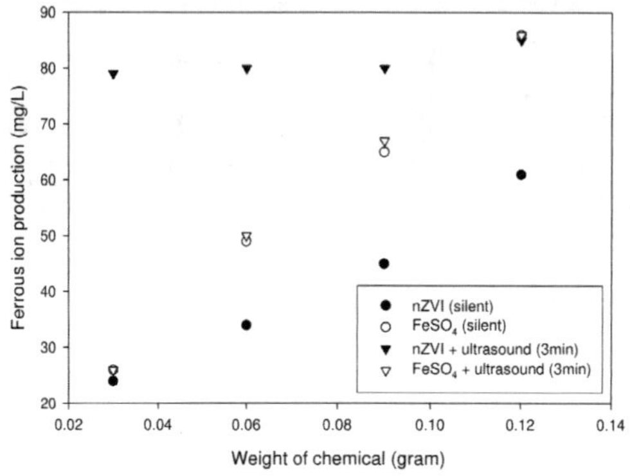

Figure 2: Fe^{2+} production by nZVI and $FeSO_4$ in aqueous solution at pH 2.

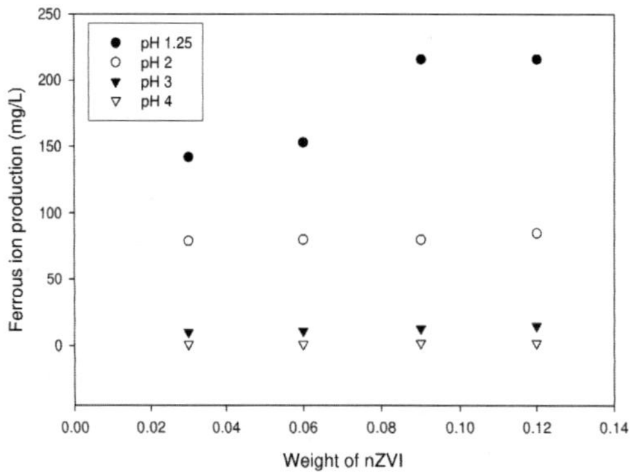

Figure 3: Fe^{2+} production by nZVI at different weight and pH.

At pH 2, the iron oxides layer was removed and Fe^0 was oxidized. The reaction can be represented as;

$$Fe^0 \rightarrow Fe^{2+} + 2e^- \quad (2)$$

At lower pH (i.e. pH 1.25), more Fe^{2+} was produced. This can be explained through excessive H^+ present in the aqueous solution. The reaction occured can be written as;

$$H_2SO_4 \rightarrow H^+ + HSO_4^- \quad (3)$$

$$H_2SO_4 \leftrightarrow H^+ + SO_4^- \quad (4)$$

$$2Fe^0 + 4H^+ + O_2 \rightarrow 2Fe^{2+} + 2H_2O \quad (5)$$

Besides pH of the aqueous solution, the ultrasound intensity and its duration also play some role in the production of Fe^{2+}. However their impact were not as crucial as pH of the solution. Figure 4, 5 and 6 shows the correlation of pH, ultrasound intensity and ultrasound duration on Fe^{2+} production by nZVI.

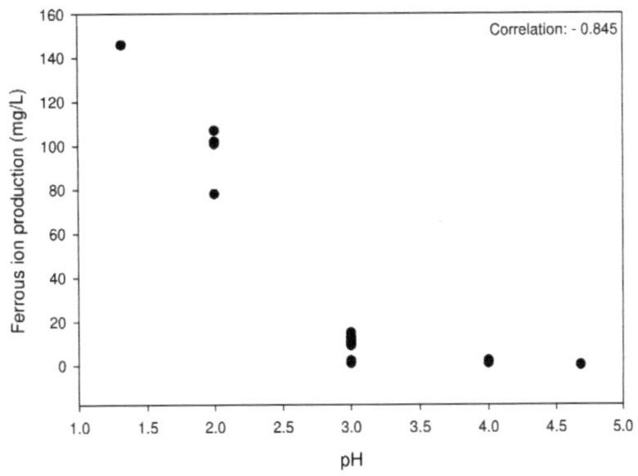

Figure 4: Correlation between Fe^{2+} production and pH of aqueous solution.

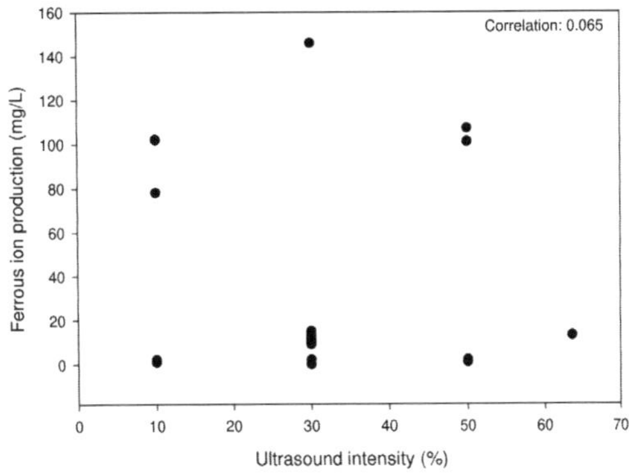

Figure 5: Correlation between Fe^{2+} production and ultrasound intensity.

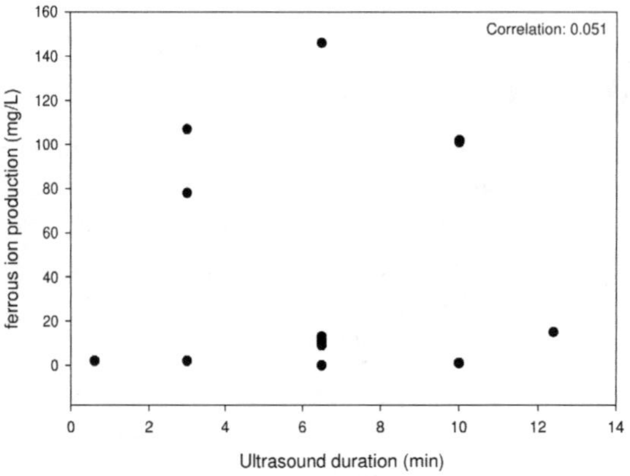

Figure 6: Correlation between Fe^{2+} production and ultrasound duration.

The effect of pH, ultrasound intensity and its duration on Fe^{2+} production from nZVI can be modeed as a linear model. The model demonstrated very low *p*–values (<0.0001) in the F- test which indicated the significant of the model was very high. Determination coefficient (R^2) of the model showed that 88.3% of the total variability could be explained by the model. Fe^{2+} production by nZVI could be explained by independent variables in term of coded as:

$$\ln(ferrous\ ion + 1.46) = 2.53 - 1.60A + 0.24B + 0.16C \quad (6)$$

A, B and C in equation (6) referred to the coded values as is in table 1. Figure 7 shows the predicted vs actual values for Fe^{2+} production by nZVI. The model fit very well especially when the aqueous solution was pH 2 and pH 3.

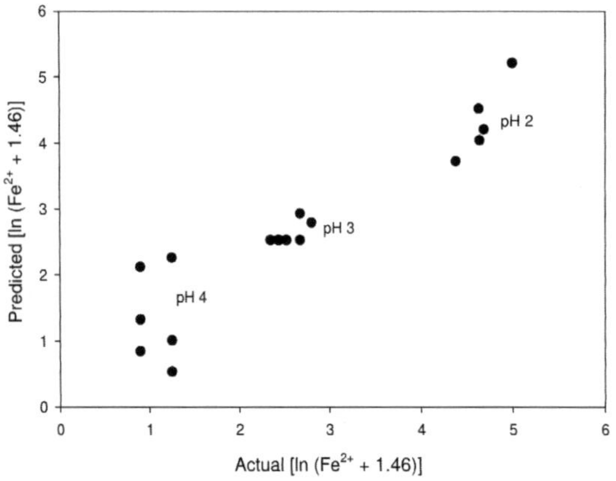

Figure 7: Actual vs. predicted values for Fe^{2+} production by nZVI.

Figure 8(a) – 8(c) demonstrate the effect of ultrasound intensity and pH on Fe^{2+} at three different ultrasound duration (3, 6.5 and 10 min respectively). The graphs clearly show that nZVI produced large amount of Fe^{2+} at pH 2 particularly with the aid of higher intensity and longer ultrasound time.

On the other hand, intensity and duration of ultrasound did not affect the production of Fe^{2+} by nZVI at pH 4. It was due to the presence of iron oxides layer that protected the Fe^0 from oxidation process.

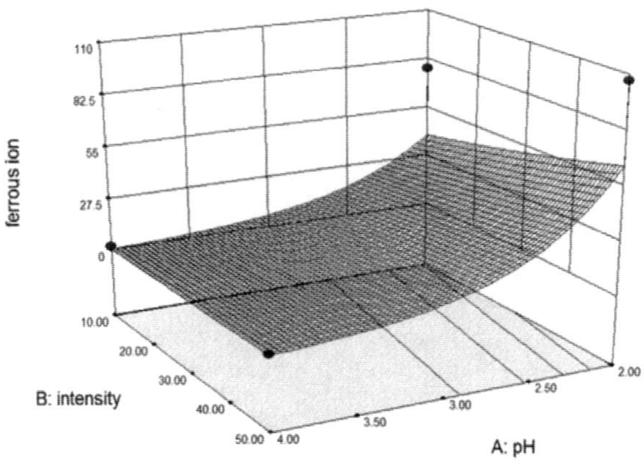

Figure 8(a): The effect of pH and ultrasound intensity on Fe^{2+} production after 3 min of ultrasound.

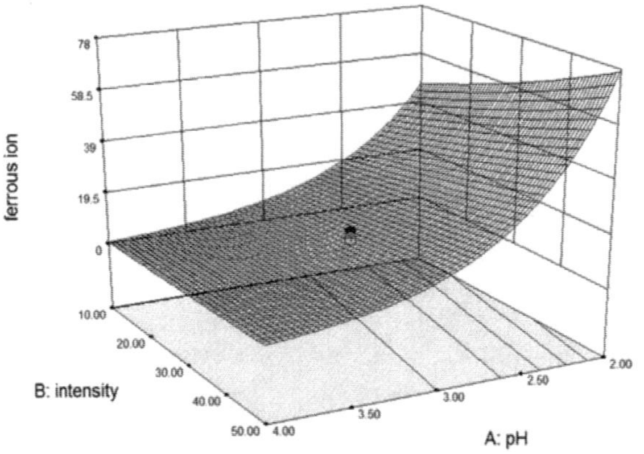

Figure 8(b): The effect of pH and ultrasound intensity on Fe^{2+} production after 6.5 min of ultrasound.

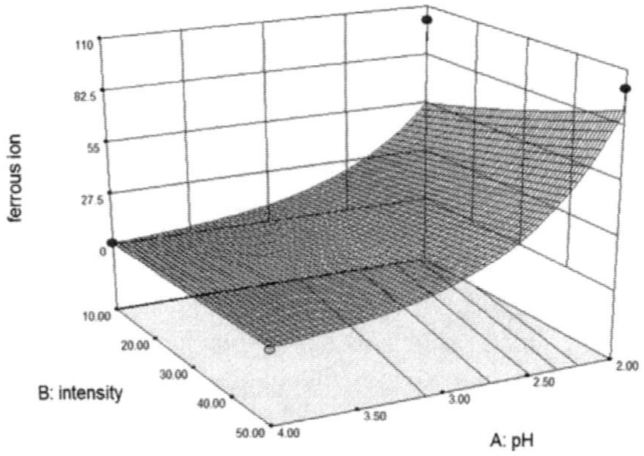

Figure 8(c): The effect of pH and ultrasound intensity on Fe^{2+} production after 10 min of ultrasound.

During COD removal test using diluted palm oil mill effluent (POME), COD readings in all samples were increased immediately after the treatment process except for experiment 4. The results of all experiments setting is represents in Figure 9.

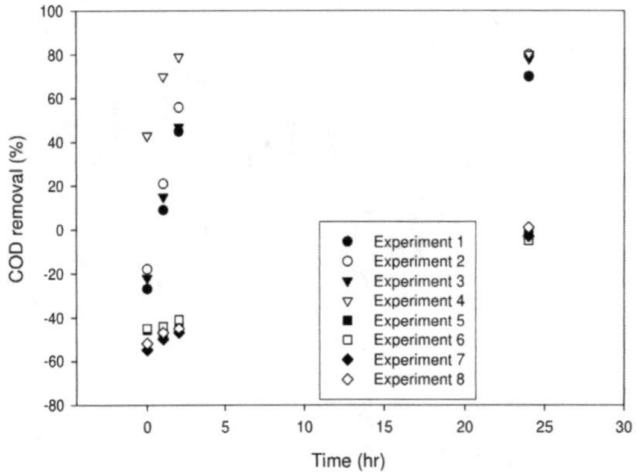

Figure 9: COD removal from diluted POME at different treatment setting.

Increment of COD reading was due to the presence of remaining H_2O_2 in the sample. During COD test, the remaining H_2O_2 was oxidized as well as the organic pollutants in the diluted POME [13]. Hence the COD content in POME arised from the initial value. This scenario clearly happened in samples which were treated at pH 4. As discussed above, obviously very few Fe^{2+} was produced at pH 4. Due to that, production of OH• was impeded due to the absence of Fe^{2+} and contributed to the excessive amount of remaining H_2O_2 in sample solution.

On the other hand, results showed that COD was removed from POME when they were treated at pH 2. In addition, the removal percentage also increased by time. In this case, the presence of Fe^{2+} in the samples reacted with H_2O_2 to produce $OH^{\cdot}$. The produced $OH^{\cdot}$ then attacked the organic pollutants and directly reduced the COD content in solution. After 24 hours of silent degradation, up to 80% of COD was removed from the POME samples.

In contrast, after 24 hours of silent degradation, COD reading in samples treated at pH 4 went back to their initial COD concentration. These indicated that no COD was removed from the diluted POME samples. In these samples, all added H_2O_2 were decomposed into water and oxygen. The decomposition of H_2O_2 as represented in the following equation:

$$2H_2O_2 \rightarrow 2H_2O + O_2 \quad (7)$$

On the other part, ultrasound intensity or ultrasound time gave no significant effect on COD removal efficiency of diluted POME at pH 4. Figure 9 shows that samples which were treated at pH 4 at various ultrasound intensity and ultrasound time resulted in almost the same COD removal. Similar trends were also noticed in experiment 1, 2, and 3. COD removal efficiency for these three treatment conditions was about the same.

However, combination of high intensity and longer ultrasound time can shorten the treatment process time. This clearly happened in experiment 4 where maximum COD removal (80%) was achieved within 2 hours instead of 24 hours of silent degradation. This can be explained by the presence of excessive Fe^{2+} in the solution due to longer exposure to high sonication. As Fe^{2+} determined the rate of the conversion of H_2O_2 to $OH^{\cdot}$, higher concentration of Fe^{2+} increased the efficiency of the process.

4. Conclusion

Production of Fe^{2+} by nZVI particles only occurred once the iron oxides layer was removed. Removal of iron oxides layer could be done by adjusting pH of the aqueous solution to pH 2. Furthermore, with the aid of ultrasound, nZVI particles were dispersed homogenously in the solution at higher speed. Rapid coalition of nZVI particles had increased its chemical reactivity hence more Fe^{2+} was produced. Thus, nZVI particles could be considered as an alternative source for Fe^{2+} in sono-Fenton process. In addition, utilization of nZVI particles in sono-Fenton process seems to be a promising option for POME treatment. COD contents can be removed faster with the presence of nZVI in sono-Fenton process particularly at higher ultrasound intensity and longer sonication time.

References

[1] Kallel M, Belaid C, Mechichi T, Ksibi M and Elleuch B 2009 Removal of organic load and phenolic compounds from olive mill wastewater by Fenton oxidation with zero-valent iron *Chem. Eng. J.* **150** 391–395
[2] Nieto L M, Hodaifa G, Rodríguez S, Giménez J A and Ochando J 2011 Degradation of organic matter in olive-oil mill wastewater through homogeneous Fenton-like reaction *Chem. Eng. J.* **173** 503–510
[3] M S Lucas, J A Peres 2009 Removal of COD from olive mill wastewater by Fenton's reagent: Kinetic study *J. Hazard. Mater.* **168** 1253–1259
[4] Y W Kang, K Y Hwang 2000 Effects of reaction conditions on the oxidation efficiency in the Fenton process *Water. Res* **34** 2786–2790
[5] R Bauer, H Fallmann 1997 The Photo-Fenton Oxidation — A cheap and efficient wastewater treatment method *Res. Chem. Intermediat.* **23** 341–354
[6] A Durán Moreno, B A Frontana-Uribe and R M Ramírez Zamora 2001 Electro-Fenton as a feasible advanced treatment process to produce reclaimed water *Water. Sci. Technol.* **50** 83–90
[7] J Beltrán-Heredia, J Torregrosa, J García, J R Domínguez and J C Tierno 2001 Degradation of olive mill wastewater by the combination of Fenton's reagent and ozonation processes with an aerobic biological treatment *Water. Sci. Technol.* **44** 103-108
[8] J Zhang, J Li, R Thring and L Liu 2013 Application of Ultrasound and Fenton's Reaction Process for the Treatment of Oily Sludge *Procedia Environmental Sciences.* **18** 686–693
[9] L Chu, J Wang, J Dong, H Liu and X Sun 2011 Treatment of cooking wastewater by an advanced Fenton oxidation process using iron powder and hydrogen peroxide Chemosphere doi: 10.1016/j.chemosphere.2011.09.007
[10] Hach 2010 *Water Analysis Handbook*, fifth ed., Hach Company, Colorado.
[11] H Zhang, H J Choi, C P Huang 2005 Optimization of Fenton process for the treatment of landfill leachate *J. Hazard. Mater.* **125** 166–174
[12] Y P Sun, X Li, J Cao, W Zhang, H P Wang 2006 Characterization of zero-valent iron nanoparticles *Adv. Colloid. Interface Sci.* **120** 47–56
[13] W G Kuo 1992 Decolorizing dye wastewater with Fenton's reagent *Water. Res.* 26 881 – 886
[14] X Zhu, J Tian, R Liu and L Chen 2011 Optimization of Fenton and electro-Fenton oxidation of biologically treated coking wastewater using response surface methodology *Sep. Purif. Technol.* **81** 444–450

Designing Process Improvement of Finished Good On Time Release and Performance Indicator Tool in Milk Industry Using Business Process Reengineering Method

M Dachyar, E Christy

Department of Industrial Engineering
University of Indonesia, Depok 16424, Indonesia

E-mail: mdachyar@yahoo.com

Abstract. To maintain position as a major milk producer, the Indonesian milk industry should do some business development with the purpose of increasing customer service level. One strategy is to create on time release conditions for finished goods which will be distributed to customers and distributors. To achieve this condition, management information systems of finished goods on time release needs to be improved. The focus of this research is to conduct business process improvement using Business Process Reengineering (BPR). The deliverable key of this study is a comprehensive business strategy which is the solution of the root problems. To achieve the goal, evaluation, reengineering, and improvement of the ERP system are conducted. To visualize the predicted implementation, a simulation model is built by Oracle BPM. The output of this simulation showed that the proposed solution could effectively reduce the process lead time and increase the number of quality releases.

1. Introduction

Increasing number of baby birth has made many FMCG (Fast Moving Consumer Good) companies for the baby willing to expand their market share by meeting the demand of baby products, one of the products is the baby milk. The demand for baby milk is increased by 6.6% (in volume) annually; from 37.300 tonnes in 2004 to 47.140 tons in 2008 [1].

Market growth, market competitive, and increased demand urged dairy industry to implement some strategies to maintain the position as a major dairy. One of the strategies is to increase customer service levels by increasing performance on quality releases – known as the finished good on time release. However, in one of multinational dairy companies its release has not been stable and on time. There is also no mapping of process information flow and no measurement for on time release process.

In this strategy, quality division has the most important role. The performance of QFS (Quality Food Safety) Division is needed to be improved in order to achieve the strategy.

2. Literature Review

2.1. Business Process Reengineering

Business process reengineering (BPR) is a business management strategy that focuses on the analysis and design of material and information workflows and processes in an organization [2]. It recommended that the old process system should be removed and replaced with a new system which is more innovative and effective.

There are several opinions about Business Process Reengineering. BPR is radical rethinking and redesign of business processes to achieve dramatic improvements modern measures of performance [3]. Business process reengineering introduces of new work strategies, actual process design, and implementation of the change in all its complex dimensions [4]. Reengineering process is a radical scrutiny, questioning, redefinition, and redesign of business processes with the aim of eliminating all activities not central to the process goals [2]. BPR is the methodology focused on the process and directly performed by the top management by performing the changes radical in order to make a better performance [5]. The four main reasons why companies need to develop business process reengineering are to fix and improve the customer service, to reduce cycle time by reducing non-added value activity, to reduce production costs and/or services, and to improve quality [3].

2.2. BPR Methods

Variuos methods of BPR have been introduced to facilitate the process of the reengineering. There are four methods that could be applied in the implementation of BPR [3]. Below are the list of BPR methods could be applied:

- Mayer Method: Motivating reengineering, Justifying reengineering, Planning reengineering, Setting up for reengineering, As-is description and validation, To-be design and validation, and Implementation
- Feldman Method: Develop vision and strategy, Create desired culture, Integrate and improve enterprise, and Develop technology solutions
- Harrison Method: Determine customer requirements and goals for the process, Map and measure the existing process, Analyze and modify existing process, Design a reengineered process, and Implement the reengineered process
- Muthu, Whitman, and Cheraghi Method: Prepare for Business Process Reengineering, Map and Analyze As-Is Process, Design To-Be Process, Reengineered Process, and Improve Continuously.

2.3. BPR Characteristic

Four main characteristics of BPR [6] ; fundamental, radical, dramatic, and processes.
Fundamental – The analysis is based on two aspects: "what" (focus on gaining the objectives of company) and "how" (focus on the order of activities in each process). *Radical* – The method is undertaken by obtaining the root problem and ignoring all the structures and proce-dures that already exist to create a new process to be run. In some BPR approaches, existing systems may not be changed in radical way because of considering some factors such as the need to do change management, build an integrated system, project management, and select stakeholders. *Dramatic* – The purpose of BPR is to achieve the improved performance dramatically, so that the ambitious and aggressive goal can be created. *Processes* – A process is a set of activities that requires a few input and produces an output that gives value to customer.

2.4. BPR Tools and Techniques

There are several tools and techniques of BPR [7].

- Process visualization ; The successful key of reengineering is based on the development of the process' vision.
- Process mapping and modeling ; Process flowcharting, IDEF, role activity diagram, SIPOC diagram, and others tools can be used to map and model the process.

- Change management ; Human elements are needed to be combined with reengineering process due to the big influence of human's work and work method
- Benchmarking ; This tool displays the process visually and its development in the operations of other organizations.
- Process and customer focus ; Main purpose of BPR is to redesign the process due to increase the performance by customer's perspective.
- Problem solving and diagnosis : Pareto diagram, cognitive map, and other problems solving tools can be used.
- Process prototyping and simulation ; By doing simulation, BPR project team would be easier to determine the process needed to be reengineered and the effects of its. Simulation provides the structural visual model which is easier to be understood, analyzed, and improved.
- Project management ; By this tool, budgeting, project scheduling, and others project indicator can be known.
- Process measurement ; Activity-based costing, statistic process control, and others measurement can be used.

2.5. Information Technology

The use of information technology is one of factors that give a big contribution to the company. Information technology has an important role in business process reengineering [8]. The new technology allows the creation and redesign of business processes. By using technology, the modeling and simulation of systems and processes, will ease the implementation process.

There are many situations that randomly affect the simulation so that simulations cannot be modeled by using others analytical technique. Simulation is the last tool that can be used when it is found that all methods of modeling cannot be done [8]. The technology is also needed in process mapping. A simple process mapping simply cannot give an enough explanation and understanding to know what the things needed to be redesigned are though it is done by big project [8]. Process map helps project teams to understand the framework of existing problems. However, the process should be supported by numerical analysis due to the team knows the things that need to be changed. Therefore, some programs and software can be used to make the process map and simulation of the as-is model and to-be engineered.

2.6. Enterprise Resource Planning

ERP systems have been considered as an important development in information technology in the company since the 1990s. By using the ERP system, all of the information that flows within a company can be integrated so that it will improve the efficiency and effectiveness of the organizational cross-functional [3].

ERP is a cross-functional enterprise systems that integrate and automize a lot of company's internal business processes, especially in the fields of manufacturing, logistic, distribution, accounting, finance, and human resource. ERP helps companies to achieve efficiency, agility, and responsiveness that are necessary to be success in a dynamic business environment [3]. ERP software typically consists of integrated modules that give the appearance of a real-time cross-functional in a company's core business process – such as production, order processing, and sales – and its resources – such as cash flow, raw materials, production capacity, and people.

The lack of conformity between the strategy, structure, organization process, and the selection of the ERP application often give a problem or risk. ERP literature and business process reengineering indicate that ERP system could not be alone to increase company's performance except the organization restructures the operational process. This is usually achieved through a business process reengineering [3].

3. Research Methodology

This research was conducted based on the Business Process Reengineering (BPR) method which has five steps such as Prepare for Business Process Reengineering, Map and analyze As-Is process, Design to-be process, Re-engineered process, and Improve Continuously.

In the first step – Prepare for Engineering – the start preparation of BPR was done by identifying the company's purpose or objective of the case and developing the strategies. In the second step – Map and Analyze As-Is Process – the flow of business processes related was mapped, then the investigation and evaluation of business processes were conducted. In the third step – Design To-Be Process – the best solutions and innovation for business process improvement was discussed. In the fourth step – Reengineered Process – the draft and development of implementation plans. In the fifth step – Improve Continuously – the monitoring and controlling of the new system were being run so the new system's performance can be evaluated and measured in order to make continuous improvement.
Some scenarios would be performed to test the as-is and to-be process. To do simulation, Oracle BPM software is used. The two simulations were compared to see each performance.

Data is collected by using secondary and primary data. Secondary data is data that is derived from QFS Division (Quality Food Safety) in the form of its historical record data of quality release or finished goods on time release process. Primary data is data that is gained by directly doing observation and interview with the supervisor, the staff of QFS Division, and operators.

Secondary data is the record of 418 data. The variable note is time which is called as a lead time of quality releases or release time. The lead time is determined by the production order number, the date of quality release report, and the date of sample received.

4. Analysis Result

4.1. Step 1: Prepare for Engineering

TIn the first study, general process understanding had to be drawn to understand the related supplier, input, process, output, and customer. This study was using SIPOC Diagram.

There are two subjects in this business process such as the supplier and customer. The suppliers are the production planner and production division, while the customers are the production planner and the warehouse. As supplier, the production planner gives production order needed while production division gives product sample which will be analyzed. As customer, the production planner will receive quality report release while the warehouse will receive the product that will be distributed.

There are also two objects in this business process such as input and output objects. Input objects are production order (from the production planner) and product sample of finished goods (from production division). Output object is the quality report release which is the main object of this business process.

To produce output, several processes are needed. The simple map of this process is shown in Figure 1.

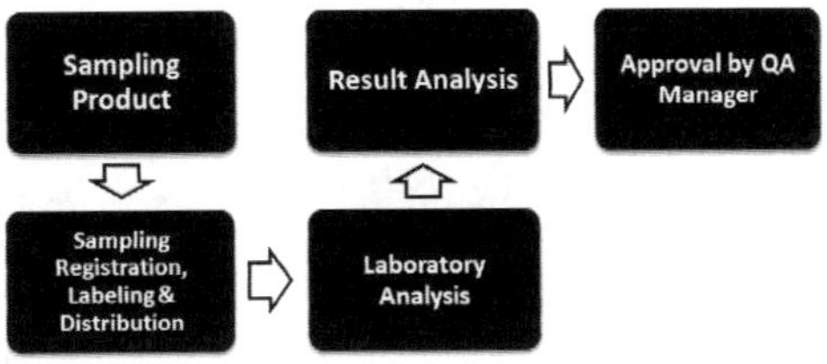

Figure 1. Simple Map of On Time Release

4.2. Step 2: Map and Analyze As-Is Process

4.2.1. *Map As-Is Process.* The main business process is Quality Process (Figure 2). There are also some sub-processes which are Sampling Process, Sample Registration and Labeling, Disposition in SAP, Input and Reporting Data/Result Analysis, and Retest Process. Table 1 shows the data of lead time of each process.

Table 1. Lead Time of Each Process

Process	Mean	SD
1. Quality Process	5d 15hr 9min 31s	7d 55min 40s
2. Sampling Process	1d 6hr 34min 34s	18hr 17s
3. Sample Registra-tion and Labeling	1hr 53min	36min 30s
4. Input and Reporting Data/ Result Analysis	16hr 20min	1d 1hr 14min 10s
5. Disposition in SAP	1hr 5min	17min
6. Retest Process	12hr 18min	1d 8min

4.2.2. *Analyze As-Is Process.*
(a) Current condition
Based on the secondary data, the average time of finished good on time release is 5.63 day. The percentage of on time release is 47.85% while the percentage of not on time release is 52.15%. The percentage of quality release weekly can be seen in Table 2. OTR is on time release, while NOT OTR is not on time release.
(b) Problem identification
Based on interview, some root problems that need to be improved, such as:

1. The information control is still conducted manually and inefficiently.
2. The ERP system implemented is not integrated and not yet meets the quality release process.
3. There is no notification information which automatically associated with quality releases.
4. The business process of quality release and its approval process are still manually, inefficiently, and ineffectively.
5. There is no good quality releases management especially during the approval process.
6. There is no performance measurement of the quality release process.
7. There are some activities which is not value-added to the quality release process.

4.3. Step 3: Design To-Be Process

To provide some solutions for solving the problems, some strategies are implemented:

1. Create a business process redesign of quality release (Problem 4 and 7)
2. Improve the existing SAP system to be more integrated and better so that it can meet the quality release process (Problem 1 and 2)
3. Make the automatic notification information so that it can show the status of quality releases (Problem 1 and 3)
4. Create a policy for the management of quality related process (Problem 5)
5. Make performance indicator tool to measure the performance of quality release process (Problem 7)

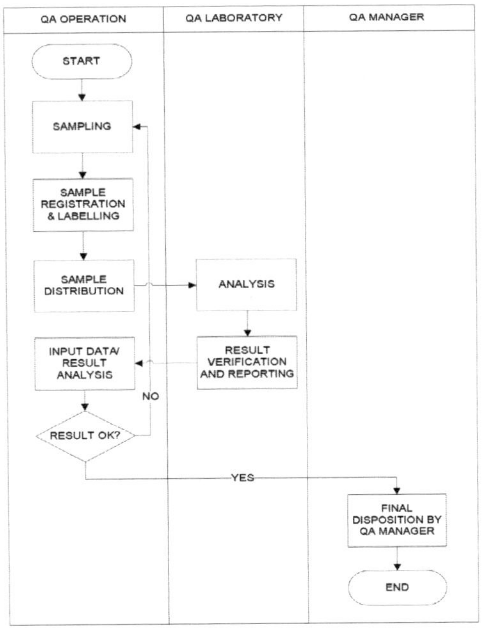

Figure 2. Quality Release Process

Table 2. On Time Release (Week 1-13)

Week	OTR	Not OTR	TOTAL	% OTR	% NOT OTR
1	2	14	16	12.50%	87.50%
2	14	14	28	50%	50%
3	19	18	37	51.35%	48.65%
4	22	14	36	61.11%	38.89%
5	23	29	52	44.23%	55.77%
6	2	19	21	9.52%	90.48%
7	13	17	30	43.33%	56.67%
8	40	11	51	78.43%	21.57%
9	11	15	26	42.31%	57.69%
10	7	12	19	36.84%	63.16%
11	17	18	35	48.57%	51.43%
12	26	12	38	68.42%	31.58%
13	22	7	29	75.86%	24.14%
Average	17	15	32	52.15%	47.85%
Total	218	200	418	52.15%	47.85%

4.4. Step 4: Reengineered Process

In this step, the strategies will be implemented. In the implementation step, there are also several phases which are needed for implementing the ERP system. There are seven phases at this step which are the initiation phase, evaluation, selection, modification, go-live, settlement, and exploitation and development phase.

4.5. Step 5: Improve Continuously
To be able to control the quality and system as well as make continuous improvements, monitoring and controlling the performance of the new system is necessary. This can be done by creating and using key performance indicator (KPI) tool, so that the performance of the system can be indicated.

4.6. Comparing Simulation of As-Is and To-Be Process
To find out the effectiveness of the new system, simulation of as-is and to-be processes are conducted to compare both processes. These simulations are run by software. These simulations use some assumptions based on interviews and primary data captured through direct observation. Visual simulation of as-is and to-be process can be shown in Figure 3 and Figure 4.

The simulation was conducted on two processes which are as-is and to-be processes process. On the to-be process, six scenarios are conducted (Table 3). Simulated results can be seen in Table 4.

Table 3. Scenario of To-Be Process

Scenario	Probability of doing retest	Probability of reject	LT of sample delivery
1	0%	10%	
2	0%	20%	
3	10%	10%	8 hours
4	10%	20%	
5	20%	10%	
6	20%	20%	

5. Discussion
The current condition shows that the process of quality releases or finished good on time release has not been run well. This is shown by the performance of quality release which the percentage of not on time quality release is 47.85%. In addition, the weekly percentage of quality release is also not consistent or always up and down. For example from the first week to the fourth week, the percentage of quality release was increasing, but after that the decreasing were happening drastically.

Besides the observation data, there were also several interviews conducted. Based on the results of observation and interview, the roots of the problem were found:

1. The information control is still done manually and not efficient. Operators still use manual way to control the information in ERP system, for example checking the system one by one. This condition certainly will reduce the performance of the operator for doing the work.
2. The implemented ERP system has not been run integratedly and has not met the needs of quality release process. The use of the existing system is not optimal. Ms. Excel is still used to process the information. This condition certainly does not support the quality release process even to make it more effective, efficient, and optimal.
3. There are some activities in the business process which does not give added-value on the quality release. Therefore, it is needed to redesign the business process to improve performance.
4. The management of quality release is still not good especially during the approval process. For example, the management of approval process still takes a long time because of the waiting activity.

Journal of Physics: Conference Series **495** (2014) 012011 doi:10.1088/1742-6596/495/1/012011

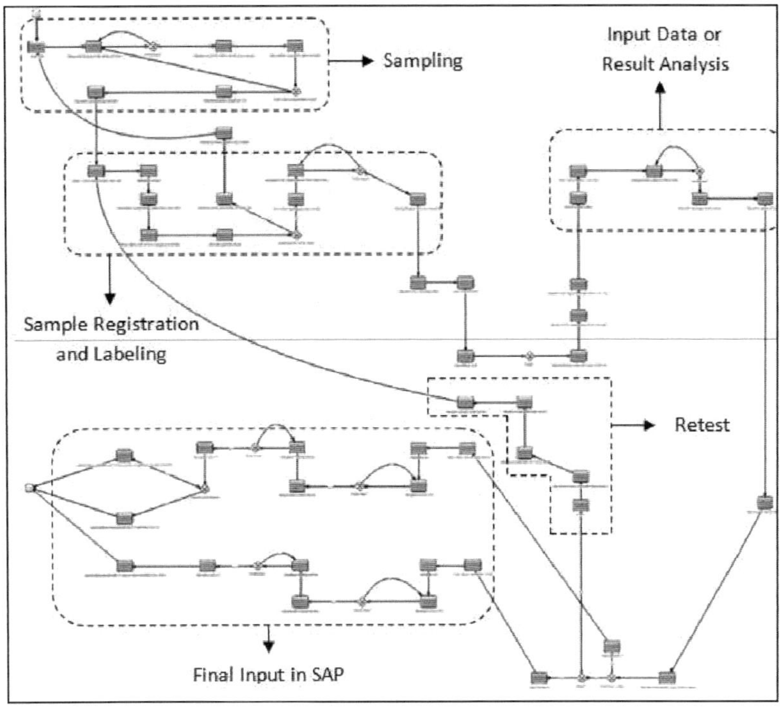

Figure 3. As-Is Process of Quality Release

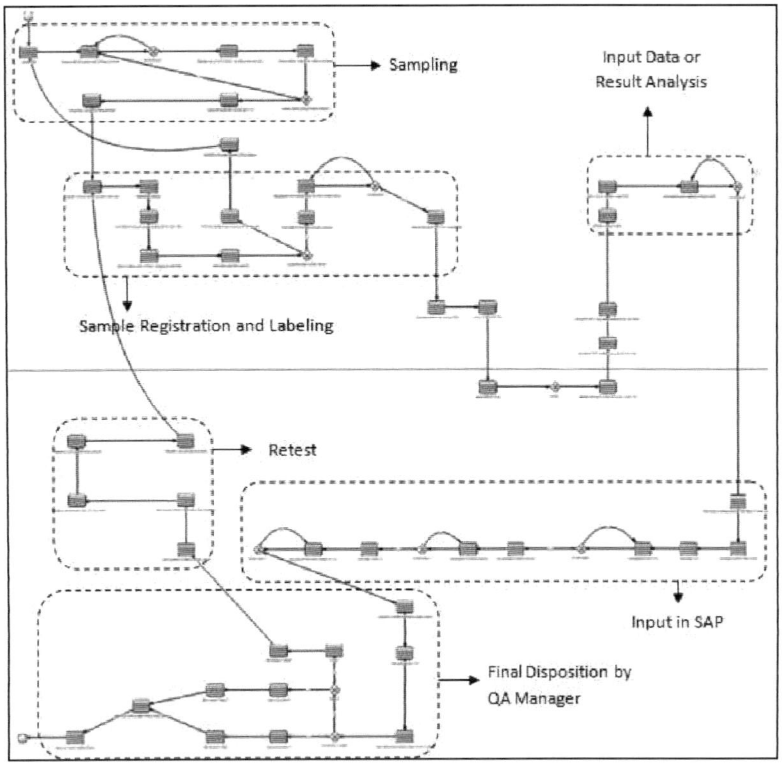

Figure 4. To-Be Process of Quality Release

Table 4. Simulation Result

Model	N	Release	Reject	Retest	LT (s)
As-Is	125	107	18	40	771,106
Scenario 1	143	135	8	0	473,164
Scenario 2	141	114	27	0	496,846
Scenario 3	139	125	14	3	505,158
Scenario 4	137	117	20	1	515,899
Scenario 5	136	122	14	6	531,505
Scenario 6	130	105	25	10	553,209

5. There is no measurement applied to measure the performance of the quality release process. Although the QFS has tools and indicators to measure the others performance, in fact there is no indicator to measure the performance of the quality release process.

In order to improve the performance of quality release process, five strategies have been designed five strategies, such as:

- Design new business process flow : In this strategy, there is one activity omitted which is the request activity of approval by QFS Manager. Besides, it is also reengineering on two activities which are sample taking process from production line and sample delivery to laboratory.
- Improve the ERP systems : The new ERP system should be designed to meet the needs of process quality release. Some functions needed are the report description, product description, stock overview, and analysis report.
- Make the system notification system : Automatic notification system needed to be installed is a system that will give a pop-up notification when there is notification information in computer. The necessary notification are the notification of data input in folder share, the notification of new quality report in the ERP system, the notification of needing retest, the notification of release or reject or retest product, and the notification if the approval process over the time limit.
- Design new policy for quality releases : This policy should be made by the director and aim to support the necessary requirements to implement the strategies which have been designed above.
- Design new performance indicator tool : Performance indicator tool can be created by using Ms. Excel. Some of the contents are (1) Measurement for Week 1-26 and 28-Week 52, KPI Graphic (2), (3) Consistency of On Time Release (OTR), and (4) the Lead Time Variability of Difference.

To show a comparison between the as-is process and to-be process, some simulations have already been conducted with several scenarios. Simulations focusing on lead time are necessary for processing the quality release. As a simulation, to-be process assessed much better compared to the as-is process. This can be proved by these points below:

- Number of completed instances : Completed instance is an operated entity in the quality release process from the early stage to the completed stage. In this case the operated entity is quality release. As the as-is process, the number of completed instances is 125. As the to-be process, the number of completed instances is 130 to 143. By implementing the to-be process, the number of quality release is increased by as much as 5 to 18. This shows that the to-be process is more effective and efficient than the as-is process.

- Average process time : Average lead time used to complete a quality release is the main focus in this research. On the as-is process, the process time used was 777.106 seconds. On the to-be process, as the optimum result of six scenarios, it was found that process time of quality release was 473.164 seconds. Process time has been declined by 43.74%. This shows that the to-be process is more effective and efficient than the as-is process.
- Number of retest and reject product : The greater the probability of retest and reject product, the amount of products released become less. This shows that, in order to increase the number of quality report which passes the test quality, the number of retest and reject has to be suppressed by improving the performance of the production process and the process of analysis in the laboratory.

In each scenario, the number of completed instances and the average process time was different. This is because of the difference scenario of retest and rejected probability. If the retest and rejected probability is smaller, the number of completed instances will be higher and the average processing time will be shorter.

6. Conclusion

Based on the research that has been conducted, there are some findings. The as-is process is ineffective and inefficient. As an analysis result, some of the root problems are the control of information is still conducted manually, the ERP system that is used is not integrated and does not meet the requirements of the process, there are some non-added-value activities in quality release process, the management of quality release is still not good, and there is no measurement applied to measure the performance and set the standard for doing improvement.

The to-be process is included the business process redesign of quality release process, implementation of ERP system to fit the needs of a new business process, implementation of the automatic notification information, creating policy related management process in quality release, and designing performance indicator tool as a reference to see the performance of the process so that it can always make improvement.

By the six scenarios, the best condition in the quality release process is on the first scenario. In this scenario, the process runs with the lowest retest and rejected probability which is 0% and 10%. As a result, the process time is reduced by 43.74% and the number of quality release is increased by 18 quality release. This scenario shows that these strategies can improve the effectiveness and efficiency of process quality release. Nevertheless, this scenario is supported by the condition of the good and qualified production process as well as a good performance during the test analysis in the laboratory.

References

[1] World Bank 2011 *Dairy industry development in Indonesia, final report* March 20, 2013 http://www.ifc.org/
[2] Harvey S and Millet B 2009 *Aust. J. Manag. Org. Behav.* **2** 30-42
[3] Erkan T 2009 *World Acad. Sci. Eng. Tech.* **30** 1508-11
[4] Dodaro G and Crowley B 1997 *Business process re-engineering assessment guide 3rd version* (United States General Accounting Office)
[5] Olalla M 2010 *Int. Adv. Eco. Res.* **6** 581-9
[6] Chan S and Choi C 1997 *Int. J. Prod. Eco.* **50** 213-24
[7] Goksoy A, Ozsoy B, and Vayvay O 2012 Int. J Busi. Manag. **7** 89-112
[8] Gupta A and Dogra P 2010 J.Theor. Appl. Inf. Tech. 102-106

Performance Evaluation of a Drilling Project in Oil and Gas Service Company in Indonesia by MACBETH Method

M Dachyar, Novandra Rhezza Pratama

Department of Industrial Engineering
University of Indonesia, Depok 16424, Indonesia

E-mail: mdachyar@yahoo.com

Abstract. Decision-making in a project is a complex undertaking. A project is a temporary organization that is surrounded by inherent uncertainties. Uncertainties that may occur in the project, among others, are the uncertainty of the time, cost, and quality. To overcome this complexity, we need a project performance evaluation. At one of the services the oil and gas in Indonesia, the performance evaluation has been done, but it is done separately and only few criteria are evaluated, so that the overall project performance is still not visible. Therefore it is necessary to evaluate the performance of the company's project with the MCDA (Multi Criteria Decision Analysis) approach using MACBETH (Measuring Attractiveness by a Categorical Based Evaluation Technique) method that can model the elements of the project into a supervisory level group to facilitate decision making by the project manager. These measurements will produce levels of monitoring classification of project element groups, i.e. absolute vigilance, strong attention, close surveillance, and normal monitoring of the project elements. This research finds that the criteria of Effectiveness, Efficiency, Time, Cost, and Complete Reports & Field Tickets require absolute vigilance so that the performance criteria can be increased.

1. Introduction

Decision-making in the project is a complex undertaking [1]. Complexity can appear in many forms and arise from a variety of sources with varying levels of intensity according to the industrial sector or the object of the project. The intensity may vary from time to time, and this variation underscores the dynamic aspect of the complexity of the project. On the one hand, the project is a temporary organization that is surrounded by the inherent uncertainty [2]. Uncertainties that may occur in, among others, the uncertainty of future projects (scheduling, duration, etc.) and uncertainty of costs (procurement, operating expenses, wages, etc.). On the other hand, the project manager can be faced with another form of the complexities involved in dealing with a lot of independent information. Another form of it is the integration and risk. To overcome this complexity, we need a performance evaluation of the project to be able to evaluate the performance of the project and address the problems that occurred on the project and get a lesson for the next projects.

At one of the services the oil and gas in Indonesia, the performance evaluation has been carried out, among others are HSE (Health, Safety, and Environment), CV (Customer Value) Metrics, and the End of Well Report. However, the evaluation is conducted separately and only a few criteria to be evaluated, so that the overall project performance is still not visible. For the purpose, it is necessary to evaluate the company's performance on the project with the MCDA (Multi-criteria decision analysis) approach using the MACBETH (measuring attractiveness by a categorical based evaluation technique) mathematical method that can model the elements of the project into a supervisory level group to facilitate decision-making by the project manager. These measurements will produce levels of monitoring classification of project element groups, i.e. absolute vigilance, strong attention, close surveillance, and normal monitoring of the project elements. This level of monitoring will be the order of priority for project managers to assist decision making in improving the performance of the project.

2. Literature Review

2.1. Project Management

A project is "a unique process, consisting of a series of coordinated and controlled activities with the initiation and completion is done to achieve the objectives in accordance with specific requirements" [3]. Project management is the planning, organizing, directing, and controlling resources company that has been established to complete specific short-term objectives [4]. Project management involves 5 process groups, i.e. project initiation, project planning, project execution, project monitoring and controlling, and project closing [5].

2.2. Performance Measurement

Performance measurement can be defined as a process for measuring the effectiveness and efficiency of an action [6] using variables that indicate the quantitative measures [7] in order to achieve those objectives derived from the company's strategic goals [8].

Performance measurement is based on the company's strategy [7]. It aims to support the implementation and monitoring of strategic initiatives. The selection of performance measures and setting targets of these measures can be seen as a concrete formulation of corporate's strategy choice. Financial or non-financial measures are necessary to translate strategy into specific objectives that provide guidance for the management of operational measures. Actual results achieved from a variety of measurements reflect how successful the company in achieving the strategy choice.

2.3. Multi-criteria decision analysis (MCDA) and MACBETH

Multi-criteria decision analysis (MCDA) is an alternative evaluation using a number of qualitative and / or quantitative criteria that have different measurement units [9] to address the complex problems which showing high uncertainty and conflicting objectives [10], in which decision makers should select, evaluate or rank the attributes according to the weight of the criteria [11] to determine the overall preference among all alternative options in which the alternatives can achieve some goals [12].

There are several commonly used methods of MCDA to assist decision making. Methods are often used are MAUT, MACBETH, AHP, ANP, ELECTRE and PROMETHEE. MCDA method for this study is the method of MACBETH. MACBETH (measuring attractiveness by a categorical based evaluation technique) is a mathematical interactive approach that uses semantic assessment of the differences in the attractiveness of various stimuli to help decision makers expressed their attractiveness relative to a quantity [13]. MACBETH is MAUT method, which is based on a comparison between different situations (which identifies context) made by decision-makers [14]. MACBETH illustrates this situation with an expression of basic performance on the one hand, and the aggregate expression on the other. The principle is to translate the qualitative information derived from the human expertise of decision-makers, into quantitative information. When the answer is inserted into the MACBETH decision support system, this method automatically verify their consistency. This then produces a numerical scale that is representative of the assessment decision-makers. Through the same process this method allows for the generation of scale weighting criteria. Moreover, this method provides a means to facilitate some kind of sensitivity analysis.

The expression of basic performance appraisal is defined by the semantics of the group situation [14]. More precisely, the concept of preference and strength of preference used and associated with the level of verbal attractiveness (none, very weak, weak, moderate, strong, very strong, extreme). Each interval is scaled so that the resulting lines with MAUT forming a linear transformation (defined by two parameters: the slope and constant). Then, to solve the problem of inter-criteria, simply by specifying, for all interval scales, two common reference points has an absolute meaning for decision makers to adjust the slope and constant. For example, "complete unsatisfactory" performance related to the values 0 and "complete satisfactory" performance related to the value of 1.

2.4. Criteria of Project Performance
There are 3 criteria of project performance: Project Performance Measurement, Project Dimension, and Project Task.

2.4.1. *Project Performance Measurement.* Project can be considered as a business process [1]. From this point of view we can associate with the Project Management Business Process Management (Business Process Management - BPM). The project can be seen as a sequence of tasks in a manner similar to the process can be viewed as a sequence of events. So, a Project Performance Management System (PPMS) can be defined as a set of metrics, or performance measures, which are used to measure both the efficiency and effectiveness of an action [6]. Performance evaluation assumes the need for a tool to analyze the measurements taken in accordance with the two dimensions of efficiency and effectiveness. This is a question to consider the impact of each component of the performance. Basically, BPM analysis adds a third component: relevance [1]. There are also two additional criteria [15], namely implementability and flexibility, so that there are five performance measures that will be used so that project performance measurement criteria: efficiency, effectiveness, relevance, implementability, and flexibility.

2.4.2. *Project Dimension.* Each project performance evaluation considers that it is necessary to analyze the measurements. The project manager must maintain a network of projects and monitoring the shift of costs, time, and quality for the duration of the project [16]. In achieving this, the project manager relies heavily on a reliable monitoring system that can provide timely signals to project issues, whether actual or potential. Project Management Body of Knowledge [5] proposed nine essential areas of knowledge management and project management to illustrate: Project Integration, Project Scope, Project Time, Project Cost, Project Quality, Project Human Resources, Project Communication, Project Risk, and Project Procurement. Other than those dimensions, there is an additional criteria, which is Project Safety [15][17][18].

2.4.3. *Project Task.* In addition to project performance measurements and project dimensions, there is one element to consider in the evaluation of project performance. The element is project tasks, which varies from one project to another. On drilling projects of an oil and gas service company, it is possible to draw the outline of each project task to be able to generalize the task that must be observed. The tasks are Business Cooperation, Prepare Resources, Mobilize Resources, Perform Services, Demobilize Resources, Complete Reports & Field Tickets, and Review Performance.

2.5. Project Performance Measurement System with Multi-Criteria
The idea consists of a combination of decision-support methods for different project contexts [1]. This method should be tailored to the needs of decision-makers in terms of tasks executed project performance measurements used, and project dimensions observed. The model consists of three dimensions: task that followed, the view that is used to analyze the performance (dimensions), and the types of performance measurement. Full details of this system can be seen in [19].

3. Research Methodology
In this section will be explained the stages of data collection and processing that will be used followed by analysis of the research results. The data collected is the secondary data from project performance

reports, as well as primary data in the form of pairwise comparisons of each of the criteria from the experts for evaluating the performance of the project.

3.1. Collecting Data

Pairwise comparison of the data is used in this study to see how important a criterion as compared with other criteria in a group of factors. This data is obtained by using the questionnaire addressed to the experts of the company to the top-level manager who has experience over 7 years in drilling project activities by the company.

3.1.1. *Questionnaire.* This questionnaire aims to filter criteria will be compared, whether the criteria are important for the project activities. If these criteria are not considered important, these criteria will not included in the pairwise comparisons. The questionnaire contained questions about the level of importance of the criteria and the project was given to 7 experts from the project team. In the questionnaire, there are 5 scales: very unimportant, unimportant, average, important, and very important. The results of this questionnaire then calculated and the Geomean values of each criterion then quantified. Geomean value obtained then filtered, if the value is less than 3.5 then the criteria are considered unimportant, so it is not included in the pairwise comparisons. Once the questionnaire processed, the criteria are decreased from 23 to 19. The criteria that are considered less important are flexibility, Project Integration, Project Procurement and Demobilize Resources.

3.1.2. *Pairwise Comparison.* This questionnaire produced the level of importance of a criterion relative to other criteria. This questionnaire contains a pairwise comparison of each sub-criterion project. This is closed questionnaire and given to the 5 experts from the project team consisting of employees of the company to the top-level managers, and decision makers in a project. In the questionnaire, there are 7 MACBETH scale. The results of this questionnaire then calculated and the Geomean values of each criterion then quantified. Goemean value obtained is the result of a comparison between the criteria with the other criteria. The result of this questionnaire is a pairwise comparison matrix and then inputted into the MACBETH model.

Table 1. Project performance data

Task	Dimension	Measurement			
		Effectiveness	Efficiency	Relevance	Implementability
Business Cooperation	Scope	4.33	4.33	4.67	5.00
	Time	3.73	3.73	4.07	4.40
	Cost	3.67	3.67	4.00	4.33
	Quality	4.00	4.00	4.33	4.67
	Human Resources	4.17	4.17	4.50	4.83
	Communication	3.83	3.83	4.17	4.50
	Risk	4.07	4.07	4.40	4.73
	Safety	4.27	4.27	4.60	4.93

3.1.3. *Project Data.* Project data is needed to obtain the value of the performance of each criterion on the project. Performance values obtained vary depending on the performance obtained and performance targets. The closer value to the target performance, the higher performance measured. Table 1 is an example of the value of the performance obtained.

4. Results and Analysis

In this section will be described the model used in the performance evaluation of the project along with the results of the evaluation and a monitoring classification of each criterion on each element to be a suggestion for the company.

4.1. Criteria Weighting

The weights of each criterion can be obtained from pairwise comparison matrices. Weighting criteria for each element is shown in Figure 1.

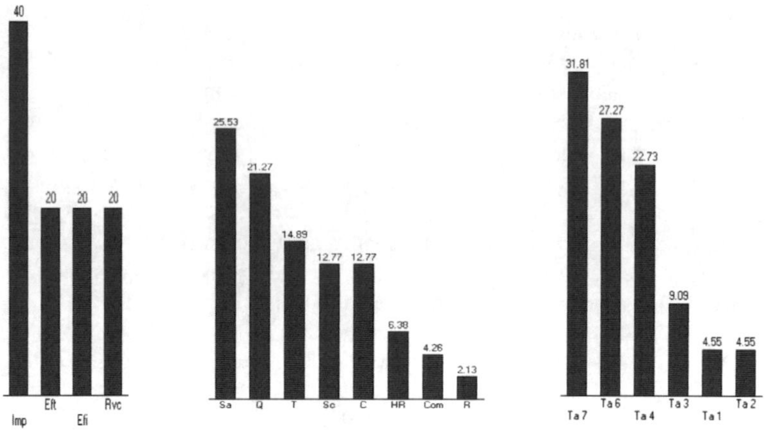

Performance Measurements Dimensions Tasks

Figure 1. The weight of the criteria for each element

Element of the Project Performance Measurements are, weights of Implementability have the highest value, higher than the weight of Effectiveness, Efficiency, and Relevance. This indicates that the implementability of the project is the most influential criteria for the success of the project, while the effectiveness, efficiency, and relevance have an equal level of importance to each other.

Element of Project Dimensions are, the weight of Project Safety criterion has the highest value, followed by Project Quality, Project Time, Project Scope and Project Cost, Project Human Resources, Project Communication, and lastly Project Risk. This indicates that the safety of the project is the most influential criteria for the success of the project, while the project risk has least impact for project success.

Element of Project Tasks are, the weight of Review Performance criterion has the highest value, followed by Complete Reports & Field Tickets, Perform Services, Mobilize Resources, and lastly Business Cooperation and Prepare Resources. This indicates that the review of the project performance is the most important activity in the project, while the business cooperation and preparation of resources is an activity which gives less influence to the success of the project.

4.2. MACBETH model design

MACBETH model consists of three orientations, namely project performance measurement-oriented, project dimension-oriented, and project task-oriented. This model is made by using M-MACBETH software to produce value for each project performance criteria. This paper will only show the task-oriented model of the project.

This project task model combines elements of the project dimension and project performance measurement criteria to be able to generate the value of each project task. This model consists of 8 nodes of project dimensions and 4 subnodes for each node of project performance measurement, as illustrated in the value tree in Figure 2. Once the value of the performance and value comparisons are included, value criterion level of each element can be generated, as shown in Figure 3.

For elements of Project Performance Measurement, can be seen that Implementability have the highest value, followed by Relevance then Effectiveness and Efficiency. This suggests that the activities in the project could be implemented well, while the effectiveness and efficiency of the project is still running poorly compared to the other criteria observed.

For elements of Project Dimensions, can be seen that Project Scope have the highest value, followed by Project Safety, Project HR, Project Risk, Project Quality, Project Communications, Project Time and lastly Project Cost. This suggests that the activities carried out in this project in accordance with project scope and safety has been maintained properly. However, it appears that poor project performance in terms of time and cost, which indicates that the project duration longer than scheduled, and the cost of the project is greater than budgeted, so improvement on both these criteria is needed.

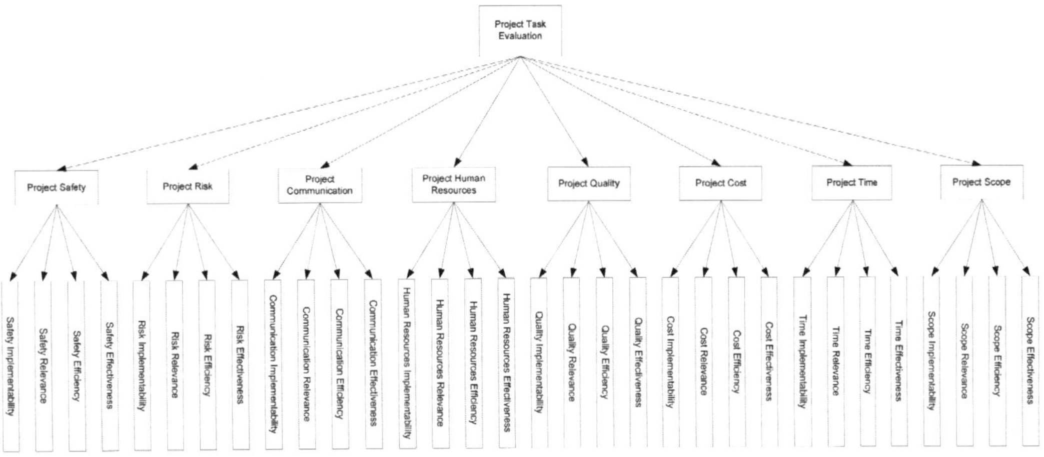

Figure 2. Value Tree of project task model

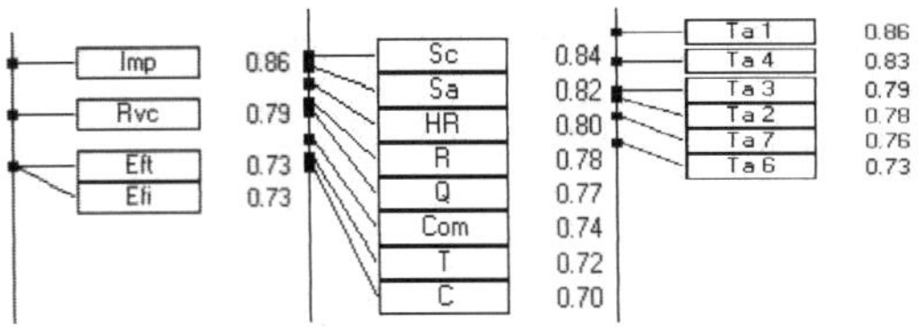

Figure 3. Project performance value

Elements of Project Tasks, can be seen that Business Cooperation have the highest value, followed by Perform Services, Mobilize Resources, Prepare Resources, Review Performance, and lastly Complete Reports & Field Tickets. This suggests that the activities of the project initiation, the business cooperation with stakeholders, run well. Meanwhile the performance of activities such as completion of the project closure report and review of performance is still not good, even though both of these activities are the most important activity.

4.3. Monitoring Classification

After getting the value of project performance, monitoring classification level of project performance can be determined. Monitoring classification is divided into 4 levels of classes [19], Absolute Vigilance, Strong Attention, Close Surveillance, and Normal Monitoring. For the element of project performance measurements, the classification can be classified as follows:

Table 2. Monitoring classification for Project Performance Measurements

Classification	Criteria
Absolute Vigilance	Effectiveness, Efficiency
Strong Attention	
Close Surveillance	Relevance
Normal Monitoring	Implementability

Table 2 shows the effectiveness and efficiency criteria require absolute vigilance to improve its performance. Companies should make sure that the project can be run more effectively and more efficiently. Meanwhile relevance requires close surveillance, so the project activities relevance can be better, and implementability only needs normal surveillance because the activities that need to be done in the project are implemented properly.

Element of the project dimensions can be classified as follows:

Table 3. Monitoring classification for Project Dimensions

Classification	Criteria
Absolute Vigilance	Time, Cost
Strong Attention	Communication, Quality
Close Surveillance	Risk, HR
Normal Monitoring	Safety, Scope

Table 3 shows that the time and cost of the project require absolute vigilance to be able to improve their performance. Companies should make sure that the project implementation time can be shortened and the project cost can be reduced. Meanwhile communication and quality of the project require strong attention, communication in the project should be smoothened and project quality should be improved. Then the project risk and HR requires close surveillance, so that the project risks can be minimized and HR projects can be better, and safety and scope of the project only need normal monitoring because their performance is considered good, the project can be run safely and in accordance with the scope of the process.

Table 4 shows that the completion of the reports requires absolute vigilance to be able to improve its performance. Companies should improve project report to be more complete and informative. Meanwhile review performance requires special supervision, it needs to be improved and the methods used need to be more varied. Then the tasks of Prepare Resources and Mobilize Resources require close surveillance, the equipment and personnel preparation and transportation to the site needs to be improved, and the tasks of Perform Services and Business Cooperation only need normal monitoring as the execution of drilling activities at the site and the process of getting the project partners are considered good.

Table 4. Monitoring classification for Project Tasks

Classification	Criteria
Absolute Vigilance	Complete Reports & Field Tickets
Strong Attention	Review Performance
Close Surveillance	Prepare Resources, Mobilize Resources
Normal Monitoring	Perform Services, Business Cooperation

5. Conclusion

Based on the research that has been done, some conclusions can be drawn according to the research objectives to be achieved. First, in the context of the drilling project, the most important criteria for evaluating the performance of the project is Implementability for project performance measurement, Project Safety for project dimensions, and Review Performance for project tasks. Then the highest value of the project performance criteria for each element is Implementability with a value of 0.86, Project Scope with a value of 0.84, and Business Cooperation with a value of 0.86. In contrast, the lowest value of the project performance criteria for each element is Effectiveness and Efficiency with a value of 0.73, Project Costs with a value of 0.70, and Complete Reports & Field Tickets with a value of 0.73. Also there are four levels of supervision for each classification criteria in the project, with the criteria of Effectiveness, Efficiency, Time, Cost, and Complete Reports & Field Tickets require absolute control so that the performance criteria can be increased.

As a refinement of materials for future research, suggestions that could be considered is the comparison of performance between the two joint performance between elements, so that the comparison between the combined performance ratings can be more accurate. The addition of project elements can be done, considering elements such as stakeholder or project sponsorship in the performance evaluation. In addition, this study can be tested on an ongoing project, so that the evaluation results can be directly applied and the results can be directlymonitored.

Conclusions from the study were expected to be able to overcome the problems of existing projects, and for companies to measure the overall performance of the various criteria to be able to see the value of the overall project performance. There are also levels of classification criteria for supervision of any of the elements of the project, so the project can be carried out surveillance activities in accordance with the classification levels so that the project will increase the performance as a whole and bring success to the implementation of future projects.

References

[1] Marques G, Gourc D, and Lauras M 2010 *Int. J. Proj. Manag.* **29** 1057-69
[2] Turner J and Muller R 2003 *Int. J. Proj. Manag.* **21** 1–8
[3] ISO 2012 ISO 21500:2012 standard: Project Management - Guide to project Management
[4] Kerzner H 2009 *Project Management: A Systems Approach to Planning, Scheduling, and Controlling* 10th ed (New Jersey: John Wiley & Sons)
[5] Project Management Institute 2008 *A Guide to the Project Management Body of Knowledge* 4th ed
[6] Neely A, Mills J, Gregory M, and Richards H 1996 *Int. J. Prod. Eco.* **46–47** 423–31
[7] Lohman C, Fortuin L, and Wouters M 2004 *Eur. J. Oper. Res.* **156** 267–86
[8] El Baz M 2011 *Exp. Sys. Appl.* **38** 6681–8
[9] Özcan T, Çelebi, N, and Esnaf Ş 2011 *Exp. Sys. Appl.* **38** 9773–9
[10] Wang J, Jing Y, Zhang C, and Zhao J 2009 *Renew. Sust. Energy Rev.* **13** 2263–78
[11] Chang T and Wang T 2009 *Infor. Sci.* **179** 355–70
[12] Herva M and Roca E 2013 *J. Cleaner Prod.* **39** 355-71
[13] Bana e Costa C and Chagas P 2004 *Eur. J. Oper. Res.* **153** 323–31

[14] Clivillé V, Berrah L, and Mauris G 2007 *Int. J. Prod. Eco.* **105** 171–89
[15] Barclay C and Osei-Bryson K 2010 *Int. J. Prod. Eco.* **124** 272-92
[16] Xiaoyi Dai C and Wells W 2004 *Int. J. Proj. Manag.* **22** 523–32
[17] Toor S and Ogunlana S 2010 *Int. J. Proj. Manag.* **28** 228-36
[18] Padhy R and Sahu S 2011 *Int. J. Proj. Manag.* **29** 1091-102
[19] Lauras M, Marques G, and Gourc D 2010 *Decis. Supp. Sys.* **48** 342-53

The selection reaction of homogeneous catalyst in soy-epoxide hydroxylation

Flora Elvistia Firdaus

Jayabaya University, Jl. Pulomas Selatan Kav 23 Jakarta -13210 Indonesia

flora_elvistia@yahoo.com

Abstract. Hydroxylation reaction of soy-epoxide has resulted soy-polyol; a prepolymeric material for polyurethane. The conversion and selectivity of soy-epoxide butanol based to hydroxylation was found higher than soy-ethylene glycol (EG) based. These reactions were performed by sulfur acid which commonly known as homogeneous catalyst. Conversion and selectivity of homogeneous catalyst compared to bentonite; a heteregeneous catalyst was lower as in fact the mixtures were more viscous. The catalysis were significantly effected to cell morphology. Foams were conducted by heterogeneous catalyst resulted an irregular form of windows while homogeneous catalyst are more ordered.

Keywords— polyurethane, conversion, selectivity, soybean, catalyst

1. Introduction

Increasing concerns about the environment and renewable resources are fueling a growing worldwide research effort which bringing significantly economic and scientific importance which impacted to reducing the dependence on fossil fuels which are rapidly being exhausted [1-2]. Vegetable oil-based products are well-known as being environment-friendly, abundant, cheap in cost, and excellent properties. The products can be developed and utilized for valuable polymeric materials [3-5].

Vegetable oils are part of larger family of chemical compounds known as fats or lipids, which made up predominantly of triesters of glycerol with fatty acids which can be processed into high value oleochemicals for various industries [6]. The composition of fatty acids contained in the vegetable oils determines the further use of oils; the conversion of oil seed crops into bioplastic could compete with plastics from petroleum chemicals [7-9]. Transformed of fatty acids into polyol through double bonds epoxidation and oxirane ring opening are used in many industries as an additives for polymers as well as composites [10], which have attracted attention for a multitude of plastic products, including polyurethanes [11-14]. The oils are characterized by their oxirane number and hydroxyl value.

Polyurethanes produced from vegetable oils; soybean-oil as well has presented some excellent properties such as hydrolytic and thermal stability.The preparation of polyols from oils have been the subject of many studies but limited attention has been paid to the effect of different alcohols to be used as the ring opening reagents for soy epoxide oil [15-16]. The hydroxyl groups are located in the middle of fatty acid chain resulting a significant steric hindrance to crosslinking. The method of using either electrophiles or nucleophiles, or catalyzed by either acids or bases. Epoxidation with peracetic acid is generally catalysed by soluble mineral acids like sulfur [17]. The reactions catalysed by clays would be an alternatives which were carried under mild conditions. The separation of the spent catalyst is achieved by filtration, and the product is recovered by mere evaporation of solvent. Catalyst used in polyol production for discontinuous batch process for a common practice is sulfur acid [18]. The effectiveness of catalyst in promoting reaction to ethylene glycol (EG) and n-butanol were found 15.6% respectively at same condition [19].

The optimum conversion of epoxide to polyols of using homogeneous catalyst were found at the previous, by temperature 50C was 150 minutes of reaction, while at 60C was 90 minutes of reaction. The best oxirane number found at this conditions was 6.7 (%) [20].

This paper report works to find out the reaction conditions that allows the addition of different alcohols to epoxidized vegetable oils in maximizing selectivity with the used of homogeneous catalyst as well as heterogeneous in polyurethane synthesis. The reason of using EG and butanol, besides performing as chain extender it is also performed as source of hydroxyl others than soy-polyol.

2.Materials and Methods

2.1Material specifications

Soy-oil based polyols used in this study were been synthesized in our laboratory using hydroxylation of epoxide methods. Sulfur acid catalysed were carried in the reaction in accordance to a known procedure to a peracetic acid *in situ* which generated from peroxide and acetic acid. The soybean-oil was purchased from grocery, is the product of Variatama Jakarta with specifications of acid value 0,024 mgr KOH/ gr sample, Iodine value 53,89 gram Iod/100 gr sample, and viscosity 443.007 cps.

3.Result and Discussion

3.1.Preparation of Flexible PU Foam

Hydroxyl groups were introduced by applying two-step methods involving epoxidation of unsaturated chain followed by epoxide ring opening. Chain extender (ethylene glycol, butanol) were taking part in the epoxidations and the amount of catalyst to be used is 1% (v/v). Bentonite was used as the comparative data. The role of these catalysts to polyurethane properties were evaluated.

Polyurethane was prepared using formula by fixing molar ratio of epoxide to chain extenders [1:1]; [1:3]; [1:5]; [1:6] and [1:10] then adding TDI (2.4): MDI (4.4') (80:20) into polyols. The mixtures were vigorously mixed to surfactant and distilled water, which then poured into an open glass mould. The foams removed from the mould and allowed to postcure for one day at room temperature. Oxirane number indicates the epoxide ring has completely reacted in hydroxylation reaction resulted hydroxyl.

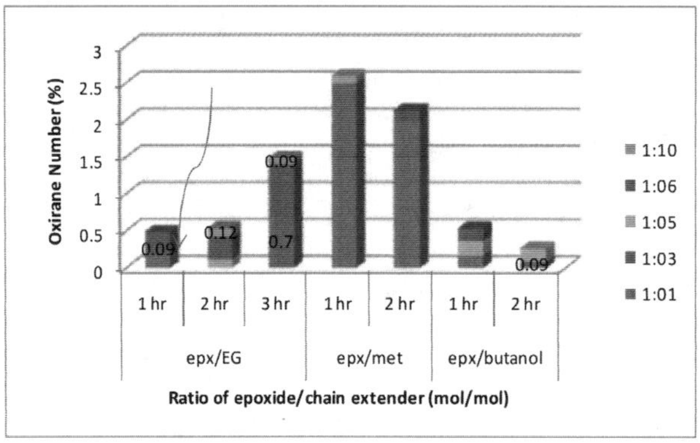

Figure 1. Time of reaction Optimization

The reduction of oxirane number of polyol would caused of increasing hydroxyl value. As can be seen in figure 1. The lowest oxirane number was achieved at concentration of epoxide/ethylene glycol (1:3) at 1 hour and three hour reaction and also at two hour for ratio of epoxide/butanol. Concentration (1:3) was used as based formula for proceed reactions. Hydroxyl value of butanol using sulfur and bentonite are 549.7 mgr KOH/gr and 710 mgr KOH/gr respectively, as compared to EG was 308.5 mgr KOH/gr. High hydroxyl value would caused high consumes of catalyst if further processed to urethane foams.

One of the apparent differences between butanol and EG, which butanol has secondary hydroxyls while EG has only primary hydroxyls. Selecting of polyols for foaming it is well understood that primary hydroxyl are preferred for molded foam process, because the speed up gelling reaction.

3.2.The Catalytic Studies

The effect of polymerization catalyst is illustrated in Figure 1, keeping the weight of sulfur acid at constant for the whole reactions. The product results were compared to as if it is being catalysed by bentonite.

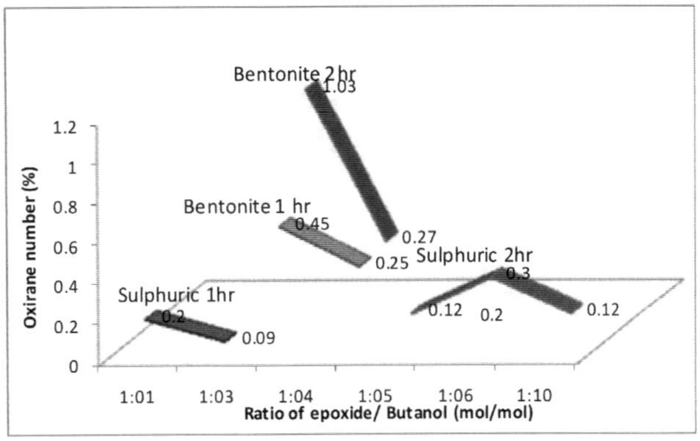

Figure 2. Comparative catalysis existence of epoxidation

Results of the epoxidation of fatty acid catalyzed by H_2SO_4 are displayed in figure 2 and figure 3, with the temperature of 60±0.5C. This figures have clearly explained the conversion and selectivity of epoxidation using butanol are higher than EG.

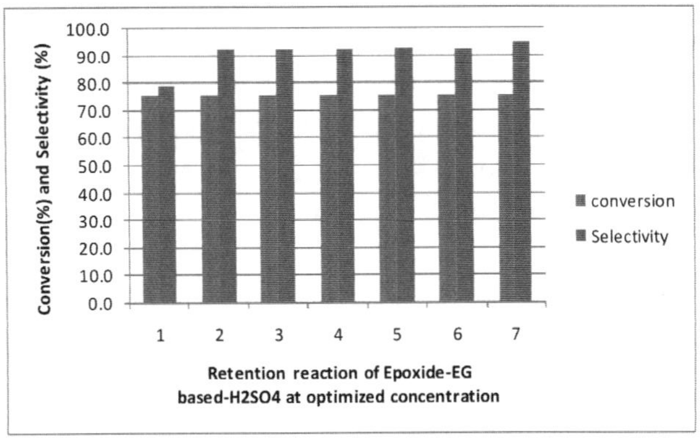

Figure 3. Epoxidation of EG- based with H_2SO_4

The glycols produced as the result reaction of water to epoxide, its formation were favoured by a lower crosslinking. Butanol although has high crosslinking will gives poor selectivity due to glycol formation. The low selectivity were assumed to happened with the decrease of cross-linking.

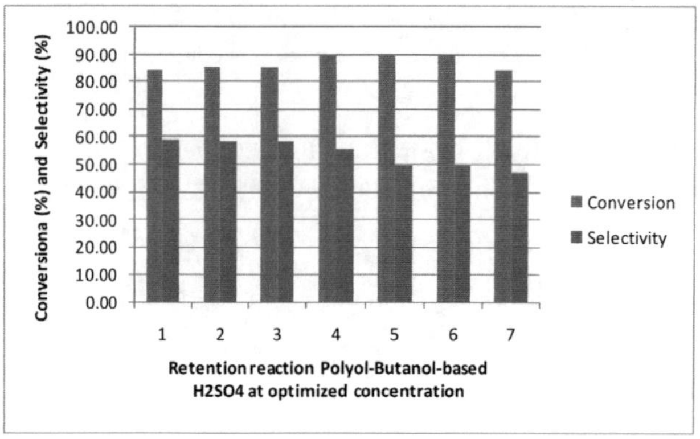

Figure 4. Epoxidation of Butanol-based with H_2SO_4

The viscosity of polyol butanol-based using sulfur acid catalyst were 569.43 cps while viscosity of polyol EG-based using sulphuric acid were 30.55 [21], logically viscosed mixture will find difficulties in the reactions, but referred to this founding the viscosity seems is not effected to conversions and selectivity. For α and β branched alcohols it was possible to improve the selectivity that shown in figure 4. Less complex alcohols it was possible to obtain 100% selectivity at 100% conversion.

The hydroxyl value of polyol butanol- based using sulphuric acid is 549.65 mgr KOH/gram and as to compared to hydroxyl value of polyol butanol-based using betonite catalyst is 710.91 mgr KOH/gram. These differences are met of reactivity and selectivity of the catalyst itself.

3.3. Characterization and Property Measurement of Polyurethane (PU) Foam

The obtained PU foams were characterized by using density measurement, the foams were cut into specimens with dimension of 1x1x1 cm. The specimens were accurately weighed using equation, density = mass/ volume. The density for each foam was ascertained using average value. Many factors and conditions can influence the foaming process of flexible cellular polyurethanes.

Natural oils polyols are significantly different from petroleum-derived polyols, the hydroxyls located in the center of polyol chains. Compared to mostly petroleum-derived polyols, which have either primary hydroxyls and secondary hydroxyls with only one covalent bond removed from the chain ends while for natural oil converted from single oil molecule are most likely to be 5-8 covalent bonds removed from the chain ends, this stated as disadvantages for natural polyol with respect to gelling reaction.

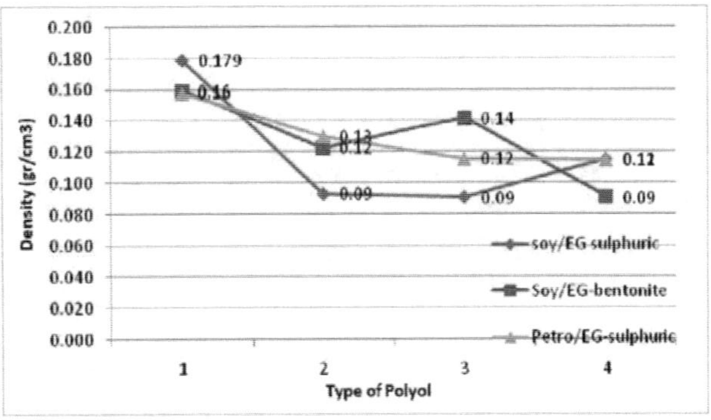

Figure 5. The density of different type of PU Foam

However the theoretical review can be voided to the finding of bentonite as displayed in figure 5. According to Chan-tu there are relationships between foams density and hydroxyl value, foams are made from vegetable oil-based polyols with hydroxyl value below 100 mgr KOH/gram, or polyol mixture having OH_m lower than 292 usually shrank within days of storage [22].

We have tried to analyse the data using outlier test, the possibility of existence density data of soy/EG using sulfur at 0.09 and 0.179, and for soy/EG using bentonite at 0.12 are not in the range distribution to the rest data by using Grubbs's (G) equation [23]. ($X_{suspect}$) values are considered to be at the extremes of the data, $\bar{x}$; mean of the data, and s standard deviation.

$$G = \underline{|\qquad\qquad|}- \qquad (1)$$

For soy/EG sulfur the $x_{suspect}$ was 0.09 with $G_{suspect}$ 0.7156 and $x_{suspect}$ 0.179 with $G_{suspect}$ 1.44 and for soy/EG bentonite the $x_{suspect}$ was 0.12 with $G_{suspect}$ 0.1142. G value then compared to tables of critical for G at $\alpha = 0.05$, If $G_{suspect} > G_{critical}$ then the suspect can be rejected. From this investigation all of the $G_{suspect} < G_{critical}$. It means that all previously $x_{suspect}$ data are in the range of foam density.

The density of using heterogeneous is seem to be much promising. It works in the surface acidity which promotes to an optimum hydroxylation reaction. Although surface area of catalyst are very critical to determine the catalytic site, but it was suitable for the soy-polyol synthesis.

3.4. Cell Morphology

SEM micrograph using of cured solid polyurethane were examined and the average cell diameter are shown in figure 5 a,b and figure 6 a,b. The foam occurred by homogeneous (sulfur) calatlyst at previous hydroxylation has large average cell size with widen cell size distribution. However the cell was in ordered form. Using heterogeneous (bentonite) catalyst in the synthesis has resulted an irregular form.

The hydroxyl value are related to gelling reaction, using heterogeneous is found to be higher than homogeneous. In extreme case the slow gelling reaction could cause foam to collapse during curing. The selectivity obtained using bentonite is poor, value below 40% is implied to incompletion of soybean-oil conversions. The foaming process evidently did not change, however the final morphology unveiled by the characterization study.

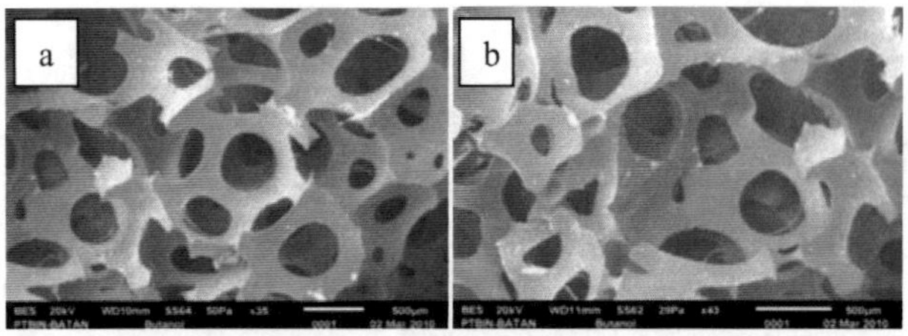

Figure 5 a,b Images of Polyurethane foam using H_2SO_4 catalyst

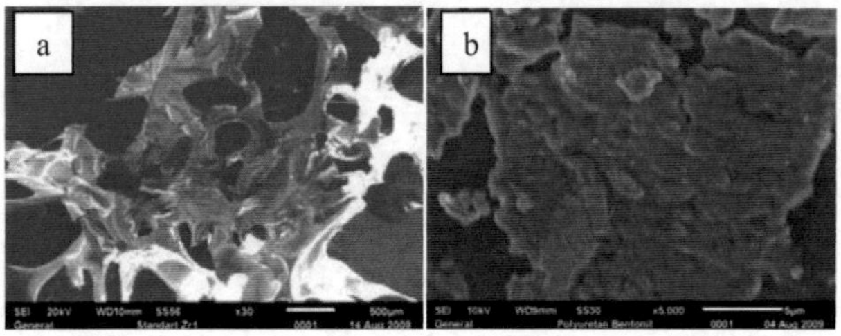

Figure 6 a,b Images of Polyurethane foam using bentonite catalyst

Conclusions

It is found the type of catalyst in polyol synthesis are effected to density, cell morphology, and to other properties of polyurethane as well. The aim of using alcohols in the polyol synthesis are synergize to the optimum existence of catalysis. Catalyst performance in natural oil can void the doubt of using soy-based for future polyurethane manufacture, which can replace petro-based polyol.

Acknowledgement

The authors gratefully acknowledge the support of the Indonesia Department of higher education (Dikti- Kementerian Pendidikan dan Kebudayaan Republik Indonesia) as funding the research.

References

[1] Sperling L.H., Manson J.A. (1983) Interpenetrating polymer networks from triglyceride oils containing special functional groups- a brief review. J Am Oil Chem Soc 60:1887-1892.

[2] Guidoti *et. al.*, (2003), Heteregeneous catalytic epoxidation of fatty acid and methyl esters on titanium grafted silicas, Green Chem, 5, pp, 421-424.

[3] Narine S.S., Kong X, Bouzidi L, Sporns (2007) Physical Properties of Polyurethanes produced from Polyols from seed oils: I. Elastomers. J.Am.Oil Chem Soc 84:55-63

[4] Narine S.S., Kong X., Bouzidi L., Sporns (2007), Physical Properties of Polyurethanes produced from Polyols from seed oils: II. Elastomers. J.Am.Oil Chem Soc 84: 65-72

[5] Lin B., Yang L., Dai H., Yi A. (2008), Kinetic Studies on oxirane cleavage of epoxidized soybean oil by methanol and characterization of polyols, J. Am Oil Chem Soc 85:113-117

[6] Pryde EH, Princern LH. Mukherjee K.D. (eds) (1981), New sources of fats and oils (American Oil Chemists Society Monograph no. 9). American Oil Chemists Society, Champaign, IL.

[7] Hong Hu Y., Gao Y., Wang D.N., Hu C.P., Zu S., Vanoverloop L. and Randall D. (2005), Journal of Applied Polymer Science, 84, p. 591

[8] Badri K.H, Ahmad S.H and Zakaria S. (2001), Journal of Applied Polymer Science, 81, p.384

[9] Zlatanic A, Lava C, Zhang W. and Petrovic Z (2004), Journal of Polymer Science, Part B: Polymer Physics, 42, p.9

[10] Nayak P.L. (2000) natural oil based polymers: opportunities and challenges. J Macromol Sci Rev Macromol Chem Phys C 40:1-21

[11] Khoe T.H., Otey F.H., Frankel E.N. (1972) Rigid urethane foams from hydroxymethylated linseed oil and polyol esters.J Am Oil Chem Soc 49: 615-618

[12] Hu Y.H., Gao Y., Wang D.N., Hu C.P., Zu S., Vanorverloop L., Randall D.(2002) Rigid polyurethane foam prepared from a rapeseed oil based polyol, J Appl Polym Sci 84: 591-597

[13] Dwan 'Isa L.J.P., Mohanty A.K, Misra M., Drzal L.T, Kazemizadeh M. (2003) Novel biobased polyurethanes synthesized from soybean phosphate ester polyols thermomechanical properties evaluations, J polym Environ 11, 161-168

[14] Guo A., Demydov D., Zhang W., Petrovic Z.S. (2002) Polyols and polyurethanes from hydroformylation of soybean oil. J Polym Environ 10: 49-52

[15] Goud V.V., Patwardhan A.V., Pradhan N.C. (2006), Studies on the epoxidation of Mahua oil (*Madhumica Indica*) by hydrogen peroxide, Bioresource Technology 97: 1364-1371

[16] Honghai D., Liting Y., Bo L., Chengsuang W., and Guag S. (2009), Synthesis and characterization of the different soy-based polyols by ring opening of propanediol, J Am Oil Chem Soc, 86: 261-267

[17] Trahanovsky N.S(1978) : Oxidation in Organic Chemistry, 5, Part C, Chapter 3, Academic Press, New York, pp. 211-294

[18] Inui K., Kurabayashi T, Sato S, Ichikawa N (2006), J. Mol Catal. A, 216; 147

[19] Lubguban A.A (2009); Synthesis and Testing of Soy-based polyols: Phospate and Glycerolysis Oligomers; Dissertation

[20] Flora Firdaus, E., (2011), Optimization of soy epoxide hydroxylation to properties of prepolymer polyurethane,World Academy of science engineering and technology, p.81

[21] Flora Firdaus, E., (2012), The type of catalyst to physical properties of soy-based polyurethane, J Chem. Chem. Eng. 6

[22] Chan Tu-Y.(2008), Polyurethane Foams from Novel based Polyols, Dissertation Univ. Of Missouri

[23] Brynn Hibbert D. and Justin Gooding J. (2006), Data Analysis for Chemistry: An Introductory Guide for Students and Laboratory Scientist, Oxford University Press, p.80

The Influence of Basic Physical Properties of Soil on its Electrical Resistivity Value under Loose and Dense Condition

M H Z Abidin[1,5], F Ahmad[2,6], D C Wijeyesekera[3,7] and R Saad[4]
[1] Faculty of Civil and Environmental Engineering, Universiti Tun Hussein Onn Malaysia, 86400 Batu Pahat Johor, MALAYSIA
[2] School of Civil Engineering, Universiti Sains Malaysia, 14300 Nibong Tebal Penang, MALAYSIA
[3] Research Center for Soft Soil, Universiti Tun Hussein Onn Malaysia, 86400 Batu Pahat Johor, MALAYSIA
[4,5] School of Physics, Universiti Sains Malaysia, 11800 USM Penang, MALAYSIA
[6] Institute of Postgraduate Studies, Universiti Sains Malaysia, 11800 USM Penang, MALAYSIA
[7] School of Architecture, Computing and Engineering, University of East London, Beckton E11 2RD London, ENGLAND

E-mail: hazreek@uthm.edu.my

Abstract. Electrical resistivity technique has become a famous alternative tool in subsurface characterization. In the past, several interpretations of electrical resistivity results were unable to be delivered in a strong justification due to lack of appreciation of soil mechanics. Traditionally, interpreters will come out with different conclusion which commonly from qualitative point of view thus creating some uncertainty regarding the result reliability. Most engineers desire to apply any techniques in their project which are able to provide some clear justification with strong, reliable and meaningful results. In order to reduce the problem, this study presents the influence of basic physical properties of soil due to the electrical resistivity value under loose and dense condition. Two different conditions of soil embankment model were tested under electrical resistivity test and basic geotechnical test. It was found that the electrical resistivity value (ERV, ρ) was highly influenced by the variations of soil basic physical properties (BPP) with particular reference to moisture content (w), densities ($\rho_{bulk/dry}$), void ratio (e), porosity (η) and particle grain fraction (d) of soil. Strong relationship between ERV and BPP can be clearly presents such as $\rho \propto 1/w$, $\rho \propto 1/\rho_{bulk/dry}$, $\rho \propto e$ and $\rho \propto \eta$. This study therefore contributes a means of ERV data interpretation using BPP in order to reduce ambiguity of ERV result and interpretation discussed among related persons such as geophysicist, engineers and geologist who applied these electrical resistivity techniques in subsurface profile assessment.

1. Introduction

Electrical resistivity technique (ERT) was a sub method from geophysical method. Geophysical method was a field that applied a principle of physics to study an earth. This field was originally championed by people from geophysicist that has a strong background and fundamental of physical sciences. Some of the physics properties used were electrical resistance, seismic velocity, density, magnetic susceptibility, etc. Recent continuous rapid development of technology has produced sophisticated innovative geophysical equipment such as electrical resistivity, seismic methods, gravity, ground penetrating radar, magnetic, etc. However, the standard performance of individual geophysical method were still depends on fundamental physical constraints, e.g. penetration, resolution, and signal to-noise ratio [1]. Traditionally, most of the data measured from field will be analyzed using utility software with an anomaly contrast. Finally, this anomaly contrast will be interpreted subjectively based on past references and expert experience. In the past, ERT has widely adopted in engineering, environmental and archeological studies. Commonly, the main objective of ERT application was for the detection and as a mapping tools for bedrock and overburden materials [2], groundwater [3] – [5], contamination plumes [6] and [7] meteorite crater [8], etc. ERT has increasingly popular in geotechnical and structural engineering works due to its good efficiency in terms of cost (lower cost), time (less time) and provides large data coverage (2D image) which is therefore able to complement the existing borehole data [9] – [13]. Conventional geotechnical drilling test can only determine information at particular drilling (1D information) point thus require soil interpolation which may be wide in contrast against ERT which can possibly provide a continuous image of the subsurface profile [14]. Field operations require less manpower while data processing and results have become quite easy and fast to be produced compared to the conventional drilling method. ERT consist of several separated set of devices and equipment is suitable to be used as an alternative tool for subsurface site investigation especially in situations of difficult accessibility for the application of conventional borehole method. Furthermore, ERT adopts surface techniques which require minimal contact to the ground thus reducing site damageability during the field measurement [15].

Ground exploration was always associated with subsurface profile characterization which relative to soil and rock characteristics. It is important to have a comprehensive site assessment which importantly used as a design input for construction of civil engineering structure. In engineering perspective, soil has being defined as material that can be worked without drilling or blasting [16]. Basically, soil consists of three natural intermixture of materials which is solid, water and air. Generally, soil can be divided into two major groups which is coarse and fine soil. Coarse soil consists of gravel and sand while fine soil consists of silt and clay particles fraction. Each type of soil has a unique characteristics derived from its nature of uncertainties. However, quantification of basic physical properties of soil may helps engineers to understand their fundamental and mechanics.

The operational process of ERM involving field measurement, data processing, interpretation and conclusion was always championed by physicists due to it being within their field of expertise. Hence, previous ongoing problem regarding the application of ERM gave rise to some lack of confidence among the engineers who were often bemused by the lack of clarity of results and justification produced by geophysicist. There is too much unclear information being covered up by geophysicists especially when they are dealing with geophysical methods related to geotechnical works. According to [15], geophysicists still possess only little appreciation from an engineer's point of view and lack the knowledge of the soil science. Furthermore as reported by [9], some geophysical results and conclusions are difficult to assimilate in sound and definitive ways as some geophysicists attempt to hide their expertise for business reasons. In the past, conventional geophysicist interpretation practice was too obsessed with qualitative anomaly approach which sometimes creates some unconvincing justification and weak results verification. Furthermore, conventional reference tables of geomaterials used for anomaly interpretation also sometimes was difficult to decipher due to its wide range of variation and overlapping values [17]. As a result, a strong verification is vital to support the interpretation outcome which otherwise have been traditionally interpreted based on a qualitative approach depending on the experience of the expert [18]. Otherwise, ERM interpretation will always be subjected to doubts arising from uncertainties and unreliability. Moreover, too many geophysical methods have been used without any reference to the geological situation thus producing disappointed results that lead to a mistrust of the geophysical method by many engineers [15].

As a result, the solutions to these challenges will require multidisciplinary research across the social and physical sciences and engineering [19]. The success at any site investigation works is based on the integration of method [20]. According to [13], studies that relate to geophysical data and geotechnical properties are much rarer and lesser known. Hence, this study proposed the influence of basic physical properties of soil (soil moisture content, densities, specific gravity, void ratio and porosity supported by grain size characteristics) with electrical resistivity value using small scale trial embankment with soil fill placed in a loose and dense condition in order to reduce some black box and ambiguities of electrical resistivity anomaly interpretation via quantitative integration analysis between electrical resistivity value and geotechnical properties. As reported by [14], the quantification of geotechnical properties has become an important factor for rigorous application of resistivity technique in engineering applications.

2. Methodology

This study was performed via fieldwork, laboratory work and data processing. Fieldwork was begun with constructing a small soil embankment model using lateritic soil in loose condition. Dimensions of soil model were 3.0 (length, m) x 1.0 (wide, m) x 0.3048 (height, m) with all sides of the model edge shaped into a gentle slope < 45°. Then, a line of electrical resistivity test was performed using a single leveled line of 2D tomography imaging on the top of each soil model based on electrode configurations using ABEM SAS 4000 equipment as shown in Table 1. Two land resistivity cables were connected to 41 steel electrodes via jumper cables. Then, both resistivity land cables were connected to the electrode selector and Terramater SAS 4000 data logger for field setup. Finally, 12 volt battery was connected to the data logger to supply direct current (DC) during the data acquisition. This study used Wenner array due to its simplicity and for good near surface data. As reported by [21], [22], Wenner array was applicable to obtain a dense near surface cover of resistivity data. Several considerations involving device and equipment setting, position of electrical resistivity line, ground condition, raw data processing etc. needed to be carefully considered and performed in order to determine the best ERV outcome. For example in order to reduce boundary effect that may reduced the ERV accuracy caused by refracted and reflected current, the electrical resistivity line was placed at the center of the soil model with additional offset (0.5m) from each end of its length. Based on [23], electrical current may propagate in geomaterials via the process of electrolysis where the current is carried by ions at a comparatively slow rate. Hence, soil models were poured with water before the electrical resistivity test was conducted. Otherwise, current will be loathed to propagate through the model due to the dry soil condition which will cause some error in the electrical resistivity readings. All raw data obtained from field measurement was transferred to the computer using SAS4000 utilities software. Then, those data was processed and analyzed using RES2DINV software of [24] to provide an inverse model that approximate the actual subsurface structure.

After that, field density test was performed on the top and center of the model using sand replacement method. Then, soil sample based on field density test was immediately taken to the geotechnical laboratory for density record, moisture content test (oven drying method), specific gravity test (gas jar method) and sieve test (wet and dry sieve). Dry and wet sieve test was performed (lateritic soil) due to its mixture of composition between coarse and fine grain of particles. Dry sieve test was conducted using mechanical shaker while hydrometer test was used for wet sieving. The entire related basic geotechnical test was based on [25]. Then, the soil model was compacted using a mechanical compacter and the whole testing process was repeated under dense condition. Both models under 2D Electrical resistivity data acquisition was given in Figure 1 while a schematic diagram representing soil sampling and electrical resistivity line alignment was given in Figure 2. As referred to in [25] and [16], the following equation (1-4 and 5-6) were used to calculate moisture content (w), bulk density (ρ_{bulk}), dry density (ρ_{dry}), specific gravity (Gs), void ratio (e) and porosity (η) of soil sample studied.

$$w = ((m_2 - m_3)/(m_3 - m_1)) \times 100 \quad (1)$$

where m_1 is the mass of container, m_2 is the mass of container and wet soil and m_3 is the mass of container and dry soil

$$\rho = (m_w/m_b)) \times \rho_a \quad (2)$$

where m_w is the mass of the wet soil from hole, m_b is mass of sand in hole and ρ_a is bulk density of sand

$$\rho_d = (100\rho)/(100 + w) \quad (3)$$

where ρ is the bulk density of soil and w is moisture content

$$G_s = (m_2 - m_1)/((m_4 - m_1) - (m_3 - m_2)) \quad (4)$$

where m_1 is the mass of empty jar, m_2 is mass of bottle + dry soil, m_3 is mass of bottle + soil + water and m_4 is mass of bottle + water only

$$e = (G_s\rho_w/\rho_d) - 1 \quad (5)$$

where Gs is the specific gravity of soil, ρ_w is density of water and ρ_d is dry density of soil

$$n = (e/1 + e) \quad (6)$$

where e is the void ratio of soil

Table 1. Configuration used in 2D electrical resistivity test.

No	Setting	Description
1	Array	Wenner
2	Electrode specification	Small steel electrodes: 6 inch of length with 2 mm of diameter
3	Electrode spacing	0.05 m (50 mm)
4	Total number of electrode	41
5	Total number of small jumper cable	42
6	Total length of 2D resistivity test	2 m (2000 mm)

Figure 1. Soil model (lateritic soil) tested by 2D electrical resistivity test: Loose condition (left) and dense condition (right).

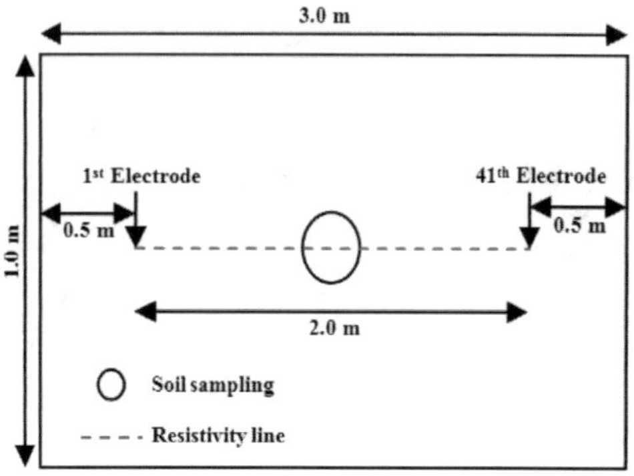

Figure 2. Schematic diagram of the soil sampling position and resistivity line alignment (drawing not to scale).

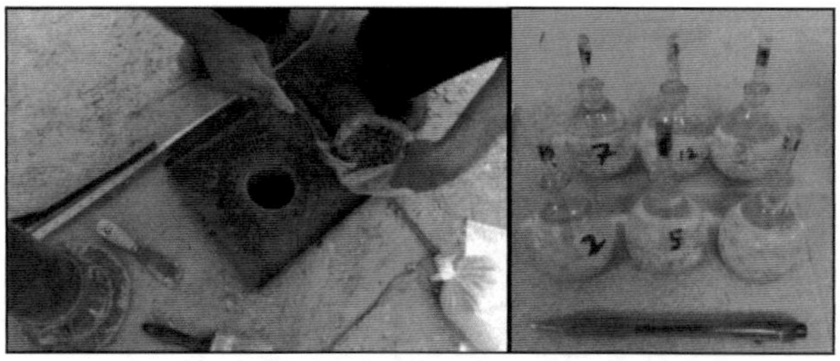

Figure 3. Field density test (left) and specific gravity test (right).

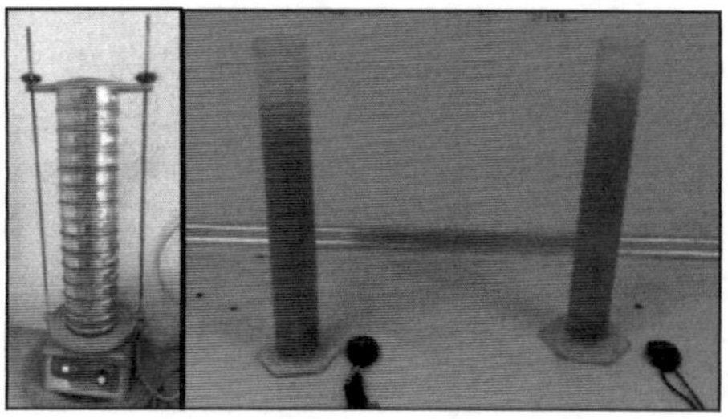

Figure 4. Mechanical sieve (left) and hydrometer test (right).

3. Results and Discussions

All results presented and discussed are based on field electrical resistivity value (ERV), basic physical properties of soil and relationship of field ERV with moisture content (w), density (ρ), void ratio (e), porosity (η) and grain size of soil (d). All results are presented in Figure 5 – 7 and Table 2.

3.1. Electrical resistivity value (ERV)

ERV was determined by measuring the potential difference at points on the ground surface which caused the propagation of direct current through the subsurface [26]. The ERV obtained in Table 2 was originally extracted from the global 2D electrical resistivity tomography section particularly at point L and D given in Figure 5 and 6. Each point of ERV was extracted at the exact location (horizontal: x and depth: y) of the soil sample tested. It was found that soil model under loose condition has higher ERV ($\rho_{(L)}$ = 48763 Ωm) compared to the dense soil model ($\rho_{(D)}$ = 1112 Ωm) due to the different composition of basic physical properties of soil such as solid, water and air. Electrical propagation in soil is largely electrolytic process by flowing in connected pore spaces and along grain boundaries of geomaterial [27].

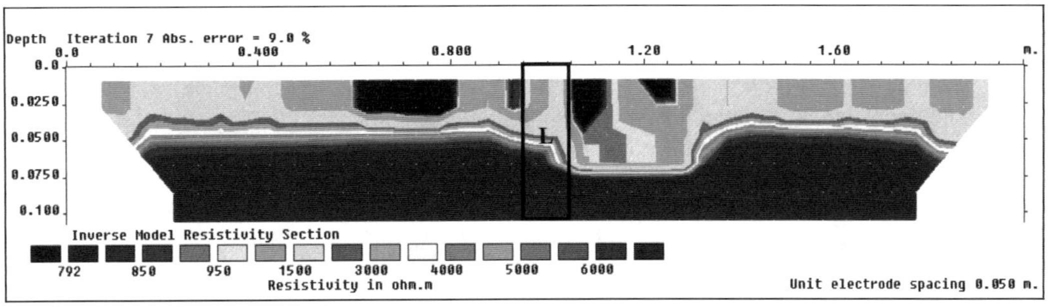

Figure 5. Global 2D electrical resistivity tomography section and localize selected point of ERV under loose condition (L).

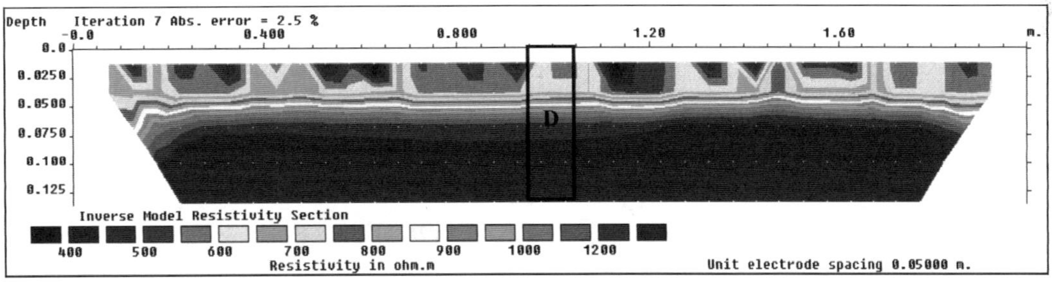

Figure 6. Global 2D electrical resistivity tomography section and localize selected point of ERV under dense condition (D).

3.2. Soil moisture content, void ratio, porosity, densities and grain size fraction

Basic physical properties of soil obtain from geotechnical analysis was given in Figure 7 and Table 2. At soil model under dense condition, it was found that the moisture content (w) value was higher compared to the loose soil model due to the random wetting process of soil model performed before the electrical resistivity data acquisition. Generally, both soil models have demonstrate a different variation of moisture content due to the dissimilarity of grain sizes present at both soil models. Even both of the soil model was built from the same lateritic soil, the variation of different particle fraction was still dissimilar due to the uncertainties nature of soil.

Specific gravity (Gs) test of each soil was conducted using 50 ml bottle. Each soil model was tested three times for averaging purposes. By knowing the Gs, void ratio and porosity of each soil model can be determined using mathematical equation as given in previous section. It was found that the lowest void ratio (e) and porosity (η) occurred at dense soil model ($e = 0.4289$ & $\eta = 0.3002$) compared to other loose soils soil model ($e = 1.8210$ & $\eta = 0.6455$). The variation of void ratio and porosity between dense and loose condition quite obvious due to the different quantity of basic physical properties of soil. These results may indicate that the soil have experienced a different degree of denseness derived from loose and dense state of model condition which can be observed and verified through the soil density results. Physically, the lower void ratio and porosity can indicate the soil was in dense condition and vice versa. The quantity of air and water in soil will be reduced due to the compaction process compared to the loose state condition. A relationship between void ratio and porosity was linear and this parameter has a big influence to the soil density variations.

In soil mechanics and geotechnical engineering, soil density was basically described using bulk density (ρ_{bulk}) and dry density (ρ_{dry}). Bulk density was defined by total mass of solids and water per total volume while dry density was defined by mass of solids per total volume. Quantities of densities provide a measure of the material quantity related to the space amount it occupies [16]. Dense soil model was found to higher density value ($\rho_{bulk} = 2.034$ Mg/m^3 & $\rho_{dry} = 1.673$ Mg/m^3) compared to the loose soil model ($\rho_{bulk} = 1.289$ Mg/m^3 & $\rho_{dry} = 0.867$ Mg/m^3) due to the geomaterials and moisture content variations. It can be observed that the densities of each soil model were relative to the moisture content variations by showing general relationship of densities was linearly proportional with the moisture content. Higher quantity of water will increased the total soil weight thus increasing its density and vice versa. Quantity of fine particles from dense soil model was also greater than loose soil model which allow more water to be absorb in order to increase its densities. Based on Table 2, quantity of clay and silt particles at dense soil model was greater than loose soil model. Loose soil model has a lower capability to absorb water due to its highly porous characteristics which consequently producing low soil densities. Hence, it was strongly believed that the grain size variation has also played some great influences to both soil model densities. Hence, it was strongly believed that the density of lateritic soil was linearly proportional to the presence of moisture content and fine gain geomaterial.

Generally, soil can be in the form of both granular and fine particle. Based on Figure 7 and Table 2, it was found that the lateritic soil was classified as Silty SAND (mixture of both granular and fine particle) based on engineering soil classification. All sieve analysis results of soil specimen tested from both models has shown some slightly variation in terms of grain size quantification due to the natural heterogeneity features of soil. Detailed results obtain in Table 2 was originally extracted from particle size distribution curve (PSD) presented in Figure 7.

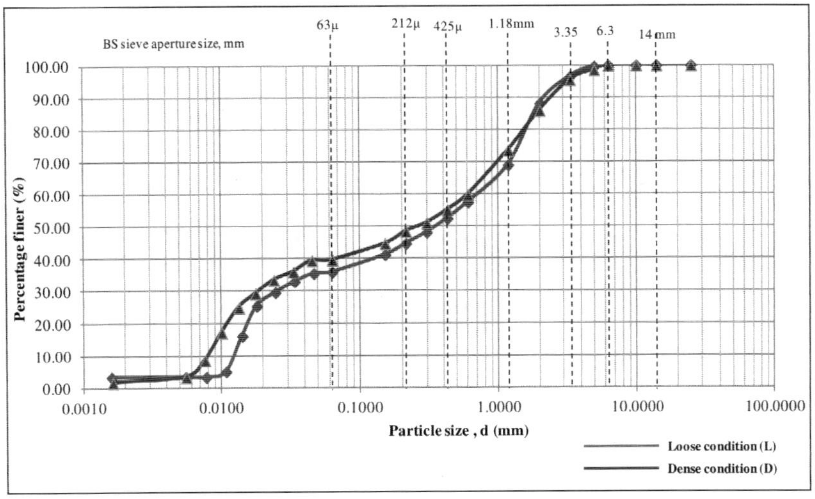

Figure 7. PSD curve for lateritic soil model at loose and dense condition.

Table 2. Extracted ERV and basic physical properties of soil model under loose and dense condition.

Soil Sample	Loose				Dense			
Electrical resistivity value, ρ (Ωm)	48763				1112			
Moisture content, *w* (%)	15.83				21.57			
Particle size analysis, d (µm – mm, %)	Clay	Silt	Sand	Gravel	Clay	Silt	Sand	Gravel
	3.38	32.62	52.23	11.77	1.72	38.05	46.24	13.99
	36.00		64.00		39.77		60.23	
Specific gravity, Gs	2.446				2.391			
Void ratio, e	1.8210				0.4289			
Porosity, η	0.6455				0.3002			
Bulk density, ρ_{bulk} (Mg/m^3)	1.289				2.034			
Dry density, ρ_{dry} (Mg/m^3)	0.867				1.673			

3.3. Relationship of Electrical Resistivity Value due to the basic physical properties of soil

All results from ERV and BPP were analyzed and presented using statistical bar chart in order to demonstrate some relationship between those properties investigated. Resistivity value was highly influenced by pore fluid and grain matrix of geomaterials [28]. As reported by [29], a soil's ERV generally varies inversely proportional to the water content and dissolved ion concentration as clayey soil exhibit high dissolved ion concentration, wet clayey soils have lowest resistivity of all soil materials while coarse, dry sand and gravel deposits and massive bedded and hard bedrocks have the highest ERV. Furthermore a decrease of ERV was results from an increased of metal ions or inorganic elements in geomaterials [30]. According to [31], soil parameters determined in grain size analysis could replicate the variety of resistivities obtained on the site very well. Hence, the field ERV can give varying values due to the variation of soil physical state. In other words, BPP can strongly influence the field ERV due to soil composition variation such as relative to the quantity of solid, air and water.

Based on Figure 8 and 9, it was found that the ERV was high due to the lower moisture content and vice versa. As illustrated in Figure 8, the highest ERV from loose soil model ($\rho_{(L)}$ = 48763 Ωm) was highly influenced by the least amount of moisture content ($w_{(L)}$ = 15.83 %). In contrast, the highest amount of moisture content ($w_{(D)}$ = 21.57 %) has influenced dense soil model ($\rho_{(D)}$ = 1112 Ωm) for having the lowest field ERV. As stated by [23], electrical current may propagate in geomaterials via the process of electrolysis where the current was carried by ions at a comparatively slow rate. The application of field ERT has theoretically stated that the water content in subsurface materials has a close positive correlation with the electrical conductivity [32]. Hence, it was shown that ERV obtained from both soil model was highly influenced by the presence of moisture content which can be established by a general relationship that the field ERV was inversely proportional to the amount of moisture content ($\rho \; \alpha \; 1/w$) since a higher moisture content will caused field ERV to be low and vice versa.

This controlled miniature model study also revealed that the soil electrical resistivity value was highly influenced by the presence of air void content. The ERV of soil model was found to be very high ($\rho_{(L)}$ = 48763 Ωm) due to the high volume of void ($v_{(L)}$ = 1.821) and porosity ($\eta_{(L)}$ = 0.6455) based on the first soil model results which focused on loose trial embankment model. After that, the results from dense soil model has revealed that the ERV was reduced ($\rho_{(D)}$ = 1112 Ωm) together with the reduction of void ratio ($v_{(D)}$ = 1.821) and porosity ($\eta_{(D)}$ = 0.6455). Due to the loose condition of soil model, it enables a higher air filled void which able to increased the ERV over the range of the previous reference charts and tables. As reported by [33], air filled void posses a higher resistivity value compared with the water filled void. Hence, careful considerations such as supported data from others need to be considered in order to interpret a reliable result from loose soil condition. Otherwise, it can be wrongly interpreted as hard rock materials.

Based on Figure 8 and 9, it was found that soil densities under dense condition (ρ_{bulk} = 2.034 Mg/m^3 and ρ_{dry} = 1.673 Mg/m^3) has the lowest ERV ($\rho_{(D)}$ = 1112 Ωm) compared to the densities of soil model under loose condition (ρ_{bulk} = 1.289 Mg/m^3, ρ_{dry} = 1.867 Mg/m^3 and $\rho_{(L)}$ = 48763 Ωm). The soil density value of dense condition was higher than loose condition since the soil quantity of compact condition require more than loose condition. During the compaction effort, volume of air contained in pore was decreased and thus require an additional soil added and compacted until it fully fit according to the soil model dimension required. Hence, the amount of soil used was higher compared to the loose conditions which contribute to a greater value of densities. Under loose condition, soil consist higher of voids which dominantly filled by air thus contribute to a lower weight which relative to the lower densities value measured. In the past, void ratio and porosity can influence the variation of soil density since a denser soil was derived from the soils with a low void ratio and porosity and vice versa. This study has demonstrated that the denser soil will decreased the ERV due to the low void ratio and porosity. The low void ratio and porosity in dense soil will assist the current propagation (electrolysis process was easily due to low porosity which contained more water) thus producing a low ERV. Hence, this study has successfully demonstrated that the highest field ERV was due to the low soil densities which associate to higher volume of void and porosity as the relationship can be established as $\rho \; \alpha \; 1/\rho_{bulk/dry}$.

According to [34], ERV can be influenced by soil grain size as a higher ERV was derived from the larger coarse soil and vice versa. Based on Figure 8 and 9, the highest field ERV was at loose soil model ($\rho_{(L)}$ = 48763 Ωm) which having the greatest amount of coarse soil (CS = 64.00 %) and lowest fine soil (FS = 36.00 %). In contrast, the lowest field ERV was at dense soil model ($\rho_{(D)}$ = 1112 Ωm) which composed of the lowest coarse soil (CS = 60.23 %) and highest fine soil (FS: 39.77 %). Hence, it was shown that the field ERV was influenced by the presence of soil grain size which can be stated by a general relationship that the ERV was linearly proportional to the amount coarse soil ($\rho \propto CS$) since the higher ERV was caused by the higher amount of coarse soil. In other case, a lower ERV also has demonstrates a significant relationship due to the higher composition of fine soil. Hence, the relationship of field ERV due to the fine soil can be established as $\rho \propto 1/FS$. Furthermore, the higher quantity of fine soil will allow more water to be absorbed thus helps the ease of current propagation and finally will reduced the ERV.

However in some cases, those general relationships presented will turn inversely especially when the properties obtained was almost similar to each other. Hence, other major non similar properties will take placed to influence the field ERV. Based on [35], detailed study related to the field condition such as porosity, degree of saturation, salt concentration in pore fluid, grain size, size gradation, temperature and activity can produce more accurate correlation performed from the laboratory experiment. Hence, it has been shown that the ERV was influenced by the BPP variations. This study can contribute to the related parties which used the electrical resistivity technique (ERT) as a strong verification of ERV interpretation. Conventional subjective anomaly interpretation of ERV can possibly being enhanced using the BPP relationship thus increasing the sense of appreciation and confidence level of an engineers to applied ERT in geotechnical site investigation (GSI). Moreover, the ERV reliability can also being increased objectively due to the strong direct data verification (BPP). According to [15], geophysical techniques offer the chance to overcome some of the problems inherent in the more conventional ground investigation techniques. Hence, further research can possibly be studied in the future such as the application of ERT as a tool to predict the BGP quantitatively. Current GSI works is growing rapidly thus require an alternative tool such as ERT in order to assist and enhanced the conventional GSI techniques (drilling method). Based on [9], it is important to quantify the BGP numerically for the purpose of analysis and design. Furthermore, BGP can further influence the geotechnical engineering properties such as shear strength and compressibility. ERT can benefit our sustainable ground investigation since it can reduce time, money and compliment others conventional method especially by its surface (non-destructive) 2D/3D surface technique of investigation.

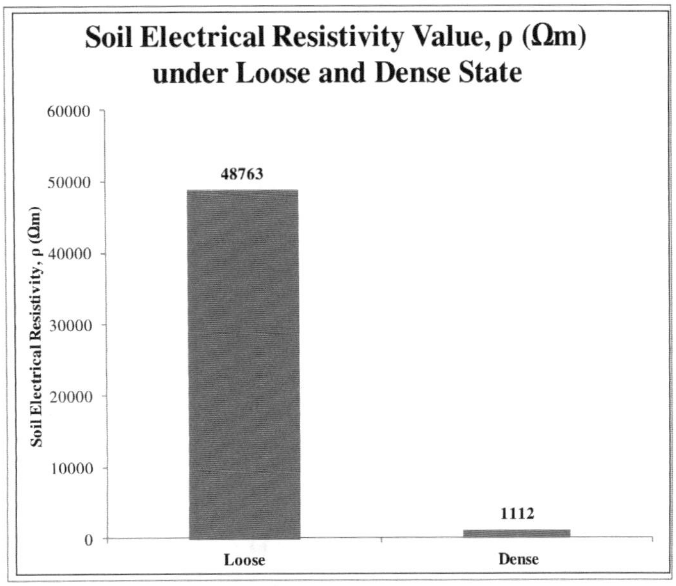

Figure 8. ERV of soil model under loose and dense condition.

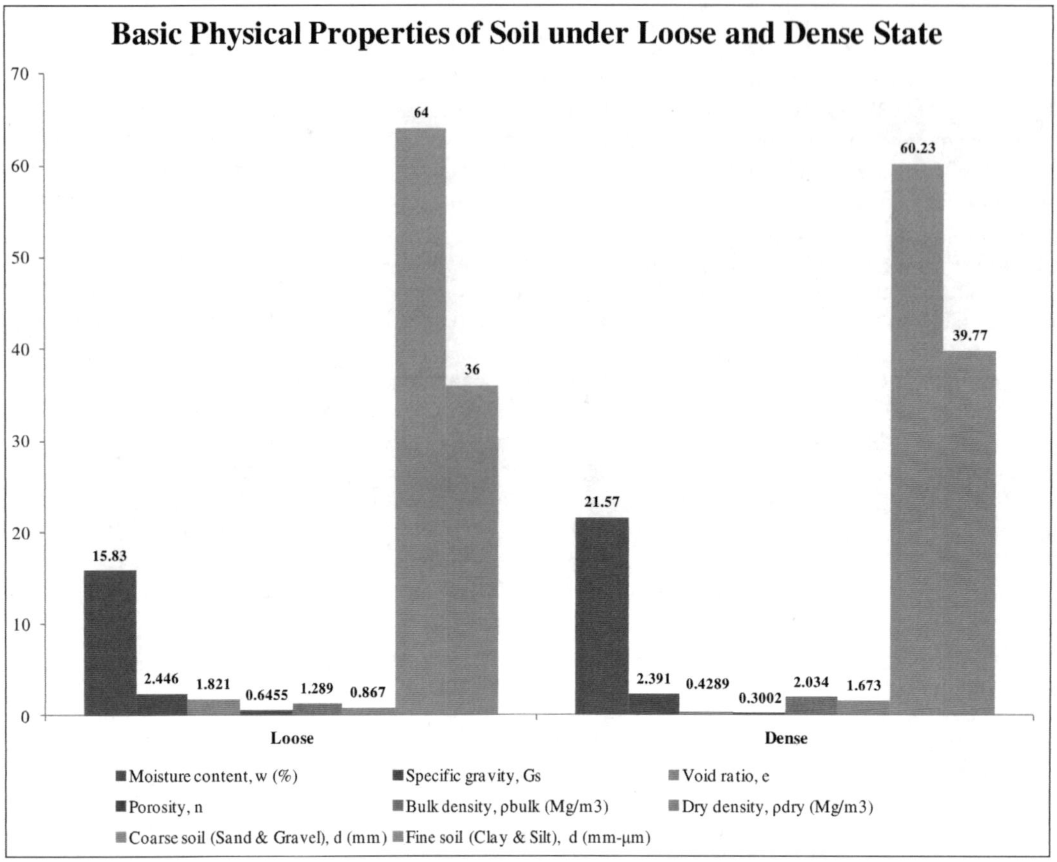

Figure 9. Basic physical properties of soil model under loose and dense condition.

4. Conclusion

The electrical resistivity value of Silty SAND was successfully performed under loose and dense state of small soil embankment model. The influence on soil resistivity value due to variations of moisture content, densities, void ratio, porosity and grain size fraction was successfully and methodically studies and presented. The ERV was influenced by the variation of soil physical state which related to the composition of water, air and solid in soil such as its sensitivity to the quantitative proportion of water, porosity and geomaterial particle fractions under different denseness condition. Different denseness level of soil has creates some variations to the basic physical properties of soil such as the dense condition has less void and porosity compared to the loose condition. The integration of geophysical results such as electrical resistivity value with basic physical properties of soil obtained from geotechnical testing and formulation provided a meaningful contribution to the geophysicist and geotechnical engineers since it applicable to minimize and explain some of the ambiguity during the data interpretation stage. Hence, the confidences level and reliability of electrical resistivity results can be increased due to the strong verification and supporting data obtained from the laboratory test.

Acknowledgment

First author wish to acknowledge gratefully to supervisors and research members for their tremendous guidance, work and cooperation.

References

[1] Mauritsch H J, Seiberl W, Arndt R, Romer A, Schneiderbauer K and Sendlhofer G P 2000 Geophysical investigations of large landslides in the Carnic Region of Southern Austria *Engineering Geology.* **56** 373-388

[2] Saad R, Muztaza N M and Mohamad E T 2011 The 2D Electrical Resistivity Tomography (ERT) Study for Civil and Geotechnical Engineering Purposes *Electronic Journal of Geotechnical Engineering.* **16** 1537-1545

[3] Saad R, Nawawi M N M and Mohamad E T 2012 Groundwater Detection in Alluvium Using 2-D Electrical Resistivity Tomography (ERT) *Electronic Journal of Geotechnical Engineering.* **17** 369-376

[4] Juanah M E, Ibrahim S, Sulaiman W and Latif P 2012 Groundwater resources assessment using integrated geophysical techniques in the southwestern region of Peninsular Malaysia *Arabian Journal of Geosciences.*1-16

[5] Kalisperi D, Soupios P, Kouli M, Barsukov P, Kershaw S, Collins P and Valliantos F 2009 Coastal aquifer assessment using geophysical methods (TEM, VES), case study: Northern Crete, Greece *Proc. of the 3rd IASME / WSEAS Int. Conf. on GEOLOGY and SEISMOLOGY (GES'09) (Cambridge, UK, 2009)* pp 158-164

[6] Samsudin A R, Rahim B E A, Yaacob W Z, Hamzah U 2006 Mapping of contamination plumes at municipal solid waste disposal sites using geoelectric imaging technique: Case studies in Malaysia *Journal of Spatial Hydrology.* **6**

[7] Moradi M, Hafizi M K, Taheri B and Kamal H S 2011 Application of Geophysical Methods to Delineation of LNAPL Contaminated Plume *Proc. of the 7th IASME/WSEAS Int. Conf. on Energy, Environment, Ecosystems and Sustainable Development and the 4th IASME/WSEAS Int. Conf. on Landscape Architecture (Angers, France, 2011)* pp 135-139

[8] Saad R, Saidin M M, Muztaza N M, Ismail N A and Ismail N E H 2011Subsurface Study Using 2-D Resistivity Imaging Method for Meteorite Impact at Bukit Bunuh, Perak *Electronic Journal of Geotechnical Engineering.* **16** 1507-1513

[9] Abidin M H Z, Ahmad F, Wijeyesekera D C, Saad R and Baharuddin M F T 2013 Soil resistivity measurements to predict moisture content and density in loose and dense soil *Applied Mechanics and Materials.* **353-356** 911-917

[10] Khatri R, Shrivastava V K and Chandak R 2011 Correlation between vertical electric sounding and conventional methods of geotechnical site investigation *Int. Journal of Advanced Engineering Sciences and Technologies.* **4** 042-053

[11] Liu C and Evett J B 2008 *Soils and Foundation* (New Jersey: Pearson International)

[12] Godio A, Strobbia C and De Bacco G 2005 Geophysical characterisation of a rockslide in an alpine region *Engineering Geology.* **83** 273-286

[13] Cosenza P, Marmet E, Rejiba F, Jun Cui Y, Tabbagh A and Charlery Y 2006 Correlations between geotechnical and electrical data: A case study at Garchy in France *Journal of Applied Geophysics.* **60** 165-178

[14] Hossain J, Khan M S, Kibria G and Hossain M S 2012 Determination of Moisture Content and Unit Weight of Clayey Soil Using Resistivity Imaging (RI) *GeoCongress.* pp 3398-3407

[15] Clayton C R I, Matthews M C and Simons N E 1995 *Site Investigation* (UK: Blackwell Science Ltd)

[16] Whitlow R 2001 *Basic Soil Mechanics* (Dorset: Prentice Hall)

[17] Solberg I L, Hansen L, Rønning J S, Haugen E D, Dalsegg E and Tønnesen J 2011 Combined Geophysical and Geotechnical Approach to Ground Investigations and Hazard Zonation of a Quick Clay Area, Mid Norway *Bulletin of Engineering Geology and the Environment.* **71** 119-133

[18] Abidin M H Z, Wijeyesekera D C, Saad R and Ahmad F 2013 The Influence of Soil Moisture Content and Grain Size Characteristics on its Field Electrical Resistivity *Electronic Journal of Geotechnical Engineering.* **18/D** 699-705

[19] Fragaszy R, Santamarina J, Amekudzi A, Assimaki D, Bachus R, Burns S, Cha M, Cho G, Cortes D, Dai S, Espinoza D, Garrow L, Huang H, Jang J, Jung J, Kim S, Kurtis K, Lee C, Pasten C, Phadnis H, Rix G, Shin H, Torres M and Tsouris C 2011 Sustainable development and energy geotechnology — Potential roles for geotechnical engineering *KSCE Journal of Civil Engineering.* **15** 611-621
[20] Benson R C, Yuhr L and Kaufmann R D 2003 Some Considerations for Selection and Successful Application of Surface Geophysical Methods *Proc. The 3rd Int. Conf. on Applied Geophysics (Orlando, Florida. 2003)*
[21] Baharuddin M F T, Taib S, Hashim R, Abidin M H Z and Rahman N I 2012 Assessment of seawater intrusion to the agricultural sustainability at the coastal area of Carey Island, Selangor, Malaysia *Arabian Journal of Geosciences.* 1-20
[22] Hamzah U, Yaacup R, Samsudin A R and Ayub M S 2006 Electrical imaging of the groundwater aquifer at Banting, Selangor, Malaysia *Environmental Geology.* **49** 1156- 1162
[23] Telford W M, Geldart L P and Sheriff R E 1990 *Applied Geophysics* (Cambridge: Cambridge University Press)
[24] Loke M H, Acworth I and Dahlin T 2003 A comparison of smooth and blocky inversion methods 2-D electrical imaging surveys *Explor Geophys.* **34** 182–187
[25] British Standard 1377 1990 *Methods of test for Soils for Civil Engineering Purposes*
[26] Burger H R, Sheehan A F and Jones C H 2006 *Introduction to Applied Geophysics* (New York: W.W. Norton & Company)
[27] Anita T, 2005 Water content and porosity estimated from ground-penetrating radar and resistivity *Journal of Applied Geophysics.* **58** 99-111
[28] Griffiths D H and King R F 1981 *Applied Geophysics for Geologist and Engineers-The Element of Geophysical Prospecting* (Oxford: Pergamon Press)
[29] Liu C and Evett J B 2008 *Soils and Foundation* (New Jersey: Pearson International)
[30] Jung Y, Lee Y and Ha H 2000 Application of Electrical Resistivity Imaging Techniques to Civil and Environmental Problems *Use of Geophysical Methods in Construction*
[31] Friedel S, Thielen A and Springman S M 2006 Investigation of a slope endangered by rainfall-induced landslides using 3D resistivity tomography and geotechnical testing *Journal of Applied Geophysics.* **60** 100-114
[32] Abidin M H Z, Saad R, Ahmad F, Wijeyesekera D C and Baharuddin M F T 2012 Integral Analysis of Geoelectrical (Resistivity) and Geotechnical (SPT) Data in Slope Stability Assessment *Academic Journal of Science.* **1** 305-316
[33] Jusoh Z 2010 Application of 2-D resistivity imaging and seismic refraction technique in subsurface investigation for civil engineering, *M.S. thesis*, Sch. Physics, Science Univ., Penang, Malaysia
[34] Chik Z and Islam S M T 2012 Finding Soil Particle Size through Electrical Resistivity in Soil Site Investigations *Electronic Journal of Geotechnical Engineering.* **17** 1867-1876
[35] Rinaldi V A and Cuestas G 2002 Ohmic Conductivity of a Compacted Silty Clay *Journal of Geotechnical and Geoenvironmental Engineering.* **128** 824-835

Aitken delta-squared generalized Juncgk-type iterative procedure

M. De la Sen and R. Nistal
Institute of Research and Development of Processes. University of Basque Country Campus of Leioa (Bizkaia) - PO Box. 644- Bilbao, 48080- Bilbao. SPAIN email: manuel.delasen@ehu.es

Abstract. This paper discusses a general Aitken delta-squared generalized Jungck-modified S-iterative scheme. The study applies generalized versions of Aitken delta-squared procedure and Venter theorem to discuss positivity and global stability of the generalized Jungck iterative scheme which is of interest in numerical methods and its acceleration of convergence

Keywords. Aitken delta- squared, Jungck iteration, Venter theorem, stability.

1. Introduction

Iterative schemes are of interest in numerical computations and its properties related to acceleration of convergence to solve scientific and engineering problems. A very common iterative scheme is the so-called Jungck iterative scheme, which involves the use of two coupled mappings, and its various extensions, [1-13]. Such a scheme is useful also in fixed point theory to find common fixed points of both mappings. In this paper, we extend such an iterative scheme and we study its stability and positivity under certain parametrical restrictions. Also, we study the acceleration of convergence by proposing a type of delta-squared Aitken procedure for the acceleration of convergence and to combine the results with the generalization of a background Venter stability theorem result, whose basic form is well-known for discrete parametrical recursive identification, [14].

1.1 Notation

The real sequences $\{y_n\}$and $\{x_n\}$ are equivalent, denoted by $\{y_n\}\approx\{x_n\}$, if both have the same limit.

2. The iterative sequence and some preliminaries

The following Aitken delta-squared generalized Jungck- modified S-iterative scheme is of interest in order to accelerate the convergence of the modified sequence to the limit of the unmodified one provided that such a limit exists:

$$Sz_{n+1}=(1-a_n)T^n z_n+a_nT^n y_n \ ; \ z_{-1}=0, \quad z_0\in C \tag{2.1.1}$$

$$ASz_{n+1}=Sz_{n+1}-\mu_{n+1}\frac{(\Delta Sz_{n+1})^2}{\Delta^2 Sz_{n+1}} \tag{2.1.2}$$

$$Sy_n=(1-b_n)Sz_n+b_nT^n z_n \tag{2.1.3}$$

$$ASy_n=Sy_n-\nu_n\frac{(\Delta Sy_n)^2}{\Delta^2 Sy_n} \tag{2.1.4}$$

; $\forall n\in N$, where $S,T:C\to C$ are two mappings on a nonempty subset C of a Banach space $(X,\|\ \|)$ subject to $T(C)\subseteq S(C)$, $\{a_n\}$and $\{b_n\}$are real sequences in $[0,1]$, $\{\mu_n\}$ and $\{\nu_n\}$ are binary sequences taking values 0 or 1; and Δ is a finite difference operator which defines the correcting Aitken-type terms of the above iterative procedure as follows:

$$\Delta Sz_{n+1}=Sz_{n+2}-Sz_{n+1} \ ; \quad \Delta Sy_n=Sy_{n+1}-Sy_n \tag{2.2.1}$$

$$\Delta^2 Sx_{n+1}=Sz_{n+1}-2Sz_{n+2}+Sz_{n+3} \ ; \quad \Delta^2 Sy_n=Sy_n-2Sy_{n+1}+Sy_{n+2} \tag{2.2.2}$$

; $\forall n \in N$. The binary sequences $\{\mu_n\}$ and $\{v_n\}$ have two objectives, namely:

a) To remove the Aitken correction if it would imply division by zero at some iteration (i.e. if $\Delta^2 S z_{n+1}$ or $\Delta^2 S y_n$ is zero)

b) To decide if implementing the Aitken correction or not at any particular iteration which does no imply division by zero.

The subsequent two technical preliminary results reflect the features that the sequences $\{S x_n\}$ and $\{S y_n\}$ can have different limits if such limits exist and that under mild conditions the Aitken correction leads to identical limits like its standard Jungck iterative process provided that such limits exist.

Lemma 2.1. The following properties hold:

(i) Assume that $\{a_n\} \subset [0, 1)$ and $\{S z_n\}$ converges to $(Sz)^*$. Thus, the following limit exists:

$$\lim_{n\to\infty} \left\{ [1+a_n(b_n-1)](Sz)^* - (1-a_n)S y_n - a_n b_n T^n y_n \right\} = 0$$

(ii) Assume that $\{b_n\} \subset (0, 1]$ and $\{S y_n\}$ converges to $(Sy)^*$. Thus, the following limit exists:

$$\lim_{n\to\infty} \left\{ (1-a_n)(Sz)^* - b_n S z_{n+1} - (1-a_n)(1-b_n) S z_n + a_n b_n T^n y_n \right\} = 0$$

(iii) Assume that $\{a_n\} \subset [0, 1)$, $\{b_n\} \subset (0, 1]$, $\{S z_n\}$ converges to $(Sz)^*$ and $\{S y_n\}$ converges to $(Sy)^*$.

Thus, the following limit exists:

$$\lim_{n\to\infty} \left\{ [1+a_n(b_n-1)](Sz)^* - (1-a_n)(Sy)^* - a_n b_n T^n y_n \right\} = 0$$

If, in addition, $(Sx)^* = (Sy)^*$ then

$$\lim_{n\to\infty} \left\{ a_n b_n \left((Sz)^* - T^n y_n \right) \right\} = 0$$

If $(Sz)^* = (Sy)^*$ and $\liminf_{n\to\infty} a_n b_n > 0$ then $\exists \lim_{n\to\infty} T^n y_n = (Sz)^*$.

Proof: One obtains explicitly $T^n z_n$ from (2.1.1), since $a_n \neq 1$, and after its replacement into (2.1.3), one gets:

$b_n S z_{n+1} + (1-a_n)(1-b_n) S z_n = \left((1-a_n) S + a_n b_n T^n \right) y_n$; $\forall n \in N$

Property (i) follows since $\{S z_n\} \to (Sz)^*$. To prove Property (ii), one obtains explicitly $T^n z_n$ from (2.1.3), since $b_n \neq 0$, and then one gets that

$(1-a_n) S y_n - b_n S z_{n+1} - (1-a_n)(1-b_n) S z_n + a_n b_n T^n y_n = 0$, $\forall n \in N$

by replacing it into (2.1.1) . Property (ii) follows since $\{S y_n\} \to (Sy)^*$. Property (ii) follows from Properties (i) -(ii). □

Lemma 2.2. The subsequent properties follow:

(i) If $\{Sx_n\}$ converges to a finite limit $(Sz)^*$ then $\{ASz_n\} \to (Sz)^*$ provided that for any $n \in N_0 = N \cup \{0\}$

$$\Delta^2 S z_{n+1} = (1-a_n)T^n z_n + a_n T^n y_n - 2(1-a_{n+1})T^{n+1} x_{n+1} - 2a_{n+1}T^{n+1} y_{n+1}$$
$$+ (1-a_{n+2})T^{n+2} x_{n+2} + a_{n+2}T^{n+2} y_{n+2} = 0 \Rightarrow \mu_{n+1} = 0$$

Furthermore, if $\{\mu_n\} \to 1$ then $\{ASz_n\} \approx \{Sz_n\}$ and $\{ASz_n\} \to (Sz)^*$ faster than $\{Sz_n\} \to (Sz)^*$.

(ii) If $\{S y_n\}$ converges to a finite limit $(Sy)^*$ then $\{ASy_n\} \to (Sy)^*$ provided that for any $n \in N$

$$\Delta^2 S y_n = S y_n - 2S y_{n+1} + S y_{n+2} = (1-b_n)Sz_n + b_n T^n x_n - 2(1-b_{n+1})Sx_{n+1} - 2b_{n+1}T^{n+1} x_{n+1}$$
$$+ (1-b_{n+2})Sz_{n+2} + b_{n+2}T^{n+2} x_{n+2} = 0 \Rightarrow \nu_n = 0$$

Furthermore, if $\{\nu_n\} \to 1$ then $\{ASy_n\} \approx \{Sy_n\}$ $\{ASy_n\} \to (Sy)^*$ faster than $\{Sy_n\} \to (Sy)^*$.

Proof: Note from direct computations that

$$\sum_{j=1}^{n} (A-I)(S z_{j+1} - Sz_j) = \mu_{n+1} \frac{(S x_{n+2} - S z_{n+1})^2}{S z_{n+1} - 2S z_{n+2} + S z_{n+3}} - \mu_1 \frac{(\Delta Sz_1)^2}{\Delta^2 Sz_1}$$

; $\forall n \in N$ and, since $\{S y_n\}$ converges to a finite limit $(Sy)^*$, one gets:

$$\lim_{n\to\infty} \left[\sum_{j=1}^{n} (A-I)(S z_{j+1} - Sz_j)\right] = -\mu_1 \frac{(\Delta Sz_1)^2}{\Delta^2 Sz_1} = \lim_{n\to\infty} (A-I)(S z_{n+1} - Sz_1)$$
$$= \lim_{n\to\infty} (A-I)S z_{n+1} - (A-I)Sz_1 = A(Sz)^* - (Sz)^* - (A-I)S z_1$$
$$= \lim_{n\to\infty} AS z_{n+1} - (Sz)^* - (A-I)S z_1 = -\mu_1 \frac{(\Delta Sz_1)^2}{\Delta^2 Sz_1}$$
$$= (A-I)S z_1 = ASz_1 = -\mu_1 \frac{(Sz_2)^2}{S z_3 - 2S z_2}$$

since $Sz_1 = 0$ from (2.1.1), $\Delta^2 S z_{n+1} = 0 \Rightarrow \mu_{n+1} = 0$ and

$$(A-I)S z_1 = -\mu_1 \frac{(\Delta S z_1)^2}{\Delta^2 S z_1} = -\mu_1 \frac{(Sz_2 - Sz_1)^2}{S z_1 - 2S z_2 + S z_3} = -\mu_1 \frac{(Sz_2)^2}{Sz_3 - 2S z_2}$$

Then, $\{ASz_n\} \approx \{Sz_n\}$ and $\{AS z_n\} \to (AS z)^* = (Sz)^*$. On the other hand, simple calculations with (2.1.2) yield:

$$ASz_n - (Sz)^* = Sz_n - (Sz)^* - \mu_n \frac{(\Delta S z_n)^2}{\Delta^2 S z_n}$$

since $(ASz)^* = (Sz)^*$ so that if $\{\mu_n\} \to 1$ then $\{ASz_n\} \to (Sz)^*$ faster than $\{Sz_n\} \to (Sz)^*$ since $\left(\frac{(\Delta S z_n)^2}{\Delta^2 S z_n \left(S z_n - (Sz)^* \right)} \right) \to 1$ as $n \to \infty$ so that the subsequent limit exists:

$$\lim_{n\to\infty} \frac{AS z_n - (Sz)^*}{S z_n - (Sz)^*} = 1 - \lim_{n\to\infty} \left(\frac{(\Delta S z_n)^2}{\Delta^2 S z_n \left(S z_n - (Sz)^* \right)} \right) = 0$$

Property (i) has been proved. Property (ii) is proved "mutatis-mutandis" from the evaluation of $\sum_{j=1}^{n} (A - I)(S y_{j+1} - Sz_j)$; $\forall n \in \mathbf{N}$ so that

$$\lim_{n\to\infty} \left[\sum_{j=1}^{n} (A - I)(S y_{j+1} - Sy_j) \right] = -v_1 \frac{(\Delta Sy_1)^2}{\Delta^2 Sy_1}$$

$$= \lim_{n\to\infty} AS y_{n+1} - (Sy)^* - (A - I)S y_1 = (A - I)S y_1 = ASy_1 = -v_1 \frac{(Sy_2)^2}{S y_3 - 2S y_2} \qquad \square$$

The following result proves the properties of stability and convergence for the case when $T, S : X \to X$ are linear and the sequences $\{a_n\}$ and $\{b_n\}$ are bounded while non-necessarily nonnegative.

Lemma 2.3. Assume that $\{a_n\}$and $\{b_n\}$ are real bounded sequences and that $T, S : X \to X$ are linear and $S : X \to X$ is, furthermore, one-to-one and of closed range. Then, the following properties hold:

(i) If, for some $k_{1,2} \in \mathbf{R}_{0+}$, one has

$$\limsup_{n\to\infty} \left[|a_n| \left\| T^n \right\| - k_1 \right] \le 0, \quad \limsup_{n\to\infty} \left[|b_n| \left\| T^n \right\| - k_2 \right] \le 0$$

$$\limsup_{n\to\infty} \left[|1 - a_n| \left\| T^n \right\| - k_1' \right] \le 0, \quad \limsup_{n\to\infty} \left[|1 - b_n| - k_2' \right] \le 0$$

$$k_2' + \mu^{-1}(S) k_2 \le 1; \quad \mu^{-1}(S) \left[k_1' + k_1 \left(k_2' + k_2\, \mu^{-1}(S) \right) \right] \le 1$$

with the inequality $\mu^{-1}(S) \left[k_1' + k_1 \left(k_2' + k_2\, \mu^{-1}(S) \right) \right] \le 1$ being guaranteed if

$$k_2' + \mu^{-1}(S) k_2 \le 1; \quad \mu^{-1}(S) \left(k_1' + k_1 \right) \le 1$$

Then, $\{z_n\}$ and $\{y_n\}$are bounded. Furthermore, $\{Az_n\}$ and $\{Ay_n\}$ are also bounded under the conditions of Lemma 2.2.

(ii) If $\|T\| \le 1$ and if

$$\mu^{-1}(S) \left(|1 - b_n| \|S\| + |b_n| \right) \le M; \quad \mu^{-1}(S) \left[|1 - a_n| + \mu^{-1}(S) |a_n| \left(|1 - b_n| \|S\| + |b_n| \right) \right] \le 1; \quad \forall n \in \mathbf{N}$$

Then, $\{x_n\}$ and $\{y_n\}$are bounded. Furthermore, $\{Az_n\}$ and $\{Ay_n\}$ are also bounded under the conditions of Lemma 2.2.

(iii) If $\|T\| \le 1$ and
$\mu^{-1}(S)\left(|1-b_n|\|S\| + |b_n|\right) \le M$; $\left[|1-a_n| + |a_n| M\right] \le \mu(S)$; $\forall n \in \mathbf{N}$
Then, $\{x_n\}$ and $\{y_n\}$are bounded. Furthermore, $\{Az_n\}$ and $\{Ay_n\}$ are also bounded under the conditions of Lemma 2.2.

(iv) If $\|T\| < 1$ and $\{a_n\} \to 1$ then $\{z_n\}$and$\{y_n\}$ are bounded, equivalent and converge to zero. Under the conditions of Lemma 2.2, $\{Az_n\}$ and $\{Ay_n\}$ are bounded, equivalent and they converge to zero faster than $\{z_n\}$and$\{y_n\}$.
(v) If $\|T\| < 1$ and $\{b_n\} \to 1$ then $\{z_n\} \approx \{y_n\}$ are bounded and they converge to zero. Under the conditions of Lemma 2.2, $\{Ax_n\}$ and $\{Ay_n\}$ are bounded, equivalent and they converge to zero faster than $\{x_n\}$and$\{y_n\}$.

Proof: The various properties for the Aitken-modified sequences$\{Az_n\}$ and$\{Ay_n\}$ which follow from the corresponding ones for $\{z_n\}$ and $\{y_n\}$ are direct under Lemma 2.2 so that no specific proofs are needed. Thus, only proofs for the generalized Jungck-modified S-iterative scheme are now given. One gets from (2.1.1) and (2.1.3)

$$\|z_{n+1}\| \le \mu^{-1}(S)\left(|1-a_n|\left\|T^n\right\|\|z_n\| + |a_n|\left\|T^n\right\|\|y_n\|\right) \tag{2.5}$$

$$\|y_n\| \le \mu^{-1}(S)\left(|1-b_n|\|S\| + |b_n|\left\|T^n\right\|\right)\|z_n\| \tag{2.6}$$

where the minimum modulus of S is positive, i.e. $\mu(S) > 0$, since $S : X \to X$ is linear, one-to-one and of closed range then

$$\|z_{n+1}\| \le \mu^{-1}(S)\left[|1-a_n| + \mu^{-1}(S)|a_n|\left(|1-b_n|\|S\| + |b_n|\left\|T^n\right\|\right)\right]\left\|T^n\right\|\|z_n\|$$

; $\forall n \in \mathbf{N}$ and one gets under the given assumptions and recursive calculations that

$$\|y_n\| \le \left(k_2' + \mu^{-1}(S)k_2\right)\|z_n\| \le \|z_n\|;\ \forall n \in \mathbf{N}$$

$$\|z_{n+1}\| \le \mu^{-1}(S)\left[k_1' + k_1\left(k_2' + k_2\,\mu^{-1}(S)\right)\right]\|z_n\| \le \|z_n\| \le \|z_1\|;\ \forall n \in \mathbf{N}$$

which proves the boundedness result of Property (i) for the unmodified sequences and also for the Aitken delta-squared corrected sequences under the conditions of Lemma 2.2. Property (ii) follows from the subsequent relations which follow from the given assumptions:

$$\|z_{n+1}\| \le \|z_n\| \le \|z_1\|;\ \|y_n\| \le \mu^{-1}(S)\left(|1-b_n|\|S\| + |b_n|\right)\|z_n\| \le M\|z_n\| \le M\|z_1\|;\ \forall n \in \mathbf{N}$$

Property (iii) follows under sufficient conditions to prove the boundedness of the sequences under Property (ii) holds, by combining both given assumptions. Properties (iv)-(v) follow from (2.5), (2.6) with $\left\{\left\|T^n\right\|\right\} \to 0$ as follows:

$\{z_n\}\to 0$, $\limsup_{n\to\infty}\left[\|y_n\|-\mu^{-1}(S)\left(|1-b_n|\|S\|\right)\|z_n\|\right]\le 0$, so that $\{y_n\}\to 0$ if $\{a_n\}\to 1$, $\{y_n\}\to 0$ and $\{z_n\}\to 0$ if $\{b_n\}\to 1$. □

3. Extended Venter theorem and its use in the modified Juncgk iterative scheme

The subsequent preliminary result is then used:

Theorem 3.1. Consider the following iterative scheme parameterized by real sequences and constants:

$$x_{n+1}\le(1-\alpha_n+\gamma_n)x_n+\omega_n+\sigma\ ;\ x_0\ge 0 \tag{3.1}$$

where

$$\sigma\ge 0,\alpha_n\in(0,1],\ \omega_n\ge 0,\ \{\gamma_n\}\text{ is bounded, }\gamma_n\ge 0;\ \forall n\in \boldsymbol{N}_0=\boldsymbol{N}\cup\{0\} \tag{3.2.1}$$

$$\sum_{i=0}^{\infty}\omega_n<+\infty,\ \sum_{i=0}^{\infty}\alpha_n=+\infty,\ \alpha_n\to 0\ \text{ as } n\to\infty \tag{3.2.2}$$

and define the real constant $K\in[0,1)$ as $K=\lim_{n\to\infty}\frac{1}{n+1}\sum_{i=0}^{\infty}(1-\alpha_n)$ (which exists since $\alpha_n\to 0$ as $n\to\infty$).Then, the following properties hold:

(i) If $\sigma=\gamma_n=0$; $\forall n\in\boldsymbol{N}_0=\boldsymbol{N}\cup\{0\}$ then $\{x_n\}\to 0$ (Venter theorem [14]). Furthermore,

$$\sum_{i=0}^{n}K^{n-i}\left[\omega_i-(K+\alpha_i-1)^i x_i\right]\ge -K^{n+1}x_0;\ \lim_{n\to\infty}\sum_{i=0}^{n}K^{n-i}\left[\omega_i-(K+\alpha_i-1)^i x_i\right]=0 \tag{3.3}$$

(ii) If $\sigma=\gamma_n=0$; $\forall n\in\boldsymbol{N}_0$ then $\{x_n\}$ is bounded, $\sum_{n=0}^{\infty}\alpha_n x_n<+\infty$ and $\sum_{n=0}^{\infty}x_n<+\infty$.

(iii) If $\sigma=0$; $\forall n\in\boldsymbol{N}_0$ then $\{x_n\}$ is bounded and $\sum_{n=0}^{\infty}(\alpha_n-\gamma_n)x_n<+\infty$.

(iv) If $\limsup_{n\to\infty}(\gamma_n-\alpha_n)\le 0$ then $\{x_n\}$ is bounded. If, furthermore, $\sup_{n\in\boldsymbol{N}_0}(\gamma_n-\alpha_n)<0$ then

$$\sup_{n\in\boldsymbol{N}_0}x_n\le\frac{1}{\inf_{n\in\boldsymbol{N}_0}(\alpha_n-\gamma_n)}\left((1-K)x_0+\sigma+\sup_{0\le i\le n}\omega_i\right)<+\infty$$

Proof: If $\sigma=\gamma_n=0$; $\forall n\in\boldsymbol{N}_0$ then $\{x_n\}\to 0$ from Venter theorem. The second part of Property (i) follows by rewriting (3.1) for $\sigma=\gamma_n=0$; $\forall n\in\boldsymbol{N}_0$ in the equivalent form

$$x_{n+1}\le Kx_n+(1-\alpha_n-K)x_n+\omega_n+\sigma\ ;\ x_0\ge 0$$

which leads to

$$x_{n+1}\le Kx_n+(1-\alpha_n-K)x_n+\omega_n\ ;\ x_0\ge 0$$

; $\forall n\in\boldsymbol{N}_0$ so that one gets via recursive calculations:

$$0\le x_n\le K^n x_0+\sum_{i=0}^{n-1}K^{n-i-1}\left[(1-\alpha_i-K)x_n+\omega_i\right]$$

with $K<1$ which leads to (3.3) and Property (i) is proved. On the other hand, one gets directly from (3.1) for $\sigma=0$

$$\lim_{n\to\infty}\sum_{i=0}^{n}\left(x_{i+1}-x_i\right)=-x_0=\sum_{i=0}^{\infty}\left(\omega_i+\left(\gamma_i-\alpha_i\right)x_i\right)$$

which yields

$\sum_{n=0}^{\infty}\left(\alpha_n-\gamma_n\right)x_n=x_0+\sum_{n=0}^{\infty}\omega_n<+\infty$ and hence Property (iii) follows. If, in addition, $\gamma_n=0$; $\forall n\in \boldsymbol{N}_0$ then $\sum_{n=0}^{\infty}\alpha_n x_n<+\infty$, $0\le\left(\inf_{n\in \boldsymbol{N}_0}\alpha_n\right)\left(\sum_{n=0}^{\infty}x_n\right)\le\sum_{n=0}^{\infty}\alpha_n x_n<+\infty, \sum_{n=0}^{\infty}x_n<+\infty$ and $\{x_n\}$ bounded. Hence, Property (ii) follows. Note that, since $K<1$, Equation (3.1) is equivalently rewritten as follows:

$$\begin{aligned}x_{n+1}&\le Kx_n+\left(1+\gamma_n-\alpha_n-K\right)x_n+\omega_n+\sigma\\&=K^{n+1}x_0+\sum_{i=0}^{n}K^{n-i}\left[\left(1+\gamma_i-\alpha_i-K\right)x_i+\omega_i+\sigma\right]\\&\le K^{n+1}x_0+\frac{1-K^{n+1}}{1-K}\sup_{0\le i\le n}\left[\left(1+\gamma_i-\alpha_i-K\right)x_i+\omega_i+\sigma\right] \qquad (3.4.1)\\&\le x_0+\frac{1-K^{n+1}}{1-K}\sup_{0\le i\le n+1}\left[\left(1+\gamma_i-\alpha_i-K\right)x_i+\omega_i+\sigma\right] \qquad (3.4.2)\end{aligned}$$

; $\forall n\in \boldsymbol{N}_0$, with $x_0\ge 0$, which implies

$$\sup_{0\le i\le n}x_i\le x_0+\sup_{0\le i\le n}\left[\frac{1-K^n}{1-K}\left(\left(1+\gamma_i-\alpha_i-K\right)x_i+\omega_i+\sigma\right)\right] \qquad (3.5)$$

or

$$\left[1-K-\sup_{0\le i\le n}\left(1-K^i\right)\sup_{0\le i\le n}\left(1+\gamma_i-\alpha_i-K\right)\right]\sup_{0\le i\le n}x_i\le\left(1-K\right)x_0+\sigma+\sup_{0\le i\le n}\omega_i \qquad (3.6)$$

Then $\{x_n\}$ is bounded if $0\le\lim_{n\to\infty}\sup_{0\le i\le n}x_i<+\infty$ which is guaranteed from (3.6) if $\lim_{m>n}\sup_{n\to\infty}\left[\left(1-k^m\right)\left(\alpha_n-\gamma_n\right)\right]>0$ which is, in turn, guaranteed if $\lim\sup_{n\to\infty}\left(\gamma_n-\alpha_n\right)\le 0$. Thus, one gets from (3.4.1) that

$$\lim\sup_{n\to\infty}x_{n+1}\le\frac{1}{1-K}\lim\sup_{n\to\infty,}\left(\sup_{0\le i\le n}\left[\left(1+\gamma_i-\alpha_i-K\right)x_i+\omega_i+\sigma\right]\right)$$

If, furthermore, $\sup_{n\in \boldsymbol{N}_0}\left(\gamma_n-\alpha_n\right)<0$ then, it follows from (3.4.2) that

$$0\le\inf_{n\in \boldsymbol{N}_0}\left(\alpha_n-\gamma_n\right)\sup_{n\in \boldsymbol{N}_0}x_n\le\left(1-K\right)x_0+\sigma+\sup_{0\le i\le n}\omega_i<+\infty$$

with its first inequality being strict if $x_0 > 0$. Thus, Property (iv) has been proved.

Theorem 3.2. Consider the iterative scheme (2.1.1)-(2.1.4) under the constraints:

1) $\{b_n\} \subset (0,1]$, $\{b_n\} \to 1$

2) $\{a_n\} \subset [0,1]$, $\{a_n\} \to 1$, $\|T^n\| = o(|y_n|)$ or $\{a_n\} \subset [0,1]$, $\{a_n\} \to 0$, $\|T^n\| = o(|z_n|)$

3) $T, S : X \to X$ are linear and $S : X \to X$ is one-to-one with closed range with the sequence $\{S^{-1}T_n\}$consists of composite positive operators.

4) Define nonnegative real sequences $\{\gamma_n\}$ and $\{\alpha_n\}(\subseteq [0,1]) \to 0$, of general terms satisfying $\gamma_n = \frac{2-\alpha_n}{b_n}$, such that $\sum_{i=0}^{\infty} \alpha_i = \sum_{i=0}^{\infty} (2 - b_i \gamma_i) < \infty$.

Then, the iterative scheme (2.1.1)-(2.1.4) is globally stable for any $z_1 \geq 0, y_1 \geq 0$ while it fulfils $\{S z_n\} \to 0$, and it is a nonnegative sequence , $\{S y_n\} \to 0$, $\{T^n y_n\} \to 0$, $\{T^n z_n\} \to 0$; and $\{AS z_n\} \to 0$, respectively, $\{AS y_n\} \to 0$ faster than $\{S z_n\} \to 0$, respectively, faster than $\{S y_n\} \to 0$.

Proof: The iterative scheme (2.1.1)–(2.1.4) verifies Theorem 3.1 with the definitions $S z_n \equiv x_n$; $n \in \mathbf{N}_{0+} = \mathbf{N} \cup \{0\}$ under the above constraints 1-4 with $\{\omega_n\}$ of general term

$$\begin{aligned} \omega_n &= b_n^{-1} S y_n + a_n T^n (y_n - z_n) = b_n^{-1}(1-b_n) S z_n + a_n T^n (y_n - z_n) + T^n z_n \\ &= b_n^{-1}(1-b_n) S z_n + a_n T^n S^{-1}(S y_n) + (1-a_n) T^n S^{-1}(S z_n);\ \forall n \in \mathbf{N}_{0+} \end{aligned}$$

which is nonnegative for $z_1 \geq 0$ and $y_1 \geq 0$, of finite summation, and convergent to zero. Then, the result follows from Theorem 3.1 using Lemmas 2. 2-2.3. □

Acknowledgements
The authors are very grateful to the Spanish Government for its support of the research through Grant DPI2012-30651, and to the Basque Government by its support of this research trough Grants IT378-10 and SAIOTEK S-PE12UN015. They are also grateful to the University of Basque Country for the financial support through Grant UFI 2011/07.

REFERENCES

[1] I. Inchan,"Viscosity iteration method for generalized equilibrium points and fixed point problems of finite family of nonexpansive mappings", *Applied Mathematics and Computation*, Vol. 219, Issue 6, pp. 2949-2959, 2012.

[2] M. De la Sen, "Stable iteration procedures in metric spaces which generalize a Picard-type iteration", *Fixed Point Theory and Applications*, Vol. 2010, Article ID 572057, 15 pages, 2010.doi:10.1155/2010/953091.

[3] M. De la Sen, "Stability and convergence results based on fixed point theory for a generalized viscosity iterative scheme", *Fixed Point Theory and Applications*, Article IDF 314581, Vol. 2009, doi: 10.1155/ 2009/314581, 2009.

[4] M. Delasen, "Online optimization of the free parameters in discrete adaptive -control systems", *IEE Proceedings-D Control Theory and Applications*, Vol. 131, Issue 4, pp. 146-157, 1984.

[5] J.J. Minambres and M. De la Sen, "Application of numerical-methods to the acceleration of the convergence of the adaptive control algorithms-The one-dimensional case", *Computers & Mathematics with Applications-Part A*, Vol. 12, Issue 10, pp. 1049-1056, 1986.
[6] M. Abbas, T. Nazir and S. Romaguera, "Fixed point results for generalizad cyclic contraction mappings in partial metric spaces", *Revista de la Real Academia de Ciencias Exactas, Fisicas y Naturales, Serie A: Matematicas,* Vol. 106, No. 2, pp. 287-297, 2012.
[7] M. De la Sen, "Linking contractive self-mappings and cyclic Meir-Keeler contractions with Kannan self-mappings", *Fixed Point Theory and Applications*, Vol. 2010, Article ID 572057, 23 pages, 2010.doi:10.1155/2010/572057.
[8] W.S. Du, "New cone fixed point theorems for nonlinear multivalued maps with their applications", *Applied Mathematics Letters*, Vol. 24, No. 2, pp. 172-178, 2011.
[9] W. Laowang and B. Panyanak, "Common fixed points for some generalized multivalued nonexpansive mappings in uniformly convex metric spaces", *Fixed Point Theory and Applications*, Article number 20, doi:10.1186/1687-1812-2011-20, 2011.
[10] H.K. Nashine and M.S. Khan, "An application of fixed point theorem to best approximation in locally convex space", *Applied Mathematics Letters*, Vol. 23, Issue 2, pp. 121-127, 2010.
[11] A. Latif and M.A. Kutbi, "Fixed points for w-contractive multimaps", *International Journal of Mathematics and Mathematical Sciences*, Vol. 2009, Article ID 769467, 7 pages, 2009, doi:10.1155/2009/769467.
[12] T. Husain and A. Latif, "Fixed points of multivalued nonexpansive maps", *International Journal of Mathematics and Mathematical Sciences*, Vol. 14, No.3, pp. 421-430, 1991.
[13] M. S. Khan, "Common fixed point theorems for multivalued mappings", *Pacific Journal of Mathematics*, Vol. 95, No. 2 , pp. 337-347, 1981.
[14] J.M. Mendel, Discrete Techniques of Parameter Estimation: The Equation Error Formulation, Marcel Dekker Inc., 1973.

Instantaneous and scale-versatile gourdron theory: pair momentum equation, quasi-stability concept, and statistical indeterminacy revealing masses of elementary, bio-molecular, and cosmic particles

Ken Naitoh

Waseda University, 3-4-1 Ookubo, Shinjuku, Tokyo, 169-8555 Japan

k-naito@waseda.jp

Abstract. Flexible particles, including hadrons, atoms, hydrated biological molecules, cells, organs containing water, liquid fuel droplets in engines, and stars commonly break up after becoming a gourd shape rather than that of a string; this leads to cyto-fluid dynamics that can explain the proliferation, differentiation, and replication of biomolecules, onto-biology that clarifies the relationship between information, structure, and function, and the gourd theory that clarifies masses, including quark-leptons and Plank energy. The masses are related to the super-magic numbers, including the asymmetric silver ratio and symmetric yamato ratio, and reveal further mechanisms underlying symmetry breaking. This paper gives further theoretical basis and evidence, because the gourd theory reported previously is a little analogical and instinctive.

1. Introduction

Each of subatomic, biomolecular, and liquid fossil fuel particles, and stars in the cosmos is flexible and breaks up. [1-13] Although the nature is with this common feature, traditional particle models, Schroedinger equation, and Newton mechanics are useful only for narrow scales.

There are so many approaches, which try to find the universal theory describing temporal history of physical phenomena from subatomic to cosmic systems, i.e., four dimensional (unsteady three-dimensional) feature. [14]

However, one of our most important targets is to know the masses of various particles in the universe. It will not be necessary for predicting particle masses to know all the temporal developments before and after breakup of particles. Thus, we think about the most important timing at which masses are determined. The important timing is rate-determining step, just before breakup.

Let us think about the shape of particle just before breakup for various cases including subatomic, bio-molecular, and cosmic particles. The common shape will be like gourd, which has two sphere-like or spheroid-like particles connected.

Therefore, our previous models based on gourd shape assumption [1-13] show a large possibility describing the masses of various particles in the universe.

Cyto-fluid dynamic theory [1-4] proposed as the first stage of model clarifies a similarity underlying breakups of liquid droplets from millimeter to meter range, biological molecules hydrated, biological cells, and organs. An important feature is fusion of symmetric and asymmetric divisions, which are related to the golden, silver, and yamato ratios. The theory is derived based on momentum conservation law and a new quasi-stability concept as weakest one, neighboring to absolute instability.

By including a new indeterminacy concept into the cyto-fluid dynamic theory, we proposed gourd theory. [5-13] The gourd theory clarifies the masses of subatomic particles including quarks, leptons, bosons, and hadrons. It is stressed that the gourd theory is based on the quasi-stability concept, whereas the traditional elementary particle theories are related to variation principle.

However, the indeterminacy concept included in the gourd theory was instinctive and qualitative. Thus, the present report gives a further theoretical explanation on the indeterminacy.

2. Cyto-fluid dynamic theory: based on pair momentum equation quasi-stability concept [1-4]

For collision and breakup during replication processes of biological molecules, we first define "parcel" as flexible lump which consists of a core molecule (such as nitrogenous base or amino acid) and water molecules hydrated to the core molecule, while "particle" implies the core molecule. For breakup phenomena of liquid droplets of water and fuel, "parcel" is identical to particle of liquid droplet. For cell division process, "parcel" includes nucleus as core and other biomolecules.

The following assumptions are also employed for deriving the model.

(1) Parcel with three-dimensional flexible axi-symmetric shapes such as spheroid- or circular-cone deformed in time,

(2) One-dimensional connection of two parcels along the axis in a free space.

(3) Unified form for various surface forces coming from coulomb force, surface tension, and gravity as the form of $1/r^m$, while r is local curvature of parcel. (This assumption of force is effective, because we observe only the short timing around breakup as rate-determinant stage.)

(4) Potential flow [15, 16] inside the particles generated by high impact speeds of water molecules. (It is well known that the potential flow occurs for a condition of impulsive flow.)

(5) Particle size proportional to parcel size.

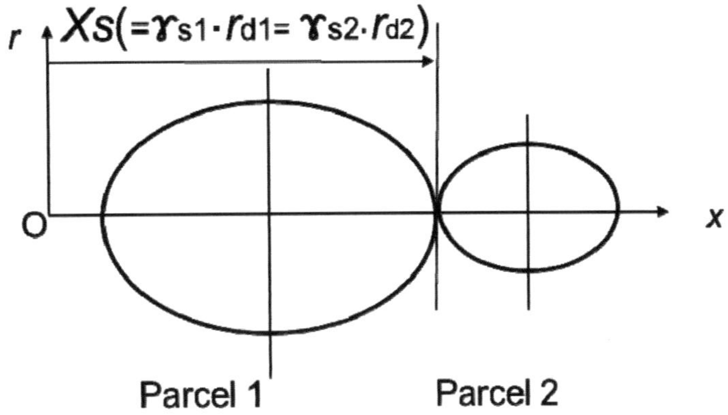

Fig. 1 Two flexible parcels connected

When there is no outer force to the two parcels connected, the time-dependent relation between dimensionless deformation rate $\gamma_k (\equiv a_k / b_k\ [k = 1, 2])$ of each spheroid-like parcel and the size ratio of the two connecting parcels $\varepsilon = r_{di} / r_{dj}\ [i, j = 1,2, i \neq j]$ with equivalent radius $r_{di} (= [a_i b_i^2]^{1/3})$ at equilibrium condition, i.e., the $\gamma_k - \varepsilon$ momentum equation, is described as

$$\frac{d^2\gamma_i}{d\bar{t}_i^2} = Bo_i(\gamma_i)(\frac{d\gamma_i}{d\bar{t}_i})^2 + Co_i(\gamma_i)\gamma_i^{\frac{5}{3}-\frac{2}{3}m} + (3-2i)Eo_i(\gamma_i)\frac{d^2\gamma_{Si}}{d\bar{t}_i^2}$$

[(*i*=1 and *j*=2) or (*i*=2 and *j*=1)] (1a)

$$\frac{d^2\gamma_{Si}}{d\bar{t}_i^{\,2}} = \frac{d^2\gamma_{Sj}}{d\bar{t}_j^{\,2}}\varepsilon^{1+m} \tag{1b}$$

$$\begin{aligned} &\varepsilon^{2-m}\frac{d^2\gamma_{Si}}{d\bar{t}_i^{\,2}} - \varepsilon^{2-m}\frac{2}{3}\gamma_i^{-1/3}\frac{d^2\gamma_i}{d\bar{t}_i^{\,2}} + \varepsilon^{2-m}\frac{2}{9}\gamma_i^{-4/3}\left(\frac{d\gamma_i}{d\bar{t}_i}\right)^2 \\ &\quad + \left[\frac{d\gamma_{Sj}}{d\bar{t}_j^{\,2}} + (3-2i)\frac{2}{3}\gamma_j^{-1/3}\frac{d^2\gamma_j}{d\bar{t}_j^{\,2}} - (3-2i)\frac{2}{9}\gamma_j^{-4/3}\left(\frac{d\gamma_j}{d\bar{t}_j}\right)^2\right] = 0 \end{aligned} \tag{1c}$$

where $\gamma_{Si} = \dfrac{X_S}{r_{di}}$ denotes dimensionless contact position of two parcels, while being

$B_{0k} = \dfrac{1}{3\gamma_k}\dfrac{\gamma_k^{\,2}-2}{\gamma_k^{\,2}-1/2}$, $C_{0k} = \dfrac{3}{8}\dfrac{2\gamma_k^{\,2m} - 1/\gamma_k^{\,m} - \gamma_k^{\,m}}{\gamma_k^{\,2}-1/2}$, $E_{0k} = 3\dfrac{\gamma_k^{\,7/3}}{\gamma_k^{\,2}-1/2}$, $\bar{t}_k = \varpi_k t$, and

, $\varpi_k^{\,2} = \dfrac{8\sigma}{\rho_L r_{dk}^{\,2+m}}$ [for k = 1, 2]

with $\sigma, \rho,$ and m denote surface force coefficient, parcel density, and a constant for extending surface force of the form of $1/r^m$ with local curvature r, respectively. [Equation 1 is derived only under the assumptions above and by purely mathematical transformation. The long derivation of Eq. 1 is in Ref. 2 and has been confirmed by the referees.]

Equation 1a for i = 1 and 2 describes the momentum one for each individual parcel, while the third and fourth ones, Eqs. 1b and 1c are the definition of contact surface between two parcels and overall momentum conservation law for two parcels connected, respectively.

Eliminating γ_{Si} in Eqs. 1a-1c and taking the first order of approximation in the Taylor series for the equation result in Eq. 2.

$$\frac{d^2 y_i}{d\bar{t}_i^{\,2}} = [-\frac{2}{3}(3-\varepsilon^3 - 2\varepsilon^{2+m})\left(\frac{dy_i}{d\bar{t}_i}\right)^2 + 3(3-\varepsilon^3)m y_i - 4\varepsilon^{1+m}\left(\frac{dy_j}{d\bar{t}_j}\right)^2 + 12\varepsilon^{1+m} m y_j]/[3(\varepsilon^3+1)] \tag{2}$$

where the deviation from a sphere is defined as y_i, which is equal to $\gamma_i - 1$. [See references 2 and 3 on the detail of derivation of Eqs. 1 and 2.]

Equation 2 shows that a symmetric ratio of 1.0 (ε =1) makes the first term on the right-hand side of the equation zero, while an asymmetric ratio of $\sqrt[3]{3}$ ($\approx$ 1.44) around 1.5 (ε^3=3) makes the second term zero for each value of m.

As we define a system as being quasi-stable when only one term on the right-hand side of the differential equation system governing the phenomenon is zero [1-13], the system of two connected parcels shown above is relatively quasi-stable when the size ratio of connected parcels takes the values of ε=1 or ε^3=3.

Let us look at biological systems containing water flows. We can classify the five bases of adenines (A), guanines (G), cytosines (C), thymines (T), and uracils (U) into two groups: purines and pyrimidines. Purines, i.e., A and G, have a relatively large size, while pyrimidines, i.e., C, T, and U, are small. This grouping specifically refers to the asymmetric size ratio of purines and pyrimidines of around 1.50 in their hydrogen bonds within DNA and RNA, i.e., that of the Watson-Crick type, although a symmetric size ratio of 1.00 is often observed in RNA. (Fig.2a) These ratios correspond to the quasi-stable ones derived from Eq. 2. Symmetric and asymmetric size ratios are also observed at the cell level of microorganisms such as yeast. (Fig. 2b)

A neutron impacting uranium 235 produces smaller child atoms that often have an asymmetric weight ratio of about 2:3 between the golden and silver ratios. [17] In contrast, varying the impact speed of neutrons results in a nearly symmetric division of uranium 235, which has the yamato ratio of 1:1. (Fig. 2c.)

The concept of quasi-stability is necessary for living beings because stronger stability cannot bring variations, i.e., adaption to environmental change and evolution. This quasi-stability is also possible for nonliving systems such as atoms because atomic systems such as uranium 235 also vary over time. It is stressed that the quasi-stable ratios of 1: $\sqrt[n]{n}$ for n=1 and 3 appear for each m. (See Eq. 2)

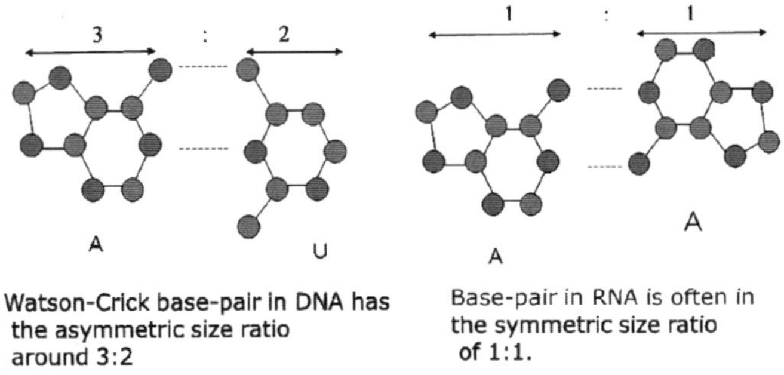

a. Asymmetric and symmetric base pairs in DNA and RNA

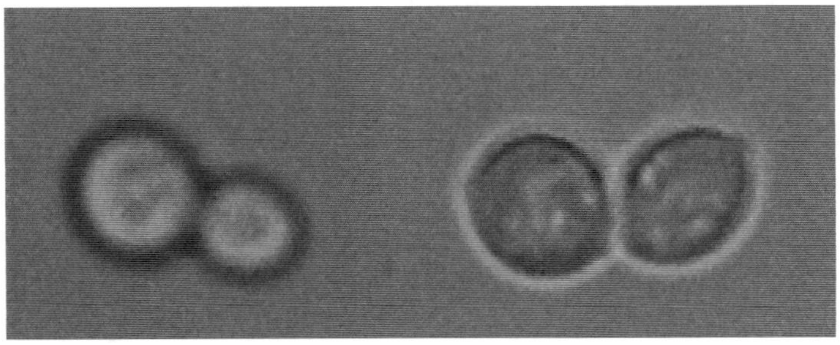

b. Asymmetric and symmetric pairs of Yeast cells

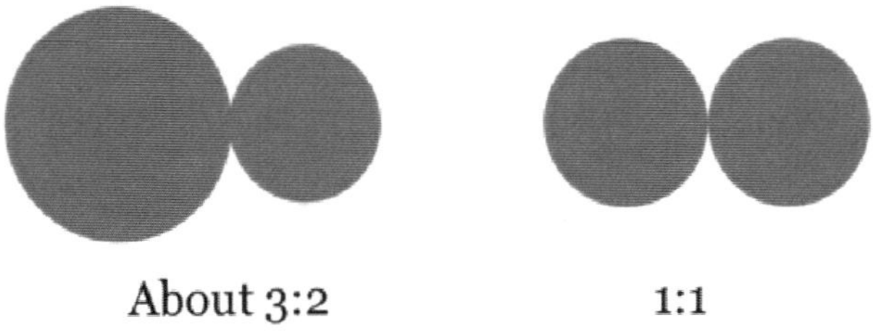

c. Asymmetric and symmetric breakups of uranium 235 obtained by varying impact speeds of neutron

Fig. 2 Asymmetry and symmetry observed in breakup of parcels

Let us consider the reason why the energy conservation law and variation principle such as the Bohr model and some extensions [14] do not explain the fusion of symmetry of 1:1 and asymmetry of around 2:3 in size and density (number density). The reason is that previous models based on energy conservation have tended to eliminate the relative motion between parcels, whereas the momentum conservation in Eqs. 1 and 2 models all of the relative motion between two parcels, nonlinear convections inside the parcels, interfacial forces at the parcel surfaces, and collisions with smaller molecules.

When we set $m = 1$, the second term in the Taylor series results in a quasi-stable ratio of about 1.27, the third term in ratios of about 3.58 and 1.35, the fourth term in ratios of about 2.47 and 1.39, the fifth term in ratios of about 2.1, and 1.4, the sixth term in a ratio of about 1.41, and the seventh in ratios of about 1.79 and 1.42. [6-13] Thus, these ratios over 2:3 reveal the molecular weights of the twenty types of amino acids of a threefold variations between 240 of cysteine as the maximum and 75 of glycine as the minimumn. The halo structures such as H10 and M32 have number ratios of neutrons and protons of about 1:3.5. The higher-order analysis for $m = 1$ clarifies the ratios over 2:3 in several systems. [6-13]

3. Four types of indeterminacy

Basically, there are two types of theoretical indeterminacies, i.e., those appearing in quantum mechanics related to Kennard's consideration [18] and those in statistical mechanics related to the Liouville equation [19]. The discontinuity of particles in nature essentially leads to statistical indeterminacy for several stages of phenomena from subatomic to cosmic, whereas very small particles such as electron particles described by the Schrodinger and Klein-Gordon equations [20] have the other type of indeterminacy indicated by quantum mechanics (quantum indeterminacy).

An important point is that the level of statistical indeterminacy, i.e., degree of variation, varies according to the window scales used for averaging (stochastic determinism window) and the number of particles involved. Statistical mechanics based on the Liouville and Boltzmann equations [19] tells us that a very large window for averaging the aggregation of a huge number of particles results in deterministic continuum mechanics, whereas a small window for a small number of particles leads to a stochastic differential equation. When a small window is used for averaging, physical quantities such as mass, size, and velocity are defined with indeterminacy, i.e., vagueness. (Fig. 3)

Baryons and mesons are constructed of only three and two quarks, respectively. There are only two electrons inside the smallest orbit around a He atom having only a few baryons. The number of carbon, oxygen, and nitrogen particles inside a nitrogenous base is not enough for a continuum, even if the nitrogenous base is hydrated, because of the order of ten or one hundred. These small numbers of particles will lead to statistical indeterminacy in the governing equations, even in cases where the indeterminacy in quantum mechanics does not work. On the other hand, a biological cell or a liquid droplet of over one millimeter in size includes a lot of molecular particles that result in deterministic governing equations. Theoretical model considering this statistical indeterminacy, i.e., an approach lying between the Langevin and Boltzmann equations, revealed even the 100-year mystery in the transition to turbulence in closed pipes. [21]

Third indeterminacy is Heidelberg's one, which is related to observation.

We proposed the forth indeterminacy, which appears only for two connected particles (parcels) shown in the above section. An important point is that, when the statistical indeterminacy appears for small number of sub-particles in a parcel and for small window of averaging, the contact-surface position, i.e., relative distance and relative velocity between particles, also becomes indeterminant, i.e., vague.

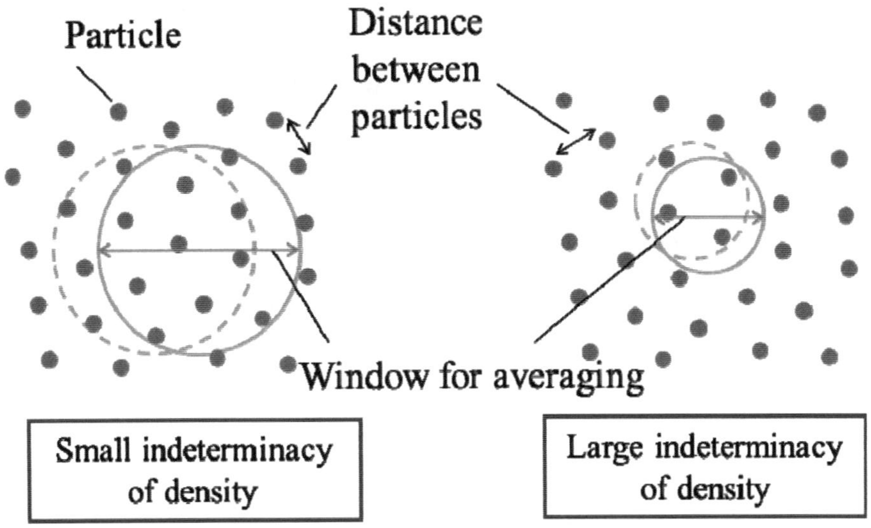

Fig. 3 Statistical indeterminacy

4. Gourd theory [5-13]

Equation 1 shown in the cyto-fluid dynamic theory is extended with the following assumptions (1) and (2) related to the indeterminacies on statistical mechanics and contact surface position depicted in Section 3 and also the other ones of (3) and (4).

(1) Stochastic random disturbance due to stochastic indeterminacy is included in momentum equations.
(2) Relative distance and relative velocity between two parcels connected are also indeterminant.
(3) Potential flow inside the particles because of high impact speeds of subatomic particles,
(4) Unified form for various surface forces coming from, strong- and weak- nucleic forces, coulomb force, surface tension, and gravity as the form of $1/r^m$, while r is local curvature of parcel.

The $\gamma_k - \varepsilon$ equation of Eqs. 1a-1c is extended as

$$\frac{d^2\gamma_i}{d\bar{t}_i^{\,2}} = Bo_i(\gamma_i)\left(\frac{d\gamma_i}{d\bar{t}_i}\right)^2 + Co_i(\gamma_i)\gamma_i^{\frac{5}{3}-\frac{2}{3}m} + (3-2i)Eo_i(\gamma_i)\frac{d^2\gamma_{Si}}{d\bar{t}_i^{\,2}} + \delta_{st\,i}$$

[(i=1 and j=2) or (i=2 and j=1)] (3a)

$$\frac{d^2\gamma_{Si}}{d\bar{t}_i^{\,2}} = \frac{d^2\gamma_{Sj}}{d\bar{t}_j^{\,2}}\varepsilon^{1+(m-\Delta m)} + \delta_{st-cs} \tag{3b}$$

$$\begin{aligned} &\varepsilon^{2-(m-\Delta m)}\frac{d^2\gamma_{Si}}{d\bar{t}_i^{\,2}} - \varepsilon^{2-(m-\Delta m)}\frac{2}{3}\gamma_i^{-1/3}\frac{d^2\gamma_i}{d\bar{t}_i^{\,2}} + \varepsilon^{2-(m+\Delta m^*)}\frac{2}{9}\gamma_i^{-4/3}\left(\frac{d\gamma_i}{d\bar{t}_i}\right)^2 \\ &\quad + \left[\frac{d\gamma_{Sj}}{d\bar{t}_j^{\,2}} + (3-2i)\frac{2}{3}\gamma_j^{-1/3}\frac{d^2\gamma_j}{d\bar{t}_j^{\,2}} - (3-2i)\frac{2}{9}\gamma_j^{-4/3}\left(\frac{d\gamma_j}{d\bar{t}_j}\right)^2\right] = \delta_{st-om} \end{aligned} \tag{3c}$$

with $\sigma, \rho, m,$ and δ_{st} denote surface force coefficient, parcel density, a constant for extending surface force of the form of $1/r^m$ with local curvature r, and indeterminacy of density and velocity due to less sub-particles inside each individual parcel, respectively, while Δm and Δm^* denote indeterminacy of the contact position of two parcels and relative velocity between two parcels connected, respectively.

After eliminating γ_{Si} in Eqs. 3a-3c and when $\Delta m = \Delta m^* = \lambda(1-m)$ and $\lambda = 1/2$ are assumed, Eqs. 3a-3c result in Eq. 4.

$$\frac{d^2}{d\bar{t}_i^{\,2}}\gamma_i = \{m_{ci}(\frac{d}{d\bar{t}_i}\gamma_i)^2 + m_{cj}(\frac{d}{d\bar{t}_j}\gamma_j)^2 + m_{si}\,\gamma_i^{\frac{5}{3}-\frac{2}{3}m} + m_{sj}\,\gamma_j^{\frac{5}{3}-\frac{2}{3}m}\}/Det + \delta_{st}$$

$$[\text{for } i, j \in \Omega, (i=1 \text{ and } j=2) \text{ or } (i=2 \text{ and } j=1), i \neq j] \tag{4}$$

with
$$m_{ci} = [(-\varepsilon - \varepsilon^4 + \frac{2}{3}\varepsilon E_{0j}\gamma_j^{-1/3})B_{0i} + \frac{2}{9}\varepsilon^{4-2\Delta m}E_{0i}\gamma_i^{-4/3}]$$

$$m_{cj} = [\frac{2}{3}\varepsilon^{2+m-\Delta m}E_{0i}\gamma_j^{-1/3}B_{0j} - \frac{2}{9}\varepsilon^{2+m-\Delta m}E_{0i}\gamma_j^{-4/3}]$$

$$m_{si} = (-\varepsilon - \varepsilon^4 + \frac{2}{3}\varepsilon E_{0j}\gamma_j^{-1/3})C_{0i}$$

$$m_{sj} = \frac{2}{3}\varepsilon^{2+m-\Delta m}E_{0i}\gamma_j^{-1/3}C_{0j}$$

$$Det = -\varepsilon - \varepsilon^4 + \frac{2}{3}\varepsilon^4 E_{0i}\gamma_i^{-1/3} + \frac{2}{3}\varepsilon\, E_{0j}\gamma_j^{-1/3},\; B_{0k} = \frac{1}{3\gamma_k}\frac{\gamma_k^{\,2}-2}{\gamma_k^{\,2}-1/2},$$

$$C_{0k} = \frac{3}{8}\frac{2\gamma_k^{\,2m} - 1/\gamma_k^{\,m} - \gamma_k^{\,m}}{\gamma_k^{\,2}-1/2},\text{ and } E_{0k} = 3\frac{\gamma_k^{\,7/3}}{\gamma_k^{\,2}-1/2}$$

It is stressed that this system is not the simple two-body problem of a rigid body because of the flexible nonlinear deformations of the parcels.

Equation 4 is derived only under the assumptions above and by purely mathematical transformation. The long derivation of Eq. 4 is in Ref. 2 and has been confirmed by the referees, although only the stochastic terms including δ_{st} are not in Ref. 2.

Moreover, we will consider a higher order of the Taylor series, for Eq. 4. For various values of m, the q-th order of the Taylor series for parcel 1 results in

$$\frac{d^2}{d\tau_1^2}y_1 = \sum_{q=1-\infty} a_q(\varepsilon,m)y_1^{q-1}[\frac{dy_1}{d\tau_1}]^2 + \sum_{q=1-\infty} b_q(\varepsilon,m)y_1^q + \sum_{q=1-\infty} c_q(\varepsilon,m)y_2^{q-1}[\frac{dy_2}{d\tau_1}]^2 + \sum_{q=1-\infty} d_q(\varepsilon,m)y_2^q + \delta_{st}'' \tag{5}$$

where the last term on the right-hand side δ_{st}'' is the stochastic one due to random fluctuation, while

$a_k(\varepsilon,m), b_k(\varepsilon,m), c_k(\varepsilon,m),$ and $d_k(\varepsilon,m)$ are four coefficients related to the size ratio ε and m. [8]

Then, the theory having only one arbitrary constant m, i.e., Eqs. 3-5, lead to the various and discrete quasi-stable ratios, which reveal the masses of subatomic particles such as hadrons, quarks, leptons, bosons including Higgs-like one, and Plank mass [6-13], whereas the traditional standard theory based on mathematics includes many parameters, which should be determined by experiments.

Figure 4a for m=1.03 shows the prediction of some bosons and plank mass, with the corresponding experimental and other theoretical ones, while Fig. 4b for m=1.03 also includes photon, because photon has a very small mass at quantum entangle, i.e., during interaction of two particles. Thus, the theory proposed predicts both the masses of W-, Z-, and Higgs-like bosons close between 80 and 130 GeV/c^2 and photon < 10^{-25} GeV/c^2. It is also stressed that theoretical mass ratios for m = 1.03 may also predict plank mass and are close to the prediction due to the SUSY-GUT theory [14]. This result may lead to a breakthrough for solving the mysterious hierarchy problem [14] in elementary particle physics. Optimizations of m and λwill bring better agreement with prediction and experiment. The present theory also prognosticates a new particle between 10 TeV and 1,000 TeV, which will be indicated by the other theoretical studies.

As the present paper considers only the short interval around breakup, the Einstein's relativity effect will be small. Even if the relativity effect on light speed appears, the mass ratio of two parcels connected will not be changed very much, because the effect works on both.

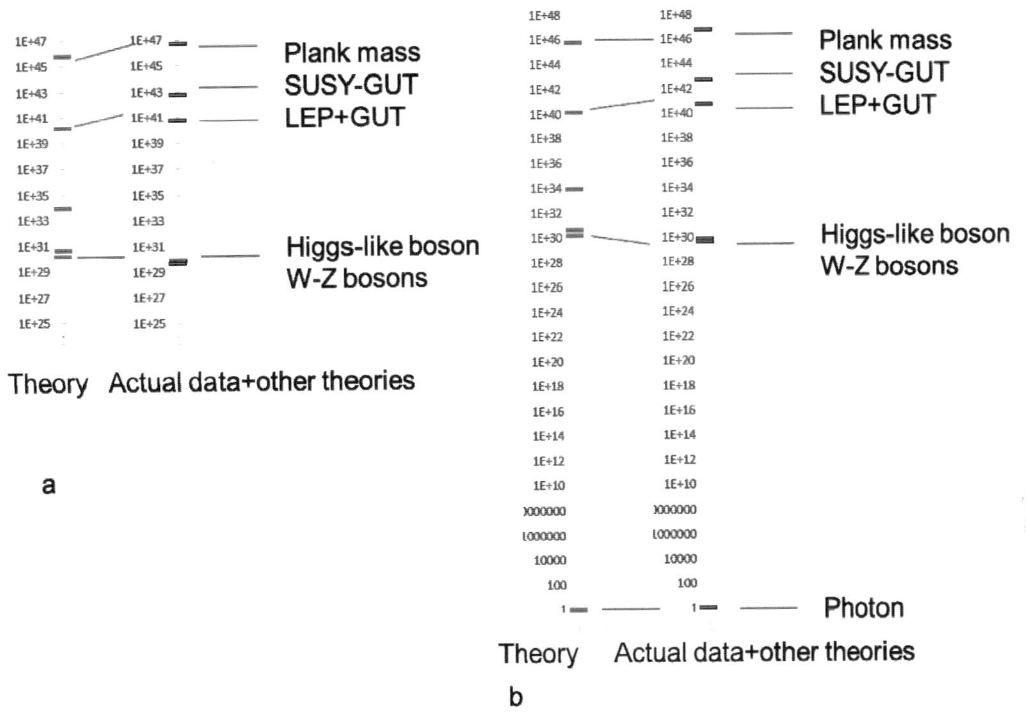

Fig. 4 Prediction of masses for bosons. (We assume that photon during interaction of quantum entanglement or single ≃ 10^{-28} GeV/c^2 and plank mass ≃ 10^{+19} GeV/c^2)

5. Further theoretical analysis on validity of indeterminacy Δm^* in gourd theory

Let us think about the indeterminacy of contact surface and of relative velocity in Eq. 3c, which is described with quantities ofΔm andΔm^* further. Question is the reason why Δm has a negative effect for m in the terms of acceleration $\frac{d^2\gamma_{Si}}{d\bar{t}_i^2}$ and $\frac{d^2\gamma_i}{d\bar{t}_i^2}$, while $\delta m*$ is positive in the term of momentum energy $(\frac{d\gamma_i}{d\bar{t}_i})^2$.

An important point is that $(\frac{d\gamma_i}{d\bar{t}_i})^2$ is nonlinear, while $\frac{d^2\gamma_{Si}}{d\bar{t}_i^2}$ and $\frac{d^2\gamma_i}{d\bar{t}_i^2}$ are linear. Thus, when $\gamma \to \gamma + \gamma'$ with the indeterminacy γ', the third term on the left hand side of Eq. 3c becomes $(\frac{d\gamma}{d\bar{t}})^2 \to (\frac{d\gamma}{d\bar{t}})^2 + 2(\frac{d\gamma}{d\bar{t}})(\frac{d\gamma'}{d\bar{t}}) + (\frac{d\gamma'}{d\bar{t}})^2$. Then, as value γ' is often comparable to γ, two terms of $2(\frac{d\gamma}{d\bar{t}})(\frac{d\gamma'}{d\bar{t}}) + (\frac{d\gamma'}{d\bar{t}})^2$ cannot be eliminated.

For thinking about $2(\frac{d\gamma}{d\bar{t}})(\frac{d\gamma'}{d\bar{t}})$ further, we can see Fig. 5 demonstrating small oscillation of parcel pair, while indeterminant (fluctuation) velocities are defined as those observed from the contact surface. Figure 5 demonstrates two types of situations of two parcels: tearing and oncoming. When two parcels tear, indeterminant velocity of $(\frac{d\gamma'}{d\bar{t}})$ can relatively take larger values, because of less collisions between smaller particles inside a parcel. However, while two parcels come near, the indeterminant velocity is small on average during a certain time, because more collisions result in plus and minus values of velocity. Therefore, this leads to the fact that parcel shape cannot be a little left-right symmetric, as observed in megascopic liquid droplet. This is apparent because left- and right-boundary surface conditions of each parcel among two ones connected are not identical. [Fluctuations of gravity center and particle surface cannot be left-right symmetric due to existence of the other particle connecting on one side.] Therefore, we use $\Delta m*$ for this nonlinear term of momentum energy, which differs from Δm of the linear term. It is also emphasized that this is not dissipation in continuum mechanics, because of fluctuation-dissipation theorem.

Next, let us think about the sign of Δm. The condition of $m \geq 1.0$ evident in our previous researches [1-13] leads to $\Delta m \leq 0$. This indeterminacy corresponds to larger size ratio of ε, which is possible because ragged parcel shape due to indeterminacy permits connection at longer distance.

Let us think about the two limits of $m=1$ and $m=\infty$. When $m = 1$, $\Delta m = \Delta m^* = \lambda(1-m)$ results in $\Delta m = \Delta m^* = 0$, whereas $m = \infty$ leads to $\Delta m / m = \Delta m^* / m = -\lambda$. This implies that Δm and Δm^* are finite. This may explain the mysterious feature that, as m becomes larger than 2.0, quasi-stable ratios are again close to those for m=1 shown in references [6-13].

6. Conclusion and further possibility

The present theory may also lead to that for ultimate particle such as prion. As the first step, we can conclude that the number of prions in a quark or lepton will be decided by values of $\Delta m / m$ and $\Delta m^* / m$. If m for breakup of a quark and lepton is less than 1.03, the number of sub-particles (prions) may be large because of smaller values of $\Delta m / m$ and $\Delta m^* / m$.

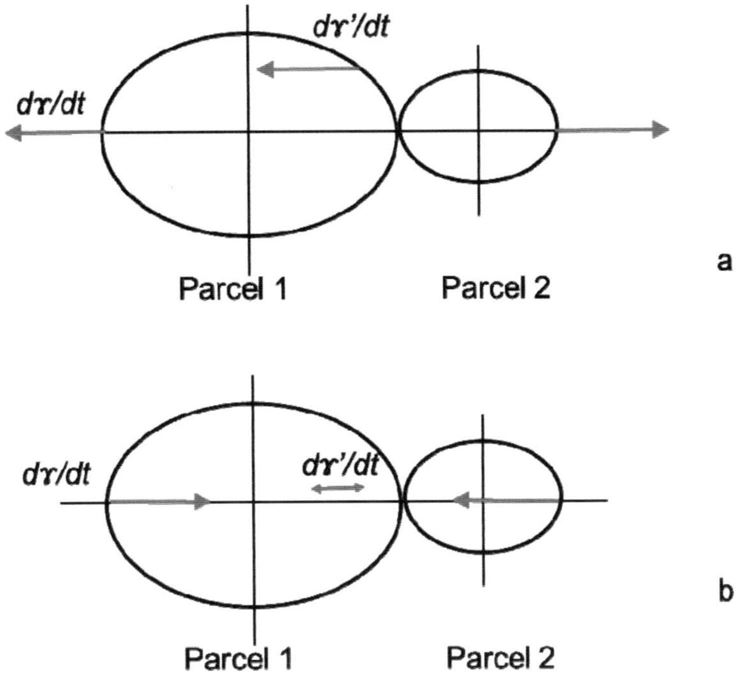

Fig. 5 Two types of patterns of indeterminant velocities.

Acknowledgements

The author sincerely thanks Prof. Dr. Ichiro Ooba of Waseda University for his advice on mass of photon. Sincere thanks are also due to Mr. Kenji Hashimoto, Mr. Hirofumi Inoue, and Mr. Shuntaro Iyoda in the Naitoh laboratory in Waseda University for their check on calculations using MATHEMATICA.

References

[1] Naitoh K 1999 Oil & Gas Science and Technology **54** 205.
[2] Naitoh K 2001 Japan Journal of Industrial and Applied Mathematics 18-1 75.
[3] Naitoh K 2006 Gene Engine and Machine Engine, Springer-Japan, Tokyo. 1-251.
[4] Naitoh K 2008 Artificial Life and Robotics **13** 10-17.
[5] Naitoh K 2010 Artificial Life and Robotics 15 117.
[6] Naitoh K 2011. Quasi-stability theory: explaining the inevitability of the Magic numbers at various stages from subatomic to biological, Proc. of 15th Nordic and Baltic Conf. on Biomedical Engineering and Biophysics, IFMBE proceedings 34, Denmark. pp.211-214.
[7] Naitoh K 2011. The inevitability of biological molecules connected by covalent and hydrogen bonds. Proc. of European IFMBE Conference, IFMBE Proceedings 37, Pp. 251-254.
[8] Naitoh K 2012 Spatiotemporal structure: common to subatomic systems, biological processes, and economic cycles. J. of Physics, C. S. 344.
[9] Naitoh K 2012. Quasi-stability, Proceedings of 17th International Conference on Cold Fusion.
[10] Naitoh K 2012. Quasi-stability, Soryushiron Kenkyu YITP-W-11-25, pp.276-294.
[11] Naitoh K 2012. Hyper-gourd theory. Artificial Life and Robotics. Vol.17, pp.275-286.
[12] Naitoh K, 2013. Gourd theory: rather than string and hyper-symmetry theories, Proceedings of JSST2013 International Conference on Simulation Technology, Tokyo.

[13] Naitoh K 2013 Gourdron theory. Artificial Life Robotics.
[14] Henley E M and Garcia A 2007 Subatomic physics. World Scientific.
[15] Tatsumi T 1982 Fluid dynamics, Baifukan, Tokyo.
[16] Lamb H 1945. Hydrodynamics, 6th ed. New York: Dover.
[17] EL-WAKIL M M 1971 Nuclear Heat Transport, Doubun Shoin.
[18] Kennard E H 1927 ZeitschriftfürPhysik 44 (4–5): 326.
[19] Hirschfelder J.O., Curtiss C.F., and Bird R.B., 1964 Molecular Theory of Gases and Liquids, (John Willey & Sons, Inc. New York).
[20] Landau E.D. and Lifshitz E.M., 1981 Quantum mechanics. 3rd edition (Butterworth-Heinemann Elsevier, Oxford).
[21] Naitoh K. and Shimiya H 2011. Japan Journal of Industrial and Applied Mathematics, Vol.28, No.1, pp.3-14.
[22] List of largest known stars: http://en.wikipedia.org/wiki/List_of_largest_known_stars
[23] Walser H 2001. The golden section. 2nd edition. The Mathematical Association of America, the publisher B.G. Teubner Stuttgart.
[24] Landau, E.D. and Lifshitz, E.M., 2004 Fluid Mechanics. 2nd Edition, (Butterworth-Heinemann Elsevier, Oxford).

Appendices

A: Inevitability of chromosomes and proteins

Equations 1 and 3 may also reveal the masses for human chromosomes and proteins. (Fig. A1 and A2)

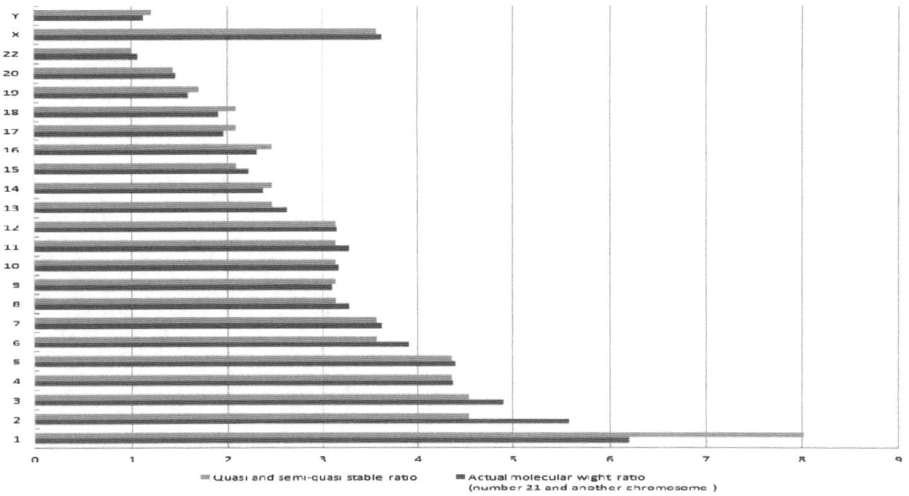

Fig. A1 Masses of human chromosomes and the corresponding predictions

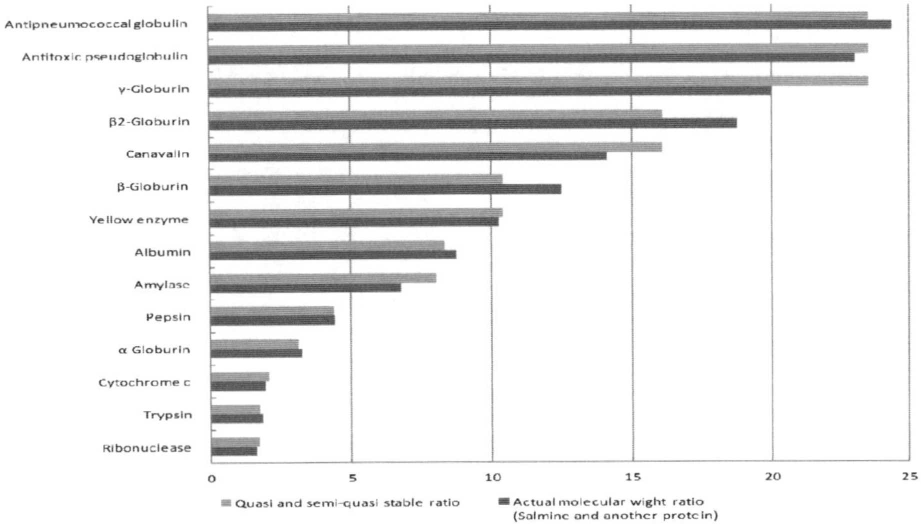

Fig. A2 Masses of proteins and the corresponding predictions

B. Inevitability of two *m* for small scales

For small subatomic particles, the prediction is possible if two values of m are simultaneously used. For example, we employ $m = 1$ and 2 for hadrons, whereas $m = 1.1$ and 1.2 for leptons. [8-13]

There are some possibilities for explaining the inevitability of these two.

(1) Strong and weak forces: Strong, weak, electric, and gravity forces work hadrons, while electric and weak forces work quarks, leptons, and bosons. [14]
(2) Two terms inside nucleic force
(3) Exponential function of distance for one term in nucleic force
(4) Attractive and repulsive forces
(5) Stabilization and un-stabilization forces in fluid mechanics
(6) Spherical- and string- like shapes of particles inside a parcel

Journal of Physics: Conference Series **495** (2014) 012017 doi:10.1088/1742-6596/495/1/012017

Optimal Cycle Time and Preservation Technology Investment for Deteriorating Items with Price-sensitive Stock-dependent Demand Under Inflation

Nita H. Shah and Arpan D. Shah
Department of Mathematics, Gujarat University,
Ahmedabad, Gujarat, India

E-mail: nitahshah@gmail.com, arpan.math@gmail.com

Abstract. The article analyzes economic order quantity for the retailer who has to handle imperfect quality of the product and the units are subject to deteriorate at a constant rate. To control deterioration of the units in inventory, the retailer has to deploy advanced preservation technology. Another challenge for the retailer is to have perfect quality product. This requires mandatory inspection during the production process. This model is developed with the condition of random fraction of defective items. It is assumed that after inspection, the screened defective items are sold at a discounted rate instantly. Demand is considered to be price-sensitive stock-dependent. The model is incorporating effect of inflation which is critical factor globally. The objective is to maximize profit of the retailer with respect to preservation technology investment, order quantity and cycle time. The numerical example is given to validate the proposed model. Sensitivity analysis is carried out to work out managerial issues.

1. Introduction

In any firm, deterioration of items is an important concern. There are various factors that affect a manufacturing process. So the deterioration of items is inevitable and plays an important role as utility of an item does not remain constant. To decrease deterioration, manufacturers have started investing in preservation technology. Many researchers have worked on controlling deterioration and derived models on the same. Ghare and Schrader (1963) developed the first EOQ model for inventory following exponential decay. Hariga (1995) Developed EOQ model for deteriorating items with shortages and time varying demand. Jaggi and Mittal (2007) developed EOQ model for deteriorating items with time dependent demand under inflationary conditions. Jaggi and Mittal (2011) developed an EOQ model for deteriorating items with imperfect quality. Jaggi et al. (2012) developed a model for effect of inspection for deteriorating items with time dependent demand under inflationary conditions. Other articles by Raafat et al. (1991), Widyadana et al. (2011), Yong and Huang (2013), Chen and Dye (2013), Hseih (2013) and Hseih and Dye (2013) are available in the same direction.

During a production process, it is impossible to achieve 100% quality products. Therefore, it is assumed that random fraction of defective items is produced at the end of production cycle and such items are sold at discounted rate instantly. On this concept, many researchers have developed various mathematical models. Porteus (1986) suggested model for optimal lot sizing, process quality improvement and setup cost reduction. Rosenblatt and Lee (1986) provided model for economic production cycles with imperfect production processes. Cardenas-Barron (2000) developed an EPQ model for items with imperfect quality. Goyal and Cardenas-Barron (2002) provided an EPQ model for items with imperfect quality. Other interesting articles are by Schwaller (1988), Zhang and Gerchak (1990), Salamah and Jaber (2000), Cardenas-Barron (2009-a, 2009-b), Cardenas-Barron et al. (2010).

Many countries are affected by inflation on a large scale and most of the models have not taken inflation rate into account. Inflation rate play an important role in deciding inventory policy for any firm. So, for determining optimum inventory policy, inflation rate can not be ignored. Also, there are many researchers who developed variants of inventory models with different constraints under inflation. Misra (1975) provided analysis of inflationary effects on inventory systems. Buzacott (1975) provided EOQ model with inflation. Datta and Pal (1991) suggested inventory model with linear time dependent demand rate and shortages under the effect of inflation and time value of money. Hariga and Ben-Daya (1996) derived optimal time varying lot sizing models under inflationary conditions. Ray and Chaudhari (1997) derived an EOQ model with stock-dependent demand, shortages, inflation and time discounting. Other interesting articles by Sarkar and Pan (1994) and Sana and Chaudhari (2003) are available in the same direction.

Inspired by this aspect, this paper proposes a model for deteriorating item with price-sensitive stock-dependent demand under inflation. Defective items are sold at a discounted rate after screening process. Demand is considered as price-sensitive and taken as a power function of selling price under inflation. Model does consider preservation technology investment to reduce the proportion of deteriorating items. This study doesn't allow any shortages during the cycle time. The objective is to maximize the profit of the firm with respect to optimal selling price, optimal length of replenishment cycle and optimal preservation technology investment.

The rest of the paper is organized as follows. In section 2, the notations and assumptions are presented. In section 3, mathematical model to maximize the total profit per unit time is developed. In section 4, solution procedure is given. Numerical examples are presented in section 5 and conclusion is drawn in section 6.

2. Notations and Assumptions:

A	Replenishment cost per order
C	Purchase cost per unit
P	Selling price per unit, $P > C$ (a decision variable)
h	Unit inventory holding cost per unit time
T	Cycle length (a decision variable)
$I(t)$	Inventory level at time t
Q	Order quantity per cycle (a decision variable)
$D(P, I(t))$	Demand rate
η	Price elasticity Markup cost, $\eta > 1$
C_s	Screening cost per unit
P_s	Selling price of defective unit
θ	Deterioration rate $(0 \le \theta \le 1)$
u	Preservation technology investment per unit to reduce deterioration rate (a decision variable)
$f(u)$	Proportion reduced deterioration rate, $f(u) = 1 - e^{-mu}, 0 < f(u) < 1$
r	Discount rate representing the time value of money
k	Inflation rate per unit time
R	Net inflation rate = $(r - k)$
λ	Proportion decrease in selling price (defective items)
$\in$	Proportion of defective items
t_1	ξT, where $0 < \xi < 1$
$\pi(P, T, u)$	Profit function

Assumptions

1. The inventory system involves a single deteriorating item over an infinite planning horizon.
2. Existence of deterioration items assumed with the initiation of production and the proportion reduced because of preservation technology are taken to be uniform.
3. There is no repair or replacement of deteriorating items during the period under consideration.
4. Shortages are not allowed.
5. Defective items are sold at a discounted rate after screening process.
6. Screening process ends at time t_1 which is pre-specified.
7. The demand rate,

   $D(P,I(t))=\int_0^T(\alpha+\beta I(t))P^{-\eta}\mathrm{d}t$, $\alpha>0,\beta>0,\eta>1$ is a non-negative power function of selling price under inflation.
8. Replenishment occurs instantly.
9. Effect of inflation is incorporated.

3. Mathematical model:

Consider a firm running a single item with a price-sensitive stock-dependent demand $D(P,I(t))$ and deterioration rate (θ). In order to reduce the deterioration rate, the firm is investing amount (u) in the preservation technology. The proportion of reduced deterioration rate is denoted by $f(u)$ .Inventory level reaches to zero at the end of each replenishment cycle due to deterioration rate and demand. The differential equation representing inventory status during interval $[0,T]$ is given by,

$$\frac{dI(t)}{dt}=-\theta(1-f(u))I(t)-D(P),\ 0\le \mathrm{t}\le T \tag{1}$$

with initial boundary condition $I(T)=0$

Solving the differential equation (1), the inventory level at any instant of time t is given by,

$$I(t)=\frac{\alpha P^{-\eta}}{\theta(1-f)+\beta P^{-\eta}}\left[e^{(\theta(1-f)+\beta P^{-\eta})(T-t)}-1\right],\ 0\le t\le T \tag{2}$$

The ordering quantity over the replenishment cycle (or the initial inventory on hand), is given by ,

$$Q=I(0)=\frac{\alpha P^{-\eta}}{\theta(1-f)+\beta P^{-\eta}}\left[e^{(\theta(1-f)+\beta P^{-\eta})T}-1\right] \tag{3}$$

Preservation technology investment for a cycle time T is given by uT. For convenience, $f(u)$ will be denoted by f in the model. Other costs involved in profit function are mentioned below.

The replenishment cost is A.

The preservation technology investment under inflation is given by uTe^{-Rt}.

As defective items are sold after screening process ,holding cost cannot be applied to "$\in Q$ "quantity of the items and hence, the inventory holding cost under inflation is given by,

$$HC=\int_0^T hI(t)\,\mathrm{e}^{-Rt}dt-h\in Q\mathrm{e}^{-Rt_1}$$, where t_1 = time at which screening process ends

The purchase cost of an item is given by,

$$CQ=\frac{C\alpha P^{-\eta}}{\theta(1-f)+\beta P^{-\eta}}\left[e^{(\theta(1-f)+\beta P^{-\eta})T}-1\right]$$

Screening cost : $C_s Q e^{-Rt_1}$

The sales revenue is $P\int_0^T (\alpha+\beta I(t))P^{-\eta}e^{-Rt}dt + P_s Q \in e^{-Rt_1}$, $P_s = P(1-\lambda)$

Hence, the profit per unit time is given by,

$$\pi(P,T,u) = \frac{1}{T}\begin{cases} \text{Sales revenue} - \text{purchase cost} - \text{Ordering cost} \\ -\text{Preservation cost} - \text{Screening cost} - \text{Holding cost} \end{cases}$$

$$= \frac{1}{T}\left\{ \begin{aligned} & P\int_0^T (\alpha+\beta I(t))P^{-\eta}e^{-Rt}dt + P_s Q \in e^{-Rt_1} - \frac{C\alpha P^{-\eta}}{\theta(1-f)+\beta P^{-\eta}}\left[e^{(\theta(1-f)+\beta P^{-\eta})T} - 1\right] \\ & -\left\{\int_0^T hI(t)\, e^{-Rt}dt - h \in Qe^{-Rt_1}\right\} - C_s Q e^{-Rt_1} - uTe^{-RT} - A \end{aligned} \right\} \tag{4}$$

4. Solution methodology:

The objective of this study is to maximize the profit with respect to optimal selling price, optimal length of replenishment cycle and optimal preservation technology investment.

By solving $\frac{\partial \pi}{\partial P} = 0, \frac{\partial \pi}{\partial T} = 0, \frac{\partial \pi}{\partial u} = 0$, we obtain the optimum value of (P,T,u). Generated equations are solved by mathematical software.

5. Numerical example:

To illustrate the solution procedure, in this section we give the following example:

Example: In this example we consider the reduced deterioration rate $f(u) = 1-e^{-mu}, m > 0$ and parametric values are as follows:

$\theta = 30\%$, $\alpha = 10000$ units/year, $\beta = 20\%$, $h = \$3$ /unit/year, $m = 5\%$, $C = \$10$ /unit/year, $A = \$50$ /order, $\eta = 1.25$, $C_s = \$3$ /unit/year, $\in = 10\%$, $\lambda = 30\%$, $r = 12\%$, $k = 6\%$, $\xi = 0.1$. We obtain optimal cycle time $T^* = 0.4521$ years, optimal selling price $P^* = \$79.07$ per unit and optimal investment $u^* = \$7.92$ per unit to reduce deterioration in the inventory system. Hence, seller's optimal profit is $\pi(P^*,T^*,u^*) = \$3566.2$ with purchase of $Q^* = 24.9$ units. Concavity of profit function with respect to two variables by taking optimum value of third variable are exhibited in Figs. 1 – 3.

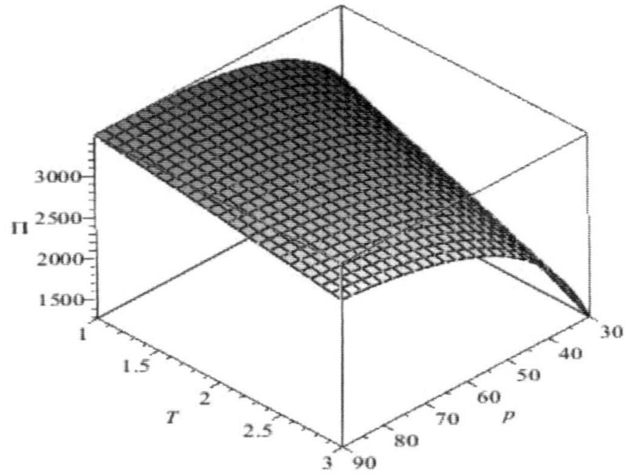

Figure 1: Change in profit with respect to T and P

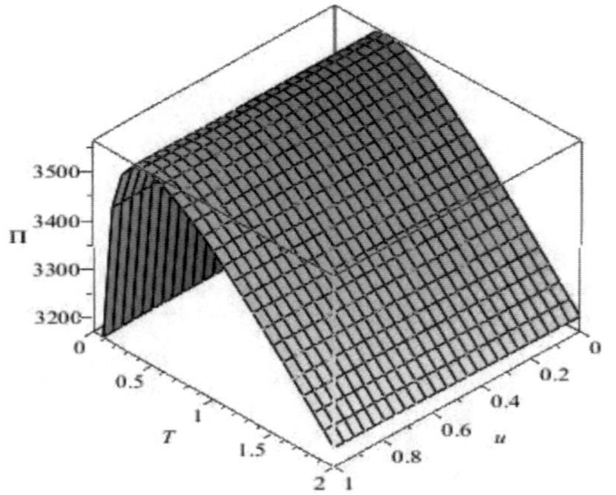

Figure 2: Change in profit with respect to T and u

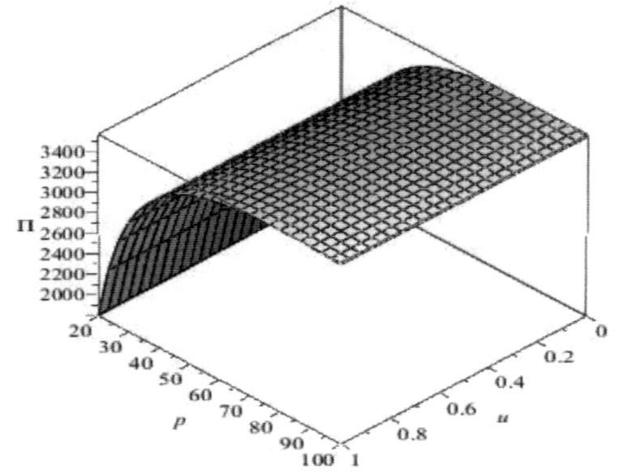

Figure 3: Change in profit with respect to u and P

Effects and results of changing the values of the system parameters on decision variables and profit are shown in Figs 4 to 7 and Table 1.

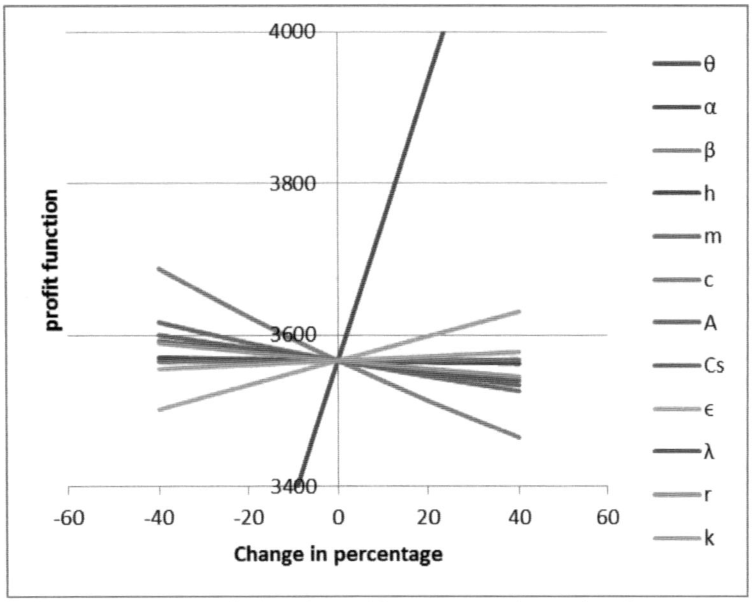

Figure 4: Change in profit function with respect to change in other parameters

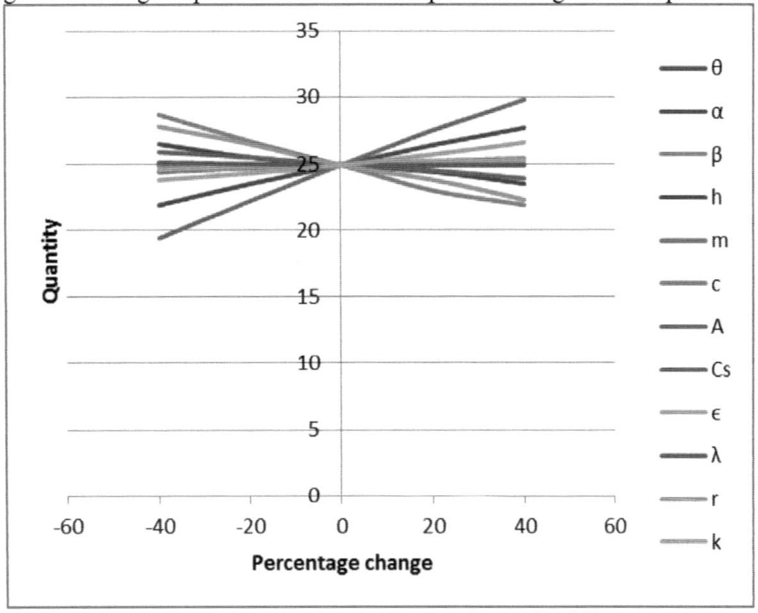

Figure: 5 Change in quantity with respect to change in different parameters

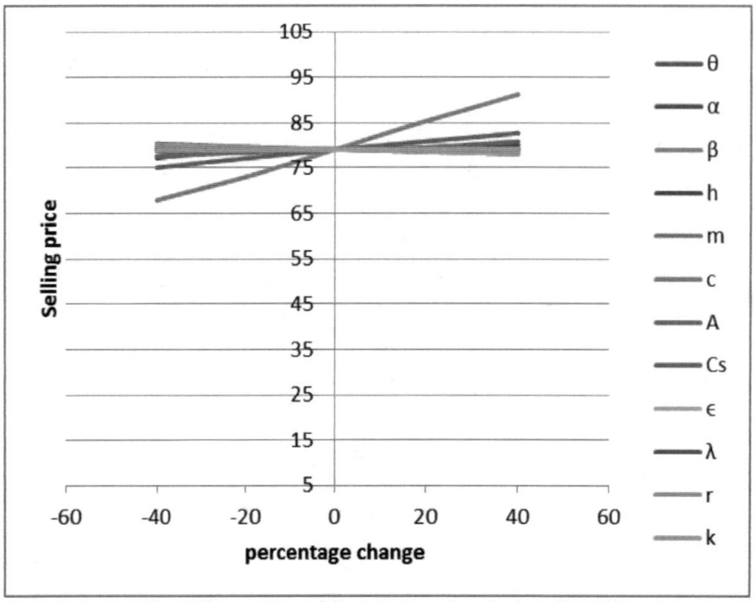

Figure 6: Change in selling price with respect to variation in different parameters.

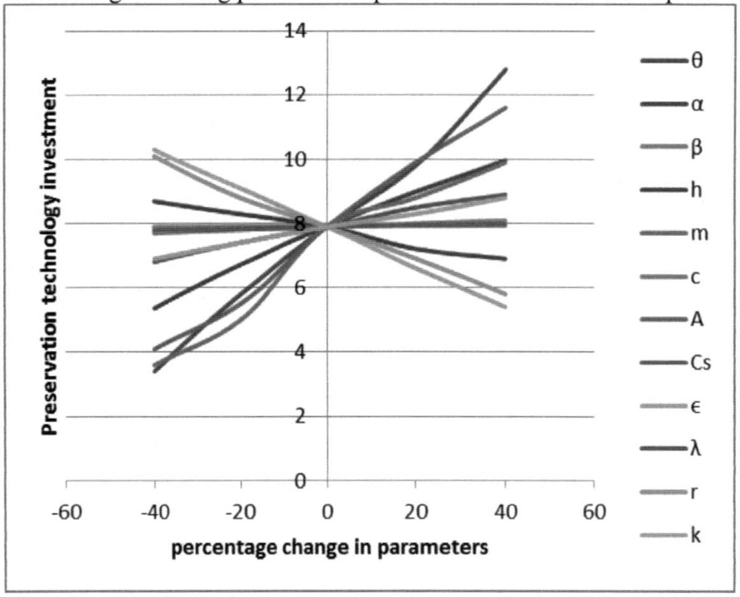

Figure 7: Percentage change in parameters versus preservation technology investment

Decision variable	Sensitively positive	Sensitively negative	Ineffective	Negligible
π	$\alpha, \in, k$	$\theta, h, C, A, C_s, \lambda, r$	NA	β, m
Q	$\alpha, m, A, \in, k$	h, C, C_s, r	θ	β, λ
P	h, C, A, C_s	α, m	θ	$\beta, \in, \lambda, r, k$
u	$\theta, \alpha, m, A, \lambda, k$	$h, \in, r$	-	β, C, C_s

Table 1: Effectiveness of different parameters on decision variables

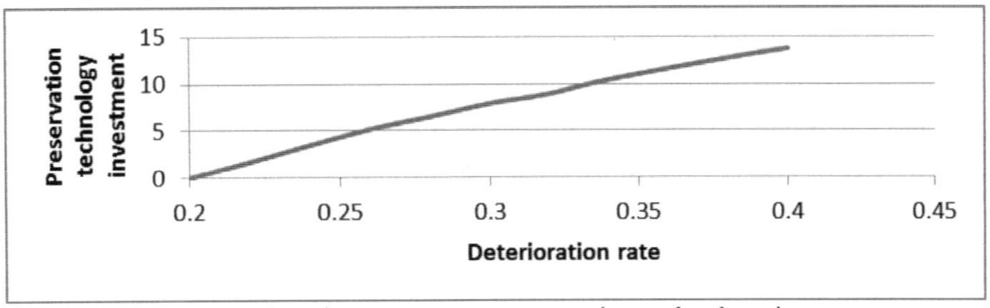

Figure 8: Deterioration rate versus preservation technology investment

Fig.8 shows the relation between deterioration rate and preservation technology investment. The trend suggests that higher investment is required for high deterioration of items in inventory system and investment is not required when the rate is marginal.

6. Conclusion:

The main objective of this paper to optimize the profit function with respect to replenishment cycle time, preservation technology investment and selling price when demand is price-sensitive and stock-dependent with inflationary effects. All the above graphs show optimality obtained in each case. It is observed that the profit function is positively affected by scale demand (α), proportion of defective items ($\in$), inflation rate per unit time (k) and negatively affected by deterioration rate (θ), inventory holding cost (h), purchase cost (C), screening cost (C_s), proportion decrease in selling price for defective items (λ), discounted rate deciding the time value of money(r). Also, positive effect of scale demand (α), proportion of defective items ($\in$), inflation rate per unit time (k) and replenishment cost per order (A) seen on quantity. Quantity is negatively affected by holding cost (h), purchase cost (C), screening cost (C_s) and discounted rate deciding the time value of money (r). Overall summary is given in table 1.
Deterioration versus preservation technology suggests high investment required if product is highly deteriorated in nature. Also, investment is not required if deterioration is less or marginable.

References:

[1] Buzacott, J.A. (1975), 'Economic Order Quantities with Inflation', *Operational Research Quarterly*,26,553-558
[2] Cardenas-Barron, L.E (2000), 'Observation on Economic Production Quantity Model for Items with Imperfect Quality', *International Journal of Production Economics*, 67,201.
[3] Cardenas-Barron, L.E (2009a),'Optimal Ordering Policies in Response to a Discount Offer: Extensions', *International Journal of Production Economics*, 122,774-782.
[4] Cardenas-Barron, L.E (2009b), 'Optimal Ordering Policies in Response to a Discount Offer: Corrections', *International Journal of Production Economics*, 122,783-789.
[5] Cardenas-Barron, L.E. (2011), 'The Derivation of EOQ/EPQ Inventory Models with Two Back orders Costs Using Analytic Geometry and Algebra', *Applied Mathematical Modelling*, 35, 2394-2407.
[6] Cardenas-Barron, L.E., Smith, N.R, and Goyal, S.K. (2010), 'Optimal Order Size to Take Advantage of a One-time Discount Offer with Allowed Backorders, *Applied Mathematical Modeling*, 34, 1642-1652.
[7] Chen, Y.R. and Dye, C.Y. (2013),'Application of particle swarm optimization for solving deteriorating inventory model with fluctuating demand and controllable deterioration rate', *International Journal of Systems Science*, 44 (6): 1026-1039.

[8] Chung-Yuan Dye, Tsu-Pang Hsieh (2013),'A particle swarm optimization for solving lot-sizing problem with fluctuating demand and preservation technology cost under trade credit', *J. Global Optimization*, 55(3), 655-679.
[9] Datta, T.K., and Pal, A.K. (1991), 'Effects of Inflation and Time value of Money on an Inventory Model with Linear Time-Dependent Demand Rate and Shortages', *European Journal of Operational Research*, 52,326-333.
[10] Ghare, P.M., and Schrader, S.K. (1963), 'A Model for Exponentially Decaying Inventory', *Journal of Industrial Engineering*, 14, 38-243.
[11] Goyal, S.K., and Cardenas-Barron, L.E. (2002), 'Note on: Economic Production Quantity Model for Items with Imperfect Quality- A Practical Approach', *International Journal of Production Economics*, 77, 85-87.
[12] Hariga, M.A (1995),'An EOQ Model for Deteriorating Items with Shortages and Time-varying Demand', *Journal of the Operational Research Society*, 46,398-404.
[13] Hariga, M.A., and Ben-Daya, M. (1996), ' Optimal time varying Lot-sizing Models Under inflationary Conditions', *European Journal of Operational Research*,89,313-325.
[14] Hsieh and Dye (2013), 'A production-inventory model incorporating the effect of preservation technology investment when demand is fluctuating with time', *Journal of Computational Applied Mathematics* (239), 25-36.
[15] Jaggi and et.al. (2012),'Effects of inspection on retailer's ordering policy for deteriorating items with time-dependent demand under inflationary conditions', *International Journal of Systems Science*, 44:9, 1774-1782.
[16] Jaggi, C.K., and Mittal, M. (2007), 'An EOQ Model for Deteriorating Items with Time-dependent demand Under Inflationary Conditions', *Indian Journal of Mathematics and Mathematical Sciences*,3,139-147.
[17] Jaggi, C.K., and Mittal, M. (2011), 'Economic Order Quantity Model for Deteriorating Items with Imperfect Quality', *Revista Investigation Operational*, 32,107-113.
[18] Misra, R.B. (1975), ' A study of Inflationary Effects on Inventory Systems', *Logistic Spectrum*, 9, 260-268.
[19] Porteus, E.L. (1986), 'Optimal Lot Sizing, Process Quality Improvement and Setup Cost Reduction', *Operations Research*, 34,137-144.
[20] Raafat, F., Wolfe, P.M., and Eldin, H.K.(1991), ' An Inventory Model for Deteriorating Items', *Computers and Industrial Engineering*,20,89-94.
[21] Ray, J., and Chaudhuri, K.S. (1997), 'An EOQ Model with Stock-dependent Demand, Shortage, Inflation and Time discounting', *International Journal of Production Economics*, 53,171-180.
[22] Rosenblatt, M., and Lee, H. (1986), 'Economic Production Cycles with Imperfect Production Processes', *IIE Transactions*, 18, 48-55.
[23] Salameh, M.K., and Jaber, M.Y. (2000), 'Economic Production Quantity Model for Items with Imperfect Quality', *International Journal of Production Economics*, 64, 59-64.
[24] Sana, S., and Chaudhuri, K.S. (2003), 'An EOQ Model with time-dependent Demand, Inflation and Money value for a Warehouse Enterpriser', *Advanced Modern Optics*, 5, 135-146.
[25] Sarkar, B.R., and Pan, H. (1994), 'Effects of Inflation and Time value of Money on Order Quantity and Allowable Shortage', *International Journal of Production Economics*, 34, 65-72.
[26] Schwaller, R.L. (1988), 'EOQ under Inspection Costs', *Production and Inventory Management*, 29, 22-24.
[27] Widyadana, G.A., Cardenas-Barron, L.E., and Wee, H.M. (2011), 'Economic Order Quantity Model for deteriorating Items and Planned Backorder Level', *Mathematical and Computer Modeling*, 54, 1569-1575.

[28] Yong, H. and Huang, H. (2013),'Optimizing Inventory and Pricing Policy for Seasonal Deteriorating Products with Preservation Technology Investment', *Journal of Industrial Engineering*, vol.2013, Article ID 793568, 7 pages, doi:10.1155/2013/793568
[29] Zhang, X. and Gerchak, Y. (1990), 'Joint Lot-Sizing and Inspection Policy in an EOQ Model with Random Yield', *IIE Transaction*, 22, 41-47.

REVIEW ARTICLE: Recent Advances in Nanomaterial Fabrication

Kitsakorn Locharoenrat[1]
Department of Physics, King Mongkut's Institute of Technology Ladkrabang (KMITL), Bangkok 10520, Thailand

E-mail: klkitsak@kmitl.ac.th

Abstract. As nanostructures with well-controlled dimension, composition, and crystallinity are expected to be a new class of intriguing system for investigating structure-property relations, this review article provides a comprehensive review of researches of these materials and related applications.

1. Introduction

Technology of making small structures, nanofabrication, has evolved greatly over the past decade from a reliance upon clever tricks appropriate for simple structures to a broadly based set of technologies applicable to make complex devices with dimension in the range of one to hundreds of nanometres. Variety of techniques has been used to produce this kind of nanostructures. In this article, the author introduces recent advances in nanofabrication techniques in order to harvest the final nanostructures. The related applications of these nanomaterials are also presented in this review article.

2. Nanofabrication

Relief structures present on the surface of a solid substrate serve as a class of natural templates for generating supported nanostructures. In this regard, decoration of these templates provides a powerful route to the formation of nanowires made of various metals and semiconductors [1, 2, 3, 4, 5, 6, 7]. It can be subsequently transferred onto the surfaces of other substrates. Sugawara et al. have fabricated the arrays of Fe nanowires on the (110) surfaces of NaCl crystals by a shadow deposition method (figure 1) [7]. When the NaCl (110) templates is annealed in a vacuum conditions, the surface becomes faceted with (010) and (100) planes in order to minimize the surface energy. Periodic macrosteps parallel to the [001] direction are then formed. After iron is deposited at a certain degree from the template normal, approximately sixty percent of (100) terraces are exposed to iron flux, and the iron nanowires are formed only near the edge of the ridges.

1 Corresponding Author: Kitsakorn Locharoenrat (klkitsak@kmitl.ac.th)

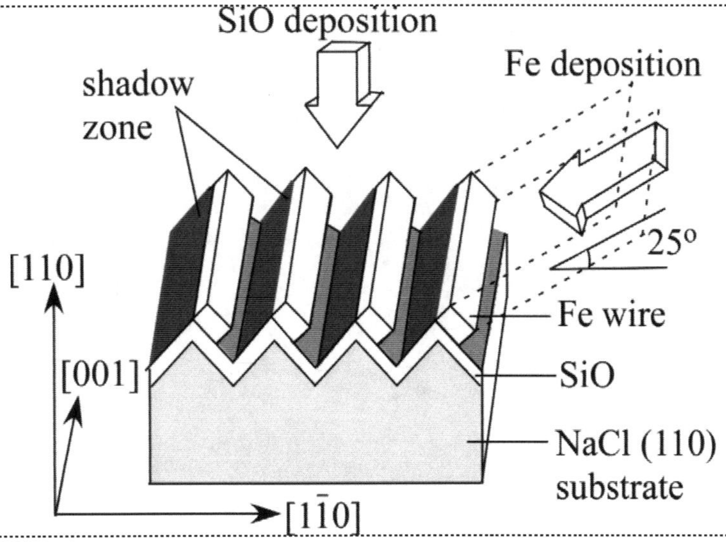

Figure 1. Schematic illustration of a shadow deposition method.

By using the same procedure, they have fabricated cobolt dot arrays sandwiched in gold nanowires onto stepped NaCl (110) surfaces [8]. As demonstrated by Kitahara et al., it is also possible to fabricate arrays of gold nanowires onto the surface of NaCl (110) substrates [9]. The author also uses this simple technique to prepare copper nanowires [10]. Briefly, NaCl (110) single crystals are dipped in distilled water before being mounted on the substrate heating stage in an ultrahigh vacuum deposition chamber. A certain thick of homoepitaxial layers of NaCl are deposited after the substrate annealing. A certain thick SiO layers are then deposited as passivation layers. Copper is deposited a certain deposition rate onto the substrate at a certain flux angle with respect to the surface normal to form copper nanowires. On the arrays of nanowires, a certain thick of SiO layers are deposited as protection layers. For Transmission Electron Microscopy (TEM) observation, the copper nanowire arrays sandwiched by SiO layers are separated from the NaCl substrates by dissolving the substrates in distilled water, and mounted these nanowires on copper grids for TEM observation. Using this approach, the author obtains wide-area arrays suitable for both TEM and optical measurements in order to study microstructures and optical properties of self-organized arrays of the nanowires.

Penner et al. have demonstrated the growth of the nanowires by templating against the step edges present on a highly oriented pyrolytic graphite (figure 2) [11, 12]. The noble metal nanowires are found to preferentially nucleate and grow along the step edges present on a graphite surface into a two-dimension parallel array that can be transferred onto the surface of a cyanoacrylate film supported on a glass slide.

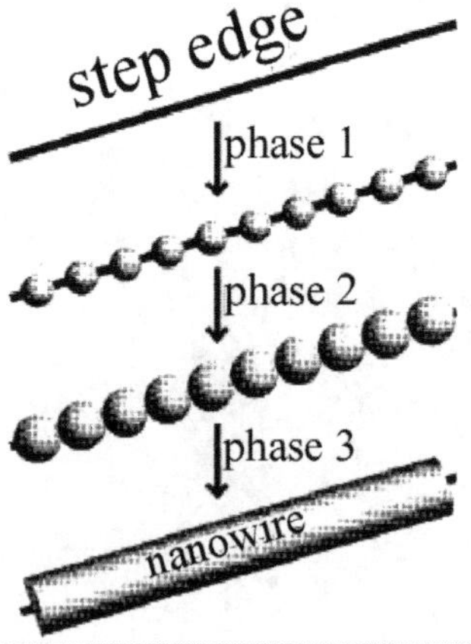

Figure 2. Three phases of nanowires electrodeposition at step edges [10]. 1. Metal nuclei forms along step edges. 2. Nucleation ceases; hemispherical nuclei grows to coalescence with nearest neighbours forming a rough or beaded nanowires. 3. Beaded nanowires grow.

The advantage of this technique is that it is applicable to the mass production of high quality metallic nanostructures and applications requiring large areas without breaking vacuum. One drawback of this technique is that the nanowires cannot be peeled off from the substrate to make freestanding form.

Electron Beam Lithography (EBL) process in figure 3 allows one to deposit metal nanowires of any shape, size, and orientation in any arrangement patterns on the flat substrate as described by Schider et al. [13]. In this process, a master image (resist) is produced onto the thin layer of metals, and it is used in transferring a pattern from a mask to surface of the silicon wafer. Finally, the resist is removed by etching in liquid or gaseous form.

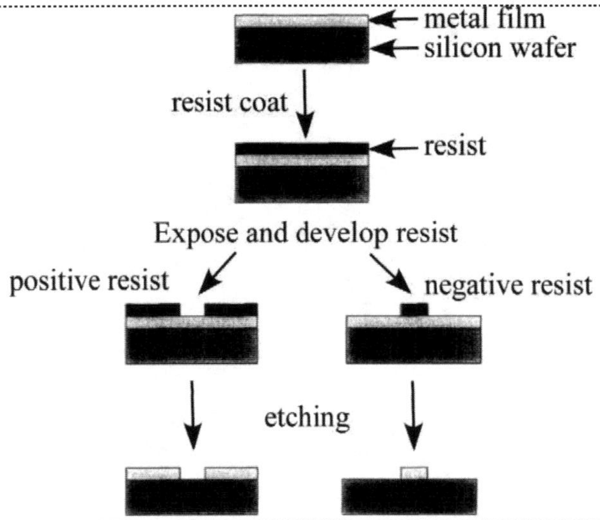

Figure 3. Schematic of main steps of electron beam lithography process.

EBL shows certain benefits over conventional photolithography techniques. It is capable of very high resolution, almost to the atomic level. Typically, EBL has a three orders of magnitude better resolution, although this is limited by the forward scattering of electrons in the resist layer, and back scattering from the underlying substrate. It is a flexible technique that can function with a wide variety of materials and an almost infinite number of patterns. On the other hand, EBL has certain drawbacks. It is slow in operation, being one or more orders of magnitude slower than optical lithography. In addition, it is expensive and complicated, with EBL systems costing many millions of dollars to purchase and require frequent servicing to maintain performance.

Channel in porous membranes provides another class of template for using in the synthesis of nanostructures (figure 4). The nanopores in the template are formed by anodizing aluminum films in an acidic electrolyte. Individual nanopore in the alumina is ordered into close-packed honeycomb structure. The separation between two adjacent pores and the diameter of each pore can be controlled by changing the anodizing conditions. Using this membrane template, nanowires of various types of metals and semiconductors can be fabricated. These nanostructures can be deposited into the pores to form continuous nanowires with large aspect ratio (length-to-diameter ratio) by either electrochemical deposition or other methods such as chemical vapour deposition. When freely standing nanowires are desired, one has to remove the template hosts after forming the nanowires in the templates. This task is accomplished by dissolving away the template materials in suitable solvent. Sandrock et al. have exploited this membrane-based template to grow gold nanowires with a well-defined dimension [14].

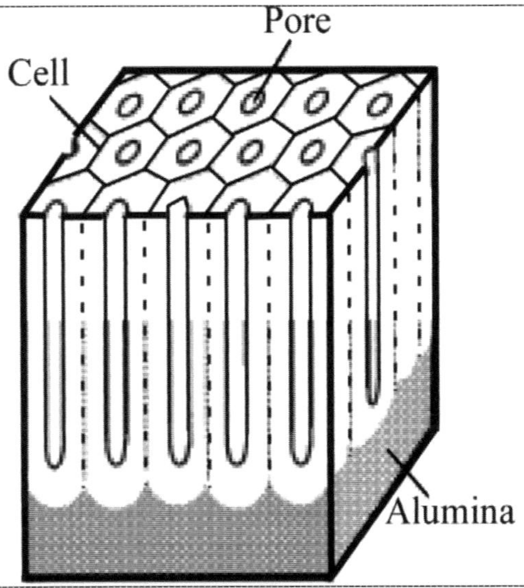

Figure 4. Schematic drawing of an anodic porous alumina template used to form the nanowires by filling the pores with the desired materials.

One advantage of this technique is the possibility of fabricating multilayered structures within nanowires. By varying cathodic potentials in the electrolyte, layers of different compositions can be controllably deposited. This method provides a low-cost approach to prepare multilayered 1-D nanostructures. One disadvantage of this approach for large-scale applications is that anodic porous alumina is a brittle ceramic film grown on a soft aluminum metal substrate. Great care needs to be exercised in the preparation of the aluminum substrate and in the manipulation of the anodic film to produce pure defect-free porous alumina films that are required to achieve uniform filling of the pores with the nanowires.

Mechanical methods can fabricate the arrays of the atomic-scale metal wires supported on solid substrates by mechanically separating two electrodes in contact. During the separation process, a metal neck is formed between the electrodes due to strong metallic cohesive energy in which is stretched into an atomically thin wire before breaking. One such method is based on a scanning tunnelling microscopy (STM) in which the STM tip is driven into the substrates and the conductance is recorded while the tip is gradually pulled out of the contact with the substrates.

Kawai's Group has employed Au (111), Au (455) and Au (788) crystals as substrates for the growth of the periodic arrays of manganese nanostructures [15]. The gold substrates are chemically etched and mechanically polished to obtain good metal surfaces. Once introducing into an ultrahigh vacuum chamber, the samples are prepared by the certain cycles of sputtering and annealing processes. After cooling the samples to room temperature, manganese is dosed with an electron beam evaporator. A typical deposition rate was about 0.05 millilitres per minute. The coverage of 1 millilitre is defined as one surface layer of manganese crystal.

The advantage of this technique is that one can fabricate a single or an array of stable nanowires. The method can be automated such that a nanowire with a preset quantized conductance can be produced at will. The nanowires supported on a solid substrate can be removed from the fabrication set-up and used as a stand-alone device or for further investigation using various experimental probes. However, the use of STM in the set-up makes it difficult for mass production. The lifetime of the nanowire is typically less than a few seconds because the gap between the STM tip and the metal surface drifts due to acoustic noise, thermal expansion and mechanical vibrations, which is also undesirable for practical applications.

3. Applications

The nanostructures have received steadily growing interests as a result of their peculiar and fascinating properties, as well as applications superior to their bulk counterparts [16]. The ability to generate such nanostructures is essential to modern science and technology when making new types of nanostructures, or downsizing the existing microstructures into 1 - 100 nanometres regimes. The most success example is provided by microelectronics, where “smaller” has meant greater performance ever since the invention of integrated circuits: more components per chip, lower cost, faster operation, and less power consumption [17]. Miniaturization represents the trend in a range of other technologies. In information storage, for instance, there are many efforts to develop magnetic and optical storage components with critical dimensions as small as tens of nanometres [18]. It is also clear that a new phenomenon is associated with nanometre-sized structures, with the well-established examples including size-dependent excitation of the quantum dot structures [19], and quantization of conductance in the metal contacts [20]. In addition, quantum confinement of electrons by the potential wells of nanoscale structures may provide one of the most powerful means to manipulate the optical, electrical, magnetic, and thermoelectric properties of a solid-state functional material.

So far, nanostructured surfaces such as nanowires, nanorods, and nanotubes become the focus of intensive research owing to their fabrication of nanodevices [21]. These nanostructures provide a good system to investigate the dependence of electrical and thermal transport or mechanical properties on the size reduction. In addition, they are expected to play an important key in the interconnection applications and the functional units in fabrication of electronic and optoelectronic devices.

Although nanostructures can now be produced using a number of advanced nanolithography techniques [22, 23], such as electron-beam, these methods are still not conveniently available to practical routes, and it is also challenging to fabricate nanowires with the widths of a few nanometre. Exploration and development of new techniques for fabrication of large quantities of nanostructures from a diversified range of materials, rapidly, and at reasonably low costs, are desirable. Many deposition techniques described herein are not intended to replace the existing methods, but rather to provide a simple alternative for generating nanostructures with a well-controlled dimension, and to help the study of their optical properties.

4. Conclusions

Recent advances in nanofabrication techniques have made it possible to produce well-defined nanostructures and each method has its specific merits and inevitable weakness. For instance, some researchers devote the most attention to the method of generating nanostructures based on anisotropic growth [24] directed or confined by the templates. The template-directed methods provide a good control over the uniformity and dimension in which the nominal thicknesses of the nanowires are measured with the sensor plane of the thickness monitor perpendicular to the metal beam direction. However, removal of the template through a post-synthesis process may cause damage to the final nanostructures' product. Most nanostructures produced using this method is polycrystalline in structure and they may limit their uses in device fabrication and fundamental studies. Therefore, judging against these aspects, some methods described here still need to be improved before it finds widespread use in commercial applications.

5. References

[1] Hu L, Wu H and Cui Y 2011 *MRS Bullet.* **36** 760
[2] Neubrech F, Kolb T, Lovrincic R, Fahsold G and Pucci 2006 *Appl. Phys. Lett.* **89** 253104
[3] He H and Tao N J 2003 *Encyclopedia Nanosci. Nanotech.* **4** 1
[4] Wu Y, Xiang J, Yang C, Lu W and Lieber A 2004 *Nature* **430** 61
[5] Tian M, Wang J, Kurta J, Mallouk T E and Chan M H W 2003 *Nano Letter* **3** 919
[6] Vossen J L and Kern W 1978 *Thin Film Processes* (New York: Academic)
[7] Sugawara A, Coyle T, Hembree G G and Scheinfein M R 1997 *Appl. Phys. Lett.* **70** 1043
[8] Kitahara T, Sugawara A, Sano H and Mizutani G J 2004 *J. Appl. Phys.* **95** 5002
[9] Kitahara T, Sugawara A, Sano H and Mizutani G 2003 *Appl. Surf. Sci.* **219** 271
[10] Locharoenrat K, Sugawara A, Sano H and Mizutani G 2007 *Surf. Sci.* **601** 4449
[11] Penner R M 2002 *J. Phys. Chem.* **106** 3339
[12] Song H H, Jones K M and Baski A A 1999 *J. Vacuu. Sci. Technol.* A **17** 1696
[13] Schider G, Krenn J R, Gotschy W, Lamprecht B, Ditbacher H, Leitner A and Aussenegg F R 2001 *J. Appl. Phys.* **90** 3825
[14] Sandrock M L, Pibel C D, Geiger F M and Foss Jr C A 1999 *J. Phys. Chem.* B **103**2668
[15] Shiraki S, Fujisawa H, Nantoh M and Kawai M 2004 *Surf. Sci.* **552** 243
[16] Nalwa H S 2000 *Handbook of Nanostructured Materials and Nanotechnology* (New York: Academic)
[17] Luryi S, Xu J and Zaslavsky A 1999 *Future Trends in Microelectronics* (New York: John Wiley & Sons)
[18] Ross C 2001 *Annu. Rev. Mat. Sci.* **31** 203
[19] Alivisatos A P 1996 *Sci.* **271** 933
[20] Krans J M, Rutenbeek J M, Fisun V V, Yanson I K and Jongh L J 1995 *Nature* **375** 767
[21] Wang Z L 2000 *Adv. Mat.* **12** 1295
[22] Krenn J R, Weeber J C, Bourillot E, Goudonnet J P, Schider B, Leitner A, Aussenegg F R and Girard C 1999 *Phys. Rev.* B **60** 5029
[23] Xia Y, Rogers J A, Paul K E, Whitesides G M 2000 *Chem. Rev.* **99** 1823
[24] Growth anisotropy describes the condition when growth rates are not equal in all directions. In contrast, when growth rates are the same rate in all directions, the growth is isotropic.

Journal of Physics: Conference Series **495** (2014) 012019 doi:10.1088/1742-6596/495/1/012019

Low wireless power transfer using Inductive Coupling for mobile phone charger

M Fareq[1], M Fitra[1], M Irwanto[1], Syafruddin Hasan[1], M Arinal[1]
[1]School of Electrical System Engineering Universiti Malaysia Perlis, Kangar 01000, Perlis, Malaysia

E-mail: mhdfitra@gmail.com

Abstract. A wireless power transfer (WPT) using inductive coupling for mobile phone charger is studied. The project is offer to study and fabricate WPT using inductive coupling for mobile phone charger that will give more information about distance is effect for WPT performance and WPT is not much influenced by the presence of hands, books and types of plastics. The components used to build wireless power transfer can be divided into 3 parts components, the transceiver for power transmission, the inductive coils in this case as the antenna, receiver and the rectifier which act convert AC to DC. Experiments have been conducted and the wireless power transfer using inductive coupling is suitable to be implemented for mobile phone charger.

1. Introduction

In recent years, there has been increasing interest in research and development of wireless power technology to eliminate the "last cable" after Wi-Fi becomes widely accepted [1]. Imagine sitting down for a cup of coffee, and placing your mobile phone on the table. The phone lights up, and starts to automatically charge without connectors or cables. You could simply grab your mobile phone on your way out in the morning, and charge it wherever you needed to at home, the office, the library, the local coffee shop. It would be even better if you did not ever need a charger. We could simply forget about USB cables, chargers and, when traveling, adapters. Inductive power transfer is being used in numerous applications for transferring power wirelessly. Inductively coupled chargers are being used for wireless charging of mobile phones, MP3 Player and other handheld devices as proposed in [2-4].

WPT can be divided into 3 parts components. First, transceiver. The transceiver electromagnetically transfers power via inductive coils which supply a wireless transfer of power to receiver units. Second, inductive coupling, the inductive coupling in this case as the antenna and forward to the bridge rectifier which act convert the induced AC voltage to the DC voltage. Third, rectifier. The rectifier finally will supplies DC voltage to the load. Block diagram of wireless power transfer as shown in Figure 1.

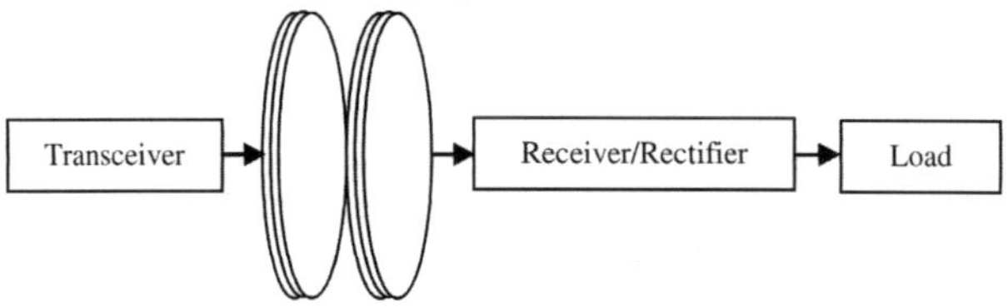

Figure 1. Block Diagram of Wireless Power Transfer

Currently WPT is requirements to be implanted for gadget charger such as mobile phone, iPad, MP3 Player and other handheld devices, for example we could simply forget about USB cables, chargers and, when traveling, and will be efficiency.

2. Methodology

2.1. Simulation.

Simulation wireless power transfer has been done simulation used multisim software, as shown in Figure 2.

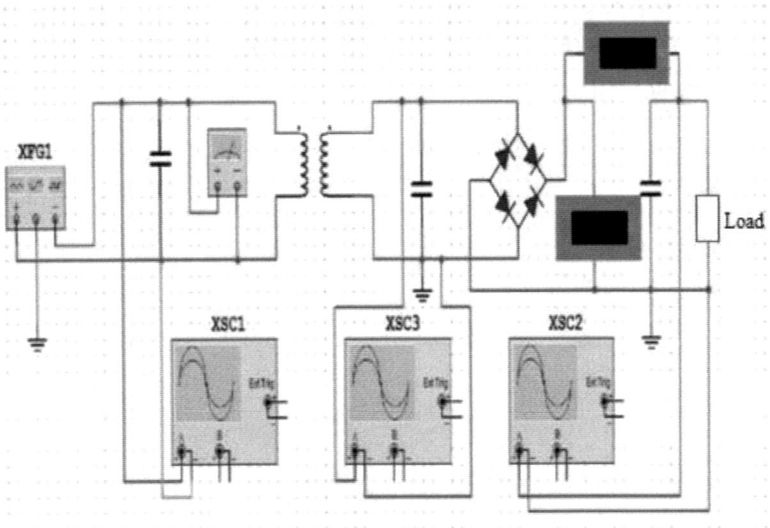

Figure 2. Simulation wireless power transfer using multisim software.

Multisim is an industry standard, best in class spice simulation environment. It is the cornerstone of the NI circuits teaching solution to build expertise through practical application in designing, prototyping, and testing electrical circuits.

2.2. Principle of wireless power transfer using inductive coupling.

Experimental setup has been studied to get the performance of WPT. In this experiment a mobile phone devices is used as an load . The experimental setup of transceiver unit has been conducted. Energy transfer by electromagnetic induction to the receiver via inductive coupling. The voltage sources to the transceiver was provide by function generator with 1.2 MHz frequency and 0.01 of Vp-p (AC). The inductive coupling is used as the antenna to wireless power delivered from the transmitting to the input of a receiver. Receiver unit, the bridge rectifier is used convert AC voltage to produces DC voltage and produce DC output. A capacitor is included in the circuit to act as a filter to reduce ripple voltage. The experimental setup as shown in Figure 3.

Figure 3. Experimental setup and collect data

2.3. Obstacle between Transmitter and Receiver

2.3.1. A presence of hands

As shown in Figure 3. Experiment has been test using obstacle hands between the transmitter and receiver inductive coupling.

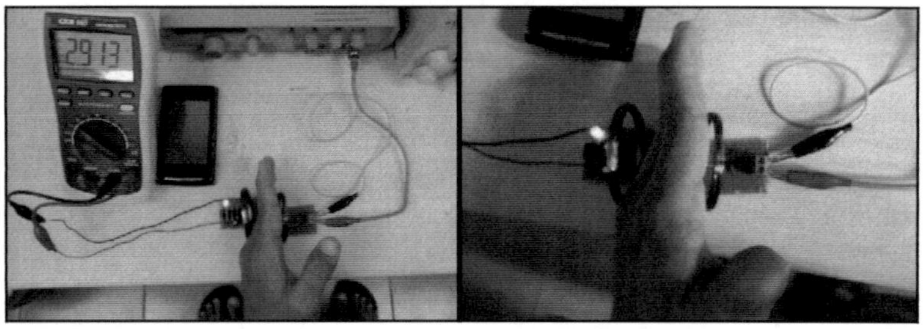

Figure 4. Experiment has been test using obstacle hands

2.3.2. A presence of book

As shown in Figure 4. Experiment has been test using obstacle a Book between the transmitter and receiver inductive coupling.

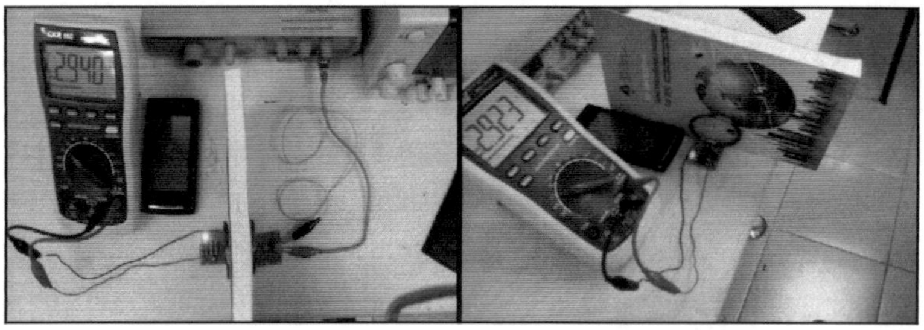

Figure 5. Experiment has been test using obstacle a book

2.3.3. A presence types of plastics

As shown in Figure 5. Experiment has been test using obstacle types of plastics (Handphone casing and Tupperware) between the transmitter and receiver inductive coupling.

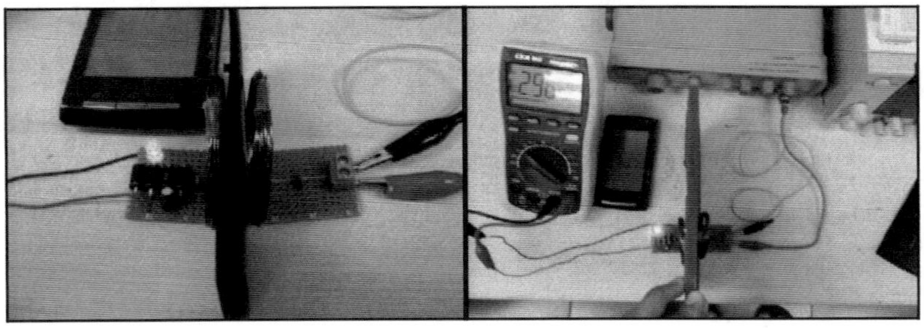

Figure 6. Experiment has been test using obstacle types of plastics (Handphone casing and Tupperware)

3. Results

An experiments has been conducted to get the WPT efficiency. The transmission inductive coupling was developed with diameter 4.5 cm number of turn is 20 while the receiver using diameter 4.5 cm number of turn is 20. The distance between these two inductive coils are varied to obtain the optimum distance for wireless power transmission.

Table 1. Varied distance data vs DC output voltage and V_{p-p}(AC)

Distance (cm)	V_{p-p}(AC)	f (MHz)	DC output voltage (Volt)
0.0	0.01	1.2	4.50
0.5	0.01	1.2	4.36
1.0	0.01	1.2	4.21
1.5	0.01	1.2	3.96
2.0	0.01	1.2	3.93
2.5	0.01	1.2	3.91
3.0	0.01	1.2	3.89
3.5	0.01	1.2	3.88
4.0	0.01	1.2	3.86
4.5	0.01	1.2	3.84
5.0	0.01	1.2	3.79

Table 1 shows the result of varied distance and the voltage differences when the distance is varied. The input source used from function generator and for wireless power transmission used frequency 1.2 MHz.

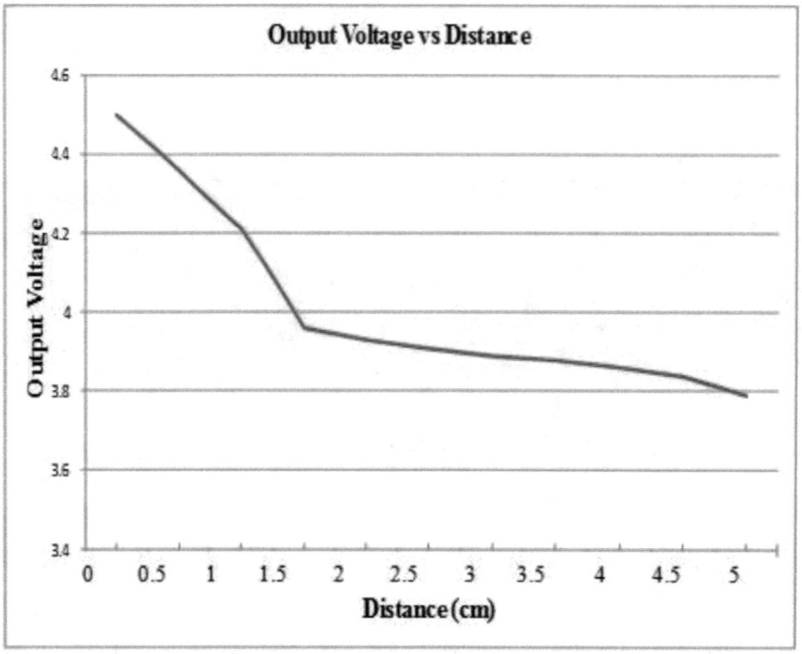

Figure 7. Graph of output voltage vs distance

Figure 7 shows the graph that has been created according to table 1. The DC output voltage getting lower as the distance are higher. From this graph we can conclude that the wireless power transmission is higher when the distance is nearer.

To observe the exact waveform of an electrical signal in this study used oscilloscope. Oscilloscopes are electronic measuring instrument that serves to project the form of an electrical signal that can be seen and studied.

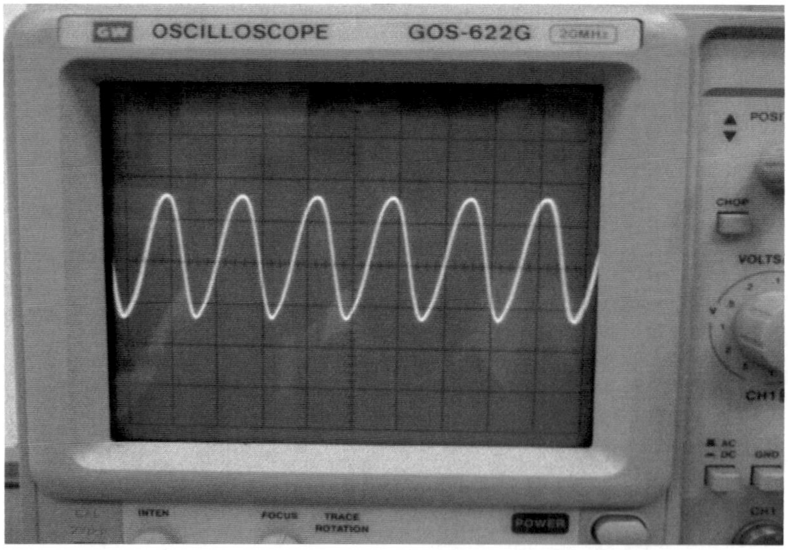

Figure 8. Sine wave at output function generator

Figure 8 shows the sine waveforms at output function generator (XSC1 refer to Figure 2) with 1.2 MHz and Vp-p (AC) 0.01 V.

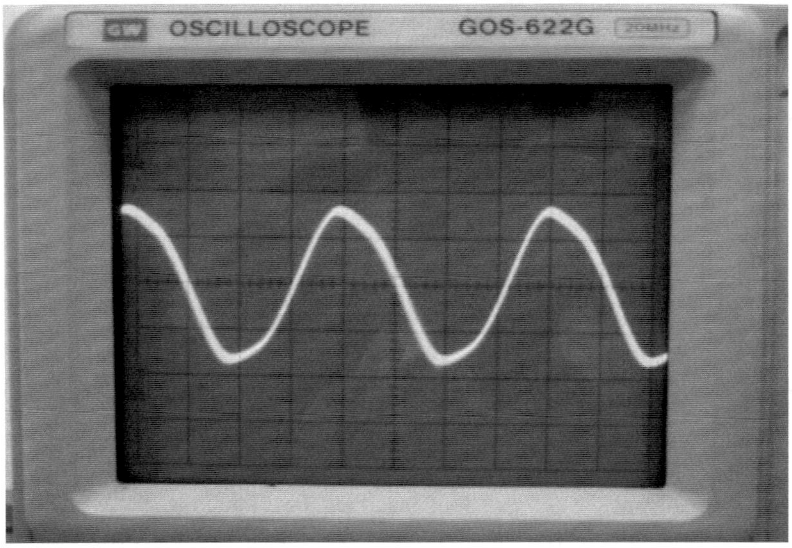

Figure 9. Voltage waveform at the inductive coupling receiver

Figure 9 shows the voltage waveform at the inductive coupling receiver terminals (XSC3 refer to Figure 2) before pass to bridge rectifier as convert AC to DC voltage.

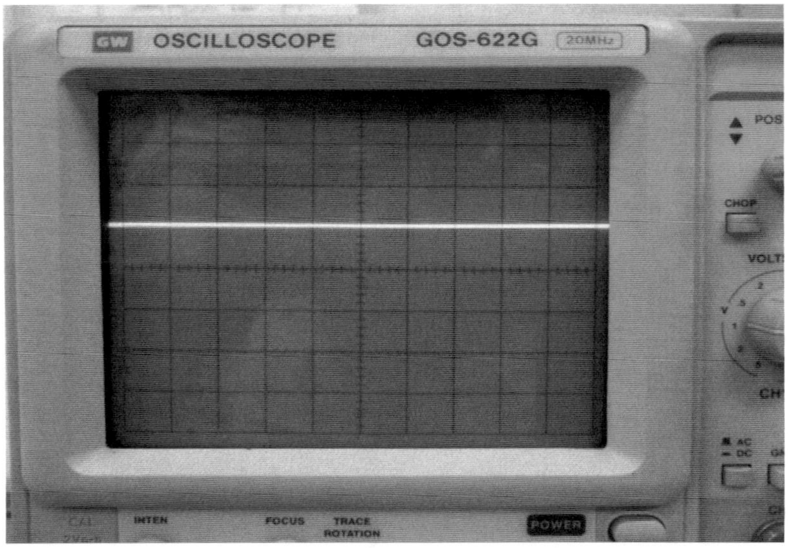

Figure 10. DC output waveform

Figure 10 displays the DC output waveform at bridge rectifier (XSC2 refer to Figure 2) as convert AC voltage to DC voltage.

Besides those experiments that has been measured above, the tests using an barrier (shielding) between the transmitter and receiver inductive coupling are also conducted. A presence of hands, books and plastics (handphone casing and tupperware) to be implanted for mobile phone charger. This results shows that the wireless power transfer is not much affected by the presence of hands, books and plastics.

Conclusions

From the overall experiment conducted from wireless power transfer using inductive coupling for mobile phone charger below conclusions are deduce :

1. Base on experimental result, the study on wireless power transfer using inductive coupling has much aspect in terms distance, range of frequency and result show the nearer the distance, the voltage transferred is higher.
2. From the experimental result in table 1, the distance 0 cm is the most efficient wireless power transmission.
3. Base on experimental result, the distance that the power can be transmitted is up to 5 cm with 3.79 Volt, because of in this condition, the mobile phone is still can charging.
4. The wireless power transfer is not much affected by shielding materials such as the presence of hands, books and types of plastics.
5. The result shows that the wireless power transfer is suitable to be implemented for mobile phone charger.

Acknowledgement

This work was supported by Associated Professor Mohd Fareq, Muhammad Irwanto Bin Misrun, Syafruddin Hasan and Mohd Arinal is gratefully acknowledged.

References

[1] Zhen Ning Low, "Design and Test of a High Power High Efficiency Loosely Coupled Planar Wireless Power Transfer System" IEEE transactions on industrial electronics, vol. 56, no. 5, May 2009.

[2] K. Oguri, "Power supply coupler for battery charger", US patent 6 356049, 2000.

[3] Y. Yang and M. Jovanavic, "Contactless electrical energy transmission system",US Patent 6 301 128, 2000.

[4] H. J. Brockmann and H. Turtiainen, "Charger with inductive power transmission for batteries in a mobile electrical device", US Patent 6 118249, 1999

Effects of Taiwan Roselle anthocyanin treatment and single-walled carbon nanotube addition on the performance of dye-sensitized solar cells

C S Chou[1], P J Tsai[2], P Wu[3], G G Shu[4], Y H Huang[1] and Y S Chen[1]

[1] Department of Mechanical Engineering, National Pingtung University of Science and Technology, Pingtung 912, Taiwan
[2] Department of Food Science, National Pingtung University of Science and Technology, Pingtung 912, Taiwan
[3] Engineering Product Development, Singapore University of Technology and Design, 20 Dover Drive, Singapore 138682
[4] Nuclear Materials Joint Lab, Tsinghua University Graduate School, Tsinghua University City Park (518055), Shenzhen, China

E-mail: cschou@mail.npust.edu.tw, wuping@sutd.edu.sg

Abstract. This study investigates the relationship between the performance of a dye-sensitized solar cell (DSSC) sensitized by a natural sensitizer of Taiwan Roselle anthocyanin (TRA) and fabrication process conditions of the DSSC. A set of systematic experiments has been carried out at various soaking temperatures, soaking periods, sensitizer concentrations, pH values, and additions of single-walled carbon nanotube (SWCNT). An absorption peak (520 nm) is found for TRA, and it is close to that of the N719 dye (518 nm). At a fixed concentration of TRA and a fixed soaking period, a lower pH of the extract or a lower soaking temperature is found favorable to the formation of pigment cations, which leads to an enhanced power conversion efficiency (η) of DSSC. For instance, by applying 17.53 mg/100ml TRA at 30 for 10 h, as the pH of the extract decreases to 2.00 from 2.33 (the original pH of TRA), the η of DSSC with TiO_2+SWCNT electrode increases to 0.67% from 0.11% of a traditional DSSC with TiO_2 electrode. This performance improvement can be explained by the combined effect of the pH of sensitizer and the additions of SWCNT, a first investigation in DSSC using the natural sensitizer with SWCNT.

Keywords: Dye Sensitized Solar Cell, Natural Sensitizer, Taiwan Roselle Anthocyanin, Single-Walled Carbon Nanotube

1. Introduction

Several alternatives to silicon-based photovoltaic (PV) cells have recently become available, and they may facilitate reduction of the cost of electricity generation via a PV cell. Among these alternatives, the dye-sensitized solar cell (DSSC), proposed by O'Regan and Grätzel [1], is an attractive alternative because of its properties, such as low production cost, the various choices of substrates that can be used, and low environmental impact during fabrication [2-3]. The dye sensitizer, which is used to harvest solar energy and generate excitons [2], is one of the major DSSC components. Therefore, novel sensitizers were synthesized and applied in DSSCs to promote the absorption of the visible spectrum [4-11].

For the sake of reducing environmental pollution (or production cost) during fabrication of synthesized sensitizers, research on natural sensitizers extracted from plants has received substantial attention. For example, natural sensitizers such as anthocyanins extracted from Calafate and Jaboticaba [12], black rice [13], hibiscus flower [14], mulberry fruit [15], Delonix regia and Indian blackberry [16], chlorophylls extracted from pomegranate leaf [15] and shiso leaf [17], and betanin extracted from beet root [18] were prepared and applied in DSSCs. Moreover, Zhou et al. observed that for photo-electrochemical performance, rutin was the most effective sensitizer for DSSC among 20 natural extracts [19]. Wongcharee et al. investigated the effects of extracting temperature, extracting solvent, and pH of the extract solution on the performance of a DSSC [20].

Interestingly, Senthil et al. indicated that the anthocyanin pigment has hydroxyl groups in the molecule, and it can effectively chelate to the surface of a TiO_2 film [16]. Aside from this, the molecular structure of the Taiwan Roselle anthocyanin (TRA) (Figure 1(a)) [21] extracted from Hibiscus Sabdariffa Linnaeus (Figure 1(b)) which are planted in Taitung, Taiwan, lacks an electron, and this natural sensitizer may easily capture the electron from an electrolyte as it is applied in a DSSC. Therefore, research on the probable factors affecting anthocyanin stability during extraction (or soaking) and on enhancing the performance of a DSSC sensitized using anthocyanin from local Roselle are worthy of continued study. These are the motivations of this study.

The TRA was extracted from Hibiscus Sabdariffa Linnaeus as a natural sensitizer of the DSSC. The effects of the concentration of TRA in the natural sensitizer, the pH of the natural sensitizer solution, the soaking temperature, the soaking period, and the addition of the single-walled carbon nanotube (SWCNT) on the performance of the DSSC were investigated.

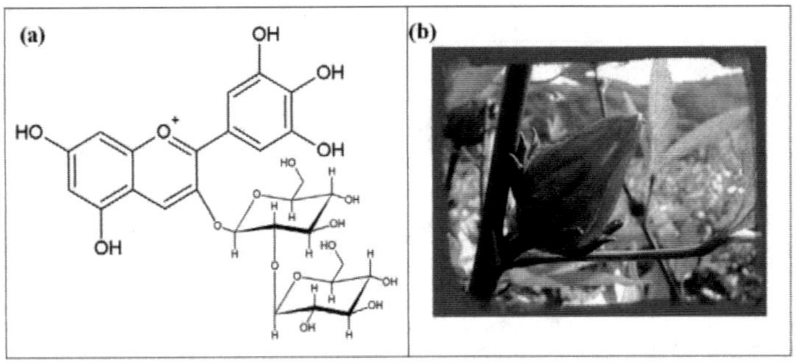

Figure 1. Schematic of the molecular structure of the TRA (a), the image of Hibiscus Sabdariffa Linnaeus (b).

2. Experimental

2.1. Preparing and measuring the natural sensitizer

The experiments of the natural sensitizer included: (1) preparing the powder of TRA (P_{RA}) and (2) preparing and measuring the solution of P_{RA} with different concentrations and pH values. The procedure of preparing the P_{RA} included the following steps: (1) the petals of Hibiscus Sabdariffa Linnaeus (Roselle) of Malvaceae, which had been planted in Taitung, Taiwan, were collected and cleaned using tap water; (2) these cleaned Roselle petals were placed in a freeze dryer (CT-5000D, Panchum Scientific Corp. Kaoshiung, Taiwan), and they were then ground into powder (P_{RP}); (3) 1 g P_{RP} was mixed with 50 ml de-ionized (DI) water in a beaker, and this mixture was stirred until it was uniformly mixed; (4) this uniformly mixed solution was placed in a fridge at a temperature of 5 ℃ for 24 h, so that the TRA could be fully extracted; (5) the TRA extract was separated from the other contaminants in the solution using a high-speed micro centrifuge (CF15RXII, Hitachi Koki Co., Ltd., Japan), and this extract was then placed in a freeze dryer to obtain P_{RA}.

The procedures of preparing and measuring the solution of P_{RA} included the following steps: (1) the P_{RA} with a preset mass was mixed with 200 ml DI water in a beaker, and this mixture was stirred until the P_{RA} was fully dissolved in the DI water; (2) a filter paper was used to filtrate this solution of P_{RA} to obtain the natural sensitizer for a DSSC; (3) the pH of the natural sensitizer could be changed by adding the hydrochloric acid (HCl), and its pH value (Table 1) and absorbance of light were obtained using a pH meter (MP220, Mettler toledo, Switzerland) and a spectrophotometer (U-2001, Hitachi, Tokyo, Japan), respectively. Table 1 lists the test conditions of preparing the natural sensitizers.

2.2. Preparing TiO_2 (P-25) and TiO_2 (P-25)+SWCNT working electrodes

Titanium dioxide (TiO_2, Uniregion Biotech P-25) particles of 20% rutile and 80% anatase with an average particle size of 21 nm were used in this study. Furthermore, a SWCNT with an inside diameter of 1.1 nm, a length of 0.5-100 μm , a density of 1.7-1.9 g/ml , and a resistance of 0.34×10^{-4} $\sim 10 \times 10^{-4}$ ohm/cm was procured from Sigma-Aldrich, USA, and it consisted of carbon nanotubes (> 90 VOL%), amorphous carbon (< 5%), Co (0.6%), Mg (1.2%), Mo (0.1%), and silicate (0.1%).

The procedure of preparing TiO_2 (P-25) and [TiO_2 (P-25)+SWCNT] working electrodes included the following steps: (1) a FTO (Fluorine doped tin oxide, SnO_2:F) glass substrate was immersed in the 0.05 M solution of $TiCl_4$ at a temperature of 70 for 30 min; (2) the colloids of the TiO_2 (P-25) and TiO_2 (P-25)+SWCNT were respectively prepared (Table 2), and these colloids were then homogenized in an ultrasonic homogenizer for 30 min; (3) the colloid of the TiO_2 (P-25) or [TiO_2 (P-25)+SWCNT] was deposited on the top of a FTO-glass substrate with the $TiCl_4$ treatment; (4) this substrate was then sintered at a temperature of 450 for 1 h (Table 3) in a high-temperature furnace (Thermolyne, 46100); (5) the surface of the TiO_2 (P-25) or [TiO_2 (P-25)+SWCNT] film on the FTO-glass substrate was modified by immersing this substrate in a 0.05 M solution of $TiCl_4$ at a temperature of 70 for 30 min; (6) the FTO-glass substrate with a TiO_2 (P-25) or [TiO_2 (P-25)+SWCNT] film was immersed into the solution of P_{RA} at a preset soaking temperature for a preset period; and (7) this substrate was then rinsed using ethanol. The area of the dye-covered TiO_2 (P-25) or [TiO_2 (P-25)+SWCNT] electrode of DSSC was 0.25 cm^2. Table 4 lists the test conditions for soaking TiO_2 (P-25) and [TiO_2 (P-25)+SWCNT] electrodes in the natural sensitizer of TRA.

2.3. Assembling and testing the DSSC

A dye-covered working electrode with a layer of TiO_2 (P-25) or [TiO_2 (P-25)+SWCNT] and a counter electrode with a platinum (Pt) film were assembled to become an open cell. A liquid electrolyte was injected into the space between the two electrodes; the liquid electrolyte used herein was identical to that used in our previous work [3]. An AM 1.5 solar simulator (XES-310S, San-Ei Pioneer of Light Technology) was used to spotlight the DSSC, and a digital source-meter (Keithley 2400) was used to obtain the open-circuit photovoltage and the short-circuit photocurrent of the DSSC. Detailed calculations of the power conversion efficiency of the DSSC were described in our previous works [22-23].

Table 1. Test conditions of preparing the natural sensitizers.

	Material	Concentration (mg/100ml)	Solution	pH Value
A1	Taiwan Roselle anthocyanin	8.52	200 mL DI water	2.33
A2		17.53		
A3		23.98		
A4		32.24		
A5		43.45		
A6		8.52		2.00
A7		32.24		
A8		17.53		2.00
A9				2.20
A10		32.24		2.00
A11				2.20

Table 2. Test conditions of preparing the colloids.

	Solute		Solution		
	Particle	Mass (g)	Ethanol (mL)	Acetylacetone (mL)	Triton X-100 (mL)
B0	TiO_2	1	4	0.4	0.05
B1	TiO_2 + SWCNT	1 + 0.002			

Table 3. Test conditions of preparing the TiO_2 and TiO_2+ SWCNT electrodes.

	Colloid	No. of layer daubed by spin coating	Sintering	
			Temp. (℃)	Time (h)
C0	B0	5	450	1
C1	B1			

Table 4. V_{OC}, J_{SC}, FF, η , and standard deviation of efficiency

	Substrate	Dye	Soaking Temp. (°C) /Time (h)	*Voc* (V)	*Jsc* ($mAcm^{-2}$)	*FF* (%)	η (%)	Standard Deviation of Efficiency (%)
D1	C0	A1	30/10	0.43	0.40	53	0.09	0.0083
D2		A2		0.41	0.53	51	0.11	0.0071
D3		A3		0.42	0.53	58	0.13	0.0122
D4		A4		0.40	0.52	68	0.14	0.0083
D5		A5		0.41	0.58	65	0.15	0.0109
D6		A6	30/10	0.43	0.79	55	0.19	0.0083
D7			50/10	0.41	0.72	62	0.19	0.0071
D8			70/10	0.41	0.35	52	0.08	0.0083
D9		A7	30/10	0.44	0.98	51	0.22	0.0071
D10			50/10	0.41	0.90	51	0.19	0.0043
D11			70/10	0.40	0.15	53	0.03	0.0050
D12	C1	A8	50/10	0.33	3.55	33	0.39	0.0109
D13		A8	30/20	0.31	4.88	39	0.59	0.0192
D14		A9	50/20	0.31	2.91	30	0.27	0.0112
D15		A8	30/10	0.33	4.52	44	0.67	0.0071
D16		A9		0.33	3.38	48	0.54	0.0109
D17		A4		0.42	0.96	53	0.22	0.0071

3. Results and Discussion

3.1. UV–visible absorbance of the natural sensitizer

Figure 2 shows the variations in the absorbance of the natural sensitizer of TRA with the wavelength of light in different concentrations of 8.52 (Test A1), 17.53 (Test A2), 23.98 (Test A3), 32.24 (Test A4), and 43.45 (Test A5) mg/100ml. Further, the pH value of these natural sensitizers is 2.33, which is the original pH of the TRA used in this study. The concentration of TRA influences the light absorbance of the natural sensitizer. For example, in the wavelength range of 300–700 nm, the absorbance in Test A5 was the largest among the five tests. This can be attributed to the fact that the natural sensitizer with a higher concentration of TRA has a better capability of absorbing visible light.

In the absorbance spectra, the absorption peaks of TRA were at 320 and 520 nm (Figure 2). Chang and Lo showed that the absorption peaks of N719 dye and anthocyanin natural dye extracted from mulberry were at 518 and 543 nm, respectively [15]. Therefore, the absorption peak of TRA in the region of visible light obtained in this study is closer to that of the N719 dye, compared with the absorption peak of anthocyanin extracted from mulberry presented in [15]. Wongcharee et al. observed that in the region of visible light, the absorption peak of rosella flower extract is about 520 nm while those of blue pea flower extract are about 580 and 620 nm [20]. This difference is due to the fact that the anthocyanin obtained from blue pea flower is the ternatin structure, and the anthocyanin obtained from rosella flower is the complex of delphinidin and cyaniding [20].

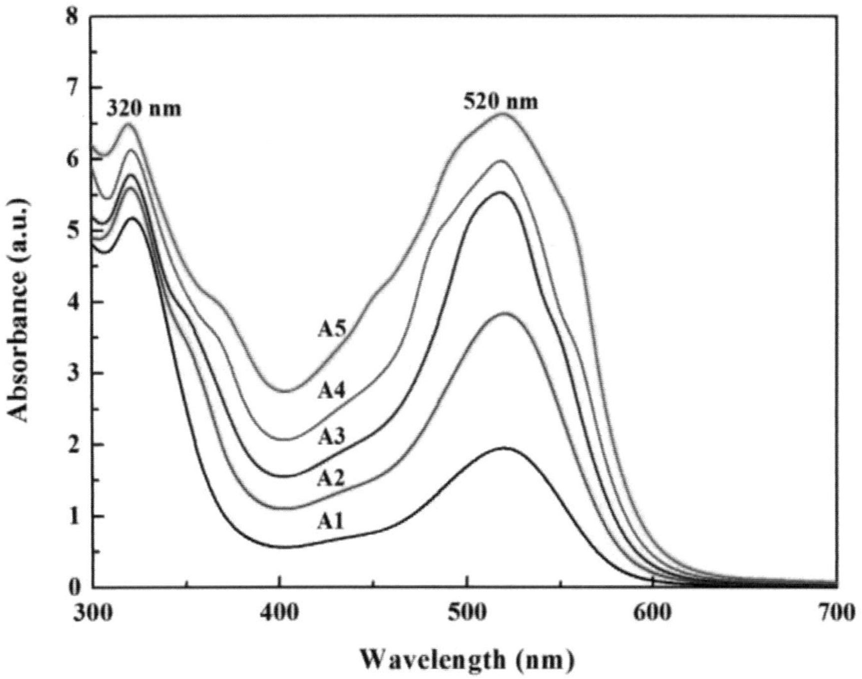

Figure 2. Variations in absorbance of TRA with the wave-length of light in Tests A1 to A5.

3.2. UV–visible absorbance of dye-covered working electrodes

Figure 3(a) shows the variations in absorbance of the TiO_2+SWCNT electrodes of DSSCs in Tests D15 and D16 with the wavelength of light. The pH values of the natural sensitizer used to soak the working electrodes of DSSCs in Tests D15 and D16 were 2.0 and 2.2, respectively. These two working electrodes were soaked in the natural sensitizer with a concentration of 17.53 mg/100ml at 30 ℃ for 10 h. In the wavelength range of 400 - 800 nm, the absorbance of the working electrode of DSSC in Test D15 exceeded that of the working electrode of DSSC in Test D16. This can be attributed to the fact that the Roselle anthocyanin solution with a lower pH is more favourable for forming the pigment cations [24], and this may stabilize the natural sensitizer, in which more red color is observed.

Figure 3(b) shows variations in absorbance of the dye-covered TiO_2+SWCNT electrode of DSSCs in Tests D12 and D15 with the wavelength of light. In Test D15, its working electrode was soaked in the nat-ural sensitizer at 30℃, and the working electrode of DSSC in Test D12 was soaked in the natural sensitizer at 50℃. These two working electrodes were soaked in the natural sensitizer with a concentration of 17.53 mg/100ml for 10 h, and the pH value of this natural sensitizer was 2.00. In the wavelength range of 400 – 800 nm, the absorbance of the working electrode with a soaking temperature of 30℃ significantly exceeded that of the working electrode with a soaking temperature of 50℃. This can be attributed to the fact that the higher soaking temperature may cause the structural modification and color change of the natural sensitizer due to a rapid reactivity of the aglycone moiety in the Roselle anthocyanin [24]. Further, increasing the possibility of losing pigment cations due to raising the soaking temperature may also reduce the ability of the natural sensitizer in capturing electrons from an electrolyte.

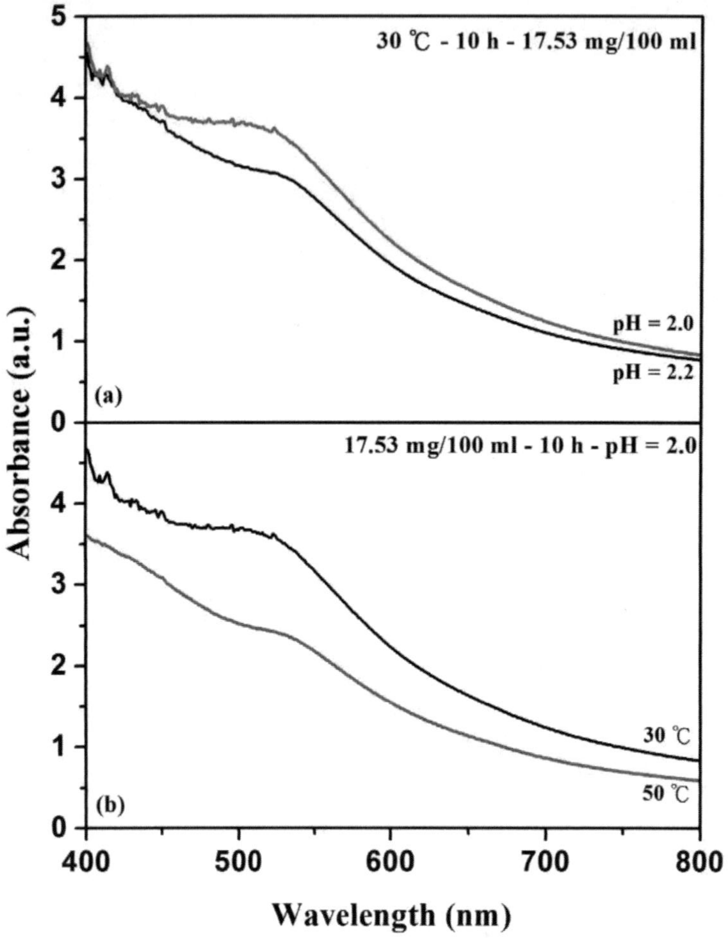

Figure 3(a)-(b). Variations in absorbance of the dye-covered working electrodes [TiO_2 (P-25)+SWCNT] of DSSCs with the wavelength of light. (a) Working electrodes of DSSCs in Tests D15 and D16; (b) working electrodes of DSSCs in Tests D12 and D15.

3.3. Voc, Jsc, η, and J-V Curve of DSSC

Table 4 shows the open-circuit photovoltage (Voc), the short-circuit photocurrent per unit area (Jsc), the fill factor (FF), the power conversion efficiency (η), and the standard deviation of efficiency of the DSSC in Tests D1 to D17. Further, Figure 4 shows the variations in photocurrent density with photovoltage (i.e., J-V characteristics of DSSC) in Tests 12, 15, 16, and 17.

3.3.1 DSSC with a dye-covered electrode of TiO_2 (P-25)

The η of DSSC increased with an increase in the concentration of Roselle anthocyanin in the natural sensitizer. For example, at a fixed pH of the extract (2.33), as the concentration of TRA increased from 8.52 to 43.45 mg/100ml, the η of DSSC, whose TiO_2 (P-25) electrode had been soaked in the natural sensitizer for 10 h at 30℃, increased from 0.09% (Test D1) to 0.15% (Test D5). However, at a fixed concentration of 8.52 mg/100ml, a fixed soaking temperature of 30℃, and a fixed soaking period of 10 h, as the pH of the extract decreased to 2.00 from the original pH of the TRA (2.33), the η of DSSC substantially increased from 0.09% (Test D1) to 0.19% (Test D6), and the Jsc of the DSSC increased from 0.40 to 0.79 $mAcm^{-2}$. Apparently, a lower pH of the extract corresponds to a favorable condition to create more pigment cations [24], which may capture more electrons and enhance both Jsc and η of a DSSC.

Aside from this, increasing the soaking temperature may be unfavorable to the performance of DSSC. For example, at a fixed concentration of 8.52 mg/100ml, a fixed pH of the extract (2.00), and a fixed soaking period of 10 h, as the soaking temperature increased from 30 to 70℃, the η of DSSC substantially decreased from 0.19% (Test D6) to 0.08% (Test D8), and the Jsc of the DSSC decreased from 0.79 to 0.35 $mAcm^{-2}$. Therefore, at a fixed concentration of TRA and a fixed soaking period, either a lower pH of the extract or a lower soaking temperature can stabilize the natural sensitizer and then enhance the Jsc of the DSSC. Wongcharee et al. observed that the efficiency and stability of DSSCs can be enhanced by adjusting the extracting temperature and pH of the extracts [20].

3.3.2 DSSC with a dye-covered electrode of TiO_2 (P-25) +SWCNT

With the original pH of the TRA (2.33), a fixed soaking temperature of 30℃, a fixed soaking period of 10 h, and a fixed concentration of 32.24 mg/100ml, the η of DSSC substantially increased from 0.14% in Test D4 [DSSC with a TiO_2 (P-25) electrode] to 0.22% in Test D17 [DSSC with a TiO_2 (P-25)+SWCNT electrode], and the Jsc of the DSSC increased from 0.52 $mAcm^{-2}$ (Test D4) to 0.96 $mAcm^{-2}$ (Test D17). This result is due to the addition of SWCNT, whose good electric conductivity may enhance the Jsc of the DSSC. Zeng et al. observed that the carbon nano tubes (CNTs) provided a good electrical connection between the ZnO overlayer and the stainless steel substrate [25].

Interestingly, decreasing the pH of the TRA can significantly enhance the η of DSSC with a TiO_2 (P-25)+SWCNT electrode. For example, at a fixed soaking temperature of 30℃ and a fixed soaking period of 10 h, although the concentration of natural sensitizer in Test D15 (17.53 mg/100ml) was smaller than that in Test D17 (32.24 mg/100ml), the η of DSSC substantially increased from 0.22% in Test D17 with the original pH of the TRA (2.33) to 0.67% in Test D15 with a pH of the natural sensitizer of 2.00. Moreover, the Jsc of the DSSC increased from 0.96 $mAcm^{-2}$ (Test D17) to 4.52 $mAcm^{-2}$ (Test D15). This can be attributed to the fact that the effect of pH might dominate over that of the concentration of TRA in the natural sensitizer. Aside from this, a natural sensitizer prepared under either a lower pH of the extract or a lower soaking temperature corresponds to a higher stability, which is favourable for creating a larger Jsc as the DSSC sensitized by the natural sensitizer is exposed to light.

At a fixed concentration of 17.53 mg/100ml, if the soaking process is under a lower pH of the TRA (such as 2.00) and a lower soaking temperature (such as 30), the η of DSSC slightly decreased from 0.67% (Test D15) to 0.59% (Test D13) as the soaking period increased from 10 to 20 h. However, if the soaking process is under a higher pH of the TRA (such as 2.20) and a higher soaking temperature (such as 50℃), a longer soaking period causes a rapid decrease in the Jsc and η of the DSSC. For example, in Test D14 the Jsc and η of the DSSC substantially decreased to 2.91 $mAcm^{-2}$ and 0.27%, respectively. Markakis indicated that the Roselle anthocyanin is the delphinin-sambubiose, which is a glycosylated polyhydroxy derivative of the flavylium salts, and the high reactivity of the aglycone moiety leads to the structural modification and the color change in the Roselle anthocyanin especially under an improper pH during a heat processing [24].

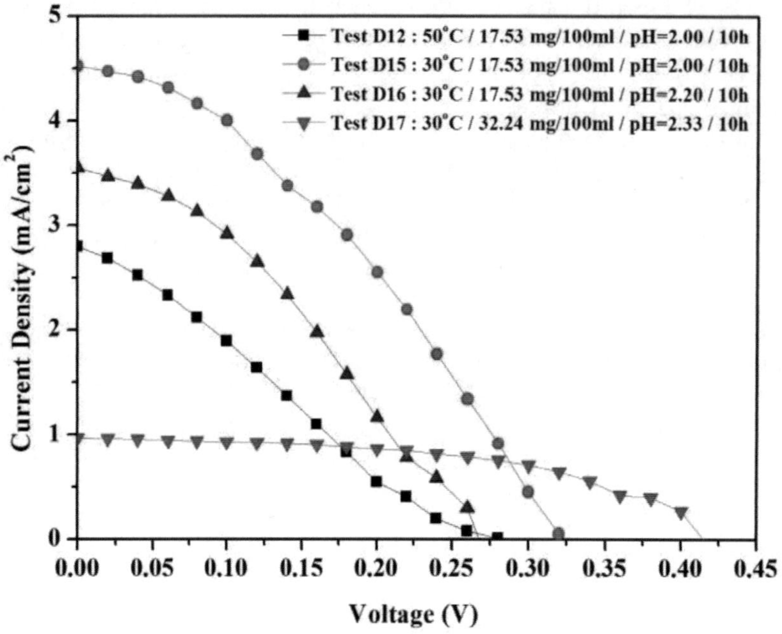

Figure 4. Variations in current density with voltage in Tests 12, 15, 16, and 17.

4. Conclusion

The TRA extracted from Hibiscus Sabdariffa Linnaeus planted in Taitung, Taiwan, was used as a natural sensitizer for sensitizing a DSSC. An absorption peak of TRA (520 nm) is close to that of the N719 dye (518 nm). At a fixed concentration of TRA and a fixed soaking period, the natural sensitizer of TRA prepared under either a lower pH of the extract or a lower soaking temperature corresponds to a higher stability of natural sensitizer, and it is favourable for creating a larger Jsc as the DSSC sensitized by the natural sensitizer is exposed to light. But, as the soaking process is under a higher pH of the TRA and a higher soaking temperature, a longer soaking period causes a rapid decrease in the Jsc and η of the DSSC. Importantly, the addition of SWCNT dramatically increases the η in a DSSC sensitized using TRA with a lower pH. Therefore, the optimum for preparing a DSSC with a TiO_2+SWCNT electrode, which is sensitized using a natural sensitizer, is worthy of continued study.

Acknowledgement

The authors would like to thank the National Science Council, R.O.C., Taiwan for financially supporting this research under Contract No. NSC 102-2221-E-020-025. The energy dispersive spectrometer (EDS) (Horiba EX-200) was performed in the Precision Instrument Center of National Pingtung University of Science and Technology.

References

[1] O'Regan B and Grätzel M 1991 *Nature* **353** 737-39.

[2] Thavasi V, Renugopalakrishnan V, Jose R and Ramakrishna S 2009 *Mater. Sci. Eng. R* **63** 81-99.
[3] Chou C S, Guo M G, Liu K H and Chen Y S 2012 *Appl. Energ.* **92** 224-33.
[4] Chen J G, Chen C Y, Wu S J, Li J Y, Wu C G and Ho K C 2008 *Sol. Energy Mater. Sol. Cells* **92** 1723-7.
[5] Ito S, Miura H, Uchida S, Takata M, Sumioka K, Liska P, Comte P, Pechy P and Grätzel M 2008 *Chem. Commun.* **41** 5194-6.
[6] Xu M, Li R, Pootrakulchote N, Shi D, Guo J, Yi Z, Zakeeruddin S M, Grätzel M and Wang P 2008 *J. Phys. Chem. C* **112** 19770-6.
[7] Yang C H, Chen H L, Chuang Y Y, Wu C G, Chen C P, Liao S H and Wang T L 2009 *J. Power Sources* **188** 627-34.
[8] Pei J, Peng S, Shi J, Liang Y, Tao Z, Liang J and Chen J 2009 *J. Power Sources* **187** 620-6.
[9] Kandavelu V, Huang H S, Jian J L, Yang T C K, Wang K L and Huang S T 2009 *Sol. Energy* **83** 574-81.
[10] Lee C W, Lu H P, Lan C M and et al. 2009 *Chem. Eur. J.* **15** 1403-12.
[11] Wang X F, Kitao O, Zhou H, Tamiaki H and Sasaki S I 2009 *J. Phys. Chem. C* **113** 7954-61.
[12] Polo A S and Iha N Y M *Sol. Energy Mater. Sol. Cells* **90** 1936-44.
[13] Hao S, Wu J, Huang Y and Lin J 2006 *Sol. Energy* **80** 209-14.
[14] Raturi A and Fepuleai Y 2010 *Renew. Energy* **35** 1010-3.
[15] Chang H. and Lo Y J 2010 *Sol. Energy* **84** 1833-7.
[16] Senthil T S, Muthukumarasamy N, Velauthapillai D, Agilan S, Thambidurai M and Balasundaraprabhu R 2011 *Renew. Energy* **36** 2484-8.
[17] Kumara G R A, Kaneko S, Okuya M, Onwona-Agyeman B, Konno A and Tennakone K 2006 *Sol. Energy Mater. Sol. Cells* **90** 1220-6.
[18] Sandquist C and McHale J L 2011 *J. Photochem. Photobiol. A* **221** 90-7.
[19] Zhou H, Wu L, Gao Y and Ma T 2011 *J. Photochem. Photobiol. A* **219** 188-94.
[20] Wongcharee K, Meeyoo V and Chavadej S 2007 *Sol. Energy Mater. Sol. Cells* **91** 566-71.
[21] Sato K, Goda Y, Yoshihira K and Noguchi H 1991 *J. Food Hyg. Soc. Jpn* **32** 301-7.
[22] Chou C S, Yang R Y, Yeh C K and Lin Y J 2009 *Powder Technol.* **194** 95-105.
[23] Chou C S, Lin Y J, Yang R Y and Liu K H 2011 *Adv. Powder Technol.* **22** 31-42.
[24] Markakis P 1982 An-thocyanins as food colors **6** 163-78.
[25] Zeng G Y, Nian K S and Lee K Y 2010 *Diam. Relat. Mater.* **19** 1457-60.

SOLUTION OF ONE PHASE INVERSE STEFAN PROBLEM BY IEF METHOD

M M Sarsengeldin[1,2], S Guvercin[3], S Kassabek[4]
[1]Department of mathematics and natural sciences, Suleyman Demirel University, Abylaikhan street 1/1,Kaskelen, Almaty, Kazakhstan, 040900
[2]National Academy of Science of Republic of Kazakhstan, institute of Mathematics and Mathematical Modelling, 125 Pushkina street, Almaty, Kazakhstan, 050010, corresponding author
[3]Department of mathematics and natural sciences, Suleyman Demirel University, Abylaikhan street 1/1,Kaskelen, Almaty, Kazakhstan, 040900
[4]Department of mathematics and natural sciences, Suleyman Demirel University, Abylaikhan street 1/1,Kaskelen, Almaty, Kazakhstan, 040900

E-mail: sarsengeldin.merey@sdu.edu.kz

Abstract:
A solution for Inverse Stefan Problem is represented in explicit analytic form. The developed method is based on use of Integral Error Functions and its properties which enables it to solve heat and mass transfer problems with domains that include moving boundaries. The method can be effectively used in the field of engineering, which requires consideration of phenomena with phase transformations, such as heat and mass transfer, low temperature plasma, filtration. The main idea of this method is to find coefficients of linear combination of Integral Error Functions which a priori satisfy the heat equation.

1.Introduction:

A wide range of transient phenomena in the fields like heat and mass transfer, low-temperature plasma, filtration and other evolutionary processes associated with phase transformations of substance, lead to solve Stefan type problems [1], [2], [3] which describe above phenomena with unknown free interphase boundaries or with known moving boundaries but unknown heat flux function which also has to be found (inverse Stefan Problem). Extensive bibliography [4] of existing methods for solving boundary value problems of heat and mass transfer in domains with free boundaries based on their reduction to the systems of integral equations [5], originally accepted non degenerate at the initial time [6],[7]. In the case of degeneracy of the integral equations, singularity occurs and development of new methods for the solution required.

Tracking answers of these questions will be as following. In the first section solution of Heat Equation by Integral Error Functions and properties that are necessary for elaborating different methods (e.g. elaborated polynomial form) are represented. In the second section one phase Inverse Stefan problem stated. In the third section by the help of properties of Integral Error Functions solution represented in the form of power series where coefficients and heat flux function P(t) has to be found. Fourth section devoted to discussion and conclusion.

1.1 Integral Error Functions

Heat equations are solved by the help of so called IEF method (Integral Error Functions or Hartree functions method) and properties of Integral Error Functions [8],[9] which were introduced by Hartree in 1935 and reasonably sometimes called Hartree functions.

The integral error functions determined by recurrent formulas

$$i^n erfcx = \int_x^\infty i^{n-1} erfcv dv, \quad n=1,2,\ldots \qquad i^0 erfcx \equiv erfcx = \frac{2}{\sqrt{\pi}} \int_x^\infty \exp(-v^2) dv \tag{1}$$

Where $$erfx = 1 - erfcx = \frac{2}{\sqrt{\pi}} \int_0^x \exp(-v^2) dv \tag{2}$$

One can obtain from

$$i^n erfcx = \frac{2}{\sqrt{\pi}} \frac{1}{n!} \int_x^\infty (v-x)^n \exp(-v^2) dv \tag{3}$$

Expressions (1) satisfy the differential equation

$$\frac{d^2}{dx^2} i^n erfcx + 2x \frac{d}{dx} i^n erfcx - 2n i^n erfcx = 0 \tag{4}$$

and recurrent formulas

$$2n i^n erfcx = i^{n-2} erfcx - 2x i^{n-1} erfcx \tag{5}$$

Integral Error Functions are very useful for investigation of heat transfer, diffusion and other phenomena which can be described by the equation

$$\frac{\partial u}{\partial t} = a^2 \frac{\partial^2 u}{\partial x^2} \tag{6}$$

in a region $D(t > 0, 0 < x < \alpha(t))$ with free boundary $x = \alpha(t)$, since the functions

$$u_n(\pm x, t) = t^{\frac{n}{2}} i^n erfc \frac{\pm x}{2a\sqrt{t}}$$

suffice the equation (6) as well as their linear combination or even series

$$u(x,t) = \sum_{n=0}^\infty [A_n u_n(x,t) + B_n u_n(-x,t)]$$

For any constants A_n, B_n.We can choose these constants to satisfy the boundary conditions at $x = 0$ and $x = \alpha(t)$, if given boundary functions can be expanded into Taylor series with powers t or $\sqrt{t}$.

1.2 Properties of Integral Error Functions

It is possible to derive properties of Integral Error Functions.

1. If n is an integer, then

$$i^n erfc(-x) + (-1)^n i^n erfcx = \frac{1}{2^{n-1} n! i^n} H_n(ix) = \frac{1}{2^{n-1} n!} e^{-x^2} \frac{d^n}{dx^n} e^{x^2} \text{ with } i = \sqrt{-1}$$ and Hermite polynomials $H_n(x)$ in the right side. Indeed, using formula (1) one can write

$$i^n erfc(-x)+(-1)^n i^n erfcx=\frac{2}{\sqrt{\pi}}\frac{1}{n!}\int_{-x}^{\infty}(v+x)^n\exp(-v^2)dv+$$

$$\frac{(-1)^n 2}{n!\sqrt{\pi}}\int_{x}^{\infty}(v-x)^n\exp(-v^2)dv=\frac{2}{n!\sqrt{\pi}}\int_{-\infty}^{\infty}(v+x)^n\exp(-v^2)dv=\frac{1}{2^{n-1}n!i^n}H_n(ix) \quad (7)$$

2. Using formula for Hermite polynomials one can derive

$$i^n erfc(-x)+(-1)^n i^n erfcx=\sum_{m=0}^{\left[\frac{n}{2}\right]}\frac{x^{n-2m}}{2^{2m-1}m!(n-2m)!} \quad (8)$$

If $n=2k$, then

$$i^{2k}erfc\ x+i^{2k}erfc\ (-x)=\sum_{m=0}^{k}\frac{x^{2(k-m)}}{2^{2m-1}m!(2k-2m)!}$$

In particular

$$erfc\ x+erfc\ (-x)=2,$$

$$i^2erfc\ x+i^2erfc\ (-x)=\frac{1}{2}+x^2\ ,$$

$$i^4erfc\ x+i^4erfc\ (-x)=\frac{1}{8}+\frac{1}{4}x^2+\frac{1}{12}x^4.$$

If $n=2k+1$, then

$$i^{2k+1}erfc(-x)-i^{2k+1}erfc\ x=\sum_{m=0}^{k}\frac{x^{2(k-m)+1}}{2^{2m-1}m!(2k-2m+1)!}$$

(9)

In particular

$$ierfc(-x)-ierfc\ x=2x\ ,$$

$$i^3erfc(-x)-i^3erfc\ x=\frac{1}{2}x+\frac{1}{3}x^3\ ,$$

$$i^5erfc\ (-x)-i^5erfc\ x=\frac{1}{2^3\cdot 2!}x+\frac{1}{2\cdot 2!\cdot 3!}x^3+\frac{2}{5!}x^5\ .$$

The proof of the formula

$$i^n erfc(-x)-(-1)^n i^n erfcx=\frac{1}{2^{n-1}n!}e^{-x^2}\frac{d^n}{dx^n}(e^{x^2}erfx) \quad (10)$$

where

$$erfx=1-erfcx=\frac{2}{\sqrt{\pi}}\int_0^x\exp(-v^2)dv$$

can be obtained by mathematical induction method using recurrent formula (5).

3. Differentiating the right side of formula (8), we obtain

$$i^n erfc(-x)-(-1)^n i^n erfcx=P_n(x)erfx-Q_n(x)\frac{2}{\sqrt{\pi}}\exp(-x^2), \quad (11)$$

where polynomials $P_n(x)$ and $Q_n(x)$ are defined by formulas

$$P_n(x) = \sum_{m=0}^{\left[\frac{n}{2}\right]} \frac{x^{n-2m}}{2^{2m-1} m!(n-2m)!}, \qquad Q_n(x) = \sum_{k=0}^{n-1} \frac{(-1)^{n-k} H_{n-k-1}(x)}{2^{n-k}(n-k)!} P_k(x)$$

4. From (8), (9) we can obtain the explicit expressions for Integral Error Functions of an integer index

$$i^n erfcx = \frac{(-1)^n}{2}[P_n(x) erfcx + Q_n(x)\frac{2}{\sqrt{\pi}}\exp(-x^2)] \tag{12}$$

$$i^n erfc(-x) = \frac{1}{2}[P_n(x) erfc(-x) - Q_n(x)\frac{2}{\sqrt{\pi}}\exp(-x^2)] \tag{13}$$

5. Using L'Hopital rule and representation (1), it is not difficult to show that

$$\lim_{x\to\infty} \frac{i^n erfc(-x)}{x^n} = \frac{2}{n!} \tag{14}$$

6. Using property 2 one can derive following formula

$$u(x,t) = \sum_{n=0}^{k} \left\{ A_{2n} \sum_{m=0}^{n} x^{2n-2m} t^m \beta_{2n,m} + A_{2n+1} \sum_{m=0}^{n} x^{2n-2m+1} t^m \beta_{2n+1,m} \right\}$$

Where u(x,t) is a solution of Heat Equation in polynomial form and $\beta(n,m) = \frac{1}{2^{n+m-1} \cdot m! \cdot (n-2m)!}$

2. Problem Statement

It is required to find the solution of Heat Equation with moving (known) boundary that satisfy to so called Stefan condition. Moreover it is necessary to find function P(t) which is also unknown.

$$\frac{\partial u}{\partial t} = a^2 \frac{\partial^2 u}{\partial x^2}, \qquad 0 < x < \alpha(t), \quad t > 0 \tag{15}$$

I.C: $u(x,0) = 0, \quad 0 < x < \alpha(t)$ (16)

B.C: $(\rho u + \theta \frac{\partial u}{\partial x})|_{x=0} = P(t), \qquad t > 0$ (17)

$u(\alpha(t),t) = \psi(t), \quad t > 0$ (18)

Stefan condition:

$$(\lambda u - \frac{\partial u}{\partial x})|_{x=\alpha(t)} = \alpha^2(t)\frac{d\alpha(t)}{dt} \tag{19}$$

$$u(0,0) = 0 \tag{20}$$

3. Solution method

Solution is represented in the polynomial form of Integral Error Functions:

$$u(x,t) = \sum_{n=0}^{k}\{A_{2n}\sum_{m=0}^{n} x^{2n-2m} t^m \beta_{2n,m} + A_{2n+1}\sum_{m=0}^{n} x^{2n-2m+1} t^m \beta_{2n+1,m}\} \tag{21}$$

Where A_{2k}, A_{2k+1} and function $P(t)$ has to be determined.

Let's say that $P(t)$can be represented in the following form

$$P(t) = p_1 t \quad + p_2 t^2 + p_3 t^3 + \ \dots\dots + p_{k+1} t^{k+1} \tag{22}$$

where$p_1, p_2, p_3, \dots, p_{k+1}$ has to be determined.

From (18)

$$\begin{aligned} u(\alpha(t),t) = {} & A_0 \hat{a}_{0,0} + A_2\left(\alpha^2 \hat{a}_{2,0} + t\beta_{2,1}\right) + \\ & + A_4\left(\alpha^4 \hat{a}_{4,0} + \alpha^2 t \hat{a}_{4,1} + t^2 \hat{a}_{4,2}\right) + \ \dots. + \\ & + A_{2k}\left(\alpha^{2k} \hat{a}_{2k,0} + \alpha^{2k-2} t \hat{a}_{2k,1} + \ \dots + \alpha^2 t^{k-1} \hat{a}_{2k,2k-1} + t^k \hat{a}_{2k,k}\right) + \\ & + A_1 \alpha \beta_{1,0} + A_3\left(\alpha^3 \beta_{3,0} + \alpha t \beta_{3,1}\right) + \\ & + A_5\left(\alpha^5 \beta_{5,0} + \alpha^3 t \beta_{5,1} + \alpha t^2 \beta_{5,2}\right) + \ \dots + \\ & + A_{2k+1}(\alpha^{2k+1} \beta_{2k+1,0} + \alpha^{2k-1} t \beta_{2k+1,1} + \ \dots + \alpha^3 t^{k-1} \beta_{2k+1,k-1} + \\ & + \alpha t^k \beta_{2k+1,k}) = \psi(t) \end{aligned} \tag{23}$$

Let $\alpha(0) = \alpha'(0) = \alpha''(0) = \alpha'''(0) = \ \dots = \alpha^{(n)}(0) = 0$ (24)

and let $\alpha(t) = \alpha$

It is easy to see that $\frac{du(\alpha(t),t)}{dt}|_{t=0} = A_2 \beta_{2,1} = \psi'(0)$

$$\frac{d^2 u}{dt^2}|_{t=0} = 2A_4 \beta_{4,2} = \psi''(0)$$

$$\frac{d^3 u}{dt^3}|_{t=0} = 3!\, A_6 \beta_{6,3} = \psi'''(0)$$

.

.

.

$$\frac{d^n u}{dt^n}|_{t=0} = n!\, A_{2n} \beta_{2n,n} = \psi^{(n)}(0) \tag{25}$$

Implies $A_{2n} = \frac{\psi^{(n)}(0)}{n! \cdot \beta_{2n,n}} = \frac{\psi^{(n)}(0) \cdot n! \cdot 2^{3n-1}}{n!} = \psi^{(n)}(0) \cdot 2^{3n-1}$ (26)

From (19)

$$(\lambda u - \frac{\partial u}{\partial x})|_{x=\alpha(t)} = \alpha^2(t) \frac{d\alpha}{dt}$$

$$\begin{aligned} & \lambda[A_0 \hat{a}_{0,0} + A_2\left(\alpha^2 \hat{a}_{2,0} + t\beta_{2,1}\right) + \\ & + A_4\left(\alpha^4 \hat{a}_{4,0} + \alpha^2 t \hat{a}_{4,1} + t^2 \hat{a}_{4,2}\right) + \ \dots. + \\ & + A_{2k}\left(\alpha^{2k} \hat{a}_{2k,0} + \alpha^{2k-2} t \hat{a}_{2k,1} + \ \dots + \alpha^2 t^{k-1} \hat{a}_{2k,2k-1} + t^k \hat{a}_{2k,k}\right) + \end{aligned}$$

$$+A_1\alpha\beta_{1,0} + A_3\left(\alpha^3\beta_{3,0} + \alpha t\beta_{3,1}\right) +$$

$$+A_5\left(\alpha^5\beta_{5,0} + \alpha^3 t\beta_{5,1} + \alpha t^2\beta_{5,2}\right) + \ldots +$$

$$+A_{2k+1}(\alpha^{2k+1}\beta_{2k+1,0} + \alpha^{2k-1}t\beta_{2k+1,1} + \ldots + \alpha t^k\beta_{2k+1,k})] -$$

$$-[2A_2\alpha\hat{a}_{2,0} + A_4\left(4\alpha^3\hat{a}_{4,0} + 2\alpha t\hat{a}_{4,2}\right) + A_6\left(6\alpha^5\hat{a}_{6,0} + 4\alpha^3 t\hat{a}_{6,1} + 2\alpha t^2\beta_{6,2}\right) +$$

$$+ \ldots + A_{2k}\left(2k\alpha^{2k-1}\hat{a}_{2k,0} + (2k-1)\alpha^{2k-3}t\hat{a}_{2k,1} + \ldots + 2\alpha t^k\hat{a}_{2k,k}\right) +$$

$$+A_1\beta_{1,0} + A_3\left(3\alpha^2\beta_{3,0} + t\beta_{3,1}\right) + +A_5\left(5\alpha^4\beta_{5,0} + 3\alpha^2 t\beta_{5,1} + t^2\beta_{5,2}\right) + \ldots +$$

$$+A_{2k+1}((2k+1)\alpha^{2k}\beta_{2k+1,0} + (2k-1)\alpha^{2k-2}t\beta_{2k+1,1} +$$

$$+ \ldots + 3\alpha^2 t^{k-1}\beta_{2k+1,k-1} + t^k\beta_{2k+1,k})] = \alpha^2\frac{d\alpha}{dt} \quad (27)$$

Thus

$$(\frac{d}{dt}\left(\lambda u - \frac{\partial u}{\partial x}\right)|_{x=\alpha(t)})_{t=0} = (\alpha^2\frac{d\alpha}{dt})'_{t=0} = 0$$

$$\lambda \cdot A_2\beta_{2,1} - A_3\beta_{3,1} = 0$$

$$\Rightarrow A_3 = \frac{\lambda \cdot A_2\beta_{2,1}}{\beta_{3,1}}$$

Taking second derivative of (27) for $t = 0$, we have

$$2\lambda \cdot A_4\beta_{4,2} - 2A_5\beta_{5,2} = 0$$

$$\Rightarrow A_5 = \frac{\lambda \cdot A_4\beta_{4,2}}{\beta_{5,2}}$$

It's easy to see that

$$A_{2k+1} = \frac{\lambda \cdot A_{2k}\beta_{2k,k}}{\beta_{2k+1,k}} = A_{2k} \cdot \frac{2^{3k} \cdot k!}{2^{3k-1} \cdot k!} = 2\lambda \cdot A_{2k} \quad (28)$$

In the same manner from (17)

$$\rho[A_0\beta_{0,0} + A_2 t\beta_{2,1} + A_4 t^2\,\beta_{4,2} + \ldots\ldots + A_{2k}t^k\,\beta_{2k,k}] +$$

$$+\theta[A_1\beta_{1,0} + A_3 t\ \ \beta_{3,1} + A_5 t^2\,\beta_{5,2} + \ldots\ldots + A_{2k+1}t^k\,\beta_{2k+1,k}] = P(t)$$

Using UC method (method of Undetermined Coefficients) and letting

$$P(t) = p_0 + p_1 t + p_2 t^2 + \ldots\ldots + p_k t^k$$

We can easily determine $p_0, p_1, p_2, p_3, \ldots, p_k$

$$p_0 = \rho A_0\beta_{0,0} + \theta A_1\beta_{1,0}$$

$$p_1 = \rho A_2\beta_{2,1} + \theta A_3\beta_{3,1}$$

$$p_2 = \rho A_4 \beta_{4,2} + \theta A_5 \beta_{5,2}$$

.

.

.

$$p_k = \rho A_{2k} \beta_{2k,k} + \theta A_{2k+1} \beta_{2k+1,k} \quad (29)$$

4. Discussion and conclusion

Main result namely coefficients of function (21) A_{2k}, A_{2k+1} and function $P(t)$ are obtained from (26),(28) and (29) respectively. It worth noting that deviation of solution can provide maximum principle. Moreover if convergence of (21) proved, wide class of heat equations that describe diverse phenomena can be solved both analytically and numerically.

Successful applications of represented method induce the question: is it possible to find and to use similar functions for solution of other equations? It can be shown that Elaborated Degenerate Hypergeometric functions and their linear combinations satisfy the equation

$$\frac{\partial \theta}{\partial t} = a^2 \left(\frac{\partial^2 \theta}{\partial z^2} + \frac{\nu}{z} \frac{\partial \theta}{\partial z} \right)$$

Where $\nu = 1,2,3$ are plane, cylindrical and spherical cases. It can be shown that mentioned elaborated Degenerate Hypergeometric Functions and their linear combinations satisfy above equation for any ν, that is the modeling of Heat transfer in solid with variable cross section.

Acknowledgments

First author would like to thank research supervisor Prof. Dr. S.N. Kharin for his great support, comments and suggestions.

[1] V. Alexiades and A. D. Solomon, *Mathematical Modeling of Melting and Freezing Processes*, Taylor and Francis, Washington, DC, 1993.
[2] M. Grassi, G. Grassi, Mathematical modelling and controlled drug delivery: matrix systems, Current Drug Delivery 2 (2005) 97–116.
[3] Kharin S.N., Sarsengeldin M. Influence of Contact Materials on Phenomena in a Short Electrical Arc. Trans Tech Publications. – Switzerland: Key Engineering Materials. – 2012. - Vol. 510-511. - P. 321-329.
[4] D. A. Tarzia, *A bibliography on moving-free boundary problems for the heat-diffusion equation. The Stefan and related problems*, MAT - Ser. A, 2 (2000), pp. 1–297.
[5] A.N. Tikhonov, A.A. Samarski, Equations of Mathematical Physics. Gostechteorizdat, 1951.
[6] A. Friedman, *Free boundary problems for parabolic equations* I. *Melting of solids*, J. Math. Mech., 8 (1959) , pp. 499–517.
[7] L. I. Rubinstein, *The Stefan Problem*, Transl. Math. Monogr. 27, AMS, Providence, RI, 1971
[8] Sarsengeldin M.M.Solution Of Heat Equation In A Domain With Moving Boundary Obtained By Hartree Functions, International scientific–practical conference // VIII international conference on Electronics and computer technologies, IKECCO-8. - Almaty, 2012. – P. 182-188.
[9] S.N. Kharin, M.M. Sarsengeldin, A.Temirkul, "Analytical Solution Of The Third Boundary-Value Problem For The Heat Equation By IEF Method", "Поиск-Ізденіс", Kazakhstan, Almaty.

Weldability of A7075-T651 and AZ31B dissimilar alloys by MIG welding method based on welding appearances

M Ishak[1, 2] and M R Islam[1]
[1]Faculty of Mechanical Engineering, University Malaysia Pahang
[2]Automotive Engineering Centre, University Malaysia Pahang
Pekan, Kuantan, Pahang, Malaysia

Email: mahadzir@ump.edu.my
rafiqme99buet@gmail.com

Abstract. It is not recommended to weld aluminium and magnesium dissimilar alloys using fusion welding method because of the formation of Al_mMg_n type intermetallic brittle compounds like Mg_2Al_3, $Mg_{17}Al_{12}$ etc. in the welding joint. These brittle compounds deteriorate the mechanical properties of the joint. But so far, insufficient researches have been attempted to stop the formation of Al_mMg_n type intermetallic brittle compounds in fusion welding method. The aim of this research work was to investigate on the weldability between A7075-T651 and AZ31B dissimilar alloys based on welding appearances and study the formation of intermetallic brittle compounds at the joint. In this research, A7075-T651 and AZ31B alloys were welded using ER5356 filler wire in MIG welding method in butt configuration. 100% argon was used as shielding gas. The results showed that, most of the welding appearances were moderate. The macroscopic investigation at all welding cross section showed that a lot of Al_mMg_n intermetallic brittle compounds were formed at the interface between weld seam and AZ31B parent metal side which caused macro cracks. A good number of macro pores were also observed at AZ31B parent metal side. These cracks and pores could easily cause the failure of the joint at very low stress.

1. Introduction

Innovative developments in dissimilar welding technologies for light weight alloys of aluminum (Al) and magnesium (Mg) would achieve weight reduction, enhance quality of components and increase fuel efficiency in automobile, marine, aviation and aerospace industries to reduce the environmental pollution. So, importance of lightweight dissimilar metal welding between Al and Mg alloys is crucial in transportation industries [1, 2]. Liu et al. stated that the welding of Mg alloy to Al alloy would improve flexibility and availability of components substantially [3].

Recently, in many industrial fields, much attention has been focused on Al and Mg alloys because not only for the above reasons but also for their various unique properties. For their realistic applications, welding technologies should be established in addition to consider issues such as alloy design, microstructure control and surface treatment [4].

Al is one of the most important industrial nonferrous metals. Al alloys are extensively used in automotive, aerospace, ship and electronic industries and communication equipment as structural material [5] because of their attractive mechanical and metallurgical properties like high specific strength, light weight, durability, low-cost maintenance [6], recovery potential, good formability, weldability, good corrosion resistance, good thermal and electrical conductivity [7], recyclability [8] etc. Magnesium, the lightest of all structural metals are also attractive, promising and has great potential in these manufacturing fields. Mg, with density two-thirds of Al alloy, has highest specific strength [6, 9], good cast ability, workability and electromagnetic shielding capability. These Al and Mg alloys have been researched extensively for practical industrial applications during the past decade. But it has been found that the process of joining Al and Mg is a very complicated one because of difference in physicochemical properties, occurrence of cracking in the joint [10] and reactions between Al and Mg atoms can yield extremely brittle intermetallic compounds specially in fusion welding process [9, 10]. So, in order to achieve further design flexibility, expanded combined structural applications [9] of the properties of Al and Mg alloys, development of reliable joints is the principal technical challenge. However to develop a reliable joining method, a variety of attempts to weld these alloys using conventional fusion welding technology showed that it is difficult to obtain sound dissimilar welds since coarse grains [1], Mg_2Al_3, $Mg_{17}Al_{12}$ etc. inter-metallic compounds near interface between two metals, which have a strong negative effect on the mechanical properties of the welding joints, are easy to form in these processes [2, 3, 5, 9, 11-15]. Thus, the problem of welding Al and Mg alloys needs to be solved to establish technologies to promote their use in industrial application. Not only in the traditional fusion welding but also even in high energy density laser welding with higher cooling rate[2], friction stir welding and laser-TIG hybrid welding processes, the detrimental intermetallic compounds could be formed easily and the joint performance could be deteriorated significantly. In additional, these methods need expensive equipment and complicated welding procedures.

So, to develop the solution in Al and Mg welding, the conventional fusion welding processes like MIG welding is very important[13]. MIG is a fusion welding method and it is easy, quick, efficient, economic, can produce very good quality welding and can be mechanized. Therefore, the aim of this research work was to investigate on the reliable ranges of welding parameters for a good welding appearance and to study the formation of Al_mMg_n type intermetallic brittle compounds at the welding joint cross section so that further research can be done to stop the formation brittle compound and thus better mechanical properties can be achieved.

2. Experimental

Commercially available 2 mm thick A7075-T651 Al and AZ31B Mg sheets were cut into 50×30 mm dimension. Then A7075-T651 sheets were welded to AZ31B sheets in butt configuration using 1mm diameter ER5356 filler wire by metal inert gas (MIG) welding method. 100% argon was used as shielding gas. The specimens were gripped by jig on an automatic welding table. The variable parameters were welding voltage *V*, welding current *I*, welding speed *WS* and gun angle *GA*. The fixed parameters were tip length (10 mm) and gas Supply rate (25 Ltr. /min).

The net Heat input, *H* can be calculated from the equation 1 given below as the ratio of the total input power, P of the heat source to its travel speed or welding speed, *v*.

$$H = P/v \quad \textbf{(1)}$$

Where, *H* is in watt-seconds or joules per mm or inch, *P* in watts and *v* in mm per second or inches per minute. For an electric arc, *P= VI* where *V* is volts and *I* is current [16].

The design of experiment (DOE) for the variable parameters was done in Box-Behnken design by response surface methodology. The first round of DOE was done with variable parameter ranges: *V* (10-25V), *I* (65-130A), *WS* (2-8 mm/s), *GA* (45-90°) and welding was performed. Based on the welding appearances the revised DOE was done with adjusted variable parameter's ranges: *V* (12-15V), *I* (55-60A), *WS* (4-6 mm/s), *GA* (45-90°) and welding was performed again. The two stages of DOE for variable parameters with ranges, heat inputs and quality of welding appearances are given in Table 1 and Table 2 respectively.

Table 1: First round design of experiment for variable parameters with ranges, heat inputs and quality of welding appearances

Sl.	Welding Voltage, V (10-25 volts)	Welding Current, I (65-130A)	Welding Speed, WS v (2-8 mm/s)	Gun Angle, GA θ (45-90°)	Heat Input, H (Joules/mm)	Welding Appearances
1	10.0	65.0	2	45.0	325.00	Burns
2	25.0	65.0	2	45.0	812.50	Burns
3	10.0	130.0	2	45.0	650.00	Burns
4	25.0	130.0	2	45.0	1625.00	Burns
5	17.5	97.5	5	45.0	341.25	Burns
6	10.0	65.0	8	45.0	81.25	Poor
7	25.0	65.0	8	45.0	203.13	Burns
8	10.0	130.0	8	45.0	162.50	Moderate
9	25.0	130.0	8	45.0	406.25	Burns
10	17.5	97.5	2	67.5	853.13	Burns
11	17.5	65.0	5	67.5	227.50	Burns
12	10.0	97.5	5	67.5	195.00	Moderate
13	17.5	97.5	5	67.5	341.25	Burns
14	25.0	97.5	5	67.5	487.50	Burns
15	17.5	130.0	5	67.5	455.00	Burns
16	17.5	97.5	8	67.5	213.28	Burns
17	10.0	65.0	2	90.0	325.00	Burns
18	25.0	65.0	2	90.0	812.50	Burns
19	10.0	130.0	2	90.0	650.00	Burns
20	25.0	130.0	2	90.0	1625.00	Burns
21	17.5	97.5	5	90.0	341.25	Burns
22	10.0	65.0	8	90.0	81.25	Poor
23	25.0	65.0	8	90.0	203.12	Burns
24	10.0	130.0	8	90.0	162.50	Moderate
25	25.0	130.0	8	90.0	406.25	Burns

Table 2: Revised design of experiment with adjusted ranges of variable parameters, heat inputs and quality of welding appearances

Sl.	Welding Voltage, *V* (12-15 volts)	Welding Current, *I* (55-60A)	Welding Speed, *WS* *v* (4-6 mm/s)	Gun Angle, *θ* (45-90°)	Heat Input, *H* (Joules/mm)	Welding Appearances
1	12.0	55.0	5	67.5	132.00	Good
2	15.0	55.0	5	67.5	165.00	Moderate
3	12.0	60.0	5	67.5	144.00	Moderate
4	15.0	60.0	5	67.5	180.00	Moderate
5	13.5	57.5	4	45.0	194.06	Moderate
6	13.5	57.5	6	45.0	129.38	Burnt
7	13.5	57.5	4	90.0	194.06	Moderate
8	13.5	57.5	6	90.0	129.38	Burnt
9	12.0	57.5	5	45.0	138.00	Moderate
10	15.0	57.5	5	45.0	172.50	Moderate
11	12.0	57.5	5	90.0	138.00	Moderate
12	15.0	57.5	5	90.0	172.50	Good

13	13.5	55.0	4	67.5	185.63	Moderate
14	13.5	60.0	4	67.5	202.50	Moderate
15	13.5	55.0	6	67.5	123.75	Burnt
16	13.5	60.0	6	67.5	135.00	Moderate
17	12.0	57.5	4	67.5	172.50	Moderate
18	15.0	57.5	4	67.5	**215.63**	Moderate
19	12.0	57.5	6	67.5	**115.00**	Good
20	15.0	57.5	6	67.5	143.75	Moderate
21	13.5	55.0	5	45.0	148.50	Moderate
22	13.5	60.0	5	45.0	162.00	Moderate
23	13.5	55.0	5	90.0	148.50	Burnt
24	13.5	60.0	5	90.0	162.00	Burnt
25	13.5	57.5	5	67.5	155.25	Moderate

3. Results and discussions

3.1 Welding Appearances

The welding appearances for first round of DOE are shown in Figure 1 and it was revealed that between heat input 162.50 to 195.00 joules/mm, the welding appearances are moderate. Below this heat input range, the welding quality is poor as shown in Figure 2 and above this heat input range, the burning of the parent metal occurs as shown in Figure 3. The poor welding appearances were due to lack of fusion and burning occurred due to high heat input. So, the parameter ranges were adjusted so that the heat input could be maintained around the range of 162.50 to 195.00 joules/mm. For this purpose, the new ranges of welding variable parameters were set as: *V* (12-15V), *I* (55-60A), *WS* (4-6 mm/s) and *GA* (45-90°). Using the new DOE with adjusted variable parameters the welding was performed again. As shown in Figure 4 (a-y), most of the welding appearances were improved. Most of the welding joints showed moderate appearances. Three types of welding appearances were attained with new DOE for adjusted variable parameters: good, moderate and poor welding appearances. The acceptable quality welding appearances were rated as good; the medium quality welding appearances were rated as moderate. The welding where burning of the parent metals occurred were rated as poor welding appearances.

(a) (b)

Figure 1: Moderate welding appearances (a) for exp. no 8 with heat input of 162.50 joules / mm and (b) for exp. no. 12 with heat input of 195.00 joules / mm

(a) (b)

Figure 2: Poor welding appearances for exp. no 6 and 22 (for same heat input of 81.25 joules / mm)

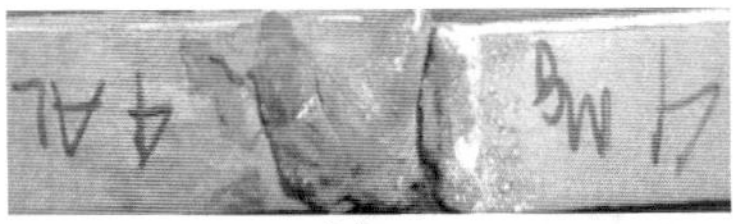

(a)

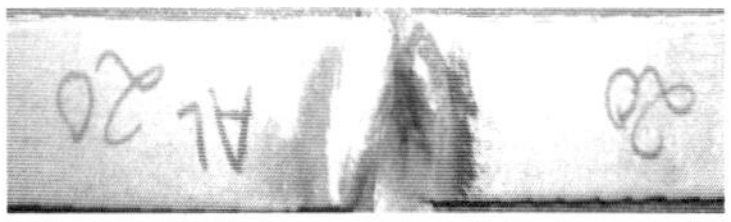

(b)

Figure 3: Burning occurs in exp. no 4 and 20 (for same heat input of 1625.00 joules / mm)

(a: good)

(b: moderate)

(c: moderate)

(d: moderate)

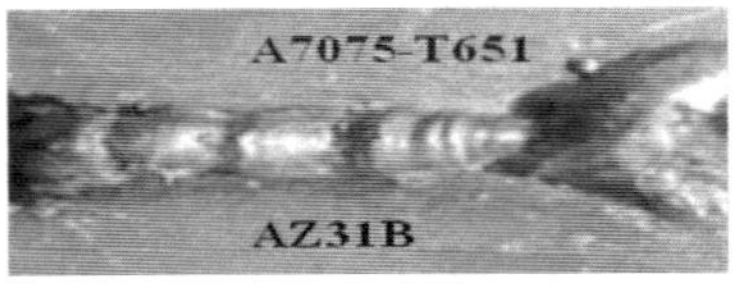

(e: moderate)

(f: burnt)

(g: moderate)

(h: burnt)

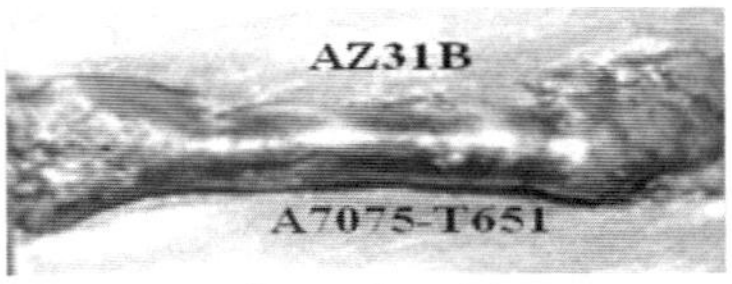

(i: moderate)

(j: moderate)

(k: moderate)

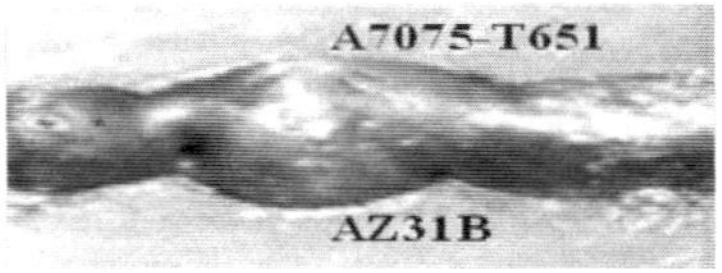

(l: good)

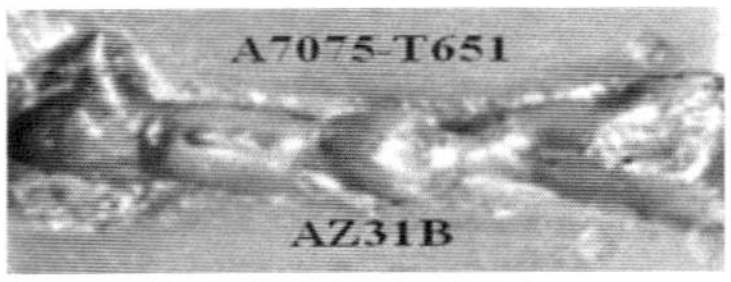

(m: moderate)

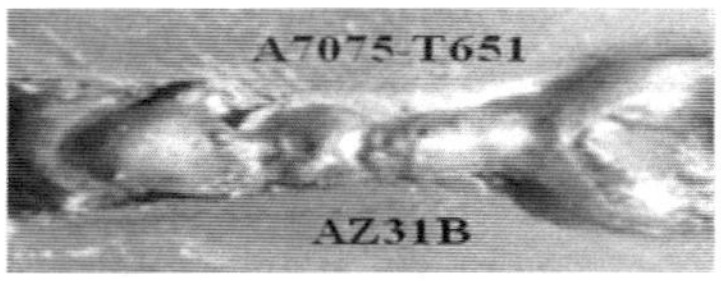

(n: moderate)

(o: burnt)

(p: moderate)

(q: moderate)

(r: moderate)

(s: good)

(t: moderate)

(u: moderate)

(v: moderate)

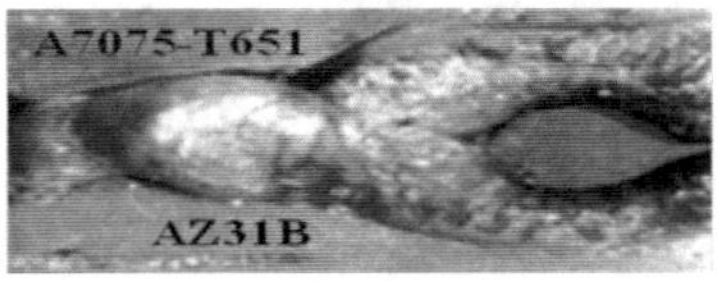

(w: burnt)

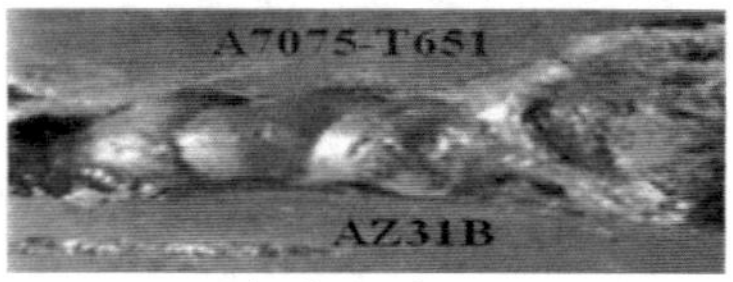

(x: burnt)

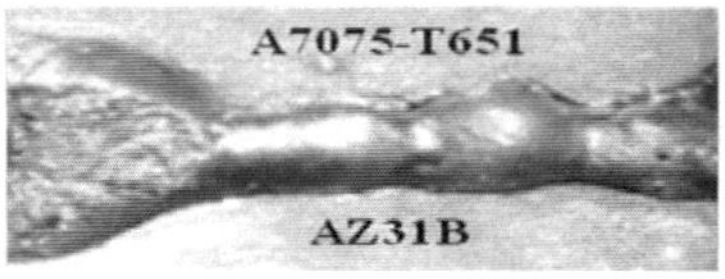

(y: moderate)

Figure 4: Welding appearances for different heat input (for exp. no 1 to exp. no 25)

3.2 Macrostructure at welding cross section

Figure 5(a) shows the optical image at the welding cross section for one of the welding joints. The welding joint contained macro cracks at the interface of AZ31B and welding seam. No cracks were observed at the interface of A7075-T651 and welding seam. The top of welding seam contained a good number of macro pores. All the welding joints depicted the same scenario.

(a) (b)

Figure 5: (a) Optical image of welding cross section showed the presence of macro cracks and macro pores, (b) Macrostructure of welding cross section showed the presence of Al_mMg_n intermetallic brittle compounds and macro pores

Figure 5(b) shows the macrostructure at welding cross section for one of the welding joints. It revealed that, a lot of Al_mMg_n intermetallic brittle compounds were formed at the interface of AZ31B and welding seam. These brittle compounds are the reason for the formation of macro cracks. A good number of macro pores were also observed at AZ31B parent metal side. No Al_mMg_n intermetallic brittle compounds were observed at the interface of A7075-T651 and welding seam. The same thing happened for all the welding joints. Al_mMg_n intermetallic brittle compounds formed because aluminium based ER5356 filler wire was used for welding which reacted with AZ31B parent. This intermetallic brittle compound is the reason for poor mechanical properties of the joint reported by all previous researchers.

4. Conclusion

a. The variable parameters range for good and moderate welding appearances are: *V* (12-15V), *I* (55-60A), *WS* (4-6 mm/s) and *GA* (45-90°).

b. The formation of Al_mMg_n intermetallic brittle compounds were at the interface of AZ31B and welding seam which produced macro cracks. A good number of macro pores were also observed at AZ31B parent metal side and at the top of welding seam. The macro cracks at the interface of AZ31B and welding seam, and macro pores at AZ31B parent metal side would definitely cause poor mechanical properties of the joint which could cause failure at very low stress.

Acknowledgement
The authors gratefully acknowledge the University Malaysia Pahang for supporting this research through the internal grant.

References

[1] Sato Y S, Park S H C, Michiuchi M and Kokawa H 2004 *Scripta. Mater.* **50(9)** 1233-1236
[2] Chang W S, Rajesh S R., Chun C K and Kim H J 2011 *J. Mater. Sci. Technol.* **27(3)** 199-204
[3] Qi X and Liu L 2012 *Mater. Design.* **33** 436-443
[4] Kwon Y J, Shigematsu I and Saito N 2008 *Mater. Lett.* **62(23)** 3827-3829
[5] Mofid M A, Abdollah Z A and Malek F G 2012 *Mater. Design.* **36** 161-167
[6] Hayat F 2011 *Mater. Design.* **32(4)** 2476 - 2484

[7] Toros S, Ozturk F and Kacar I 2008 *J. Mater. Process. Tech.* **207(1-3)** 1-12
[8] Casalino G 2007 *J. Mater. Process. Tech.* **191(1-3)** 106-110
[9] Liu C C, Chen D L, Bhole S, Cao X and Jahazi M 2009 *Mater. Charact.* **60(5)** 370-376
[10] Liu L M, Tan J H, Zhao L M and Liu X J 2008 *Mater. Charact.* **59(4)** 479-483
[11] Yong Y, Da-tong Z, Cheng Q and Wen Z 2010 *T. Nonferr. Metal. Soc.* 619-623
[12] Chen Y C and Nakata K 2008 *Scripta. Mater.* **58(6)** 433-436
[13] Zhang H T and Song J Q 2011 *Mater. Lett.* **65(21-22)** 3292-3294
[14] Yan Y B, Zhang Z W, Shen W, Wang J H, Zhang L K and Chin B A 2010 *Mat. Sci. Eng. A.* **527** 2241–2245
[15] Liu L and Ren D 2011 *Mater. Design.* **32(7)** 3730-3735
[16] Messler R W J 2004 *Principles of Welding: Processes, Physics, Chemistry and Metallurgy* (Weinheim: John Wiley & Sons Inc) chapter 5 pp 134

The Evaluation of Hydroxyapatite (HA) Coated and Uncoated Porous Tantalum for Biomedical Material Applications

Nadia Safuan[1, 2], Irza Sukmana[1, 2, 3] *, Mohammed Rafiq Abdul Kadir[1, 2], Deni Noviana[4]

[1]IJN-UTM Cardiovascular Engineering Centre (CEC), Block V01 Taman University, Johor Bahru, 81310 Malaysia, [2]MEDITEG (Medical Devices and Technolgy) Research Group, Faculty of Biosciences and Medical Engineering UTM, 81310 UTM Skudai Johor, Malaysia, [3]Department of Mechanical Engineering, University of Lampung, Jalan Prof. Soemantri Brojonegoro 1, Bandar Lampung 35143, Indonesia, [4]Divison of Surgery and Radiology, Department of Veterinary Clinic, Reproduction and Pathology, Faculty of Veterinary Medicine, Bogor Agricultural University (IPB), Dramaga Campus, Bogor, 16610, Indonesia.

***Corresponding author:** irza.sukmana@gmail.com

Abstract. Porous tantalum has been used as an orthopedic implant for bone defects as it has a good corrosion resistance and fatigue behaviour properties. However, there are some reports on the rejection of porous Ta after the implantation. Those clinical cases refer to the less bioactivity of metallic-based materials. This study aims to evaluate hydroxyapatite coated and uncoated porous Tantalum in order to improve the biocompatibility of porous tantalum implant and osseointegration. Porous tantalum was used as metallic-base substrate and hydroxyapatite coating has been done using plasma-spraying technique. Scanning Electron Microscopy (SEM) and Field Emission Scanning Electron Microscopy (FESEM) techniques were utilizes to investigate the coating characteristics while Confocal Raman Microscopy to investigate the interface and image. The effect of coating to the corrosion behaviour was assessed by employing potentiodynamic polarization tests in simulated body fluid at 37±1 °C. Based on SEM and FESEM results, the morphologies as well the weight element consists in the uncoated and hydroxyapatite coated porous tantalum were revealed. The results indicated that the decrease in corrosion current density for HA coated porous Ta compared to the uncoated porous Ta. This study concluded that by coating porous tantalum with HA supports to decrease the corrosion rate of pure porous.

1. Introduction

Bone diseases and bone fractures including osteoporosis, bone tumor and trauma count for half of all chronic diseases in the population over 50 years of age in developed countries. Based on the survey, it affects peoples about hundreds of millions across the world. Approximately, 10 million Americans have osteoporosis and more around 34 million peoples are at the risk of getting this disease. Thus, nearby future, it is estimated that it will keep increasing due to the aging of the population. These diseases cause high morbidity, and often require surgical interventions [1]. According to the latest statistical, over 200,000 hip replacements regarding load-bearing implants are performed each year in USA. However, this number expected to rise steadily due to increasing of life expectancy [2].

Metals are frequently used as orthopedic implants due their formability and mechanical properties [3-5]. However, in some cases, if metals come in contact with body fluids results in the release of toxic metallic ions that can cause prone to corrosion and damage surrounding tissue [6,7]. Besides that, most of the metallic materials and their alloys are known as non-bioactive substrates. In the in-vivo system, cell and tissue also have limited adhesion to biometallic materials [8].

Porous tantalum has been shown to be corrosion resistant in vivo. Also, it has been recognized as a good candidate for orthopedic implants and bone substitute applications due to excellent fatigue properties [6-8]. Porous tantalum also has a good biocompatibility to the bone tissue. In order to encourage natural bone growth at the interface of the prosthetic device, metal implants used to be coated with bioactive ceramic substrates, such as: Hydroxyapatite (HA) and bioglass [9]. In restorative orthopedics and dentistry, calcium phosphate and hydroxyapatite coated metallic implants are currently considered as the best bearing implants due to the possibility to take advantage of both biocompatibility of ceramics and the strength of metals [10].

Currently, biomaterials are limited due to the ever-present issues relating to biocompatibility. One of the factors that influence biocompatibility of metallic-based materials is controlled by chemical. In a specific way, it is called electrochemical interaction that results in the release of metallic ions into the surrounding tissue and the toxicology of these released substances [11]. This study aims to evaluate bioactivity of hydroxyapatite-coated and uncoated porous Tantalum for bone implant application. Corrosion behaviors of both samples were conducted in Simulated Body Fluid (SBF) solution at room temperature.

2. Experimental

2.1. Samples Preparation for corrosion test

Hydroxyapatite coating on porous tantalum was done using plasma-spraying technique at Amrec-Sirim, Kulim, Malaysia. The thickness layer of the coating is about 10-15 μm. The sample was cut by using high precision Allied wire cutter. The surface dimension of porous tantalum samples (coated and uncoated) was 7mm x 7mm, with thickness of about 0.5mm. After cutting, samples were washed in distilled water then proceed for cold mounting. Afterwards, samples were grinded with fine and coarse paper. Ultrasonic bath was then used to clean the samples before final rinsing with acetone then dried in air.

2.2. Corrosion test

Electrochemical measurements in this study were performed in a three-electrode electrochemical cell. Ag, AgCl/KCl (3.5M) (0.7V) electrode was used as reference electrode and for counter electrode, graphite was used. The electrolyte was Simulated Body Fluid (SBF) solution and all tests were performed at room temperature which is 37±1 °C. A potential scanning rate at 0.25 [mV S^{-1}] initiated at -250 [mV] below the open circuit potential and the atmosphere was open to air, recorded was the dynamic polarization curves. Polarization curves were obtained 1000s after immersion in the solution, varying the potential between -0.5 and 2.5V. The current density was measured using VersaSTAT 3 potentiostat interfaced with a computer.

2.3. Scanning Electron Microscope

Scanning Electron Microscope (SEM) technique used to analyze the HA coating quality, and to evaluate the porous microstructure and morphology of the samples. Field Scanning Electron Microscope (FESEM) was also used to estimate the ratio of Calcium over Phosphate (Ca/P) at the Hydroxyapatite coated area to measure the weight element of each samples.

3. Results and Discussions

3.1. SEM and FESEM images

The result of SEM shows the average coating thickness of deposition HA on pure porous Ta substrate. Meanwhile, the pore size before and after coating also was investigated shown on Fig.1a, the average coating thickness obtained by using plasma spray coating technique is about 157µm.

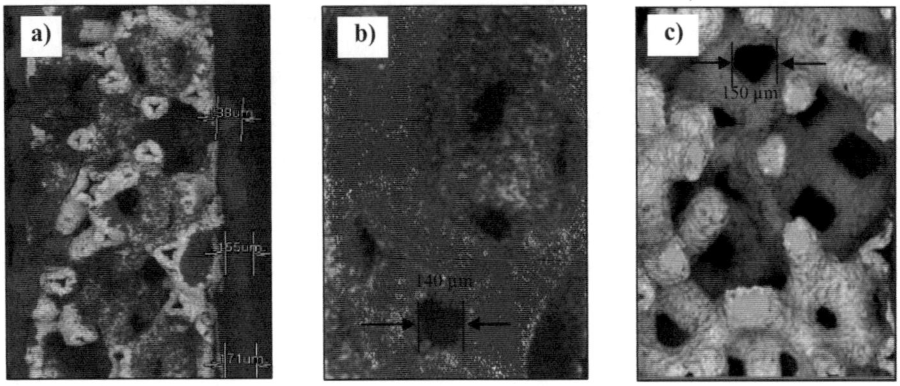

Fig. 1 SEM Image of Average Coating Thickness on HA Coated Porous Ta (a) and SEM Image of Pore Size of Porous Tantalum before (b) and after (c) coating with HA

Fig. 1b shows the pore size thickness of porous Ta before and after coating with HA. The average of pore size for pure porous tantalum is 160 µm while after coating with HA it becomes smaller which is148.67 µm as shown in Table 1.

Table 1 Pore Size of Porous Tantalum before and after coating with HA

Before coating (µm)	After coating (µm)
150	140
163	152
167	154
Average = 160	Average = 148.67

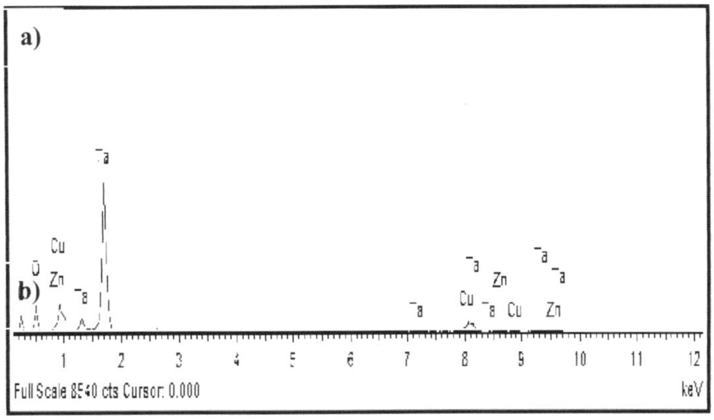

Element	Weight%
Tantalum	53.10

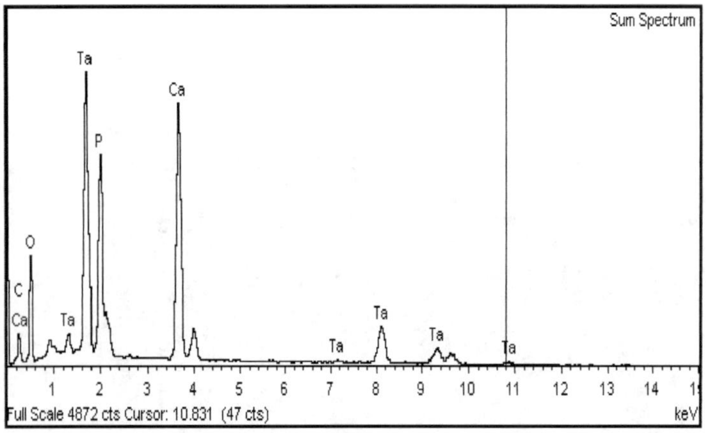

Element	Weight%
Tantalum	30.65
Calcium	19.61
Phosphorus	11.57

Fig. 3 FESEM graph and weight element for porous Tantalum (a) Uncoated (b) HA coated

As shown in Fig. 3 (b), Ta is still present in the coated sample however the weight decline since it is not pure anymore. By using only pure porous tantalum can caused adverse effect to the surrounding tissue. It is occurred due to the released of metallic ion to the surrounding tissue. In order to make metal surfaces interact to the surrounding tissue (bioactive), porous tantalum was coating with the calcium phosphate that have been modified the surface of the implants [9].

3.2. Confocal Raman Microscopy

Interface Raman microscopy was used to analyze the raman spectrum of porous tantalum coated HA. It used the concept of electromagnetic radiation due to the consequences at certain frequencies; it will correlate to the vibration of specific sets of chemical bonding within a molecule [12]. Therefore in Fig. 5, the green line showed on the graph was a constant because there was no absorption of electromagnetic radiation on the surface of porous tantalum sample. Besides, Confocal Raman Microscopy could not detect the present of pure metallic-based material such as Tantalum.

The blue line represented Phosphorus element which in the region of 10-200 cm^{-1} presented the lattice vibrations in crystals, LA modes. Thus, the Raman was strong, showed the highest peak at the beginning of testing. Therefore, it consumed the highest absorption of electromagnetic radiation initially towards Phosphorus elements.

While the red line in Fig. 5 indicated v(cc) alicyclic, aliphatic chain vibrations. The range of region is about 600-1300 cm^{-1} showed the medium peak and also indicating medium Raman for calcium elements based on the coating for the tested sample. Thus, it was proven that the coating was attached to the substrate.it shows that the green line on the graph is constant thus Confocal Raman Microscopy could not detect the present of pure metallic-based material such as Tantalum.

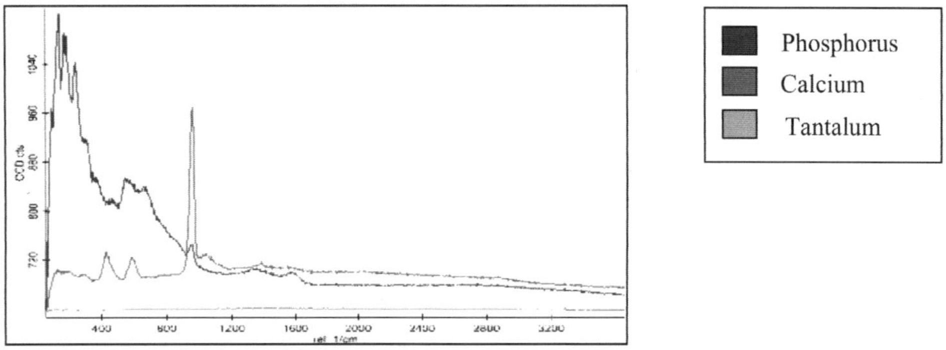

Fig. 5 Interface Raman Spectrums of Confocal Raman Microscopy

3.3 Corrosion Behavior Studies

Potentiodynamic polarization used to analyze the corrosion behavior of tantalum material as shown in Fig. 6 (a) and (b).

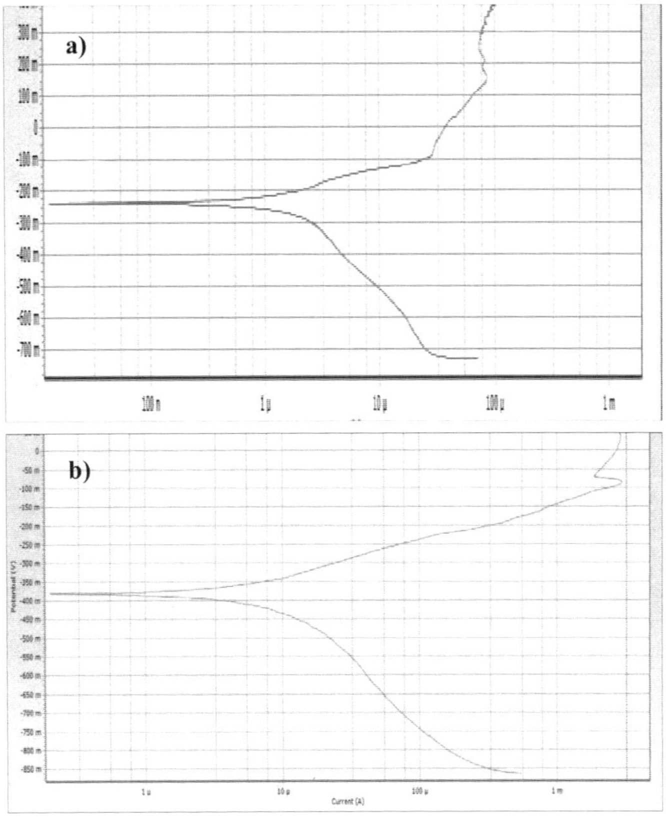

Fig. 6 Polarization Curves of Porous Ta (a) Uncoated (b) HA Coated

Table 2 Mean values (standard deviation) of corrosion current densities and corrosion potentials of uncoated porous Ta and HA coated in SBF solution at ±37° C.

Porous Ta	Ecorr (mV)	Icorr (uA)
Uncoated	-239.002	8.714
HA coated	-381.428	6.820

Uncoated porous Ta obtained higher corrosion current density [Ecorr= -239.002mV, icorr= 8.714uA, Table 2] hence lower the corrosion resistance in SBF solution when compared to HA coated porous Ta [Ecorr= -381.428, icorr= 6.820, Table 2] that shows lower corrosion current density. Based on Fig. 6 (a), the pitting corrosion potential of the uncoated porous tantalum is lower than the coated porous tantalum. Previous study showed that coating the porous tantalum with HA reduced the corrosion current density however also increased the pitting corrosion potential of the porous tantalum. This showed that the coated porous tantalum is effective to anticorrosion. These results are in agreement with the studies that have been carried out on metallic dental implants [12].

The current density corresponds to the rate of electrochemical reactions associated with corrosion processes taking place on the material surface in the electrolyte under polarization. The corrosion current density of HA coated porous Ta had declined and gave a positive effect of the metallic substrate towards corrosion resistance. It is important to have excellent corrosion resistance for biocompatible alloys such as porous tantalum so that the human body will be safe from corrosion process that take place in the human body. Surface films like HA plays important part in corrosion and osseointegration process of implants. It helps by modifying the surface of alloy implants as well improving the corrosion resistance [13]. This explained the corrosion resistance is inversely proportional to the current density. As an example, if it had low current density indicates high corrosion resistance and consequently a low level of metals release towards the surrounding. It shows it was beneficial for the biocompatibility.

This study was conducted due to the corrosion resistance may affect the cell metabolism and also cell behavior from the corrosion occurred [13]. Currently, tantalum-based materials had gain attention as one of the promising bone implant materials with higher biocompatibility and corrosion resistance. However, if the porous tantalum in pure form, it is mechanically weak as poor bone integration with the surrounding cell. To enhance the usage of Ta in biomedical application, it was coated with bioceramics such as Hydroxyapatite. With the ability to bind the bond and its biocompatibility, Hydroxyapatite (HA) had been intensively studied for bone repair and replacement applications [14].

Previous *in vitro* research showed that if the metal coated with single HA reduced the corrosion current density of 316L SS substrate in physiological solutions. However this effect was not sufficiently complete. It is a fact that the surface morphology and structure of a single HA coating consists of a porous surface and some micro cracks. It will not become a complete barrier against the release of metallic ions from metallic implant to physiological solutions just by coating with HA. The HA coating act simply as a semi-mechanical barrier. On the other hand, previous in vivo study showed that the bone osteointegration around single HA coated Vitallium alloy and single HA coated SS were almost similar. Based on the point of clinical success and osteointegration, there was no significant difference was observed [15-16]. Vitallium was originally used in endodontic implants however the biocompatibility was questioned when it was shown to undergo surface corrosion [17]. Our result is also in a good agreement with other researchers [12, 18]. By coating the porous tantalum with HA we believe it can improve the corrosion resistance and thus the biocompatibility of the tantalum material [19]. From clinical study for porous tantalum implant, the results indicated the present of swelling besides there also fluid developed around the implant within the first week of surgery. Based on the cell culture perspective, the implant also caused tissue failure happened at or beyond the implant surface [20]. This may occurred due to the questioning of biocompatibility of metallic biomaterials which had considered for medical use in term of corrosion resistance. If the corrosion resistance was excellent hence it would help to minimize the released of toxic metal-ions in the human body [21]. Therefore, Hydroxyapatite (HA) coated porous tantalum would be the best way to help solving problems regarding

4. Conclusion

Based on SEM and FESEM results, the morphologies as well the weight element consists in the uncoated and hydroxyapatite coated porous tantalum were revealed. As for pure tantalum, 53.10% weight of element was present while in HA coated samples there was just only 30.65 % consist of pure tantalum followed by calcium 19.61 % and phosphorus 11.57 %. From this study, the average thickness of coating by plasma technique had also obtained which is 160 μm. The results based on electrochemical test indicated that the decrease in corrosion current density for HA coated porous Ta compared to the uncoated porous Ta. Thus, by coating porous tantalum with HA helps to reduce the corrosion rate of pure porous tantalum and the release of metallic ions to the surrounding tissue. By releasing the toxic metallic-ion towards surrounding environment may affect the growth of the cell and it is not good for people who get the bone implant procedure. As one of the challenge in this research which is to produce biomaterial that has most characteristics similar to bone composition such as bone-bonding and mechanical properties, we had magnificently proven that by coating porous tantalum with bioceramic, Ca/P had improve the surface of the implants being optimum biocompatible. With this research, we hope that it be a helpful and contributes to the world of orthopedic field. In future research, there is needed to test in-vitro biocompatibility assessment of HA coated and uncoated porous tantalum by doing cell culture and animal studies. Besides, it also needed to study the cytotoxicity of the tantalum materials.

Acknowledgements
The authors thank to Dr. Hendra Hermawan (FBME UTM) and Dr. Jamal Siddiqui for their constructive comments. UTM Tier-1 (vote#03H12), MOHE-FRGS (vote#4F128) and UTM RU (vote#00G670) are also kindly acknowledged.

References
[1] National Osteoporosis Foundation. Available from: http://www.nof.org.
[2] Kim S. Changes in surgical loads and economic burden of hip and knee replacements in the US: 1997-2004. Arthritis Rheum 2008. 59(4). p. 481-8.
[3] D. H. Kohn, Curr. Opin. *Solid St. M.* 1998. (3). p. 309.
[4] V.K Balla. S Bodhak, S. Bose, A. Bandyopadhyay, *Acta Biomater.* 2010. (6). p. 33.
[5] G. Ryan, A. Pandit, D.P. Apatsidis, *Biomaterials.* 2006. (27). p. 265.
[6] M. Pourbaix, *Biomaterials.* 1984. (5). p. 122.
[7] F. W. Sunderman, J. *Mech. Behav. Biomed.* 1989. (13). p. 205.
[8] M.A. Mcgee, D.W. Howie, K.Costi, D.R. Haynes, C.I. Wildaneur, M.J. Pearcy, J.K. Mclean, *Wear.* 2000. (241) p.158.
[9] Lyle D. Zardia Ckas, Douglas E. Parsell, Lance D. Dillom, Darrel W. Mitchell, Laura A. Nunnery, Robert Poggle. *Structure, Metallurgy and Mechanical Properties of Porous Tantalum Foam.* 2000. p. 180-187.
[10] T. LI, J. Lee, T. Kobayashi and H. Aoki, J. *Mater. Sci., Mater. Med.* 1996. (7). p. 355.
[11] Williams David F.(ed.), 1981. Biomaterials an Biocompatibility: An introduction, in Fundamental Aspects of Biocompatibility, vol. 1, CRC press, Boca Raton, Fl, pp. 2-3
[12] Joan Coates, *Interpretation of Infrared Spectra, A Practical Approach,* Encyclopedia of Analytical Chemistry, R. A. Meyers (Ed.) 2000. p. 10815-10837
[13] M.H. Fathi, F. Azam, *Materials Letters, Novel Hydroxyapatite/Tantalum Surface Coating for Metallic Dental Implant.* 2007. 61. p. 1238-1241.
[14] M. Metikos-Hukovic, A. Kwokal, J. Piljac, *Biomaterials.* 2003. 24. p. 3765.
[15] X. Miao, *Materials Letters.* 2003. 57. p. 1848.
[16] M.H. Fathi, M. Salehi, A. Saatchi, V. Mortazavi, S. B. Mousavi, *Dent. Mater.* 2003. 19. p. 188.
[17] M.H. Fathi, M. Salehi, A. Saatchi, V. Mortazavi, S. B. Mousavi, *Surface Engineering*. 2001. 17. p. 459.
[18] Locci P, Marinucci L, Lilli C, Belcastro S, Staffolani N, Bellocchio S, et al. *Biocompatibility of alloys used in orthodontics evaluated by cell culture tests. J Biomed Mater Res* 2000. 51(4). p. 561-68.
[19] W. Suchanek, M. Yoshimura, J. *Mater. Res.* 1998. 13. p. 94.
[20] S.A. Hacking,J.D. Bobyn,K.-K. Toh, M. Tanzer,J.J. Krygier, Fibrous tissue ingrowths and attachment to porous tantalum, 2000
[21] Wallinder D., Pan J., Leygraf C. and A. Delblanc-Bauer, *Electrochemical behaviour and corrosion test,* 1999

Mathematical models for estimating earthquake casualties and damage cost through regression anaylysis using matrices.

J D Urrutia, L A Bautista and E B Baccay
Polytechnic University of the Philippines

Email: math_urrutia@yahoo.com.ph / jackieurrutia20@gmail.com / labautista1971a@yahoo.com / math_baccay@yahoo.com

ABSTRACT: The aim of this study was to develop mathematical models for estimating earthquake casualties such as death, number of injured persons, affected families and total cost of damage. To quantify the direct damages from earthquakes to human beings and properties given the magnitude, intensity, depth of focus, location of epicentre and time duration, the regression models were made. The researchers formulated models through regression analysis using matrices and used $\alpha = 0.01$.

The study considered thirty destructive earthquakes that hit the Philippines from the inclusive years 1968 to 2012. Relevant data about these said earthquakes were obtained from Philippine Institute of Volcanology and Seismology. Data on damages and casualties were gathered from the records of National Disaster Risk Reduction and Management Council.

The mathematical models made are as follows:

$$ln\hat{y}_1 = -13.405 - 0.001x_1 + 2.214x_2 - 0.013x_3 + 0.374x_4 - 0.001x_5$$
$$ln\hat{y}_2 = -13.426 + 0.360x_1 + 1.899x_2 + 0.015x_3 + 0.444x_4 - 0.001x_5$$
$$ln\hat{y}_3 = -16.905 + 0.083x_1 + 2.577x_2 - 0.043x_3 + 0.630x_4 + 0.002x_5$$
$$ln\hat{y}_4 = -17.075 + 0.234x_1 + 2.520x_2 - 0.033x_3 + 0.383x_4 + 0.001x_5$$

This study will be of great value in emergency planning, initiating and updating programs for earthquake hazard reductionin the Philippines, which is an earthquake-prone country.

KEYWORDS. *Mathematical Models, Earthquake, Regression Analysis, Matrices, Correlation Coefficient*

1. INTRODUCTION

Earthquakes or called quakes are serious natural hazard in many parts of the world. It is a sudden movement of the earth's crust caused by the release of stress accumulated along geologic faults or by volcanic activity. Throughout the past two decades, earthquakes have caused more than a million deaths, physical damages and injuries worldwide. Since the Philippines is situated within the "Pacific Ring of Fire", it causes the country to have frequent seismic and volcanic activity which may results to a volume of casualties and enormous damages. Moreover, the country sits on the Philippine Plate, which is sandwiched by the Pacific Plate and Eurasian Plate. In this tectonic setting, the country is widely known as an earthquake country.

Earthquake damage depends on what area is hit. If an unpopulated region is struck, there will be low loss of life or property. If it hits a large city, there may be many injuries and destruction. Many of the areas at risk on the Ring of Fire are largely populated. Major earthquakes hitting those areas today could produce terrible damage.

The Philippine archipelago is prone to earthquakes because of its geotectonic characteristics. It is also transacted by numerous faults, foremost of which is the Philippine Fault Zone that runs the length from Luzon through Eastern Visayas to Eastern Mindanao and bounded by several trenches: in the east by the East Luzon Trench, Philippine Trench and Davao Trench; and in the west by the Manila Trench, Negros Trench, Sulu Trench and Cotabato Trench. These tectonic features serve as the major earthquake generators in the country which hosts at least five unfelt to felt earthquakes per day.

2. OBJECTIVE OF THE STUDY

The objective of the study was to formulate mathematical models through regression analysis using matrices to physically quantify the direct damages from earthquakes to human beings and structures given the duration, magnitude, intensity and depth of focus of the earthquakes. These models will be of great value in planning, initiating and updating programs for earthquake hazard reduction and in emergency planning in the Philippines.

The researchers used the variables shown in the figure to make models that assess casualties and cost of damage. Significant relationships among variables were obtained after some verifications. Multiple Regression using matrices was then used to make models that will best estimate the casualties and damaged cost.

RESEARCH PARADIGM

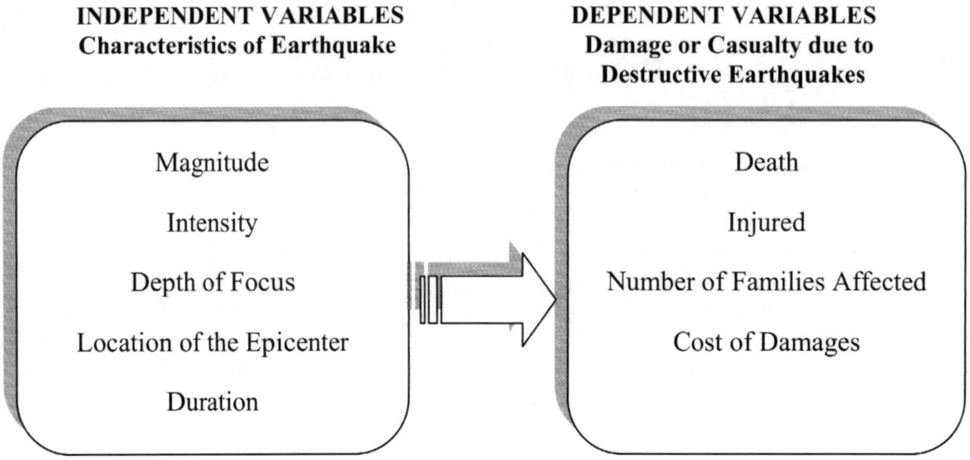

Figure 1.*Research Paradigm of the Study*

3. STATEMENT OF THE PROBLEM

This study is conducted to formulate mathematical models through regression analysis using matrices to estimate casualties and cost of damages due to destructive earthquakes. In particular, the point of the study was to answer the following questions:

1. To what extent are the major earthquakes that strike the Philippines be characterized in terms of:
   1.1 intensity;
   1.2 magnitude;
   1.3 depth of focus;
   1.4 location of the epicenter; and
   1.5 duration?

2. How much damage can earthquakes produce in terms of
   2.1 death,
   2.2 injuries
   2.3 families affected, and
   2.4 cost of damage?

3. Whatmathematical models can be formulated through regression analysis using matrices that estimately describe following due to destructive earthquakes.
   3.1 death,
   3.2 injured,
   3.3 families affected, and
   3.4 cost of damage?

4. How significant are the formulated mathematical models to assess the possible damages due to an earthquake event?

4. SCOPE AND DELIMITATION

The researchers limit this study only to the destructive earthquakes with magnitude 5 and higher. They also considered the thirty earthquake events in the Philippines that brought enormous casualties and damages. The data were gathered from the Philippine Institute of Volcanology and Seismology[8] and National Disaster Risk Reduction and Management Council[9]. The researchers formulate regression modelsusing matrices by considering independent variables like magnitude, intensity, depth of focus, location of the epicenter and duration of earthquake recorded within 1968 to 2012.

The researchers considered the zone where the epicenter is located. The different zone defined by Sergio Su[4] are the following:

Zone 1 – East Luzon subduction zone
Zone 2 – Philippine Trench
Zone 3 – Philippine fault
Zone 4 – Double forearc associated with Manila Trench
Zone 5 – Manila Trench
Zone 6 – Negros and Sulu Trench
Zone 7 – Cotabato trench and the Northern Extension of the Molucca Sea Plate

5. REVIEW RELATED LITERATURES

This portion presents a review of several literatures that would be advantageous to the study summarized from previous writings, divulging thorough facts asserted by few people and pioneer in the field of applied mathematics.

Casualty counts and damages of earthquake may depend on the location where it occurs. Most earthquake-related deaths are caused by the collapse of structures and the construction practices play a tremendous role in the death toll of an earthquake. This was proved in the web article[3].

According to Arnold E.P. [2] that on November 30, 1645, Manila experienced the most destructive earthquake in its history. A number of stone buildings were ruined, and an estimated 3,000 lives lost.This earthquake was noted to compare in magnitude with the greatest recorded in world history.

Numerous other earthquakes occurred in 1743, causing large scale destruction in thePhilippines. On January 12, in the province of Tayabas, Quezon, the church and amonastery made of stone were devastated by an earthquake and accompanying landslides. The records show that inhabitants had never previously seen such destruction, with the town rendered uninhabitable and its people forced to move to another place to build a new town.

According to Ma. Leonila P. Bautista[3], There is a research study made by that is related to the current study which estimated the magnitudes and epicenters of 74 moderate-to-large historical earthquakes of the Philippines 1589 to 1896 through relating them to modern earthquakes. This was completed by initially deriving or formulating empirical equations relating surface magnitude (Ms) of recent Philippine earthquake (1987 to 1995) with felt areas enclosed by different intensities.

The said study found that the proposed epicenters of these historical events show expected and at the same time interesting results. While most of these epicenters fall within established and known active tectonic structures, some epicenters fall within established and known active tectonic structures, some epicenters fall within some structures thought to be inactive. For example, the Philippine fault exhibited active seismicity throughout the 300-year record. The Philippine Trench-east Luzon Trench region failed to show such remarkable seismicity but this is probably due to lack of populated towns in the eastern shores of the Philippines that could have reported felt big events.

According to Sergio S. Su[4] that the knowledge of how ground acceleration or velocity attenuates with epicentral distances is essential to the estimation of seismic risk at a given site. In the virtual absence of instrumental data, this attenuation function for acceleration or velocity may be obtained indirectly in two steps: first, through a study of how felt-intensity attenuates with epicentral distance; and second, through empirical correlation of felt intensity with acceleration or velocity.

Using the first step and data from 83 earthquakes in the Philippine region, Su derived an intensity-attenuation equation in the form:

$$I(R) - I_0 = 4.01 - 0.015R - 2.40 \log R$$

Where $I(R)$ = Intensity as a function of distance
I_0 = Intensity at the epicenter
R = Epicentral distance in kilometers

Paul C. Thenhaus[5]estimated the economic loss from earthquakes to dwellings in a new housing development near Canlubang. Their estimates suggested an average annual loss per dwelling of P 576.00. Because of the lack of historic record of earthquake intensities and resulting damage at the site, the estimate is recognized as having relatively large uncertainty. Assuming $\pm$Modified Mercalli Intensity (MMI) unit variability in the intensities on which the loss estimate is band, results in bounding estimates of average annual loss per dwelling of P 1,236.00 and P 216.00. An estimate of catastrophe potential, or the largest economic loss that might reasonably be expected to occur to housing project, is estimated to be P 121,447,471.00 with a 90 % probability of not being exceeded in 250 years. This economic loss is equivalent to a 22 % physical loss to dwelling construction in the housing project.

To assess possible damages due to earthquakes, Dr.Felicisma V. Olaya,[7] proposed mathematical models with predictors intensity x_1, magnitude $x_{2,}$ depth $x_{3,}$ and epicenter $x_{4.}$ According to her, using stepwise multiple regression analysis, one of the models that predicts death is the combination of magnitude, ln(magnitude) and epicenter which gave highest $R^2 = 0.563$ and that is

$$\hat{y}_1 = 59578.966 + 11182.497x_2 - 71139.37\ln(x_2) + 167.795x_{4.}$$

The model for predicting injuries with highest $R^2 = 0.685$ is the combination of magnitude and ln(epicenter) with ln(injured) and that is

$$\ln(\hat{y}_2) = 16.697 + 2.664x_2 + 1.241\ln(x_4)$$

One of the models that predicts cost of damage is the combination of ln(intensity) and epicenter with highest $R^2 = 0.611$ and the model is

$$\ln(\hat{y}_3) = -18.787 + 9.477\ln(x_2) + 0.719x_{4.}$$

6. RESEARCH METHODOLOGY

The method of research that was used by the researchers was a descriptive type of researchthrough a statistical analysis wherein the study is focused on present and past situations.

The researchers gathered data on destructive earthquakes that hit the Philippines from 1968 to 2012. The information on the characteristics of earthquakes were collected from Philippine Institute of Volcanology and Seismology[8],which is the main institute studying earthquakes in the Philippines. Data on casualties and cost of damages due to these earthquakes were collected from the files of the National Disaster Risk Reduction and Management Council[9].

6.1. Statistical Treatment

Since there were more than one independent variables involved, multiple regression using matrices was utilized to establish the mathematical model that would be best assess the damage due to destructive earthquakes particularly the number of casualties and families affected and the cost of damage.

All calculations were made using normal equation. Matrix computations were calculated using MATLAB and regression models were verified using SPSS.

In fitting a multiple linear regression model,knowledge of matrix theory can facilitate the mathematical manipulations considerably. Using matrix notation, we can write the equation

$$y = X\beta + c$$

Where

$$y = \begin{bmatrix} y_1 \\ y_2 \\ \vdots \\ y_n \end{bmatrix}, \quad X = \begin{bmatrix} 1 & x_{11} & x_{21} & \cdots & x_{k1} \\ 1 & x_{12} & x_{22} & \cdots & x_{k2} \\ \vdots & \vdots & \vdots & \cdots & \vdots \\ 1 & x_{1n} & x_{2n} & \cdots & x_{kn} \end{bmatrix}, \quad \beta = \begin{bmatrix} \beta_0 \\ \beta_1 \\ \vdots \\ \beta_k \end{bmatrix}, \quad \epsilon = \begin{bmatrix} \epsilon_1 \\ \epsilon_2 \\ \vdots \\ \epsilon_n \end{bmatrix}$$

Then the least squares method for estimation of β, it involves finding b for which

$SSE = (y - Xb)'(y - Xb)$is minimized, then the X matrix is

$$A = X'X = \begin{bmatrix} n & \sum_{i=1}^{n} x_{1i} & \sum_{i=1}^{n} x_{2i} & \cdots & \sum_{i=1}^{n} x_{ki} \\ \sum_{i=1}^{n} x_{1i} & \sum_{i=1}^{n} x_{1i}^2 & \sum_{i=1}^{n} x_{1i}x_{2i} & \cdots & \sum_{i=1}^{n} x_{1i}x_{ki} \\ \vdots & \vdots & \vdots & \cdots & \vdots \\ \sum_{i=1}^{n} x_{ki} & \sum_{i=1}^{n} x_{ki}x_{1i} & \sum_{i=1}^{n} x_{ki}x_{2i} & \cdots & \sum_{i=1}^{n} x_{ki}^2 \end{bmatrix}; g = X'y = \begin{bmatrix} g_0 \\ g_1 \\ \vdots \\ g_k \end{bmatrix} = \begin{bmatrix} \sum_{i=1}^{n} y_1 \\ \sum_{i=1}^{n} x_{1i}y_i \\ \vdots \\ \sum_{i=1}^{n} x_{ki}y_i \end{bmatrix}$$

allows the normal equations to be put in the matrix form $Ab = g$.[6]

7. PRESENTATION, ANALYSIS AND INTERPRETATION OF DATA

7.1. Characteristics Of Destructive Earthquakes In The Philippines

7.1.1. Intensity

The table 1 shows the frequency distribution of the intensity of the thirty earthquakes in the Philippines that considered in the study.

TABLE 1
Distribution of the Intensity of the Earthquakes

Intensity	Frequency	Percentage
4	1	3.3%
5	2	6.7%
6	5	16.7%
7	15	50.0%
8	7	23.3%
	n = 30	100%

As shown in the table 1, 50% of the earthquakes were of intensity 7. Earthquakes of intensity 7 were classified by the PHIVOLCS as destructive. Moreover, 23.3% of the considered quakes were of intensity 8 and these were considered as very destructive. Subsequently, 16.7% were of intensity 6, 6.7% were of intensity 5 and 3.3% of the total considered earthquakes were of intensity 4. These were classified by the PHIVOLCS as very strong, strong, and moderately strong respectively.

TABLE 2

Summary of Means and Standard Deviations of the Independent Variables

Variable	Mean	Standard Deviation
Intensity (x_1)	6.83	0.99
Magnitude (x_2)	6.65	0.69
Depth of Focus (x_3)	23.67	15.73
Location of the Epicenter(x_4)	3.7	1.66
Duration (x_5)	481.21	734.59

7.1.2. Magnitude

The table 2 shows that the mean magnitude of the earthquakes was 6.65 Ms with standard deviation of 0.69. According to the Richter magnitude scale used by the PHIVOLCS, seismic activities with magnitude 6 to 7 are strong earthquakes and accompanied by local damages near the epicenters. First class seismological stations can observe them wherever they occur within the earth.

7.1.3. Depth of Focus

The above table 2 divulges that the average depth of considered earthquakes was 23.67 kilometers with standard deviation of 15.73. Shallow earthquakes are those quakes with depth ranges from 0 to 70 kilometers. The earthquakes with shallower depths bring more damages than those who have greater depths.

7.1.4. Location of the Epicenter

The table below shows the frequency distribution of where the considered earthquakes were located.

TABLE 3
Epicenter Locations of the Earthquakes

Zone	Frequency	Percentage
Zone 1	2	6.67%
Zone 2	6	20.00%
Zone 3	6	20.00%
Zone 4	9	30.00%
Zone 5	1	3.33%
Zone 6	4	13.33%
Zone 7	2	6.67%
	n = 30	100.00%

As shown in table 3, the most number of earthquakes had epicenters located at Zone

4. The thirty percent (30%) of the total earthquakes considered in the study had epicentersat West Luzon Trench, whereas the double forearc associated with the Manila Trench. This implies that Luzon is very prone to be stricken by the earthquakes or ay seismic activities.

7.1.5. Duration of the Earthquake

The average duration of the earthquake considered in the study was 481.21 seconds with standard deviation of 734.59 as shown in table 2. Earthquakes with longer time duration bring more damages than those with shorter one.

7.2. Damages Caused By Earthquakes

The table 4 underneath shows the mean, median and standard deviation of the different dependent variables.

TABLE 4
Summary of Means, Medians and Standard Deviations of the Dependent Variables

Variables	Mean	Median	Standard Deviation
Death (y_1)	187.03	4.5	720.65
Injured (y_2)	468.27	27.5	1732.31
Affected Families (y_3)	13,188.73	0	43687.49
Total Damage Cost (y_4)	466.65	4.98	2229.16

As shown in table 4, the mean death owing to earthquakes was 187.03 as associated to the median 4, which is more lifelike estimate of death. The most number of deaths aroused during the Moro Gulf earthquake in August 17, 1976 which left 3,792 persons dead. It was an earthquake that resulted in enormous destruction of properties and huge loss of lives. The

tsunami produced by the earthquake added immensely to the devastation. The great number of casualties during the incident could also be ascribed to the tremor that occurred just after midnight when most people were sleeping.

The mean of the number of people injured was 468.27 while the median was 27.5. The largest number of injured happened on Moro Gulf earthquake in 1976 which wounded 9,240 persons.

The cost of property damaged had a mean of P466.65M and a median of P4.98M. The largest cost of damages occurred in July 16, 1990 earthquake with epicentre at Cabanatuan City. It was of Intensity VIII at the epicentre and had a magnitude of 7.8 in the Richter scale. The earthquake left in its wake about P12.252M of damages to buildings, infrastructures and properties in Baguio-Cabanatuan-Dagupan area. It is deemed as the most devastating earthquake to strike the northern Philippines in the 19th century.

7.3. Relationships Of The Independent And Dependent Variables

The relationships of the independent to the dependent variables were ascertained using Pearson's coefficient of correlation, as shown in the table below.

Journal of Physics: Conference Series **495** (2014) 012024 doi:10.1088/1742-6596/495/1/012024

TABLE 5
Pearson's coefficient of correlation

	Intensity	Magnitude	Depth	Location	Duration
Death (y_1)	0.121	0.448*	-0.013	0.350	-0.089
Injured (y_2)	0.112	0.439*	-0.015	0.375*	-0.095
Affected Families (y_3)	0.242	0.426*	-0.056	0.086	0.126
Total Damage Cost (y_4)	0.233	0.336	-0.003	0.044	0.129

*Correlation is significant at the 0.05 level (2-tailed)

The table 5 shows that the intensity of the considered quakes was not significantly correlated with any of the dependent variables. Unlike the intensity, magnitude of the earthquake was significantly correlated with number of death, injured, and families affected at 0.05 level. The coefficients of correlations of depth and time duration were not significantly correlated with any of the dependent variables but the location of the epicenter was significantly correlated with injured.

This also shows that all the independent variables did not correlate with the total damage cost of damage and as a consequence the independent variables that had considered in the research will not estimate the said dependent.

The assumptions for multiple linear regression guided the researchers to formulate mathematical models that estimate the casualties and cost of damage of an earthquake event. The deviation from linearity guided them to know if there are linear relationships between the independent and dependent variables. Since such assumption of multiple regression did not meet, the researchers decided to transform the dependent variables.

After the transformation, all the assumptions for formulating multiple regression models were satisfied. The assumptions for normality and linearity were tested through graphical method, the test of independence was verified by Durbin-Watson Test, the homoscedasticity was tested by Levene'stest , and the multicollinearity was verified through the Variance Inflation Factor (VIF) value.

TABLE 6
Pearson's Correlations Coefficients

	ln_death	ln_injuries	ln_affected_families	ln_total_damage
intensity	.340	. 1*	.303	.469**
magnitude	.533**	.562**	.457*	.644**
depth	.016	52	-.089	-.070
epicenter	.264	42	.213	.213
duration	-.186	-.107	.509**	.320

**Correlation is significant at the 0.01 level (2-tailed).

*. Correlation is significant at the 0.05 level (2-tailed).

The table 6 shows that the transformed variable death was significantly correlated with magnitude. The transformed variable injuries were significantly correlated with intensity and magnitude. The transformed variable affected families were significantly correlated with magnitude and time duration. The transformed variable total damage was significantly correlated with intensity and magnitude.

7.4. Proposed Mathematical Models

7.4.1. Mathematical Model For Estimating Death

The least squares estimating equations $(X'X)b = X'y$

$$\begin{bmatrix} 30 & 205 & 199.6 & 710 & 111 & 14436.4 \\ 205 & 1429 & 1374.7 & 4813 & 758 & 101312 \\ 199.6 & 1374.7 & 1341.62 & 4818.8 & 735 & 100163.82 \\ 710 & 4813 & 4818.8 & 23980 & 2399 & 356200 \\ 111 & 758 & 735 & 2399 & 491 & 50919.8 \\ 14436.4 & 101312.2 & 100163.82 & 356200 & 50919.8 & 22596163.36 \end{bmatrix} \begin{bmatrix} b_0 \\ b_1 \\ b_2 \\ b_3 \\ b_4 \\ b_5 \end{bmatrix} = \begin{bmatrix} 56.56 \\ 407.78 \\ 399.52 \\ 1355.08 \\ 237.16 \\ 18521.53 \end{bmatrix}$$

and then, using the relation $b = (X'X)^{-1}X'y$, the estimated regression coefficients are obtained as:

$$b_0 = -13.405;\ b_1 = -0.001;\ b_2 = 2.214;\ b_3 = -0.013; b_4 = 0.374; b_5 = -0.001$$

The coefficients were obtained using MATLAB. Therefore, the number of persons killed in an earthquake event can be computed using the regression equation.

$$\boldsymbol{ln\hat{y}_1 = -13.405 - 0.001x_1 + 2.214x_2 - 0.013x_3 + 0.374x_4 - 0.001x_5}$$

This regression model yields the coefficient of determination $R^2 = 0.509$ which implied that 50.9% of the variation in the number of death during an earthquake is explained by the regression equation.

7.4.2. Mathematical Model For Estimating Injuries

The least squares estimating equations $(X'X)b = X'y$

$$\begin{bmatrix} 30 & 205 & 199.6 & 710 & 111 & 14436.4 \\ 205 & 1429 & 1374.7 & 4813 & 758 & 101312 \\ 199.6 & 1374.7 & 1341.62 & 4818.8 & 735 & 100163.82 \\ 710 & 4813 & 4818.8 & 23980 & 2399 & 356200 \\ 111 & 758 & 735 & 2399 & 491 & 50919.8 \\ 14436.4 & 101312.2 & 100163.82 & 356200 & 50919.8 & 22596163.36 \end{bmatrix} \begin{bmatrix} b_0 \\ b_1 \\ b_2 \\ b_3 \\ b_4 \\ b_5 \end{bmatrix} = \begin{bmatrix} 97.95 \\ 696.84 \\ 677.83 \\ 2481.22 \\ 389.79 \\ 41805.14 \end{bmatrix}$$

and then, using the relation $b = (X'X)^{-1}X'y$, the estimated regression coefficients are obtained as:

$$b_0 = -13.426;\ b_1 = 0.360;\ b_2 = 1.899;\ b_3 = 0.015; b_4 = 0.444; b_5 = -0.001$$

The coefficients were obtained using MATLAB.Therefore, the number of persons injured in an earthquake event can be computed using the regression equation

$$\boldsymbol{ln\hat{y}_2 = -13.426 + 0.360x_1 + 1.899x_2 + 0.015x_3 + 0.444x_4 - 0.001x_5}$$

This regression model yields the coefficient of determination $R^2 = 0.501$ which implied that 50.1% of the variation in the number of death during an earthquake is explained by the regression equation.

7.4.3. Mathematical Model for EstimatingAffected Families

The least squares estimating equations $(X'X)b = X'y$

$$\begin{bmatrix} 30 & 205 & 199.6 & 710 & 111 & 14436.4 \\ 205 & 1429 & 1374.7 & 4813 & 758 & 101312 \\ 199.6 & 1374.7 & 1341.62 & 4818.8 & 735 & 100163.82 \\ 710 & 4813 & 4818.8 & 23980 & 2399 & 356200 \\ 111 & 758 & 735 & 2399 & 491 & 50919.8 \\ 14436.4 & 101312.2 & 100163.82 & 356200 & 50919.8 & 22596163.36 \end{bmatrix} \begin{bmatrix} b_0 \\ b_1 \\ b_2 \\ b_3 \\ b_4 \\ b_5 \end{bmatrix} = \begin{bmatrix} 99.30 \\ 716.64 \\ 700.58 \\ 2172.53 \\ 412.56 \\ 95469.76 \end{bmatrix}$$

and then, using the relation $b = (X'X)^{-1}X'y$, the estimated regression coefficients are obtained as

$$b_0 = -16.905;\ b_1 = 0.083; b_2 = 2.577;\ b_3 = \ -0.043;\ b_4 = 0.630; \qquad b_5 = 0.002$$

The coefficients were obtained using MATLAB.Therefore, the number of families that would be affected in an earthquake event.

The number of families that affected during an earthquake event can be computed using the regression equation

$$\boldsymbol{ln\hat{y}_3 = -16.905 + 0.083x_1 + 2.577x_2 - 0.043x_3 + 0.630x_4 + 0.002x_5}$$

This regression model yields the coefficient of determination $R^2 = 0.466$ which implied that 46.6% of the variation in the number of death during an earthquake is explained by the regression equation.

7.4.4. Mathematical Model for Estimating Total Cost of Damage

The least squares estimating equations $(X'X)b = X'y$

$$\begin{bmatrix} 30 & 205 & 199.6 & 710 & 111 & 14436.4 \\ 205 & 1429 & 1374.7 & 4813 & 758 & 101312 \\ 199.6 & 1374.7 & 1341.62 & 4818.8 & 735 & 100163.82 \\ 710 & 4813 & 4818.8 & 23980 & 2399 & 356200 \\ 111 & 758 & 735 & 2399 & 491 & 50919.8 \\ 14436.4 & 101312.2 & 100163.82 & 356200 & 50919.8 & 22596163.36 \end{bmatrix} \begin{bmatrix} b_0 \\ b_1 \\ b_2 \\ b_3 \\ b_4 \\ b_5 \end{bmatrix} = \begin{bmatrix} 65.83 \\ 486.15 \\ 472.67 \\ 1471.57 \\ 271.48 \\ 50130.28 \end{bmatrix}$$

and then, using the relation $b = (X'X)^{-1}X'y$, the estimated regression coefficients are obtained as

$$b_0 = \ -17.075;\ \ b_1 = 0.234;\ \ b_2 = 2.520;\ b_3 = \ -0.033;\ b_4 = 0.383;\ b_5 = 0.001$$

The coefficients were obtained using MATLAB.Therefore, that total cost of damage in an earthquake event can be computed using Thus, the regression equation that estimates total cost of damage is

$$\boldsymbol{ln\hat{y}_4 = -17.075 + 0.234x_1 + 2.520x_2 - 0.033x_3 + 0.383x_4 + 0.001x_5}$$

This regression model yields the coefficient of determination $R^2 = 0.564$ which implied that 56.4% of the variation in the number of death during an earthquake is explained by the regression equation.

8. SUMMARY OF FINDINGS

From the results of data analysis, the following are the findings of the study:

8.1. Characteristics of Earthquakes:

1. Fifty percent of the total considered earthquake events that hit the Philippines from 1968 to 2012 were of intensity seven. These earthquakes were classified by the PHIVOLCS as destructive.

2. The average magnitude of the earthquakes that were considered in the research was 6.65 Ms with standard deviation of 0.69. According to the Richter magnitude scale used by the PHIVOLCS, those seismic activities with magnitude 6 to 7 are the strong earthquakes

3. The average depth of considered earthquakes from 1968 to 2012 was 23.67 kilometers with standard deviation of 15.73. Shallow earthquakes are those quakes with depth ranges from 0 to 70 kilometers.

4. The thirty percent of the earthquakes considered in the study had epicenters at Zone Four (4) and that is in West Luzon Trench, whereas the double forearc associated with the Manila Trench.

5. The average duration of the earthquake considered in the study was 481.21 seconds with standard deviation of 734.59.

8.2. Damages Caused By Earthquakes

1. The average death owing to earthquakes was 187.03 as associated to the median 4, which is more lifelike estimate of death. The most number of deaths aroused during the Moro Gulf earthquake in August 17, 1976 which left 3,792 persons dead.

2. The mean of the number of people injured was 468.27. The largest number of injured also happened during the Moro Gulf earthquake in 1976 which had wounded 9,240 persons.

3. The average number of affected families was 13,188 families.

4. The cost of property damaged had a mean of P466.65M and a median of P4.98M. The largest cost of damages occurred in July 16, 1990 earthquake with epicentre at Cabanatuan City.

8.3. Proposed Mathematical Models

$$ln\hat{y}_1 = -13.405 - 0.001x_1 + 2.214x_2 - 0.013x_3 + 0.374x_4 - 0.001x_5$$

$$ln\hat{y}_2 = -13.426 + 0.360x_1 + 1.899x_2 + 0.015x_3 + 0.444x_4 - 0.001x_5$$

$$ln\hat{y}_3 = -16.905 + 0.083x_1 + 2.577x_2 - 0.043x_3 + 0.630x_4 + 0.002x_5$$

$$ln\hat{y}_4 = -17.075 + 0.234x_1 + 2.520x_2 - 0.033x_3 + 0.383x_4 + 0.001x_5$$

8.4. Significance of the Models

The first model estimates number of death and was significant with p-value of 0.003 with has a strong positive linear relationship with correlation coefficient of 0.713.

The second model predicts number of injured persons and was significant with p-value of 0.004 with has a strong positive linear relationship with correlation coefficient of 0.703.

The third model predicts number of families affected and was significant with 0.007 p-value and has a positive linear relationship with correlation coefficient $R = 0.682$.

The last regression model estimates total cost of damage and was significant with 0.001 p-value. This has strong positive linear relationship with correlation coefficient $R = 0.751$

9. CONCLUSION

Based on the findings of the study, the conclusions were drawn:

The assumptions for multiple linear regression guided the researchers to formulate mathematical models that estimate the casualties and cost of damage of an earthquake event. The deviation from linearity guided them to know if there are significant relationships between the independent and dependent variables. Since the assumption for linearity did not meet, the researchers decided to transform the dependent variables and thus, the mathematical models can be used to estimate earthquake casualties and cost of damages.

10. RECOMMENDATION

The researchers propose looking for another transformation process, wherein the researchers will obtain high value of coefficient of determination and also advocate adding more independent variables like local geotectonic settings, aftershocks etc., to assess the possible damages of an earthquake event more accurate.

REFERENCES

BOOKS AND ONLINE RESOURCES

[1] Nelson, Stephen A. *"Natural Disasters"*, Tulane University,
[2]Arnold, E.P. (ed.) June 1985. *Series of Seismology*, Volume **IV** Philippines. Southeast Asia Association of Seismology and Earthquake Engineering,
[3] Bautista, Ma. Leonila P., 1996 "*Estimation of the Magnitudes and Epicenters of Historical Earthquakes of the Philippines*", (Unpublished Master's Thesis. Kyoto University,)
[4]Su, Sergio S, August 1980 "*Attenuation of Intensity with Epicentral Distance in the Philippines*", Bulletin of Seismological Society of America Vol. **70**,.
[5] Thenhaus, Paul C. et. al., 1993 "*Estimates of Economic Loss to Dwellings, Canlubang, Philippines*", (PHIVOLCS Unpublished Report,)
[6] Mayers,Mayers,Walpole.,1998 *Probability and Statistics for Engineers and Scientists,6th Edition*

DISSERTATION

[7] Olaya, Felicisima, Ph. D., Centro Escolar University, 2003,*A Proposed Mathematical Model for Estimating Possible Damages due to Earthquake.*

GOVERNMENT AGENCIES

[8]Philippine Institute of Volcanology and Seismology
[9] National Disaster Risk Reduction and Management Council

Numerical Solutions of Free Convection Boundary Layer Flow on a Solid Sphere with Convective Boundary Conditions

H T Alkasasbeh[1], M Z Salleh[1], R M Tahar[2] and R Nazar[3]

[1] Faculty of Industrial Science and Technology, University Malaysia Pahang, 26300 UMP Kuantan, Pahang, Malaysia
[2]Faculty of Technology, Universiti Malaysia Pahang, 26300 UMP Kuantan, Pahang, Malaysia
[3]School of Mathematical Sciences, Faculty of Science and Technology, Universiti Kebangsaan Malaysia, 43600 UKM Bangi, Selangor, Malaysia

Email: zukikuj@yahoo.com

Abstract. The free convection boundary layer flow on a solid sphere with convective boundary conditions has been investigated. The basic equations of boundary layer are transformed into a non-dimensional form and reduced to nonlinear systems of partial differential equations are solved numerically using an implicit finite difference scheme known as the Keller-box method. Numerical results are obtained for the wall temperature, the local heat transfer coefficient and the local skin friction coefficient, as well as the velocity and temperature profiles of the fluid. The features of the flow and heat transfer characteristics for Prandtl number, Pr = 0.7 7 and 100, the conjugate parameter γ = 0.05, 0.1, 0.2 and the coordinate running along the surface of the sphere, $0^o \leq x \leq 120^o$ are analyzed and discussed.

1. Introduction

Free convection boundary layer flow about a solid sphere represents also an important problem, which is related to numerous engineering applications. This returns to the fact that a sphere presents an important geometry for the study of the convection flow in many engineering applications (spherical storage tanks, packed beds of spherical bodies, etc.). The above mentioned investigations have been reviewed by Jafarpur and Yovanovich [1]. The boundary layer theory was first introduced by a Ludwig Prandtl, in his lecture on "Fluid Motion with Very small Friction" at the Heidelberg Mathematical Congress in August 1904 (Schlichting [2]). Using theoretical considerations together with some simple experiments, Prandtl showed that the flow past a body can be divided into two major parts. The larger part concerns a free stream of fluid, far from any solid surface, which is considered to be in viscid. The smaller part is a thin layer adjacent to the solid surface in which the effects of viscosity are felt. This thin layer where friction effects cannot be ignored is called the boundary layer (see Burmeister [3], Acheson [4]).

[1] To whom any correspondence should be addressed.

The analysis of free, forced and mixed convection about a sphere studied by Chen and Mucoglu [5]. Nazar et al. [6-8] considered the free and mixed convection boundary layer flows on a sphere in micropolar and viscous fluids, respectively. The natural convection heat and mass transfer from a sphere in micropolar fluids with constant wall temperature and concentration was presented by Cheng [9]. All the papers above considered the boundary condition of either constant wall temperature or constant heat flux.

It is worth mentioning that Newtonian heating (NH), in which the heat transfer from the surface is proportional to the local surface temperature has been first used by Merkin [1]. Salleh et al. [11-14] considered the free and mixed convection boundary layer flow on a sphere with Newtonian heating in viscous and micropolar fluids,respectively. The free convection boundary layer flow on solid sphere in viscoelastic fluid with Newtonian heating studied by Kasim et al. [15].

On the other hand, the convective boundary conditions in which the heat is supplied through a bounding surface of finite thickness and finite heat capacity recently used by Aziz [16] who obtained the similarity solution for laminar thermal boundary layer over a flat plate with a convective surface boundary condition. Subsequently, Makinde and Aziz [17] studied the magnetohydrodynamic (MHD) mixed convection flow from a vertical plate embedded in a porous medium with a convective boundary condition. Ishak et al. [18, 19] obtained the similarity solutions for flow and heat transfer over a permeable surface and the radiation effects on the thermal boundary layer flow over a moving plate with covective boundary conditions, respectively. The mixed convection boundary-layer flow past a horizontal circular cylinder embedded in a porous medium filled with a nanofluid with convective boundary condition presented by Rashad et al. [20]. The numerical solutions of the steady magnetohydrodynamic two dimensional stagnation point flow of an incompressible nanofluid towards a stretching cylinder with convective boundary condition using fourth-order Runge-Kutta-Fehlberg method with a shooting technique has been investigated by Akbar et al. [21]. Recently, the numerical solutions of stagnation point flow over a stretching surface with convective boundary conditions using the shooting method has been studied by Mohamed et al. [22].

Motivated by the above mentioned studies, therefore, the aim of the present paper is to study the free convection boundary layer flow on a solid sphere with convective boundary conditions. The governing boundary layer equations are first transformed into a system of non-dimensional equations via the non-dimensional variables, and then into non-similar equations before they are solved numerically by the Keller-box method, as described in the books by Cebeci and Bradshaw [23] and Na [24].

2. Mathematical Analyses

Consider a heated sphere of radius a, which is immersed in a viscous and incompressible fluid of ambient temperature T_∞. The surface of the sphere is subjected to a convective boundary conditions (CBC), as shown in figure 1.

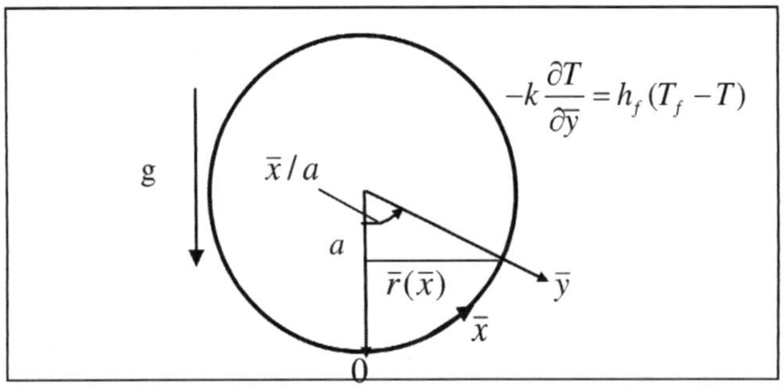

Figure 1. Physical model and coordinate system.

Under the Boussinesq and boundary layer approximations, the basic dimensional equations of the flow are

$$\frac{\partial}{\partial \bar{x}}(\bar{r}\,\bar{u}) + \frac{\partial}{\partial \bar{y}}(\bar{r}\,\bar{v}) = 0 \tag{1}$$

$$\bar{u}\frac{\partial \bar{u}}{\partial \bar{x}} + \bar{v}\frac{\partial \bar{u}}{\partial \bar{y}} = \nu \frac{\partial^2 \bar{u}}{\partial \bar{y}^2} + g\beta(T - T_\infty)\sin\left(\frac{\bar{x}}{a}\right) \tag{2}$$

$$\bar{u}\frac{\partial T}{\partial \bar{x}} + \bar{v}\frac{\partial T}{\partial \bar{y}} = \alpha \frac{\partial^2 T}{\partial \bar{y}^2} \tag{3}$$

subject to the boundary conditions

$$\bar{u} = \bar{v} = 0,\ -k\frac{\partial T}{\partial \bar{y}} = h_f(T_f - T) \text{ at } \bar{y} = 0$$

$$\bar{u} \to 0,\ T \to T_\infty \text{ as } \bar{y} \to \infty, \tag{4}$$

where $\bar{r}(\bar{x}) = a\sin(\bar{x}/a)$, $\bar{u}$ and $\bar{v}$ are the velocity components along the $\bar{x}$ and $\bar{y}$ directions, respectively. T is the local temperature, g is the gravity acceleration, β is the thermal expansion coefficient, T_f is the temperature of the hot fluid, ν is the kinematic viscosity, k is the thermal conductivity, α is the thermal diffusivity and h_f is the heat transfer coefficient.
We introduce now the following non-dimensional variables (Salleh et al. [11], Aziz [16]):

$$x = \frac{\bar{x}}{a},\ y = Gr^{1/4}\left(\frac{\bar{y}}{a}\right),\ r = \frac{\bar{r}}{a},$$

$$u = \left(\frac{a}{\nu}\right)Gr^{-1/2}\bar{u},\ v = \left(\frac{a}{\nu}\right)Gr^{-1/4}\bar{v},\ \theta = \frac{T - T_\infty}{T_f - T_\infty} \tag{5}$$

where $Gr = g\beta(T_f - T_\infty)\frac{a^3}{\nu^2}$ is the Grashof number. Substituting variables (5) into (1) to (3) then become

$$\frac{\partial(ru)}{\partial x} + \frac{\partial(rv)}{\partial y} = 0 \tag{6}$$

$$u\frac{\partial u}{\partial x} + v\frac{\partial u}{\partial y} = \frac{\partial^2 u}{\partial y^2} + \theta\sin x \tag{7}$$

$$u\frac{\partial \theta}{\partial x} + v\frac{\partial \theta}{\partial y} = \frac{1}{\Pr}\frac{\partial^2 \theta}{\partial y^2} \tag{8}$$

The boundary conditions (4) become

$$u = v = 0,\ \frac{\partial \theta}{\partial y} = -\gamma(1 - \theta) \text{ on } y = 0$$

$$u \to 0,\ \theta \to 0 \text{ as } y \to \infty, \tag{9}$$

where Pr is the Prandtl number, $\gamma = ah_f Gr^{-1/4}/k$ is the conjugate parameter for the convective boundary conditions, respectively. It is noticed that when $\gamma = 0$ is for the insulated plate and $\gamma \to \infty$ is when the surface temperature is constant. To solve equations (6) to (8), subjected to the boundary conditions (9), we assume the following variables:

$$\psi = xr(x)f(x,y),\ \theta = \theta(x,y), \tag{10}$$

where ψ is the stream function defined as

$$u = \frac{1}{r}\frac{\partial \psi}{\partial y} \text{ and } v = -\frac{1}{r}\frac{\partial \psi}{\partial x}, \tag{11}$$

which satisfies the continuity equation (6). Thus, equations (7) and (8) become

$$\frac{\partial^3 f}{\partial y^3} + (1 + x\cot x) f \frac{\partial^2 f}{\partial y^2} - \left(\frac{\partial f}{\partial y}\right)^2 + \frac{\sin x}{x}\theta = x\left(\frac{\partial f}{\partial y}\frac{\partial^2 f}{\partial x \partial y} - \frac{\partial f}{\partial x}\frac{\partial^2 f}{\partial y^2}\right) \tag{12}$$

$$\frac{1}{\Pr}\frac{\partial^2 \theta}{\partial y^2} + (1 + x\cot x) f \frac{\partial \theta}{\partial y} = x\left(\frac{\partial f}{\partial y}\frac{\partial \theta}{\partial x} - \frac{\partial f}{\partial x}\frac{\partial \theta}{\partial y}\right) \tag{13}$$

subject to the boundary conditions

$$f = \frac{\partial f}{\partial y} = 0,\ \frac{\partial \theta}{\partial y} = -\gamma(1-\theta) \text{ at } y = 0$$
$$\frac{\partial f}{\partial y} \to 0,\ \theta \to 0 \text{ as } y \to \infty \tag{14}$$

It can be seen that at the lower stagnation point of the sphere, $x \approx 0$, equations (12) and (13) reduce to the following ordinary differential equations:

$$f''' + 2ff'' - (f')^2 + \theta = 0 \tag{15}$$
$$\frac{1}{\Pr}\theta'' + 2f\theta' = 0 \tag{16}$$

and the boundary conditions (14) become

$$f(0) = f'(0) = 0,\ \theta'(0) = -\gamma(1-\theta(0)) \text{ (CBC)}$$
$$f' \to 0,\ \theta \to 0 \text{ as } y \to \infty \tag{17}$$

along with $\theta'(0) = -\gamma(1+\theta(0))$ (NH). Where primes denote differentiation with respect to y. The physical quantities of interest in this problem are the local skin friction coefficient, C_f and the local heat transfer coefficient, $Q_w(x)$ which are given by

$$C_f = x\frac{\partial^2 f}{\partial y^2}(x,0), \text{ and } Q_w(x) = \gamma(1-\theta(x,0)) \tag{18}$$

where $C_f = \tau_w/(\rho U_\infty^2)$ is the skin friction coefficient and $\tau_w = \mu(\partial\bar{u}/\partial\bar{y})_{\bar{y}=0}$ is the wall shear stress. At the lower stagnation point of the sphere, $x \approx 0$, the skin friction coefficients and the heat transfer coefficient are measured by $f''(0)$ and $-\theta'(0)$ respectively.

3. Results and Discussion

Equations (12) and (13) subject to the boundary conditions (14) were solved numerically using the Keller-box method for both cases of boundary conditions, Newtonian heating (NH) and convective boundary conditions (CBC) with several parameters considered, namely the Prandtl number Pr, the conjugate parameter γ and the coordinate running along the surface of the sphere, x. The numerical solutions start at the lower stagnation point of the sphere, $x \approx 0$, with initial profiles as given by equations (15) and (16) and proceed round the sphere up to 120° (see Nazar et al. [6, 7]). Values of Pr considered are Pr = 0.7, 7 and 100. It is worth mentioning that small values of Pr ($\ll 1$) physically correspond to liquid metals, which have high thermal conductivity but low viscosity, while large values of Pr ($\gg 1$) correspond to high-viscosity oils. It is also worth to be pointed out that specifically, the Prandtl number considered in this study, namely Pr = 0.7, 7 and 100, correspond to air, water and engine oil, respectively.

Tables 1 show the values of the wall temperature $\theta(0)$ for the case of Newtonian heating (NH). Some results reported by Salleh et al. [11] are included in these tables. It is seen that the agreement between the previously published results with the present ones is very good.

For the case of convective boundary conditions (CBC), Table 2 presents the values of the wall temperature $\theta(0)$ and the skin friction coefficient $f''(0)$ at the lower stagnation point of the sphere, $x \approx 0$, when Pr= 0.7, 7, 100 and $\gamma = 0.1$. It is found that, when Pr increases, the values of $\theta(0)$ and $f''(0)$ decrease.

Tables 3, 4 and Figures 6, 7 shows the values of the local heat transfer coefficient $Q_w(x)$ and the skin friction coefficient, C_f for various values of x when Pr = 0.7, 7, 100 and $\gamma = 0.1$, respectively. It is found that as Pr increases, the values of $Q_w(x)$ increases and C_f decrease. On the other hand, for a fixed Pr, as x increases, i.e. between the interval $0^o \le x \le 120^o$, the values of $Q_w(x)$ decrease and C_f increase. From Table 3 and Figures 6, it is seen that the value of $Q_w(x)$ is significantly higher at $x = 0^o$ than those at $0^o < x \le 120^o$, because the sphere temperature is almost equal to fluid temperature at $x = 0^o$, and has a different value when $0^o < x \le 120^o$. From Table 4 and Figures 7, it is found that the value of $C_f = 0$ at $x = 0^o$, because at this point the value of the wall shear stress τ_f is very small. On the other hand, the maximum value of C_f appears when $x = 120^o$, because in this case the value of τ_f is very high.

Figure 2 illustrates the variation of the wall temperature $\theta(x,0)$ with conjugate parameter γ when Pr = 0.7, 7 and 100. To get a physically acceptable solution, γ must be less than or equals to some critical value, say γ_c, i.e. $\gamma \le \gamma_c$, depending on Pr. It is found that, the critical value of γ, γ_c is 0.3766 when Pr = 0.7, 0.5971 when Pr = 7, and 0.6892 when Pr = 100.

The graphs of $\theta(x,0)$ for some values of the Prandtl number Pr when γ = 0.05, 0.1 and 0.2 are plotted in Figure3. It is found that, as the Prandtl number Pr increases, the wall temperature $\theta(x,0)$ decrease, and $\theta(x,0)$ increase as γ increases. For small values of $Pr << 1$, the difference value changing is higher than for large values of $Pr >> 1$, and it is seen that the surface temperature is very sensitive to the Prandtl number variations.

Figures 4 and 5 display the temperature and velocity profiles, respectively, at $x = 0^o, 60^o, 90^o$ when Pr = 0.7, 7, 100 and $\gamma = 0.1$, respectively. It is found that as when x is fixed and Pr increases, the temperature and velocity profiles decrease and also the thermal boundary layer thickness decrease. This is because for small values of the Prandtl number Pr (≪1), the fluid is highly conductive. Physically, if Pr increases, the thermal diffusivity decreases and this phenomenon leads to the decreasing manner of the energy transfer ability that reduces the thermal boundary layer. On the other hand, in the same figure it has been found that when Pr is fixed and x increases, the temperature, velocity profiles and the thermal boundary layer thickness increase.

Table 1. Values of the wall temperature $\theta(0)$ at the lower stagnation point of the sphere, $x \approx 0$, when Pr= 0.7, 1, 7 and $\gamma = 1$ (NH)

	$\theta(0)$	
Pr	Salleh et al. [11]	Present
0.7	26.4590	26.459113
1	17.2876	17.287591
7	3.3635	3.3634765

Table 2. Values of the wall temperature $\theta(0)$ and the skin friction coefficient $f''(0)$ at the lower stagnation point of the sphere, $x \approx 0$, when Pr= 0.7, 7 100 and $\gamma = 0.1$ (CBC)

Pr	$\theta(0)$	$f''(0)$
	Present	Present
0.7	0.236634	0.261164
7	0.148764	0.117331
100	0.090019	0.042963

Table 3. Values of the local heat transfer coefficient $Q_w(x)$ for various values of x when Pr = 0.7, 7, 100 and $\gamma = 0.1$ (CBC)

Pr	0.7	7	100
x	Present	Present	Present
0°	0.083615	0.089018	0.092998
10°	0.083420	0.088868	0.092982
20°	0.083366	0.088832	0.092960
30°	0.083256	0.088769	0.092920
40°	0.083098	0.088673	0.092863
50°	0.082892	0.088551	0.092786
60°	0.082638	0.088400	0.092678
70°	0.082311	0.088213	0.092549
80°	0.081906	0.088005	0.092392
90°	0.081409	0.087735	0.092204
100°	0.080795	0.087446	0.091969
110°	0.080060	0.087112	0.091704
120°	0.079085	0.086677	0.091356

Table 4.Values of the local skin friction coefficient, C_f for various values of x when Pr = 0.7, 7, 100 and $\gamma = 0.1$(CBC)

Pr	0.7	7	100
x	Present	Present	Present
0°	0.000000	0.000000	0.000000
10°	0.024339	0.010401	0.003838
20°	0.047659	0.020366	0.007516
30°	0.071418	0.030349	0.011241
40°	0.094013	0.040072	0.014760
50°	0.115443	0.049131	0.018063
60°	0.135174	0.057400	0.021113
70°	0.153510	0.064763	0.023958
80°	0.169912	0.071107	0.026411
90°	0.184092	0.076146	0.028611
100°	0.195770	0.080327	0.030327
110°	0.204355	0.083037	0.031600
120°	0.209970	0.084369	0.032472

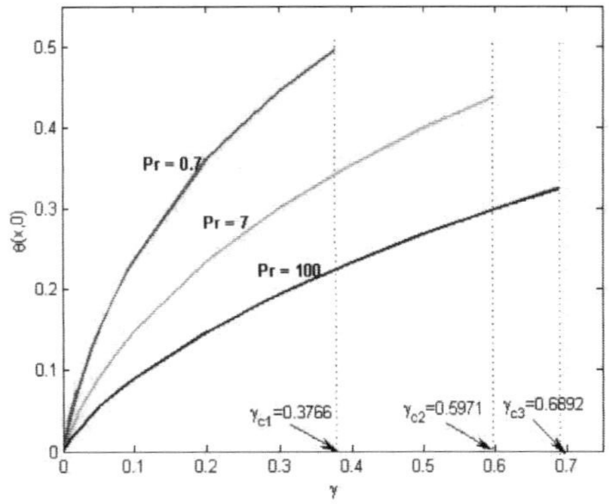

Figure 2. Variation of the wall temperature $\theta(x,0)$ with conjugate parameter γ when Pr = 0.7, 7, 100

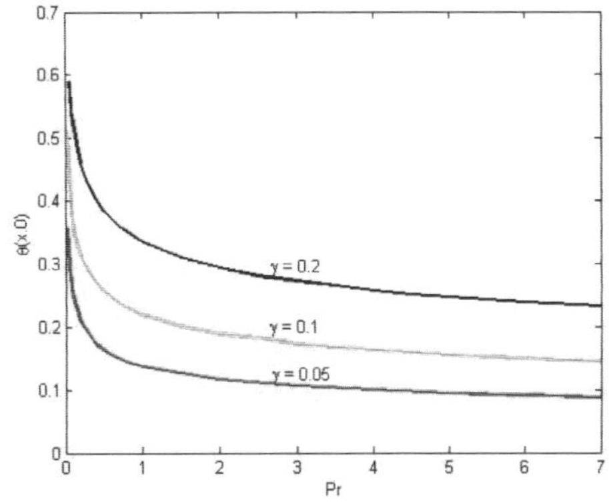

Figure 3. Variation of the wall temperature $\theta(x,0)$ with Prandtl number Pr when γ = 0.05, 0.1, 0.2

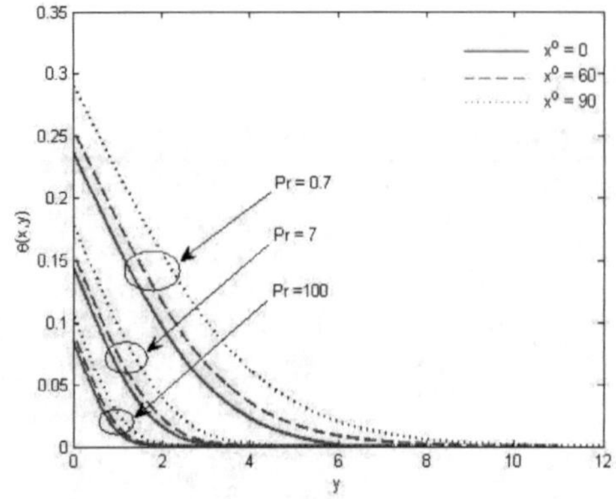

Figure 4. Temperature profiles $\theta(x, y)$ at $x = 0^o, 60^o, 90^o$ when Pr = 0.7, 7, 100 and $\gamma = 0.1$

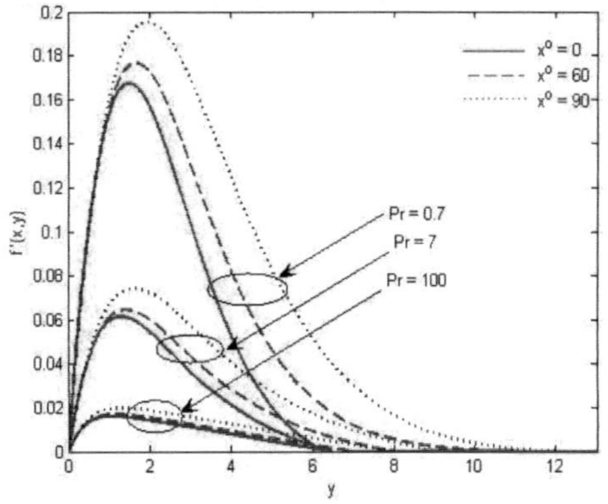

Figure 5. Velocity profiles $f'(x, y)$ at $x = 0^o, 60^o, 90^o$ when Pr = 0.7, 7, 100 and $\gamma = 0.1$

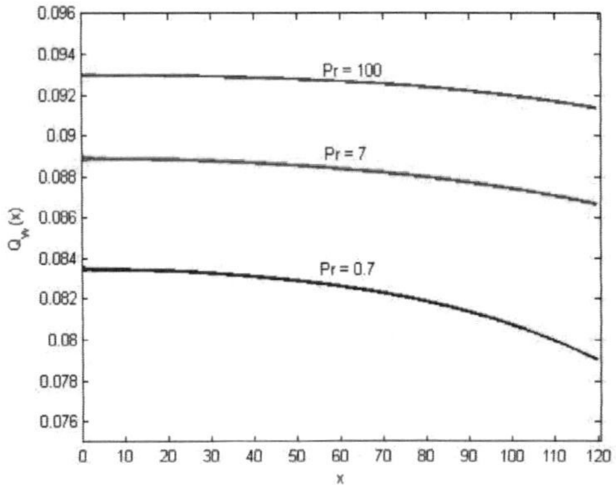

Figure 6. Variation of local heat transfer coefficient with x when Pr = 0.7, 7, 100 and $\gamma = 0.1$

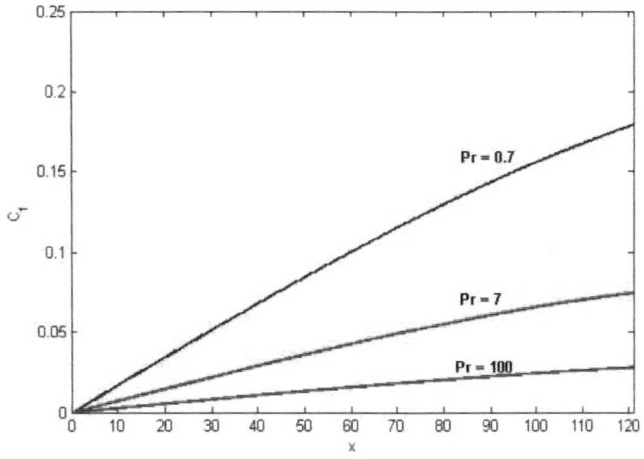

Figure 7. Variation of the skin friction coefficient with x when Pr = 0.7, 7 and 100 and $\gamma = 0.1$

4. Conclusions

In this paper, we have numerically studied the problem of free convection boundary layer flow on a sphere with convective boundary conditions. We are interested to see how the Prandtl number Pr and the conjugate parameter of convective boundary conditions $\gamma = ah_f Gr^{-1/4}/k$ affect the flow and heat transfer characteristics. We can conclude that, when Pr increases and γ is fixed, the temperature and velocity profiles, the local heat transfer coefficient $Q_w(x)$ increases and the local skin friction coefficient C_f decrease, and when Pr is fixed and the coordinate running along the surface of the sphere x increases, the temperature and velocity profiles and the thermal boundary layer thickness increase. Further, it is seen that an increase in the Prandtl number Pr, results in a decrease of the wall temperature $\theta(x,0)$. However, increasing of conjugate parameter γ leads to an increase of the wall temperature $\theta(x,0)$.

Acknowledgement

The authors gratefully acknowledge the financial supports received from the Universiti Malaysia Pahang (RDU 120390 and RDU 121302).

References

[1] Jafarpur K and Yovanovich M 1992 *Int. J. Heat. Mass. Transfer* **35** 2195-2201
[2] Schlichting H 1979 *Boundary layer theory* New York:McGraw-Hill Inc
[3] Burmeister L C 1993 *Convective heat transfer* New York:John Wiley & Sons Inc
[4] Acheson D J 1990 *Elementary fluid dynamics* Oxford University Press
[5] Chen T and Mucoglu A 1977 *Int. J. Heat. Mass. Transfer* **20** 867
[6] Nazar R, Amin N, Grosan T and Pop I 2002a *Int.Comm. Heat. Mass. Transfer* **29** 377
[7] Nazar R, Amin N, Grosan T and Pop I 2002b *Int.Comm. Heat. Mass. Transfer* **29** 1129
[8] Nazar R, Amin N and Pop I 2002c *Arab. J. Sci. Eng* **27** 117
[9] Cheng C Y 2008 *Int.Comm. Heat. Mass. Transfer* **35** 750
[10] Merkin J 1994 *Int. J.Heat. Fluid. Flow* **15** 392
[11] Salleh M Z, Nazar R and Pop I 2010 *Acta Applic Math* **112** 263
[12] Salleh M Z, Nazar R and Pop I 2010 *ArchivesMechanics* **62** 283
[13] Salleh M Z, Nazar R and Pop I 2012 *Meccanica* **47** 1261
[14] Salleh M Z, Nazar R and Pop I 2010 *SRX Physics* **62** 283

[15] Kasim A R, Mohammad N, Aurangzaib S S and Sharidan S 2012 *World Acad. Sci Eng Tech* **64** 628
[16] Aziz A A 2009 *Commun. Nonlinear Sci. Numer. Simul.* **14** 1064
[17] Makinde O and Aziz A 2010 *Int. J. Therm. Sci.* **49** 1813
[18] Ishak A 2010 *Appl. Math. Comput.* **217** 837
[19] Ishak A, Yacob N and Bachok N 2011 *Meccanica* **46** 795
[20] Rashad A, Chamkha A and Modather M 2013 *Comp. Fluids* **86** 380
[21] Akbar N S, Nadeem S, Haq R U and Khan Z 2013 *Chinese. J Aeron* doi.org/10.1016/j.cja.2013.10.008
[22] Mohamed M K A, Salleh M Z, Nazar R and Ishak A 2013 *Boundary Value Problems* **4** 1
[23] Cebeci T and Bradshaw P 1984 *Physical and computational aspects of convective heat transfer* Springer, New York
[24] Na T Y 1979 *Computational methods in engineering boundary value problem* New York: Academic Press

IDENTIFYING POTENTIAL MISFIT ITEMS IN COGNITIVE PROCESS OF LEARNING ENGINEERING MATHEMATICS BASED ON RASCH MODEL

Sh Ataei[1]
PhD Student, *Faculty of Computer and Mathematical Sciences*
Universiti Teknologi MARA Shah Alam, Selangor, Malaysia
Z Mahmud[2]
Assoc Prof Dr, *Faculty of Computer and Mathematical Sciences*
Universiti Teknologi MARA Shah Alam, Selangor, Malaysia
M N Khalid[3]
Assoc Prof Dr, *Faculty of Computer and Mathematical Sciences*
Universiti Teknologi MARA Shah Alam, Selangor, Malaysia

E-mail: sholehataei@gmail.com, zamal669@salam.uitm.edu.my, mnasir@tmsk.uitm.edu.my

Abstract. The students learning outcomes clarify what students should know and be able to demonstrate after completing their course. So, one of the issues on the process of teaching and learning is how to assess students' learning. This paper describes an application of the dichotomous Rasch measurement model in measuring the cognitive process of engineering students' learning of mathematics. This study provides insights into the perspective of 54 engineering students' cognitive ability in learning Calculus III based on Bloom's Taxonomy on 31 items. The results denote that some of the examination questions are either too difficult or too easy for the majority of the students. This analysis yields FIT statistics which are able to identify if there is data departure from the Rasch theoretical model. The study has identified some potential misfit items based on the measurement of ZSTD where the removal misfit item was accomplished based on the MNSQ outfit of above 1.3 or less than 0.7 logit. Therefore, it is recommended that these items be reviewed or revised to better match the range of students' ability in the respective course.

1. Introduction

According to Kilpatrick & Swafford [1], for students to be successful in later mathematics endeavours and to use mathematics effectively in life, they must have a sound learning of mathematics concepts. Mathematics helps to strengthen and to accelerate students understanding in many varieties of engineering courses.

There are many ways to assess learning. Among them are the traditional and conventional methods such as currently practiced by educators in the current education system. Assessment comes in various forms and methods. In the current Outcome-Based Education (OBE) system, measurement of learning is gauged against the course learning outcome while integrating Bloom's Taxonomy concept in the measurement process. However, it is still unclear as to the degree of learning that has actually taken place with the current assessment methods. Therefore, having one measurement tool that can help to measure one domain of Bloom's Taxonomy is important.

One such tool is based on the IRT (Item Response Theory) model, namely the Rasch measurement model [2]-[3]-[4]-[5], [6]. It is a scale measurement which is estimated by responses of persons and items' properties [7]. According to some researchers [8]-[9]-[10]-[11], [12], this measurement is able to measure spread and separation of the items and task areas based on Bloom's Taxonomy.

Item Response Theory (IRT) tries to apply the mathematical model in order to anticipate the probability of a person of certain ability attempting an item of certain difficulty successfully. This would depend on the log transformation of the difference between logit difficulty of item and logit ability of a person.. In addition, Person- Item Distribution Map (PIDM) in the Rasch model shows the logit transformation of raw score of person ability and item difficulty that both are calibrated against a single logit ruler within the range. [5]- [9]-[11]-[13]-[14]-[15], [16].

2. Measurement models

Assessment plays an important role in education measurement. A suitable measurement should be able to (a) provide linear measures, (b) overcome missing data, (c) give precision estimates, (d) detect misfits and (e) item or task to be completed and person must be separated from each other [5]. Therefore, the selection of proper tools for assessing mathematics learning is through the application of the Rasch measurement model which is expected to respond to all the properties above. Rasch model is a mathematical model that is applicable for ascertaining certain properties in data. The Rasch model expresses the relation between the probability of completing a task, the ability of persons and the difficulty of the task. A Rasch model transforms raw scores (ordinal data) into a linear interval scale data.

2.1. Dichotomous Rasch model

Dichotomous data is acceptable to be fitted into the Rasch model where the difference between two parameters namely, person ability (δ_i) and item difficulty (θ_n) are governed by a single logit scale as a result of the log transformation of the probability of completing a task given the ability of a person and the difficulty of the task or item given. The Rasch model is the simple logit model of IRT such as the Dichotomous Rasch model which has two options for answering either a yes (1) or no (0), correct or incorrect, agree or disagree. The Rasch measurement model function based on Rasch is illustrated as follows:

θ_n is a parameter: Person ability for responding to questions (items) and δ_i is a parameter: Difficulty of questions(items) for answering.

$$P(X=1)=\frac{\exp(\theta_n-\delta_i)}{1+\exp(\theta_n-\delta_i)} \tag{1}$$

$$P(X=0)=1-P(X=1)=\frac{1}{1+\exp^{(\theta_n-\delta_i)}} \tag{2}$$

$$\ln\left[\frac{\dfrac{e^{\theta_n-\delta_i}}{1+e^{\theta_n-\delta_i}}}{1-\left[\dfrac{e^{\theta_n-\delta_i}}{1+e^{\theta_n-\delta_i}}\right]}\right]=\theta_n-\delta_i \tag{3}$$

Correct response probability, incorrect response probability and parameter separation are achieved in Equations (1), (2) and (3). [3]- [4], [5]- [6]-[17], [18].

2.2. Simple Rasch Logistic model

The simple Rasch model (SRM; Rasch, 1960/1980) for the analysis of test data provides a way to place persons and items on a scale with a clear probabilistic interpretation of distance on the scale. Let the attitude of a person be characterized by the variable and the affective value of item. Item difficulty is 0.5 when person ability and item difficulty are equal.

The success probability is a function of the difference of person ability and item difficulty where $\log\frac{p}{1-p}=\theta_n-\delta_i$, $\theta_n-\delta_i$ is the difference between person ability and item difficulty.

If two persons are of different abilities but are attempting the same item $\delta_1=\delta_2=\delta_i$, (a) first person will have $\log\frac{p}{1-p}=\theta_1-\delta_i$, and (b) the second person will have $\log\frac{p}{1-p}=\theta_2-\delta_i$ and so on. This Rasch property is very important because we can achieve item difficulty without considering specific person or achieve person difficulty without considering specific items.

For example, in determining differences between two persons (item free), it will be

$$\log\frac{p_1}{1-p_1}-\log\frac{p_2}{1-p_2}=\theta_1-\delta_i-(\theta_2-\delta_i)=\theta_1-\theta_2 \tag{3}$$

These equation represents probability of success where probability of success $=f(\theta-\delta)$ (4) [4]-[5]-[6]-[19]-[20]-[21] and [22].

2.3Misfit

In Rasch measurement, fit statistics is able to help us expose the discrepancies between observe real data we have gathered in practice and expected data from Rasch model.

On the other hand misfit data is defined as items which fall outside certain range of acceptable fit. Green and Frantom [2] and [3] defined item fit as an index of items function which can provide a continuum of answers. In this study, we use outfit that is based on the sum of squared standardized residuals. Consider X is an observation; E is its expected value underpinning Rasch parameter estimates while σ^2 is its modelled variance on its expectation. Therefore, the standardized unweighted mean square outfit is below:

$$t_i=(u_i^{\frac{1}{3}}-1)(\frac{3}{q_i})+(\frac{q_i}{3}) \tag{5}$$

when u_i is fit mean square and q_i is fit standard error[3]-[4]-[20],[21].

Green and Frantom [2]believed that outfit happens if mean square lies between 0.7 and 1.3.The standard deviation of the standardized outfit is an index of overall misfit for persons and items when it be bigger than +2 and less than -2 [3]- [4]-[20]-[21],[22].

3. Data Analysis and Results

The results of engineering students in a Calculus III course from the Faculty of Mechanical and Electrical Engineering, Universiti Teknologi MARA are discussed in this study. In order to examine students' understanding of Calculus III, the examination questions were prepared according to the six Bloom's taxonomy cognitive levels. The questionnaire consists of 31 Calculus III multiple choice questions and 72 attitude items which were administered to 54 students.

3.1. Summary statistics

Table 1 shows the summary of 54 measured persons and 31 measured items. It shows the mean measured person at -0.24 logit and mean measured item at 0.0 logit. This indicates the test is slightly difficult for the majority of the students. Reliability index for a person is measured at 0.78 logit which indicates a moderately acceptable fit with a separation index of about 2.0. This indicates a moderately acceptable fit with a person's ability ranging from high to low and vice versa. The reliability index for the item is measured at 0.88 logit with a separation index of about 3.0. This suggests a slightly broader continuum for items than persons.

Table 1. Summary of 54 measured person.

	SCORE	TOTAL COUNT	MEASURE	MODEL ERROR	INFIT MNSQ	INFIT ZSTD	OUTFIT MNSQ	OUTFIT ZSTD
MEAN	12.4	27.5	-.24	.46	.98	.0	.95	.0
S.D.	3.9	4.1	1.07	.17	.20	.9	.34	.7
MAX.	19.0	31.0	4.13	1.30	1.36	1.9	2.79	3.6
MIN.	5.0	14.0	-1.87	.39	.25	-2.0	.06	-1.2

REAL RMSE .51 TRUE SD .95 SEPARATION 1.87 Person RELIABILITY .78
MODEL RMSE .49 TRUE SD .95 SEPARATION 1.92 Person RELIABILITY .79
S.E. OF Person MEAN = .15

Table 2. Summary of 31 measured Item.

	SCORE	TOTAL COUNT	MEASURE	MODEL ERROR	INFIT MNSQ	INFIT ZSTD	OUTFIT MNSQ	OUTFIT ZSTD
MEAN	21.6	47.9	.00	.35	1.01	.1	.97	.1
S.D.	8.5	3.4	1.19	.15	.15	.9	.16	.8
MAX.	38.0	53.0	4.69	1.13	1.58	1.9	1.34	2.3
MIN.	1.0	38.0	-1.90	.29	.83	-1.6	.69	-1.0

REAL RMSE .42 TRUE SD 1.12 SEPARATION 2.67 Item RELIABILITY .88
MODEL RMSE .38 TRUE SD 1.13 SEPARATION 2.95 Item RELIABILITY .90
S.E. OF Item MEAN = .22

3.2 Model Fit

3.2.1. Person-Item- Distribution Map

Figure 1 indicates the person-item distribution map which is the heart of Rasch analysis. The map shows that persons are distributed on the left side of the logit ruler and items on the right side of the logit ruler. In Figure 1, it can be seen that a person's mean logit falls slightly below item mean logit and also there are gaps on the left and right side in upper mean logit line that are known as person and item free respectively. These indicate the test is slightly difficult for students with limited ability.

The map shows that two students from the Faculty of Mechanical Engineering are located at 4.0 logit, seen as most able students while two from the Faculty of Electrical Engineering located at -1.8 logit are seen as least able. It is observed that item C8 of Bloom's Taxonomy comprehension level is the most difficult item while items C26 and C29 at the application level are considered easiest by the majority of the students.

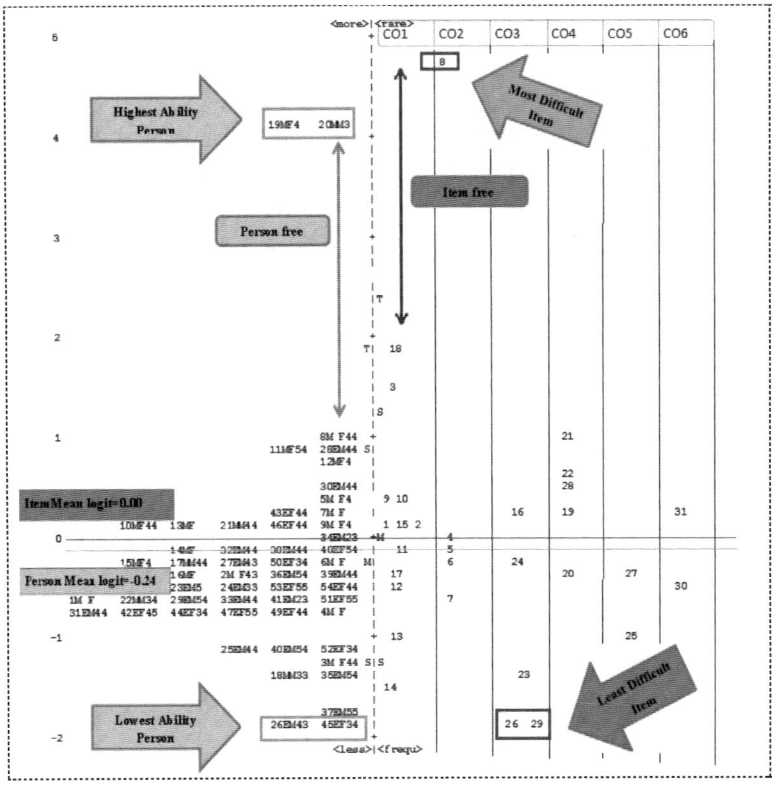

BLOOMS TAXONOMY

CO1: Knowledge

CO2: Comprehension

CO3: Application

CO4: Analysis

CO5: Synthesis

CO6: Evaluation

Figure1. Person-Item Distribution Map for learning Calculus III across Bloom's Taxonomy levels.

3.2.2. Potential misfitting item

In Figure 2, the items are shown as circles and the size of the circles indicates standard errors of measurement along the latent variable. Items were plotted to items difficulty (y-axis) and outfit ZSTD statistics (x-axis). An item is considered as potential item misfit if plot range is outside of -2 to +2 or circles are big [20], [22].

In Figure 2, all items are within the -2 and +2 range (item difficulty), except item 8 (item point :(0.58, 4.69)) and the size of points is 1.13. As can be observed measure difficulty of item 8 is $4.6 > +2$ and the largest standard error with the largest circle relates (1.13) to Item 8, indication that Item 8 is potential item misfit. As a mentioned in Figure 1, Item 8 is the most difficult item where all students were not able to answer it correctly. Figure 2 shows Items 16 (item point:(2.3,0.25), size:0.33) and also ZSTD of item 2(item point:(1.92,0.1),size:0.31) is near 2 that indicates items misfit.

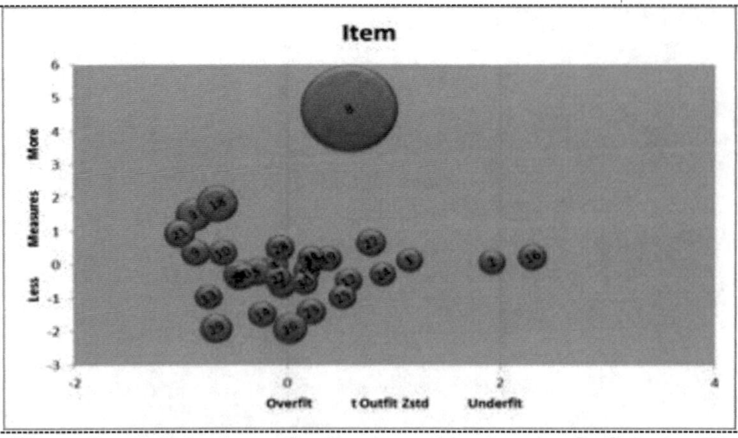

Figure 2 . Bubble chart – Students' cognitive learning In Calculus III

From Table 2, item 16 could be considered a misfit item because its standard deviation of the standardized outfit is above +2.0 and its MNSQ value is close to 1.4. Attention should be given to this item as it could allow for possible unusual responses. According to Table 2 we cannot consider item 8 as a misfit item. Similarly, Liu shows that some potential items misfit cannot be misfit items [22].

Table 2. Summary of 31 Measured Items

ENTRY NUMBER	TOTAL SCORE	TOTAL COUNT	MEASURE	MODEL S.E.	INFIT MNSQ	INFIT ZSTD	OUTFIT MNSQ	OUTFIT ZSTD	PT-MEASURE CORR.	PT-MEASURE EXP.	EXACT MATCH OBS%	EXACT MATCH EXP%	Item
8	1	45	4.63	1.13	1.58	.9	1.17	.6	A .13	.42	97.8	97.7	8
16	16	44	.25	.33	1.23	1.9	1.34	2.3	B-.13	.27	56.8	66.1	16
1	21	52	.15	.30	1.17	1.6	1.27	1.1	C .19	.36	63.5	66.7	1
2	19	48	.10	.31	1.18	1.9	1.22	1.9	D-.06	.26	52.1	63.9	2
22	13	47	.68	.34	1.12	.8	1.15	.8	E .08	.27	72.3	73.1	22
19	20	51	.21	.31	1.11	1.0	1.07	.4	F .27	.36	60.8	67.6	19
26	37	46	-1.90	.39	1.09	.5	.98	.0	G .17	.25	80.4	80.5	26
24	23	48	-.26	.30	1.06	.7	1.09	.9	H .20	.30	62.5	61.8	24
25	30	49	-.95	.31	1.03	.3	1.06	.5	I .23	.28	63.3	65.7	25
12	24	47	-.46	.31	1.05	.6	1.06	.6	J .24	.30	53.2	62.2	12
23	35	49	-1.39	.33	1.05	.4	1.03	.2	K .20	.27	73.5	72.5	23
31	18	49	.23	.31	1.05	.5	1.02	.2	L .22	.28	63.3	66.3	31
15	20	51	.16	.31	1.04	.4	1.04	.2	M .33	.36	70.6	67.8	15
11	24	51	-.17	.30	1.02	.3	1.03	.2	N .32	.35	64.7	64.2	11
30	27	51	-.53	.29	1.00	.1	1.01	.2	O .30	.30	64.7	62.3	30
7	24	43	-.58	.33	1.01	.1	.94	.0	P .33	.32	62.8	63.9	7
4	21	51	-.01	.30	1.01	.1	.98	-.1	o .29	.30	64.7	64.3	4
27	23	47	-.38	.31	.97	-.4	.99	-.1	n .33	.29	72.3	61.6	27
28	14	46	.52	.33	.95	-.3	.98	-.1	m .32	.26	78.3	70.6	28
5	18	38	-.19	.35	.96	-.4	.87	-.3	l .41	.37	65.8	64.6	5
10	18	49	.38	.32	.93	-.5	.85	-.6	k .43	.36	67.3	69.2	10
14	38	52	-1.46	.33	.93	-.4	.82	-.2	j .34	.26	78.8	74.3	14
17	27	53	-.36	.29	.93	-.9	.86	-.5	i .40	.33	66.0	63.5	17
20	25	49	-.31	.31	.91	-1.0	.86	-.4	h .41	.33	71.4	63.5	20
6	19	39	-.26	.35	.90	-1.0	.82	-.5	g .44	.36	69.2	64.2	6
9	18	49	.38	.32	.88	-1.0	.80	-.9	f .48	.36	71.4	69.2	9
3	9	49	1.51	.41	.87	-.4	.69	-.9	e .56	.44	85.7	85.1	3
29	36	46	-1.89	.37	.86	-.7	.82	-.7	d .47	.25	78.3	78.3	29
21	13	50	.95	.35	.86	-.7	.73	-1.0	c .53	.40	78.0	77.6	21
13	31	50	-.94	.31	.84	-1.6	.79	-.7	b .46	.28	76.0	66.4	13
18	7	46	1.85	.47	.83	-.4	.69	-.7	a .60	.46	89.1	88.3	18
MEAN	21.6	47.9	.00	.35	1.01	.1	.97	.1			70.2	69.8	
S.D.	8.5	3.4	1.19	.15	.15	.9	.16	.8			9.9	8.5	

4. Conclusion

This study has demonstrated that proper measurement of students' learning of Calculus III can be measured using the Rasch measurement models. Rasch models clearly have an advantage to accurately measure students' learning of Calculus III across the six cognitive levels of Bloom's Taxonomy.

Rasch measurement is also able to assess the reliability of the person and item measure and identify any misfits due to unusual responses given by the students. Calibration of items and students' responses using PIDM map has demonstrated the ability level of students against the difficulty of the items. This provides an excellent diagnosis to gauge students' learning of Calculus III. The use of misfit items is able to detect not only some dimensionality problem in the items but also determine which items of test should be deleted. Also, this study shows that potential item misfit can be different from item misfit. Generally, the empirical data moderately fit the expected Rasch model. Based on the Rasch measurement, students find that some Calculus III exam questions relating to knowledge and comprehension level are too difficult but they are can answer questions relating to application level better. The analysis also showed that some of the examination questions were either too difficult or too easy for the majority of the students.

In future study, we will investigate why engineering students are able to answer questions of application level without understanding the relative concepts. Also, having discussed about the relationship between potential item misfit and item misfit via tables and figures, the strongest arguments exists with respect to their mathematical formulas.

Acknowledgment

The authors would like to thank Universiti Teknologi MARA for funding this research under the Research Intensive Faculty grant scheme No. 600-RMI/DANA 5/3/RIF (866/2012). This research would not be possible without the university's assistance.

References

[1] Kilpatrick J, Swafford J, and Findell B 2001 *Adding it up: Helping children learn mathematics* (National Research Council (U.S.) Mathematics Learning Study Committee National Academies Press)

[2] Green K E, and Frantom C G 2002 Survey *Development and Validation with the Rasch Model. International conference on questionnaire development, Evaluation, and Testing* (Charleston, SC)

[3] Bond T G and Fox C M 2007 (2nd Ed.) *Applying the Rasch Model Fundamental Measurement in the Human Services* (Mahwah, NJ, Erlbaum)

[4] Rasch G 1960/1980 Probabilistic models for some intelligence and attainment tests Copenhagen, Danish Institute for Educational Research),ed B D Wright 1980 (The University of Chicago Press)

[5] Wu M and Adams R 2007 *Applying the Rasch Model to Psycho-Social Measurement: A Practical Approach*

[6] Alagumalai S, Curtis D D and Hungi N 2005 *Applied Rasch Measurement A Book of Exemplars: Papers in Honour of John P. Keeves* (Springer)

[7] Embretson S E and Reise S P 2000 *Item Response Theory for Psychologists* (Mahwah, NJ: Erlbaum)

[8] Othman H, Asshaari I, Ismail N A and Nopiah Z M 2011 *Evaluating the Reliability And Quality Of Final Exam Questions Using Rasch Measurement Model: A Case Study Of Engineering Mathematics Courses,* Seminar Pendidikan Kejuruteraan Kongres Pengajaran dan Pembelajaran and Alam Bina (PeKA'11)

[9] Kamsuriah A, Nazlena M A and Suhaila Z 2011 *An Improved Course Assessment Measurement for Analyzing Learning Outcomes Performance Using Rasch Model*, Procedia Social and Behavioral Sciences (Elsevier Lt) **18** 442-449

[10] Mahmud Z 2011 Diagnosis of Perceived Attitude, Importance and Knowledge in Statistics based on Rasch Probabilistic Model *Int.J. Applied Mathematics and Informatics* **5** 291-298
[11] Riccardo D N 2010 *Measure School Learning Through Rasch Analysis: The Interpretion of Results,* Procedia Social and Behavioral Sciences (Elsevier) **9** 373-377
[12] Mahmud Z, Musa N A C, Ghani N A M and Rahim A M 2012 Probabilistic Approach in Examining Quality of Survey Response Data in Statistical Education Research *International J. Mathematical Models and Methods in Applied Sciences* **6** 182-190
[13] Abdul Aziz A, Masodi M S, Mohamed Ibrahim F, Omar M Z and Zaharim A 2013 Insight into Engineering Education Learning Outcome's Assessment with Rasch Model Research *J. Applied Sciences, Engineering and Technology* **6** 3520-3526
[14] Asshaari I, Othman H, Bahaludin H, Ismail N A and Nopiah Z M 2011 *Rasch measurement in evaluation of Blooms' separation: A case study in Engineering mathematics III Course*, Seminar Pendidikan Kejuruteraan Kongres Pengajaran dan Pembelajaran and Alam bina
[15] Osman S A, Naam S I, Jaafar O, Hamidon W, Badaruzzaman W and Rahmat R A A 2012 *Application of Rasch Model in Measuring Students' Performance In Civil Engineering Design II Course* Procedia Social and Behavioral Sciences (Elsevier) **56** 59–66
[16] Osman S A, Badaruzzaman W H, Hamid R, Taib K, Khalim A R, Hamzah N and Jaafar O 2011 *Assessment on Students Performance Using Rasch Model in Reinforced Concrete Design Course Examination,* Proceedings of 10th WSEAS International Conference on Education and Educational Technology
[17] Andrich D 1988 *Rasch Model for Measurement* (Sage)
[18] Wilson M 1988 Detecting and Interpreting Local Item Dependence Using a Fannily of Rasch Models *Applied Psychological Measurement* **12** 353-364
[19] Gori E 2006 The Rasch Approach To "Objective Measurement'm The Presence Of Subjective Evaluation From"Judges" *J. Educación XXi* **9** 107-133
[20] Linacre J M 2007 *A User's Guide to WINSTEPS Rasch-model Computer Programs* (Chicago: MESA Press)
[21] Wright B D, Masters GN 1990 *Rasch Measurement Transaction (2nd ed)* (Wilmington, Delaware Wide Range, INC)
[22] Liu X 2010 *Using and developing measurement instruments in science education: A Rasch Modeling approach* (Charlotte NC: Information Age Publishing)

Journal of Physics: Conference Series **495** (2014) 012027 doi:10.1088/1742-6596/495/1/012027

Electrical Conductivity Characteristic of TiO2 Nanowires From Hydrothermal Method

Mohd Azlishah Othman[1] **Noor Faridah Amat**[1] **Badrul Hisham Ahmad**[1] **Jose Rajan**[2]

[1] Centre for Telecommunication Research and Innovation (CeTRI), Faculty of Electronic and Computer Eng., Universiti Teknikal Malaysia Melaka, Hang Tuah Jaya, 76100 Durian Tunggal, Melaka, Malaysia

[2] Faculty of Industrial Science and Technology, Universiti Malaysia Pahang, Lebuhraya Tun Razak, 26300, Gambang Kuantan, Pahang, Malaysia

E-mail: azlishah@utem.edu.my

Abstract. One dimensional nanostructures of titanium dioxide (TiO2) were synthesized via hydrothermal method by mixing TiO2 as precursor in aqueous solution of NaOH as solvent. Then, heat and washing treatment was applied. Thus obtained wires had diameter ~15nm. TiO2 nanowires will be used as a network in solar cell such dye-sensitized solar cell in order to improve the performance of electron movement in the device. To improve the performance of electron movement, the characteristics of TiO2 nanowires have been analyses using field emission scanning electron microscopy (FESEM) analysis, x-ray diffractometer (XRD) analysis and brunauer emmett teller (BET) analysis. Finally, electrical conductivity of TiO2 nanowires was determined by measuring the resistance of the TiO2 nanowires paste on microscope glass.

1. Introduction

One dimensional nanostructure become interest field to investigate in order to enhance the performance of electron movement in the dye sensitized solar cell (DSSC). Then, it will be increase the efficiency of DSSC which is new type of solar cell technology that offer low-cost material. In DSSC, light is absorbed by wide band gap of metal oxide semiconductor nanoparticle such as TiO2, ZnO, Nb2O5 and SnO[1], [2], [3] which is coating with dye-sensitive light onto the semiconductor nanoparticle surface to absorb the light. the effective surface area for dye adsorption can be greatly enhanced and efficient light absorption[4]. Figure 1 illustrate during the illumination, electrons are injected into the conduction band of semiconductor nanoparticle by excited dye molecules adsorbed on the semiconductor nanoparticle surface, which is in contact with an electrolyte. In electrolyte, triiodide will react with electron at cathode to produce iodide, then iodide will react with oxidized dye molecule to regenerate dye sensitized molecule. But there are two undesirable reactions will be occurring toward the injected electron. First, the injected electron will be recombining with the oxidized dye molecules and dye sensitized molecule will be regenerate. Second reaction is the injected electron reacts with triiodide to produce iodide. Unfortunately, from both reactions, the injected

[1] Mohd Azlishah bin Othman, Centre for Telecommunication Research and Innovation (CeTRI), Faculty of Electronic and Computer Eng., Universiti Teknikal Malaysia Melaka, Hang Tuah Jaya, 76100 Durian Tunggal, Melaka, Malaysia.

electron travels through the semiconductor to back electrode decrease, consequence the efficiency of DSSC is low.

To solve the recombination problem, the one dimensional nanostructure is the best solution to form a semiconductor network in the DSSC. The nanowires will provide direct path for electron that bring the electricity move after injected from dye in the excited state. The direct path for electron movement can avoid the electron from the recombination situation. A dense network of nanowires can provide both high surface area and direct connectivity for electron movement[4], [5], [6]. In this paper, we choose TiO2 as a material of metal oxide semiconductor to synthesis TiO2 nanowires. TiO2 has been the mostly preferred as the semiconductor compare to other semiconductor[3] because it has good mechanical flexibility, non toxic, easily available, environmental stability, low cost characteristics and its conductivity could be controlled with acid/base (doping/undoping)[3], [7]. Characteristics of TiO2 nanowires by analyzing with FESEM, BET and XRD have been reported. The electrical conductivity of TiO2 nanowires also been discussed.

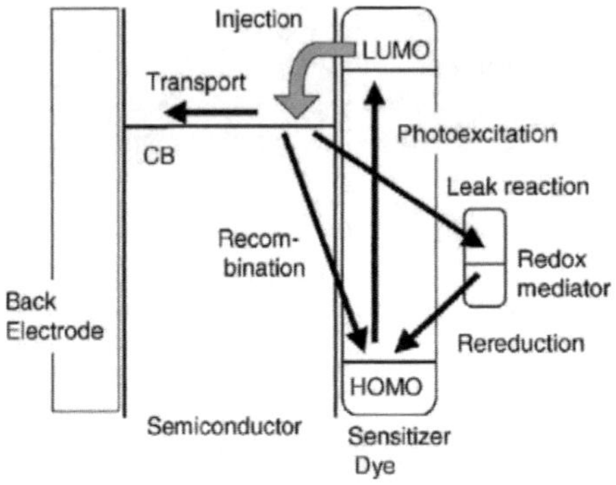

Figure 1. Primary reaction steps in dye-sensitized solar cells[8]

2. Experimental Procedure

This section is discussed all process involves to figure out the characteristics of TiO2 nanowires which is synthesizing via hydrothermal method. All the process includes procedure to synthesis TiO2 nanowires, after obtained white precipitation of TiO2 nanowires, send sample to FESEM analysis, BET analysis and XRD analysis. Next, the electrical conductivity of TiO2 nanowires have been determines by applying a formula of conductivity.

2.1. Synthesis of TiO2 Nanowires

The TiO2 nanowires were synthesized from a commercial powder of TiO2 (Degussa, P25) as a main material. A total of 2g of TiO2 powder and 50mL of 10M NaOH in aqueous solution were stirred together at room temperature until the TiO2 powder fully dissolves in the NaOH aqueous solution. The mixture was put into a Teflon-lined stainless autoclave which was heated at 150C for 5 days. After that, it was naturally cooled down to room temperature. The solution was washed with 0.1M of HCl aqueous solution by stirring the mixture for 24h. The function of HCl treatment is to remove or eliminate the Na ions in the solution[9], [10], [11], [12] . HCl solution as acid treatment that it can decrease the Na:H ratio in the solution by replacing Na+ with H- [11]. After 24h, the solution was centrifuged and white precipitation was obtained. After HCl treatment, the white precipitation was washed by deionized water until the pH value 9. Then, the white precipitation was dried at 70C to obtain TiO2 nanowires powder.

2.2. Process of coating TiO2 Paste

In order to put TiO2 nanowires onto the glass substrate, there are involved 2 methods that include preparation of TiO2 nanowires paste and the process of coating TiO2 nanowires paste onto the glass substrate.

2.2.1. *Preparation of TiO2 Paste*

TiO2 nanowires paste was prepared by mixing a different weight of TiO2 nanowires powder, such 5mg, 10mg, 20mg, and 40mg with a few drops of NMP in a mortar. NMP is used as a binder to bind the nanowires with each other. The mixture was mixed well until the paste became viscous paste. The ratio of mixture to prepare TiO2 nanowires paste as shown in the table below.

Table 1. The ratio of TiO2 nanowires paste.

Weight of TiO2, mg	Drop of NMP
5	1
10	2
20	3
40	4

2.2.2. *Coating TiO2 paste*

The TiO2 nanowires paste were pasted onto glass substrate in small area by using doctor blade technique. The area of TiO2 paste onto the glass substrate is in μm. Firstly, the glass substrate was cut into small pieces and continued with cleaning process by washing it with acetone and heating and sonicating it with deionized water for 30 minute and washed again with ethanol. After that, the small pieces of glass were dried at 70C for 10 minutes.

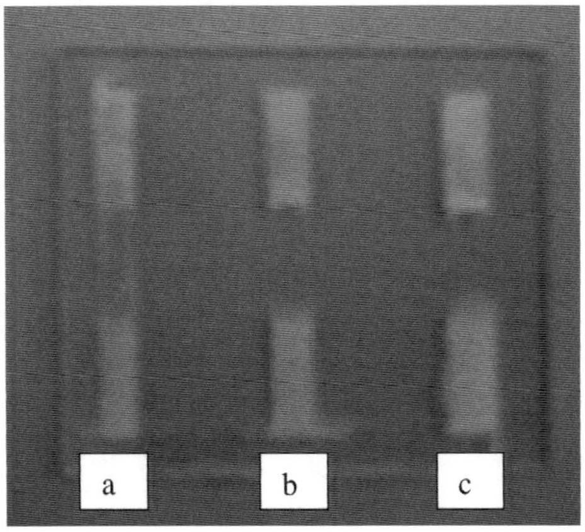

Figure 2. TiO2 nanowires paste onto the glass substrate a) 5mg,b) 10mg and c) 20mg.

2.3 Characterization Method

Finally, the TiO2 nanowires were characterized by FESEM analysis to figure out the nanostructure of TiO2, XRD analysis, BET analysis to determine the surface area and lastly, electrical conductivity. The electrical conductivity were observed by using resistance measurement by using digital multimeter equipment and by applying the resistance measurement, the conductivity of TiO2 nanowires based on different weight will be calculated. The equation below is a formula to calculate

Journal of Physics: Conference Series **495** (2014) 012027 doi:10.1088/1742-6596/495/1/012027

the conductivity where *R* is resistance (Ω), *l* is length between 2 points (mm), and *A* is area of TiO2 nanowires paste onto glass substrate (μm), ϱ is resistivity (Ωmm) and σ is conductivity ($\Omega^{-1}m^{-1}m$).

$$\rho = \frac{RA}{l} \qquad \text{Eq (1)}$$

$$\sigma = \frac{1}{\rho} \qquad \text{Eq (2)}$$

3. Result and Discussion

Figure 3 shows the FESEM analysis image of TiO2 nanowires after synthesis using hydrothermal method at 150C for 5 days. The nanowires are close-packed with each other and make a network too dense. Although, the network of nanowires is dense, the surface area of TiO2 nanowires is high which is258.2911 m^2/g. From figure 4, the XRD pattern shown that TiO2 nanowires has same as XRD pattern of TiO2 (P25) but slightly different. In TiO2 nanowires, the nanostructure is less crystalline compare to TiO2 (P25).

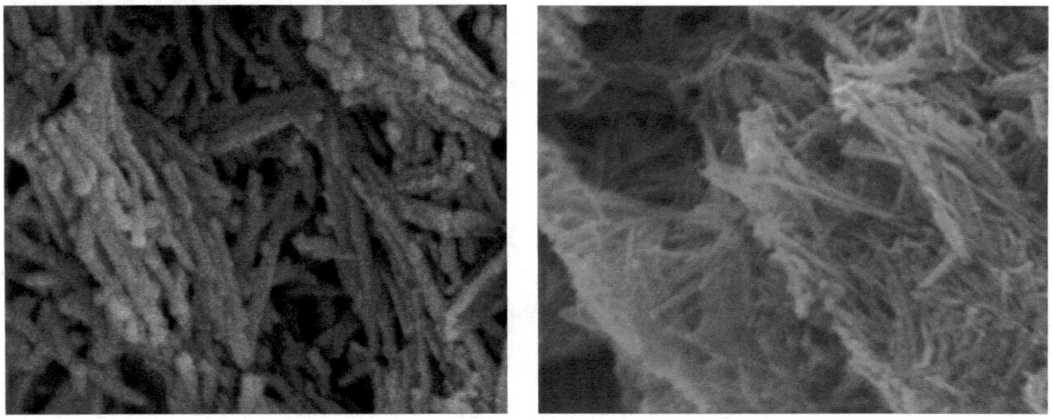

Figure 3. FESEM images of TiO2 nanowires

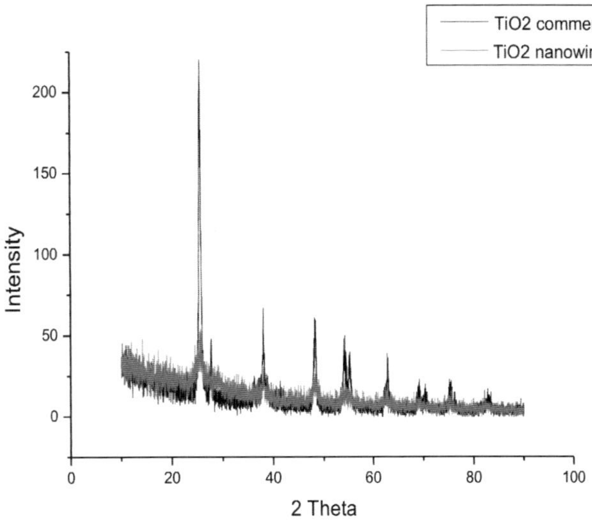

Figure 4. XRD pattern of TiO2 nanowires compared with TiO2 commercial, Degussa P25.

TiO2 nanowires also were characterized in terms of electrical conductivity. In order to improve the performance of electron movement in nanowires structures, the electrical conductivity testing is important to determine whether the nanowires has high conductivity or not by measuring the resistance values using digital multimeter. In the band theory of solids, the conduction process is whether or not there are electrons in the conduction band; more electrons on conduction band more conductive a material is[13]. Conductivity is increased when electrons in valence band supplied with external energy so that the electrons have enough energy to excite into conduction band. By applying below formula, the conductivity values will be determined as shown in figure 5. Increase the amount of TiO2 nanowires will be affected on the conductivity.

Table 2. The average values of resistance, resistivity and conductivity of TiO2 nanowires

Weight of TiO2 Nanowires, mg	Resistance, Ω	Resistivity, Ωm	Conductivity, $\Omega^{-1}m^{-1}$
5	8.26M	16.51k	6.279E-5
10	7.85M	15.70k	6.67E-5
20	2.06M	4.12k	2.2103E-4
40	2.23M	4.47k	2 2438E-4

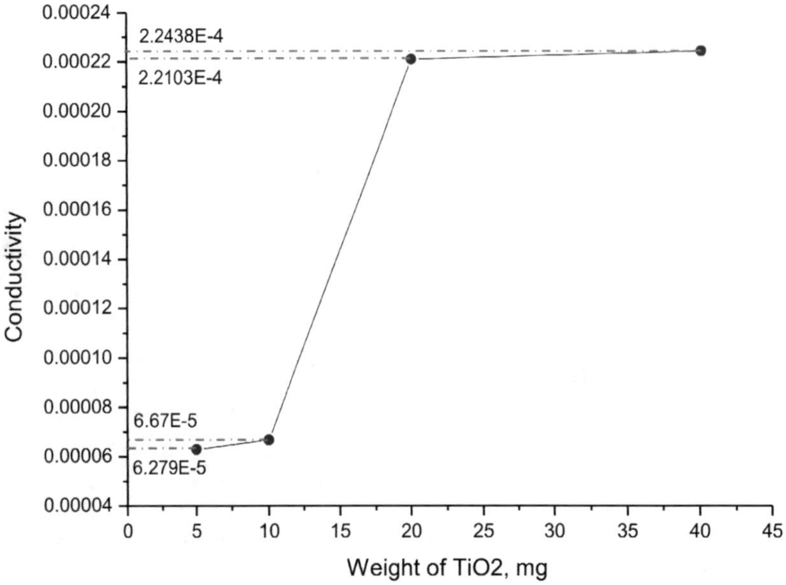

Figure 5. The graph of electrical of conductivity with different weight of TiO2 nanowires.

4. Conclusion

TiO2 nanowires were successfully synthesis by using hydrothermal method. TiO2 nanowires also have a high surface area that will be affect the performance of TiO2 nanowires especially in electrical conductivity. By analyzing the characteristics of TiO2 nanowires such XRD analysis, BET analysis, and FESEM analysis, TiO2 nanowires were successfully synthesis via hydrothermal method with high surface area. As conclusion, the electrical conductivity will be increasing as the weight of TiO2 nanowires in making paste. We achieved to show that, TiO2 nanowires have high conductivity.

5. Acknowledgment

First and foremost, the author would like to thanks Universiti Teknikal Malaysia Melaka, Malaysia (UTeM) for their equipment, financial and assistance support. This project was supported by Malaysian Technical University Network Grant; MTUN/2012/UMP-FKEKK/2-M00086. Last but not list, the author would like to thanks those involved direct and indirect in completing this project success fully.

References

[1] R. Jose, V. Thavasi, and S. Ramakrishna, "Metal Oxides for Dye-Sensitized Solar Cells," *Journal of the American Ceramic Society*, vol. 92, no. 2, pp. 289–301, Feb. 2009.

[2] K. Hara, H. Arakawa, and C. Dssc, *Dye-sensitized Solar Cells*. 2003.

[3] A. S. Karmakar and J. P. Ruparelia, "A Critical Review on Dye Sensitized Solar Cells," pp. 8–10, 2011.

[4] Q. Wang, S. Ito, M. Grätzel, F. Fabregat-Santiago, I. Mora-Seró, J. Bisquert, T. Bessho, and H. Imai, "Characteristics of high efficiency dye-sensitized solar cells.," *The journal of physical chemistry. B*, vol. 110, no. 50, pp. 25210–21, Dec. 2006.

[5] J. B. Baxter and E. S. Aydil, "Nanowire-based dye-sensitized solar cells," pp. 1–3, 2005.

[6] C. Yu and J. Park, "Thermal annealing synthesis of titanium-dioxide nanowire–nanoparticle hetero-structures," *Journal of Solid State Chemistry*, vol. 183, no. 10, pp. 2268–2273, Oct. 2010.

[7] S. G. Pawar, S. L. Patil, M. A. Chougule, B. T. Raut, D. M. Jundale, and V. B. Patil, "Scholars Research Library," vol. 2, no. 2, pp. 194–201, 2010.

[8] R. Katoh, A. Furube, A. V. Barzykin, H. Arakawa, and M. Tachiya, "Kinetics and mechanism of electron injection and charge recombination in dye-sensitized nanocrystalline semiconductors," *Coordination Chemistry Reviews*, vol. 248, no. 13–14, pp. 1195–1213, Jul. 2004.

[9] R. Yoshida, Y. S. Ã, and S. Y. Ã, "Syntheses of TiO 2 (B) nanowires and TiO 2 anatase nanowires by hydrothermal and post-heat treatments," vol. 178, pp. 2179–2185, 2005.

[10] D. T. T. Le, D. D. Vuong, and N. D. Chien, "Synthesis and LPG-sensing properties of TiO 2 nanowires," *Journal of Physics: Conference Series*, vol. 187, p. 012086, Sep. 2009.

[11] M. Chang, J. R. Deka, C. H. Lin, and C. C. Chung, "In situ mechanical property measurement of titania nanowires," pp. 187–191.

[12] A. M. Bakhshayesh, H. Dadar, and M. R. Mohammadi, "Synthesis and optical characterization of anatase-TiO 2 nanowires by hydrothermal method," no. 00, pp. 12–14, 2012.

[13] M. S. P. Sarah, M. Z. Musa, M. N. Asiah, and M. Rusop, "Electrical conductivity characteristics of TiO<inf>2</inf> thin film," *2010 International Conference on Electronic Devices, Systems and Applications*, pp. 361–364, Apr. 2010.

10 GHz Microstrip Spanar Antennas: An Experimental Analysis

Mohd Azlishah Othman[1] Hazwani Azman[1] Mohd Nor Husain[1] Mohamad Zoinol Abidin Abd. Aziz[1] Yahaya Abd. Rahim[2] Ahmad Naim Che Pee[2] Mohamad Radzi Motsidi[2] Mohd Fairuz Iskandar Othman[2]

[1]Microwave Research Group, Centre for Telecommunication Research and Innovation (CeTRI), Faculty of Electronic and Computer Engineering, Universiti Teknikal Malaysia Melaka, Hang Tuah Jaya, 76100 Durian Tunggal, Melaka, Malaysia

[2]Communication Multimedia Network and Systems Reseach Group, Centre of Advanced Computing Technology, Faculty of Information Technology and Multimedia Communication, Universiti Teknikal Malaysia Melaka, Hang Tuah Jaya, 76100 Durian Tunggal, Melaka, Malaysia

E-mail: azlishah@utem.edu.my

Abstract. This paper presents Spanar Antenna designed using CST Microwave Studio Simulation 2011. The proposed antenna was designed to operate at 10 GHz, which suggested return loss, S_{11} must be less than -10 dB and voltage standing wave ratio (VSWR) must be less than 2. The best performance of simulation of Spanar Antenna was obtained at a small size of 24.8 mm x 8.0 mm with dimension board of FR4 substrate 31.7 mm x 18.5 mm. The thickness (h) and dielectric constant (Ɛr) of substrate were 1.6 mm and 4.7. An analysis between simulation result and measurement result has been compared in order to see the antenna performance.

1. Introduction

Antenna is a one of useful component in communication era in different communication bands [1].10 GHz frequency spectrum in Malaysia mainly is used for fixed wireless access with two spectrum band; lower (10.15 –10.30 GHz) and upper (10.50 – 10.65 GHz) band [2]. The fixed wireless access is mainly used for point-to-point communication usually between two base stations as a backhaul as an option to replace wired connectivity. Besides that, antenna is being proposed at 10 GHz for bio-detecting system to detect heart and respiration rates of human within far away [3]. It is due wireless bio-signal detection is the recent tendencies in the medical industry, and research for sensing bio-signal continues to rise.

A novel spanar printed antenna design aiming for X-Band frequency spectrum at 10 GHz radar applications that has advantages such as simple structure, compact size and easy fabrication is proposed in this work. By introducing a slit to the antenna, frequency operation for the radar

[1] Mohd Azlishah bin Othman, Microwave Research Group, Centre for Telecommunication Research and Innovation (CeTRI), Faculty of Electronic and Computer Engineering, Universiti Teknikal Malaysia Melaka, Hang Tuah Jaya, 76100 Durian Tunggal, Melaka, Malaysia

application can be achieved [4]. Details of the antenna design are described, and both simulated and measured results such as return loss and radiation patterns are presented and discussed.

In this paper, the Spanar antenna design has referred from micro strip element. This antenna will use high dielectric constants such as FR4 with relative permittivity of $\mathcal{E}_r$, = 4.7 and thickness, *h* of 1.6 mm in order to get a smaller antenna size. An analysis between simulation and measurement results are conducted to investigate the effect of parameter for length by feed line on antenna performance. The thickness of copper of the patch is 0.035 mm with antenna feed by 50 Ω.

2. Antenna Design and Architecture

The dimension of patch is treated a circular patch element by the actual radius, a_e is given in [5] [6]. f_r is the resonant frequency, v_o is the free space speed of light, *h* is the thickness of substrate and $\mathcal{E}_r$ is the dielectric constant of substrate.

The Effective radius of patch was starting with:

$$= \frac{8.791\ x\ 10^9}{f_r\sqrt{\varepsilon_r}}$$

Actual/physical radius of patch,a:

$$a = \frac{}{\left[1 + \frac{2h}{\pi\varepsilon_r}\left(In\left(\frac{\pi F}{2h}\right) + 1.7726\right)\right]^{\frac{1}{2}}}$$

Effective radius of patch, a_e:

$$a_e = a\left[1 + \frac{2h}{\pi\varepsilon_r a}\left(In\left(\frac{\pi a}{2h}\right) + 1.7726\right)\right]^{\frac{1}{2}}$$

The dimension of rectangular patch is using equation as follows in [7] [8] where f_r is the resonant frequency, $\mathcal{E}_r$ is the dielectric constant of substrate and *c* is the space speed of light.

Width of patch, W_r:

$$W_r = \frac{c}{2f_r}\sqrt{\frac{2}{\varepsilon_r + 1}}$$

Length of patch, L_r:

$$L_r = \frac{0.49\,\frac{c}{f_r}}{\sqrt{\varepsilon_r}}$$

Effective of dielectric constant, ε_{reff}:

$$\varepsilon_{reff} = \frac{\varepsilon_r + 1}{2} + \frac{\varepsilon_r - 1}{2}\left[1 + 12\frac{h}{W}\right]^{-\frac{1}{2}}$$

Changes of length of patch, ΔL:

$$\Delta L = 0.412h\,\frac{\left(\varepsilon_{reff} + 0.3\right)\left(\frac{W}{h} + 0.264\right)}{\left(\varepsilon_{reff} - 0.258\right)\left(\frac{W}{h} + 0.8\right)}$$

Actual/physical length of patch,L:

$$L = \frac{1}{2f_r\sqrt{\varepsilon_{reff}}\sqrt{\mu_o\varepsilon_o}}$$

Effective length of patch, L_{eff}:

$$L_{eff} = L + 2\Delta L$$

The length of triangular patch element can be determined in [9]. f_r is the resonant frequency, *a* be the side length of the triangular and *c* is the space speed of light.
Length of side of patch,a:

$$a = \frac{2c}{3f_r\sqrt{\varepsilon_r}}$$

$$f_r = \frac{2c}{3a\sqrt{\varepsilon_r}}$$

The dimension of width and length of feed line are determined in [10]. *d* is the thickness of substrate, Z_o is the impedance 50Ω, $\mathcal{E}_r$ is the dielectric constant of substrate, Ø is the loss tangent for dielectric constant, *f* is the frequency and *c* is the space speed of light.

Width of feed line of patch, W_f:

$$A = \frac{Z_o}{60}\sqrt{\frac{\varepsilon_r+1}{2}} + \frac{\varepsilon_r-1}{\varepsilon_r+1}(0.23+\frac{0.11}{\varepsilon_r})$$

$$B = \frac{377\pi}{2\,Z_o\sqrt{\varepsilon_r}}$$

$$W_f = \frac{8\,e^A}{e^{2A}-2}\,x\,d\,if\,\frac{W}{d}<2$$

$$W_f = \frac{2}{\pi}\left[B-1-\ln(2B-1)+\frac{\varepsilon_r}{2\varepsilon_r}+0.39-\left(\frac{0.61}{\varepsilon_r}\right)\right]\,x\,d\,if\,\frac{W}{d}>2$$

Length of feed line of patch, L_f:

$$L_f = \frac{\emptyset c}{2\pi f\sqrt{\varepsilon_{reff}}}$$

The calculation value is used in the CST Microwave Studio Simulation Software. The parameter design adjusted by the parametric method is used to achieve 10 GHz operating frequency. **Figure 1** and **Figure 2** show the geometry of the Spanar antenna that has been designed to cover operating frequency of 10 GHz.

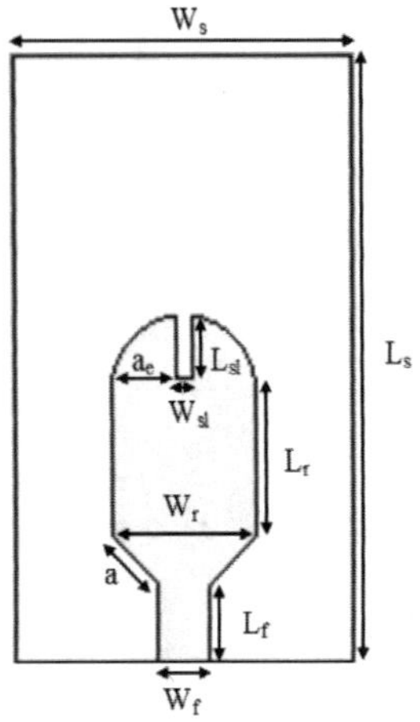

Figure 1. Front view of

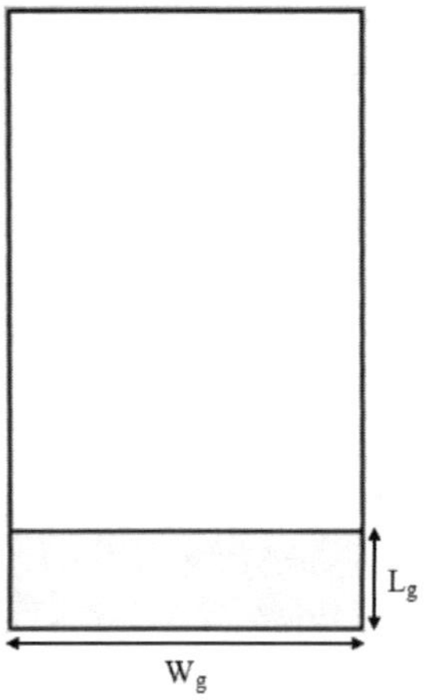

Figure 2. Back view of

geometry of the proposed antenna.

geometry of the proposed antenna.

The antenna is fabricated on a thin FR4 substrate of thickness of 1.6 mm with a relative permittivity, ε_r is equal 4.7. The thickness of copper of the patch is 0.035 mm with antenna plate fed by 50 Ω. **Figure 3** shows front view of Spanar antenna and **Figure 4** shows back view of Spanar antenna. **Table 1** is shown the parameter and dimensions of the proposed Spanar antenna.

Figure 3. Front view of Spanar antenna.

Figure 4. Back view of Spanar antenna.

Table 1. Proposed Spanar antenna parameter.

Parameter	Value (mm)
L_s x W_s	31.7 x 18.5
a_e	4.0
W_r	8.9
L_r	7.6
a	9.0
W_f	2.9
L_f	5.235
W_g	5.0
L_g	18.5
W_{sl}	1.0
L_{sl}	3.95

3. Results and Discussion

Figure 5 shows the simulation and measurement of Return Loss, S_{11} in dB of Spanar antenna. From graph, the antenna achieves the target of operating frequency at 10 GHz. The designs are simulated by using the CST Microwave Studio Simulation. The best Return Loss, S_{11} of simulation for this antenna is obtained -30.5 dB when it reaches the length of feed line antenna, which is at 5.235 mm. The measurement is obtained -17.07 dB using the Vector Network Analyzer. The good antenna must have return loss below than -10dB where is return loss is that portion of a signal that cannot be absorbed by the end of line termination, or cannot cross an impedance change at some point in the transmission system. It also show that the antenna has less loss and able to transmit 90% from the maximum power. In this return loss the good efficiency antenna is less than 90%. In this simulation the efficiency is below than 90%. This measurement result is due to the soldering the connector to at the antenna feed lines.

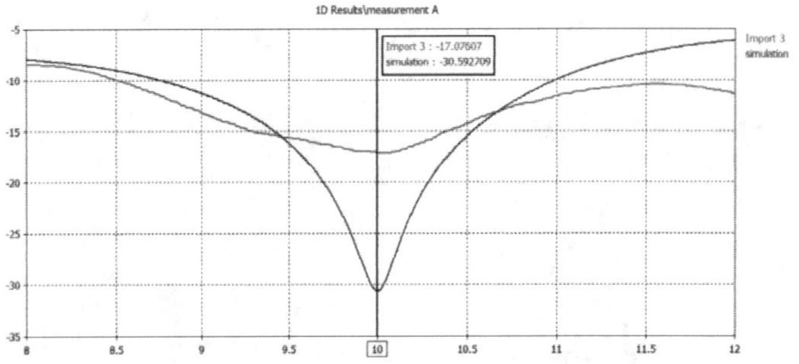

Figure 5. Return Loss, S11 in dB of Spanar antenna.

In the process of achieving this particular configuration for Spanar antenna to perform efficiently there is always a reflection of the power, which leads to the standing waves, which is characterized by the Voltage Standing Wave Ratio, VSWR. VSWR at the respective operating frequency is determined 1.06 from simulation and 1.29 from measurement that means the reflection coefficient is 0.09, which is less than 10% and plot in **Figure 6**.

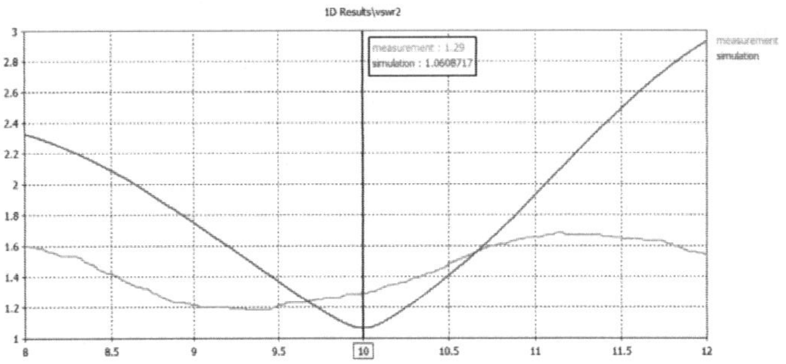

Figure 6. Voltage Standing Wave Ratio, VSWR of Spanar antenna.

The Directivity and the Gain are probably the most important parameters of an antenna. In this simulated Spanar antenna having the gain value is 4.123 dB as shown in **Figure 7** and directivity value its similar same with the value of parameter gain which is 5.587 dB as shown in **Figure 8**. Directivity is the ability of an antenna to focus energy in a particular direction when transmitting or to receive energy better from a particular direction when receiving. The relationship between gain and directivity: Gain = efficiency/Directivity. The Gain of antenna is an actual which is less than the directivity due to ohmic losses in the antenna or its radome [11]. From both the efficiency is -1.4 dB which is respectively less than -3 dB and nearly 0 dB.

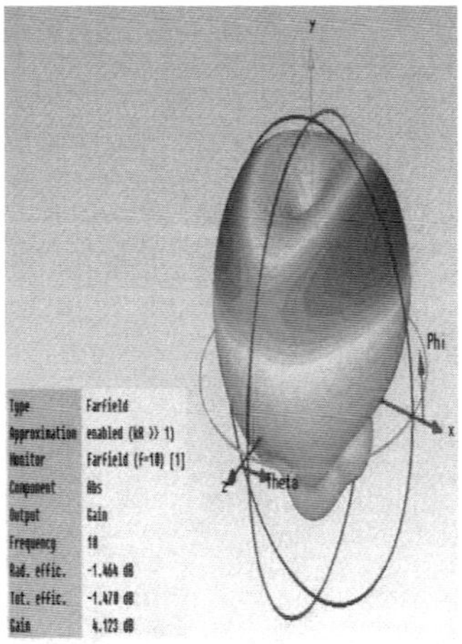

Figure 7. Gain of the proposed antenna.

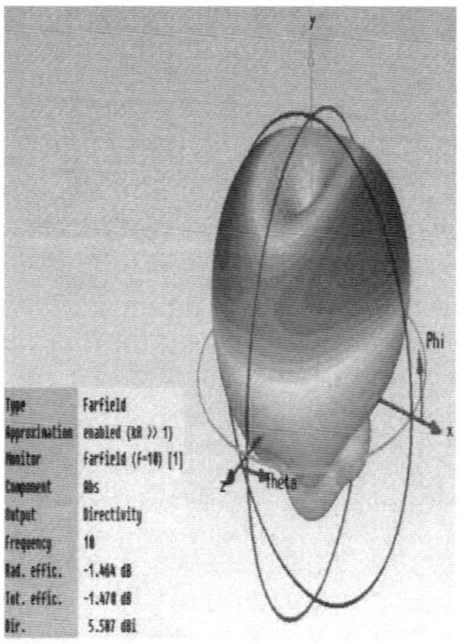

Figure 8. Directivity of the proposed antenna.

From the radiation pattern, parameter of HPBW and FNBW is obtained. **Figure 9** shows E plane radiation pattern the HPBW is 54.3^0 and FNBW is 108.6^0. **Figure 10** shows H plane radiation pattern HPBW is 131.4^0 and FNBW is 262.8^0. The HPBW are obtained by the angular width (3dB). HPBW is the angular separation in which the magnitude of the radiation pattern increases 50% from the peak of the main beam. The FNBW value usually twice the value of HPBW and for this design, the FNBW of both plane is about twice the HPBW and it is fulfil the characteristic which is the FNBW is larger than the HPBW.

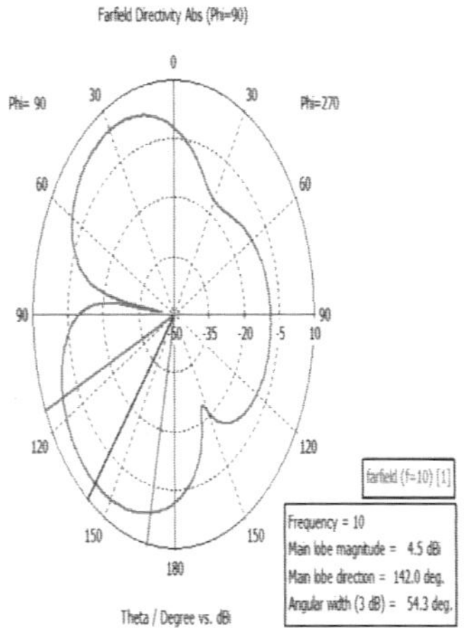

Figure 9. Radiation Pattern in Elevation (E) plane of proposed antenna.

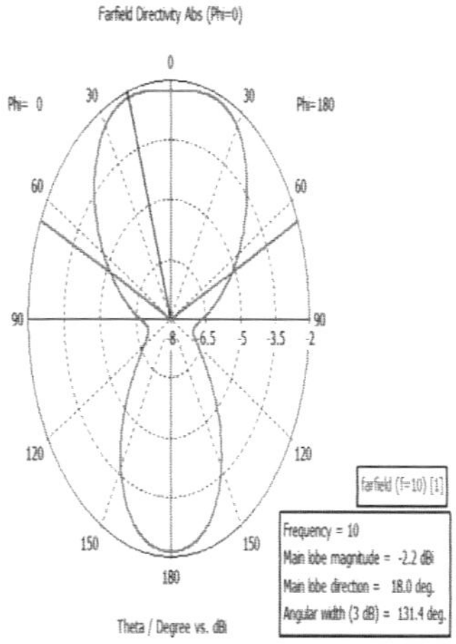

Figure 10. Radiation Pattern in Azimuth (H) plane of proposed antenna.

The parametric study has been done to see the effect by varying the lengths of the element length feed line, L_f. **Figure 11** shows the S-parameter simulations while **Figure 12** shows VSWR of the Spanar for different length, L_f by feed line is simulated by using the CST Microwave Studio Simulation. The most effective response for the antenna is at length feed line at 5.235 mm. The most challenging task in designing the antenna is to sufficiently match to its input transmission line that is approximately 10 % or less of the incident signal which is lost due to the reflection. VSWR range at the respective operating frequency is determined which is less than 2.

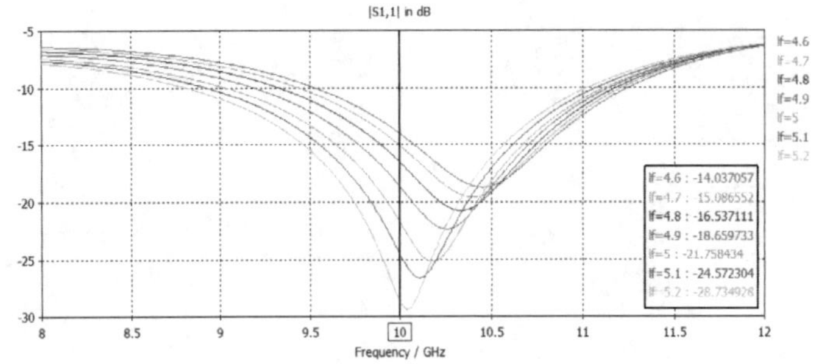

Figure 11. Return loss of the proposed antenna for various lengths of feed line.

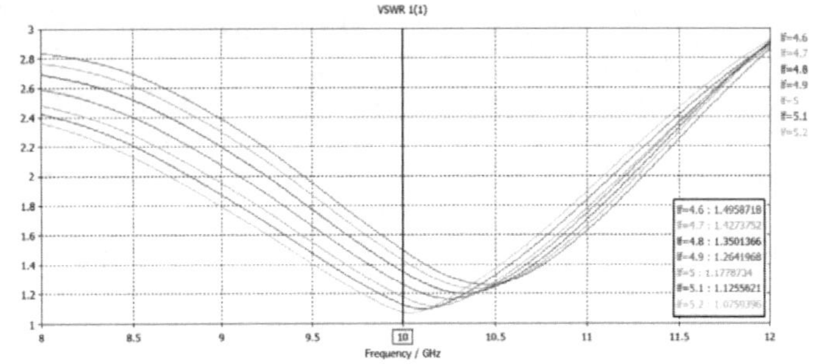

Figure 12. VSWR of the proposed antenna for various lengths of feed line.

Table 2 shows the simulated Return loss and VSWR for different length by feed line. It is obvious that the length of feed line has a lot of effect on the frequency resonance and the return loss also increases. It also displays the VSWR decrease with increases in length by feed line.

Table 2. Comparison different length of feed line, L_f at frequency 10 GHz.

L_f	S_{11}	**VSWR**
4.6 mm	-14.03 dB	1.49
4.7 mm	-15.08 dB	1.42
4.8 mm	-16.53 dB	1.35
4.9 mm	-18.65 dB	1.26
5.0 mm	-21.75 dB	1.17
5.1 mm	-24.57 dB	1.12
5.2 mm	-28.73 dB	1.07

Conclusion

A 10 GHz Spanar antenna design and concept was successfully discussed and presented. Parametric study of length of feed line that is affected a return loss and VSWR are done. The good antenna must have return loss below than -10dB and VSWR range at the respective operating frequency is determined which are less than 2. Antenna design gain is 4.123 dB and Directivity is 5.587 dB respectively. Also, the radiation patterns in E-plane and H-plane at 10 GHz frequency are obtained which are reported in the paper. The simulated and measurement results are compared and obtained the better result in desired operating frequency. With these features, this antenna is suitable for wireless communication systems operating at X-band.

Acknowledgment

First and foremost, the author would like to thank Universiti Teknikal Malaysia Melaka, Malaysia (UTeM) for financial assistance under project grant PJP/2012/FTMK (43C) S1047 and Zamalah cooperative support. Special dedication to kind-hearted project coordinating supervises lectures, student, parents, public and those involved directly and indirectly for their moral supports, contribution and cooperation in completing this paper.

References

[1] Rekha S and R. S. Meena 2013 Design Microstrip patch antenna and Role of Hexaferrite Cavity in X-band *International Conference on Communication Systems and Network Technologies* pp 7–10

[2] Azhari A, Norman J, Dzuzlindah M A and Ahmad K S 2012 Design & Development of 10GHz Slotted Waveguide Antenna *IEEE Asia-Pacific Conference on Applied Electromagnetics (Melaka, Malaysia, 11-13 December 2012)* pp 331–333

[3] Jee H L, Yun T I and Seong O P 2009 X-band Radar System for Detecting Heart and Respiration Rates *IEEE Information Technology in Biomedicine* pp 400–404

[4] Aruna R and A K Gautam 2011 Improvement in Gain and Bandwidth of Rectangular and U Slot Loaded Patch *IEEE International Journal of Computer Science Issues (November 2011)* pp 283–288

[5] B. J. Kwaha, O. N Inyang and P. Amalu 2011 The Circular Microstrip Patch Antenna – Design and Implementation *IJRRAS* pp 86–95

[6] Balanis C A 1982 *Handbook of Microstrip Antennas* (John Wiley and Sons, New York)

[7] D Murugan, M Ganesh M and S Piramasubramanian *2012 Design and Performance Prediction of 10GHz Microstrip Array Antenna Structures IEEE ICCCNT* (Coimbatore, India, 26-28 July 2012) pp *1–5*

[8] Pozar D M 2005 *Microwave Engineering* (John Wiley & Sons, United States)

[9] Anzar K, Rajesh N and Puran G 2012 Investigation Of Triangular Microstrip Patch Antenna On Six Different Substrates For X –Band Applications *International Journal of Engineering Research and Applications* (September-October 2012) pp 1578–1584

[10] Ali M T, Muhamud S, Abd Rahman N R and Ya'acob N 2011 A Microstrip Patch Antenna with Aperture Coupler Technique at 5.8 GHz *IEEE International Conference on System Engineering and Technology* pp 121–124

[11] John D K and Ronald J M 2003 *Antennas For All Application* (Kraus & Marhefka, Singapore)

Journal of Physics: Conference Series **495** (2014) 012029 doi:10.1088/1742-6596/495/1/012029

Energy monitoring based on human activity in the workplace

N.H. Mustafa[1], M.N. Husain[1], M.Z.A. Abd Aziz[1], M.A. Othman[1], F. Malek[2]
[1]Center for Telecommunication Research and Innovation (CeTRI), Fac. of Electronic and Comp. Eng., Universiti Teknikal Malaysia Melaka (UTeM), Hang Tuah Jaya, 76100 Durian Tunggal, Melaka, Malaysia

[2]School of Computer and Communication Engineering, Universiti Malaysia Perlis (UNIMAP), Perlis, Malaysia

Email: nurhanim.mustafa@yahoo.com.my, mohamadzoinol@utem.edu.my

Abstract. Human behavior is the most important factor in order to manage energy usage. Nowadays, smart house technology offers a better quality of life by introducing automated appliance control and assistive services. However, human behaviors will contribute to the efficiency of the system. This paper will focus on monitoring efficiency based on duration time in office hours around 8am until 5pm which depend on human behavior atb the workplace. Then, the correlation coefficient method is used to show the relation between energy consumption and energy saving based on the total hours of time energy spent. In future, the percentages of energy monitoring system usage will be increase to manage energy in efficient ways based on human behaviours. This scenario will lead to the positive impact in order to achieve the energy saving in the building and support the green environment.

1. Introduction

Recently, human activity has been used to monitor the energy usage performance for smart house. Smart home is one of a universal computing in which the information-perceiving and information-processing units remain invisible in the surroundings to create a pervasive environment. People spend a significant amount of time in their houses, and this has drawn researchers to promote integration of all possible services with traditional homes [1]. Besides that, smart home is a study of human psychology to create a pervasive environment employing ambient intelligence [2].

Human habit cannot be controlled because it changes according to our life. Human activity has been critical information since knowing what activities are undertaken is well important for judging if energy consumed by appliances is well spent by users [4]. Besides that, an estimation of the occupant's location can be modelled as the belief in the fact that he/she is located at position, x at the instant, t. This is depend on the time to reflect the changes in activities during the day or week. The information from human activity is an important clue for providing an energy saving service.

Wireless Enabled Electricity Manger (WEEMAN) is an intelligent control device reported in [3]. This system can monitor the energy consumption of a single device or a set of devices. Smart meter acts as the central control node in which availability based management algorithm. This algorithm learns about the usage patterns of the appliances, collects the information such as the real time average power of appliances from WEEMAN to generate efficient energy load patterns. The algorithm will provides a unique attribute for the user to set the expected monthly electricity bill amount and proactively control the operation of the all appliances according to the amount [3].

Furthermore, it is also preferable to know the relationship between a context and its energy consumption for a smart home utilizing context-aware energy saving besides the inferred contexts [4]. So, the reason by ignoring the relationship between a user context and its energy consumption often makes smart homes hard to find effective energy saving policies without sacrificing user comfort.

Smart home energy management system will monitor, manage and control the functions of the home appliances to reduce the monthly electricity bill [3]. The major benefits of smart house to consumers are the ability to incorporate energy management types through lighting, air conditioning and home appliances. Lighting is important for humans to see and do a work. The perfect lighting system when have auto control lighting so that will be dimmable light and bright light depend on situation of human activity and need.

Human activities are often discussed about the daily routine for workers/staffs at office or man and wife in the smart house which rarely discuss in Malaysia. The focuses on energy efficiency based on workers/staffs behaviour at working place or office will be investigate in this paper. Then, the correlation coefficient of energy saving performance based on total energy usage hours of time spent will be compute and analyze. In this paper the relation of energy consumption and energy saving in the workplace for every week are limited to 5 days. Furthermore, the proposed energy monitoring system will focuses mainly on lighting control, air-conditioner (HVAC) and some electrical appliances only.

2. Case Study

Human always moved around the office because each person has a different image and personality. So, there also different behaviours of energy usage can be categorized in similar working place. The energy monitoring system required various of sensors to detect a motion and temperature in monitored area. In an office environment, workers usually would like to use 3 rooms which are office room, living room and restroom. So, we proposed the monitored working space with 1 Office Room, 1 Living Room, and 2 Restroom. Each room dimension is design by using the Sweet Home Software. The figure 1 and table 1 shows the office environment and dimensions area of proposed monitored working place.

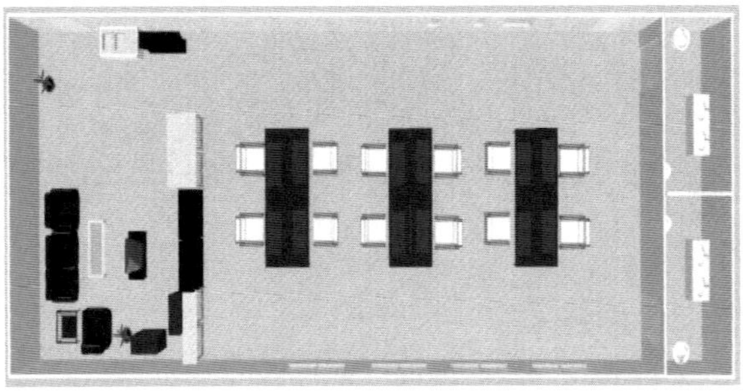

Figure 1. Proposed monitored working space environment

In a working environment, living room is used by staff to get some rest such as to make coffee for tea break. There are two restroom for male and female in order to provide comfort working environment place for workers.

Table 1. Dimension area proposed for each room

Rooms	**Office Room**	**Living Room**	**Restroom**	**Total**
Area/m^2	104.3	35.0	14.6	153.9

There are 10 unit sensors are place in selected position in order to provide an accurate information about the inhabitant's location. Occupancy sensor is proposed instead of motion sensors, since the occupancy sensor can detect the movement accurately and usually preferred for security system. Occupancy sensors can be switched automatically by detecting any movements within a coverage space. Occupancy sensors can be divided into two: Passive Infra Red (PIR) and PIR Ultrasonic. PIR will detect motion by sensing temperature profile change which is suitable for large body. PIR ultrasonic is more suitable to detect tiniest movement to ensure maximum effectiveness for almost every application in every space. Table 2 shows the quantity of sensors for each room.

Table 2. Total sensor for each room

Total	Office Room	Living Room	Restroom	Target
6	3	1	2	Light Control
4	3	1	0	HVAC

Each room is simulated by using DIALux software to calculate the power required for lighting. The DIALux software allow the users to choose any type of lamp which suitable for the simulated area. Figure 2 shows simulated lighting power in the restroom by using DIALux software. The air-conditioner and home appliance capacity were calculated according to the Consumer Guide to Home and the General Electric as proposed in [5]. The total power consumption of each room for lighting, air conditioning and home appliances are simulated in a KWH/day.

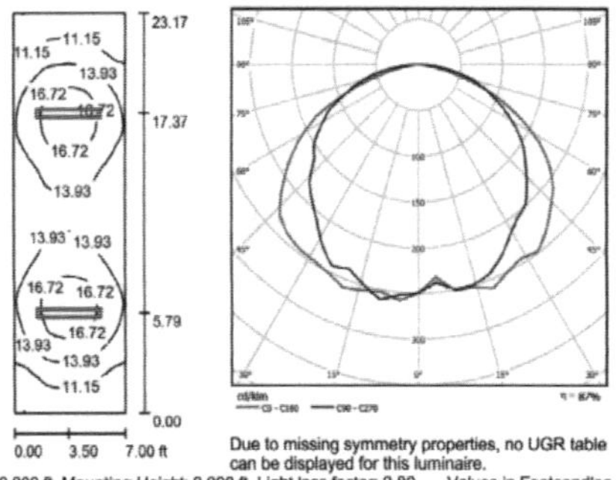

Due to missing symmetry properties, no UGR table can be displayed for this luminaire.

Height of Room: 9.000 ft, Mounting Height: 9.000 ft, Light loss factor: 0.80 Values in Footcandles, Scale 1:91

Surface	ρ [%]	E_{av} [fc]	E_{min} [fc]	E_{max} [fc]	u0
Workplane	/	14	8.20	18	0.599
Floor	20	10	7.12	12	0.683
Ceiling	80	3.79	2.40	4.93	0.633
Walls (4)	50	8.43	2.99	25	/

Workplane:
Height: 2.493 ft
Grid: 64 x 32 Points
Boundary Zone: 0.000 ft
Illuminance Quotient (according to LG7): Walls / Working Plane: 0.671, Ceiling / Working Plane: 0.277.
Proportion of points with less than 400 lx (for IEQ-7): 100.00%.

Luminaire Parts List

No.	Pieces	Designation (Correction Factor)	Φ (Luminaire) [lm]	Φ (Lamps) [lm]	P [W]
1	2	LUMICENTER CCN07-E132 (1.000)	2361	2700	36.0
			Total: 4721	Total: 5400	72.0

Specific connected load: 0.44 W/sq ft = 0.32 W/sq ft/10 fc (Ground area: 162.17 sq ft)

Figure 2. Simulated lighting power by using DIALux software

3. Proposed Scenario

The next step is to monitor the human behaviour on appliances and devices and model in term of usage time duration. Each person was divided into three categories: Person A is refer to worker which is like to do his work in an office. Person B is refer to worker which like to do his work in an office and outside of the office. While, Person C is refer to worker which is like to do his work at outside of the office for most of the time.

The input given is the time where the system will be 'start' and 'stop'. All the information about that scenario created will be stored in the microcontroller. When the Person A is entering the office area, the light and air-conditioner are automatically switch 'ON' in main office 15 minutes before to standby state. If that person has not shown in an office within next 15 minutes, the system will 'recheck' and 'call back' later at the same time for implementing the others task. The air conditioner is used in living and office room only.

Scenario management information allows users to define a set of behaviour rules. A lot of sequences can be implemented according to the habitant's age and its social status. The proposed scenario will focuses on a working time in office. Table 3 shows the proposed activity for 3 types of workers. Type A is for the busy worker, type B for normal worker and type C for off work worker. The 9 normal worker's steps and time spent for 3 types of workers is proposed in Table 3.

Table 3. Proposed Scenario

Wrong Behaviors	**Action**	**A**	**B**	**C**	**Time**	**Step**
	Office Room is ON	Enters Office			At 8 o'clock	1
Office Room is left on for 4 hours	Living Room is ON	Enters Living Room			8-8.30	2
	Living Room is OFF Office Room still ON		Enters Office		At 8.30 o'clock	
Living Room is left on for 30 minutes	Restroom is ON Office Room still ON	Enters Restroom	Enters Living Room		8.30-8.45	3
	Restroom is OFF	Start Working at her/him desk			At 8.45	
	Office Room still ON		Start Working at her/him desk	Enter Office Start Working at her/him desk	At 9 o'clock	
	Restroom is ON Living Room is ON		Enters Restroom	Enters Living Room	At 10 o'clock	4
Living Room is left on for 30 minutes	Restroom is OFF				At 10.15	
	Living Room is OFF Office Room still ON	Still Working	Continue Working	Continue Working	At 10.30 o'clock	

	Restroom is ON			Enters Restroom	11.30 - 11.45	5
	Restroom is OFF			Leave office go to lunch	At 11.45 o'clock	
	Office Room is OFF	Leave office go to lunch	Leave office go to lunch		At 12 o'clock	
	Office Room is ON	Enters Office			At 2 o'clock	6
Office Room is left on for 3 hours	Office Room still ON		Enters Office	Enters Office	At 2.15 o'clock	
	Living Room is ON			Enters Living Room	3-3.15	7
Living Room is left on for 15 minutes	Living Room is OFF	Still Working	Still Working	Continue Working	At 3.15 o'clock	
	Restroom is ON	Enters Restroom			4-4.15	8
	Restroom is OFF				At 4.15 o'clock	
	Living Room is ON		Enters Living Room	Enters Living Room	4.30-5.00	9
Living Room is left on for 30 minutes	All rooms are OFF	Leave office to go back	Leave office to go back	Leave office to go back	At 5 o'clock	

The proposed human habits which is cannot be controlled and depend on personal life is tabulated in Table 3. Thus, the difference of power consumption related to the total hours spent and saving in each room is recorded in KWH. Table 4, 5, and 6 represents the number of duration time spent in each person in every room for 9 working behaviour's steps that proposed in Table 3. Equation (1),(2) and (3) are used to get the result for calculation in table 4, 5 and 6. The all equation is very help to complete the result for power saving, power consumption and saving in cost.

$$\text{otal Power, kWh} = \text{otal Time Spent for each room (hour)x Total Power per day, kWh} \tag{1}$$

$$\text{Power Consumption, kWh} = \text{ime Spent(hour)x Total Power use per hour, kWh for each room} \tag{2}$$

$$\text{Power Saving, kWh} = \text{otal Power, kWh} - \text{Power Consumption, kWh} \tag{3}$$

$$\text{Saving in cost (RM)} = \text{Power Saving, kWh x RM 0.218 per unit (tariff rate in Malaysia)} \tag{4}$$

Table 4. Power per day (Total Power per day = 77.299 kWh)

	Power Saving, kWh	**Power Consumption, kWh**	**Saving in cost (RM)**
A	30.599	46.7	6.671
B	36.471	40.828	7.951
C	39.715	37.584	8.658

Table 5. Power per week (Total Power/week = 386.495 kWh)

	Power Saving, kWh	**Power Consumption, kWh**	**Saving in cost (RM)**
A	152.995	233.5	33.353
B	182.355	204.14	39.753
C	198.575	187.92	43.289

Table 6. Power per month (Total Power per month = 1545.98 kWh)

	Power Consumption, kWh	**Power Saving, kWh**	**Saving in Cost, RM**
AAAA	934	611.98	133.412
AAAB	904.64	641.34	139.812
AAAC	888.42	657.56	143.348
AABB	875.28	670.7	146.213
AABC	859.06	686.92	149.749
AACC	842.84	703.14	153.285
ABBB	845.92	700.06	152.613
ABBC	829.7	716.28	156.149
ABCC	813.48	732.5	159.685
ACCC	797.26	748.72	163.221
BBBB	816.56	729.42	159.014
BBBC	800.34	745.64	162.55
BBCC	784.12	761.86	166.086
BCCC	767.9	778.08	169.621
CCCC	751.68	794.3	173.157

4. Result and Discussion

The power consumption calculations in Table 7 are not considered the usage of air-conditioner and also no appliances usage in restroom. So, air –conditioner and the others appliance are usage in office room and living room. This paper also shows total power consumption calculated for home appliances including 12 Personal Computer, 1 LCD Television, 1 Water Heater, 12 Cellphone and 12 MP3 Player. All calculation of power consumption for lighting, air-conditioning and home appliances is shown in table 7. In order that, the result in table 7 is used to calculate the value of power consumption in table 4, 5, and 6 depend on time spent per day, week and month.

Based on figure 3, the result shows that the power consumption and power saving are changes depend on the human behavior. Thus, human behavior is important thing must to think first and control to get the efficient monitoring system. In the several tested data, that will help the people to know when the fixed time a human goes to the restroom and living room, and also how long their take a time in restroom, living room and go outside for off work.

Table 7. Calculation of Power Consumption for each Room

Total	Office Room	Living Room	Restroom	
153.9	104.3	35.0	14.6	**Area/m^2**
0.3040	0.0310	0.2010	0.0720	Light KWH/day
6.1000	5.0000	1.1000	0.0000	HVAC KWH/day
2.1848	1.4928	0.6920	0.0000	Home Appliances KWH/day

From previous data, the total power/month for power consumption, power saving and percentage of power saving/month have been calculated. The figure 3 is show the comparison between power consumption(in kWh), power saving (in kWh) and power saving cost (in RM).

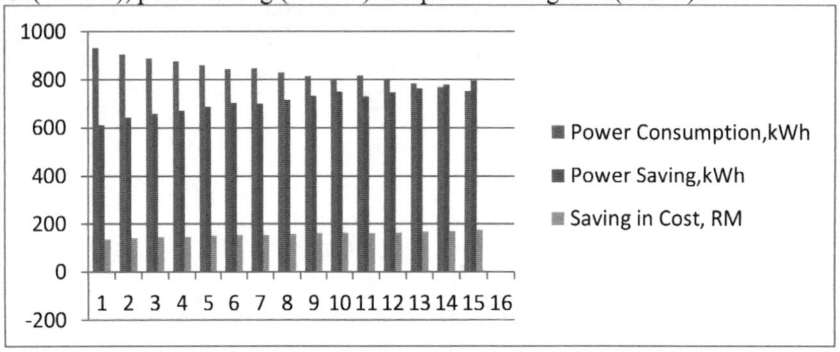

Figure 3. Comparison between power consumption, power saving and power saving in cost

5. Correlation Coefficient of Power Consumption and Power Saving

The correlation statistical method is used to evaluate the relation between the power consumption and the power saving. A correlation coefficient is a single number that describes the degree of relationship between two variables that ranges from -1 to +1, indicating a perfect negative correlation at -1, absence of correlation of zero, and perfect positive correlation at +1 [6]. Table 8 indicates the relation between two random variables (X, Y) according to the correlation coefficient [6]. While, Table 9, show the graph of power consumption versus power saving for per day, per week and per month.

Table 8. Relationship between X and Y

Relationship between X and Y		
r=+1.0	Strong – Positive	As X goes up. Y always also goes up
r=+0.5	Weak – Positive	As X goes up: Y tends to usually also go up
r=0	-No Correlation-	X and Y are not correlated
r=-0.5	Weak – Negative	As X goes up: Y tends to usually go down
r=-1.0	Strong – Negative	As X goes up. Y always goes down

Table 9. Correlation Coefficient for Power Consumption, kWh and Saving in Cost, RM

<table>
<tr><td></td><td>Power Consumption, kWh</td><td>Saving in Cost, RM</td></tr>
<tr><td>Per day</td><td>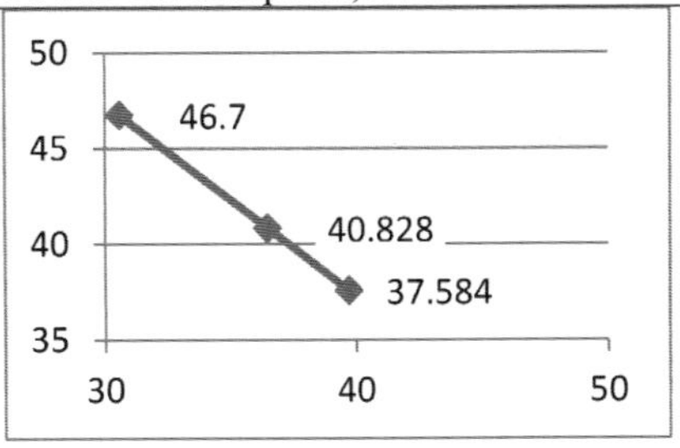

r = -1</td><td>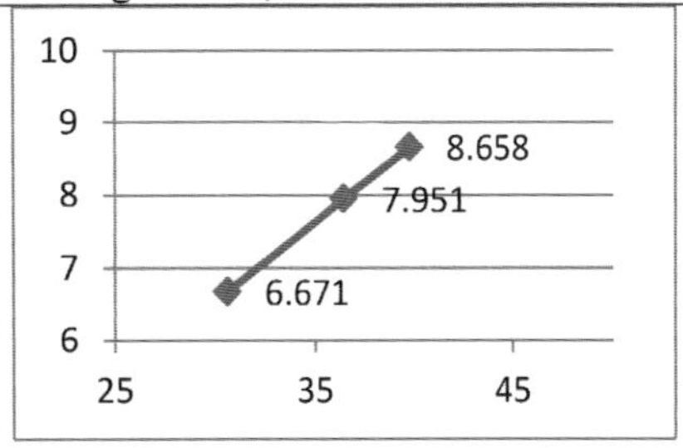

r = 0.99</td></tr>
<tr><td>Per week</td><td>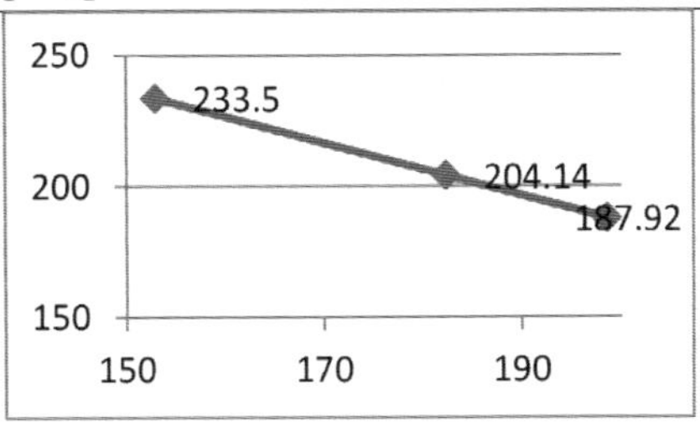

r = -1</td><td>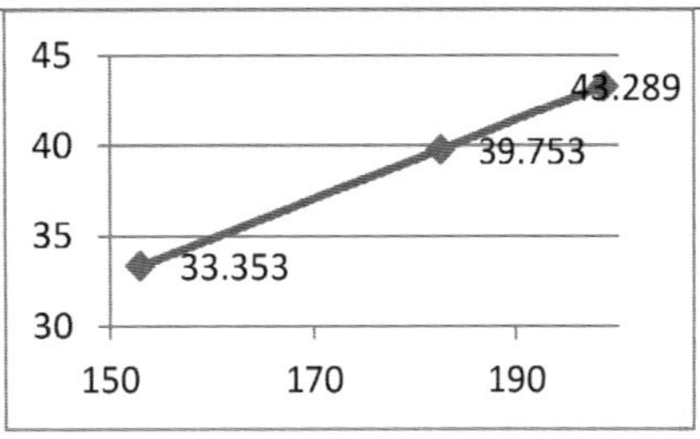

r =1</td></tr>
<tr><td>Per month</td><td>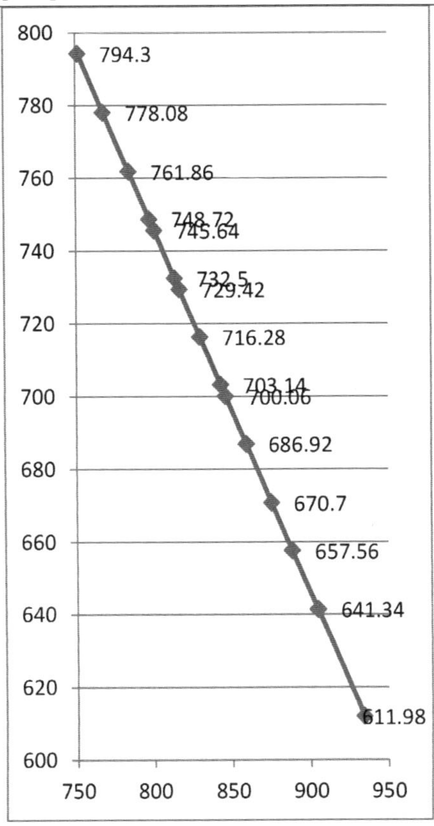

r = -1</td><td>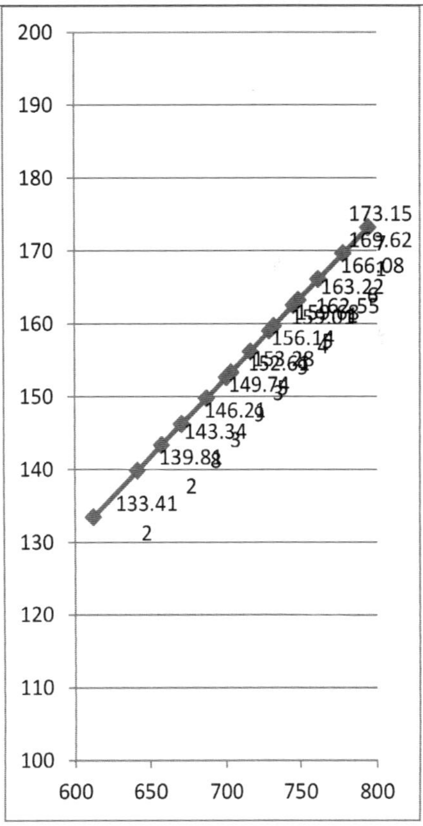

r =1</td></tr>
</table>

Thus, based on table 9, the relation between power consumption and power saving indicates a strong negative relation (r=-1). This means that when power consumption increases, the power saving will decrease.

6. Conclusion

A proposed scenario is suggested for weekday to minimize power consumption and maximize occupant's energy saving. The results indicate that smart room based on a set of sensors could perform better energy management which is not only an individual need but also for economical target. It is expected that the smart monitoring system based on human behaviour will improve energy efficiency in buildings and raise building user's awareness on green environment. The intention to install more sensitive sensors in order to sense more useful information on human behaviour in the smart room environment can provide more efficiency to the system. The result also shows a strong negative relation between power consumption and power saving in cost.

7. Acknowledgement

The authors wish to extend their thanks for the support provided by the Universiti Teknikal Malaysia Melaka (UTeM). This work was supported by the Malaysia Ministry of Education (MOE) through MTUN CoE Research Grant: MTUN/2012/UNIMAP-FKEKK/1 M00008.

8. References

[1] M. R. Alam, "SPEED: An Inhabitant Activity Prediction Algorithm for Smart Homes",*IEEE Transaction on Systems, Man, and Cybernetics-Part A:Systems and Human*, Vol. 42, No. 4, pp. 985-990, July 2012

[2] M. R. Alam, M. B. I. Reaz, F. H. Hashim and M. A. M. Ali, "Stochastic Analysis of Smart Home User Activities", *Proceedings of International Joint Conference on Neural Networks, San Jose, California, USA*, pp. 21-23, July 31- August5 2012

[3] A. R. Devidas, T. S. Subeesh, and M. V. Ramesh, "Design and Implementation of User Interactive Wireless Smart Home Energy Management System", pp. 1-6, 2012

[4] M-Y. Weng, C-L. Wu, C-H. Lu, H-W. Yeh, and L-C. Fu, "Context-aware Home Energy Saving based on Energy-Prone Context", *IEEE/RSJ International Conference on Intelligent Robots and Systems, Vilamoura, Algarve, Portugal*, pp. 5233-5238, October 7-12 2012

[5] "Wholesale Solar",*Consumer Guide to Home and the General Electric website*, http://www.wholesalesolar.com/StartHere/HowtoSaveEnergy/PowerTable.html

[6] I. I. Attia, and H. Ashour, "Energy Saving Through Smart Home", *The Online Journal on Power and Energy Engineering (OJPEE)*, Vol. 2, No. 3,pp. 223-227

[7] P. Paschke, M. Plonczak, P. Klis, and M. Grunt, "Perspectives of development of integrated monitoring system of power supply and air conditioning equipment towards technical environment equipment monitoring system of the operator", *INTELEC 2008 - 2008 IEEE 30th International Telecommunications Energy Conference*, pp. 1–6, Sep. 2008.

[8] B.H. Ng and Z. A. Akasah, "An Overview of Malaysia Green Technology Corporation Office Building: A Showcase Energy-Efficient Building Project in Malaysia", *Journal of Sustainable Development*, vol. 4, no. 5, pp. 212–228, Sep. 2011.

[9] M. S. Mathavi, D. Vanitha, S. Jeyanthi, and P. Senthil, "The Smart Home : Renewable Energy Management System for Smart Grid Based On ISM Band Communications", *International Journal of Scientific& Engineering Research* , vol. 3, no. 3, pp. 1–8, March, 2012.

[10] A. M. Carreiro, S. Member, P. S. Moura, J. I. Moreno, A. T. De Almeida, S. Member, and J. L. Malaquias, "In-House Monitoring and Control Network for the Smart Grid of the Future", *Innovative Smart Grid Technologies (ISGT Europe), 2011 2nd IEEE PES International Conference and Exhibition on*, pp. 1–7, 2011.

Journal of Physics: Conference Series **495** (2014) 012030 doi:10.1088/1742-6596/495/1/012030

A survey on human behavior towards energy efficiency for office worker in malaysia

N.H. Mustafa[1], M.N. Husain[1], M.Z.A. Abd Aziz[1], M.A. Othman[1], F. Malek[2]

[1]Center for Telecommunication Research and Innovation (CeTRI), Fac. of Electronic and Comp. Eng., Universiti Teknikal Malaysia Melaka (UTeM), Hang Tuah Jaya, 76100 Durian Tunggal, Melaka, Malaysia

[2]School of Computer and Communication Engineering Universiti Malaysia Perlis, Perlis, Malaysia

Email: nurhanim.mustafa@yahoo.com.my, mohamadzoinol@utem.edu.my

Abstract. Green environment has become an important topic around the world. This campaign can be realized if everybody understands and shares similar objectives on managing energy in an efficient way. This paper will present and analyse the survey on energy usage by office workers in Malaysia. The survey will focus on the workers in government sector. In social science surveys, it is important to support the tested data for a project. For issues related to human behaviour we must compare with real situations to verify the tested data and the results in energy monitoring system. The energy monitoring system will improve energy usage efficiency for the basic human activities in different situations and environments.

1. Introduction

A smart home management technique based on a set of sensors is presented that can be adapted according to habitant's behavior. Human habits cannot be controlled because it changes according to our life so the change of power consumption is related to the change of numbers of hours spent in each room and thus saving in KWH. The results are satisfactory and indicate that smart home based on a set of sensors could perform energy management which is not only an individual need but economical target [1]. Wireless architecture has an intelligent control device called WEEMAN (Wireless Enabled Electricity Manger) that monitors the energy consumption of a single device or a set of device [2]. This provides a unique attribute for the user to set their expected monthly electricity bill amount and proactively control the operation of all appliances according to the amount. Table 1 was shown the usages of appliance are different and different time. There will be some change for the weekend days, mainly in the time period of 10 am -4 pm (2). So, from table 1, shows that the availability management algorithm is designed to stay close to the user and learns the common behavior of human inside the house irrespective of personal identity. The algorithm will auto learn the importance of an appliance in the home based on their usage time and how frequently human uses that appliance.

Table 1. Time based usage of an appliance [2]

	Timings			
Appliances	6am – 10 am	10am- 4pm	4pm-10pm	10m-6am
Lighting	10%	5%	40%	7%
Entertainment	30%	30%	25%	10%
Kitchen	10%	5%	5%	3%
Heating	10%	5%	5%	2%
Cooling	5%	30%	10%	78%

Practical and efficient use of electricity is not only kinder to the environment, it is also kinder to get saving money and bringing more profits. Reducing electricity consumption requires surprisingly minor changes in human daily work, and perhaps some wiser decisions when choosing and installing new devices. TNB has to generate a higher amount of electricity to meet the demands of our usual work day [3]. Addition that, more electricity is required, so more power plants are fired up to generate more supply [3]. By re-scheduling the peak hours (8.00 a.m.-10.00 p.m.), that will reduce the amount of electricity required during the day, and making the most of electricity generated at night. Then, fewer power plants will need to be turn on which is good for the environment. Based on TNB Off-peak tariff Rider (OPTR) scheme, up to 20% savings for electricity usage during off-peak hours [3].

SmartCharge cuts electricity bills by storing energy during low-cost periods for use during high-cost periods. The total possible savings each day is a function of both the home's rate plan and its pattern of consumption [4]. The eGauge energy meter is used to instrument the home to measure electricity [4][5]. SmartCharge's primary benefit is that it does not require consumers to alter their electricity usage to reduce their electric bill under market-based pricing plans. Instead, SmartCharge reduces costs by determining when to switch a home between using (and storing) grid power and using previously stored power from a battery array. Since electricity prices are largely set in day ahead markets [4][5], next day prices are well-known. Then predict next day consumption by developing statistical machine learning (ML) to build a model based on important predictive metrics, such as weather, time of day, day of week, and etc. By combining SmartCharge with market based pricing is capable of reducing electricity costs for consumers over the short-term , consumers save by storing energy during low cost periods for use during high cost periods and long term, as Smart Charge penetration increases, average price will fall due to significant reduction in peak demand. In order to know better, SmartCharge's potential for savings, it is useful to consider a worst-case scenario where 100% of the home's consumption occurs during the day's highest rate period [4].

The Modern technology implicated in greenhouse environment can improve the production efficiency and security, and also reduce production cost. Besides that, green technology also can promote the improvement of agricultural automation degree; still can reduce the production strength; improve employee management level; and thus bring huge economic benefits and social benefits. Greenhouse environment monitoring has a very important role, but only for the accurate extraction of greenhouse environment information can better use modern technology to manage the greenhouse environment. Mobile phones as daily life necessary communication tools, mobile platform as the receiving system, can reliably and effectively monitor real-time data everywhere its portability is self-evident [6].

Human movement analysis is an emerging field, which involves not only medical branches, but even more technology and engineering sciences [7]. Many investigators study and analyze human movement and activities in order to improve a system especially in energy monitoring. Human behavior will analyze to get the efficient energy control. Nowadays, energy consumption and energy saving depends on human daily routine. Few papers discusses about human daily routine in case of smart house, and also at medical organization to monitor patient. This paper present the survey and analysis that have been done on the government workers especially those work in office environment. This investigation of the relationship between daily physical activities requires an objective method to assess physical activity under daily living conditions. The question in the survey form is focused on the activities that related to the energy usage which represent the worker's behavior himself. The information from survey will be used by energy monitoring system to predict and manage the energy with more efficient.

2. Survey Form

The questions in the survey form are design to acquire the feedback from workers in government sector according to their daily activities which related to the energy usage. The questionnaire is focused on the worker in workplace to know their activity during working at 8 am until 5 pm. There are following question in survey form:

2.1. Question 1

Ask about usual time comes to the office. This question will be used to record the time for staff coming to work. There are 3 categories of arriving time available for sorting purposed.

2.2. Question 2

Ask about making a coffee or tea itself in the office. There are given condition to choose whether Yes or No, if said yes must to write the proper time. This question would be ask because to see their activity and also to see whether using water heater or not in the office.

2.3. Question 3

Ask about breakfast. There also has a condition to choose whether Yes or No; If Yes must to write that time. This question would be asked because to know the time their go for breakfast and to know how many people currently like go for breakfast.

2.4. Question 4

Ask about listening to the music during do a work. This question to know whether worker like or not to listen the music, so that will shows that worker are using a device or electrical appliances.

2.5. Question 5

Ask about usage others appliances included with charging handphone, mp3, mp4, camera battery, using own laptop or tablet and also use the projector. This question because to see how many worker a like to use the devices selected and to combined for electrical appliances calculation.

2.6. Question 6

Ask about usual time go to the restroom and how long a worker spent in the restroom. This question to see the specific time them go the restroom and to categories the types of worker whether in person A, B or C.

2.7. Question 7

Ask about usual time for lunch or pray and how long a worker spent for lunch and pray. This question to see the specific time them go for lunch or pray and to categories the types of worker whether in person A, B or C.

2.8. Question 8

Ask about usual time going to the living room and how long a worker spent in the living room. This question to see the specific time them go to living room and to categories the types of worker whether in person A, B or C.

2.9. Question 9

Ask about activity worker in living room. This question to see how many workers like to read newspaper or magazines, watching TV, Make a coffee, or do small meeting. That also can be shows the electrical appliances used.

2.10. Question 10
Ask about type of lighting that workers prefer during working whether normal, dimmable or brightness. This question to know because want to choose the best lighting are worker prefer in the office.

2.11. Question 11
Ask about automatic control for lighting and air-conditional. This question will help to decide the right type of switch (manual or auto control).

2.12. Question 12
Ask about temperature level that workers prefer in office. This question will help to set the temperature and to know how many much prefer to use fan in office.

3. Data Survey Analysis
Human is unique and each person have different behaviour either at working place or at their home. Based on survey, 54 workers have respond with positive feedback from 3 government school located in Penang and Malacca, Malaysia. There are 42 and 12 respond recorded for female and male respectively. Figure 1 to Figure 15 show the analysis for each question in the survey form.

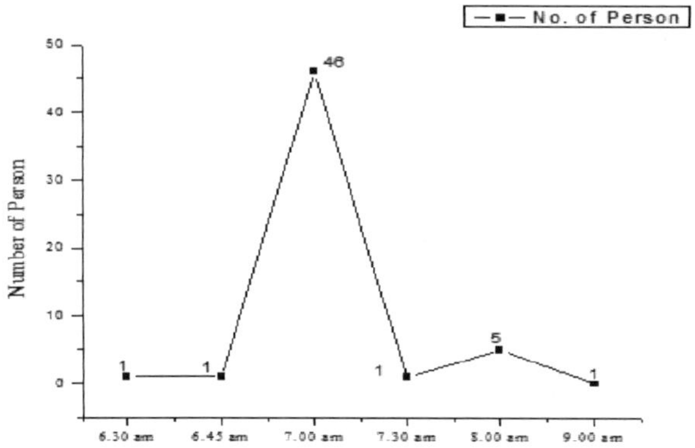

Figure 1. Time worker come to the office

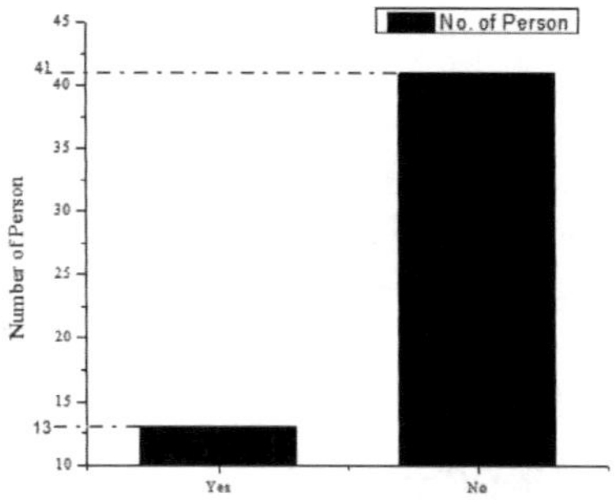

Figure 2. Workers like to make a coffee/ tea itself

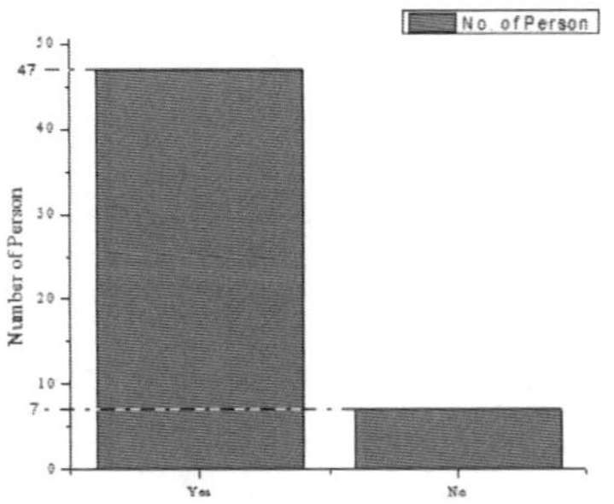

Figure 3. Worker like go to the breakfast

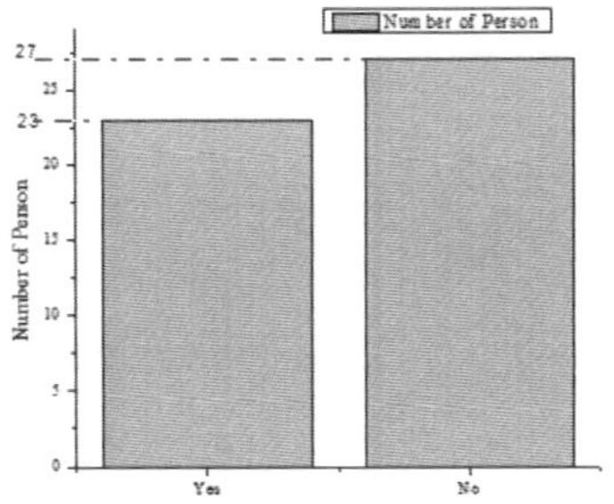

Figure 4. Worker don't like to listen music during do a work and 23 worker like to listen music

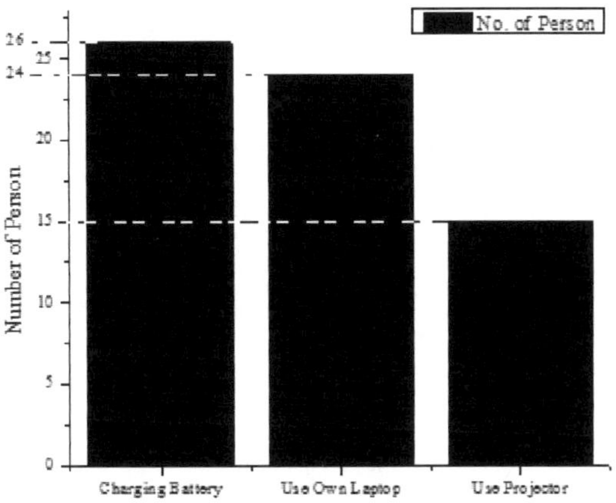

Figure 5. Worker is charging battery, use their own laptop and the projector in office

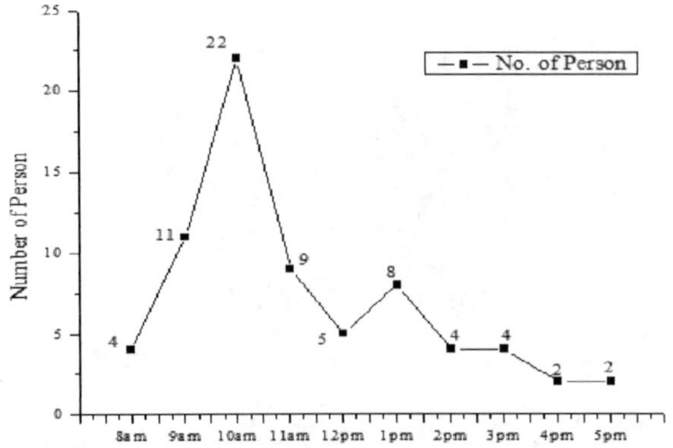

Figure 6. Worker like go to restroom at 10.00 am

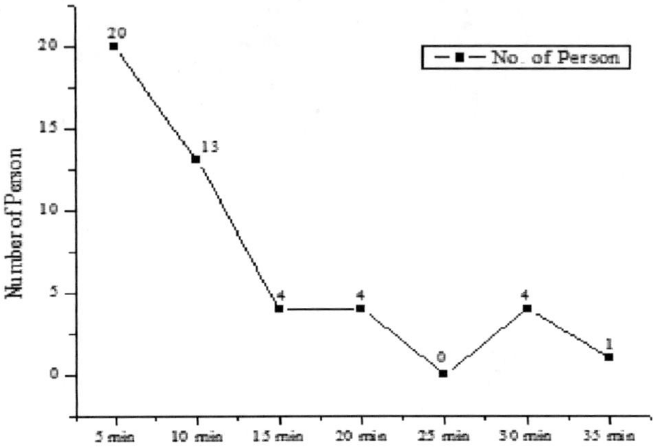

Figure 7. Time spent in the restroom

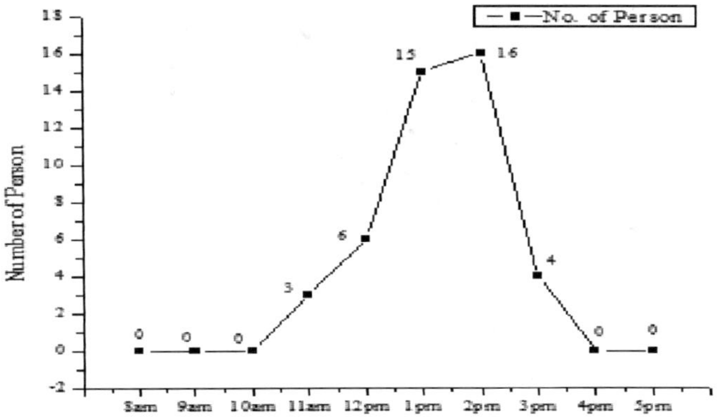

Figure 8. Time for lunch and pray

IOP Publishing
doi:10.1088/1742-6596/495/1/012030

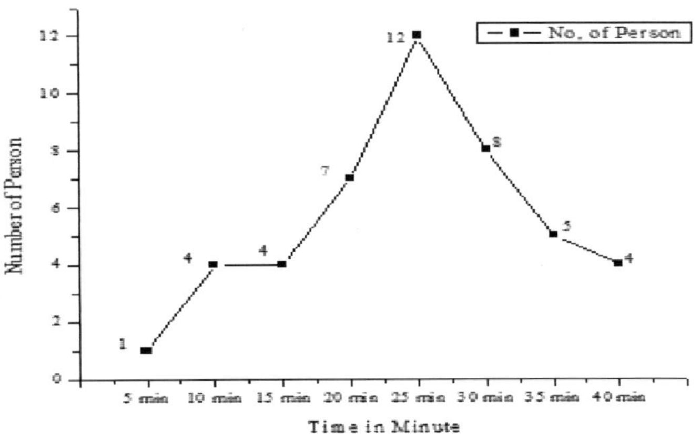

Figure 9. Time spent for lunch and pray

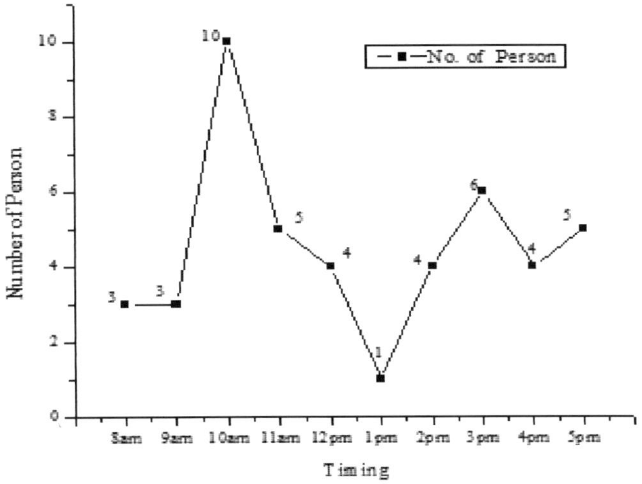

Figure 10. Time used to living room

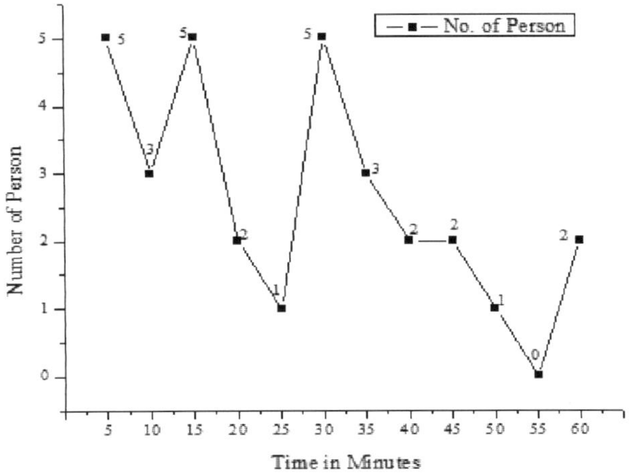

Figure 11. Time spent in living room

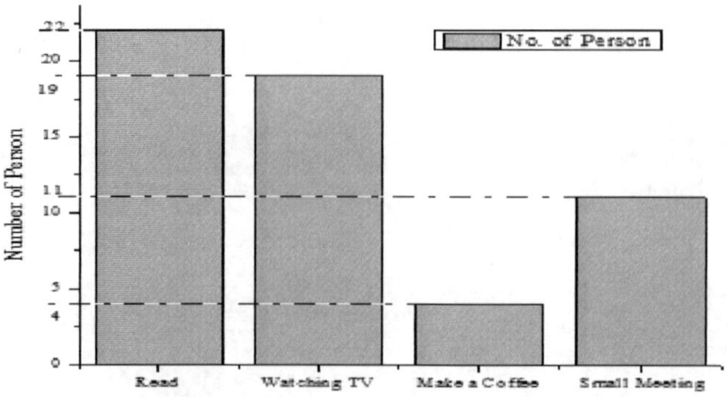

Figure 12. Activities in living room

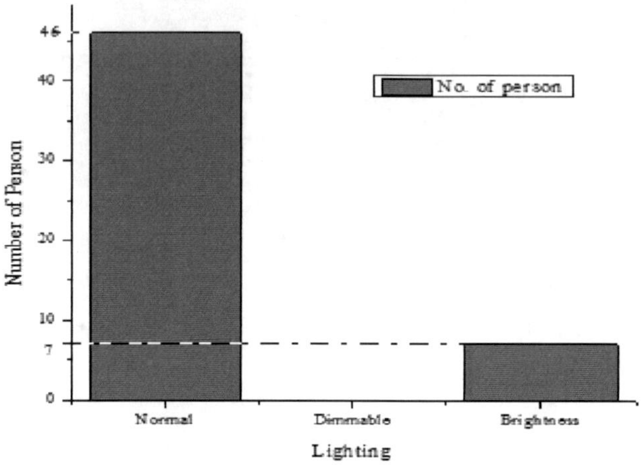

Figure 13. Lighting condition

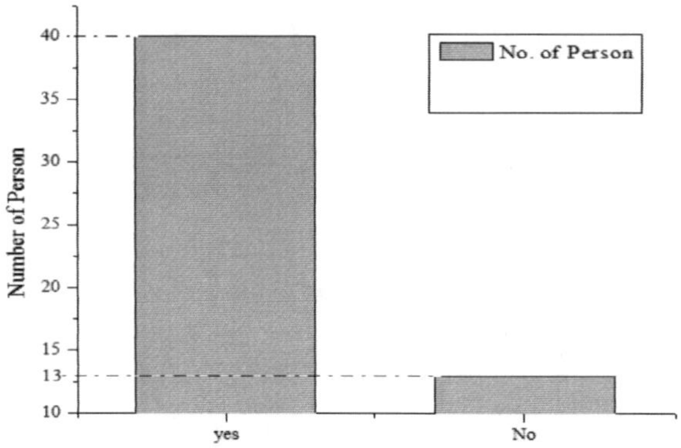

Figure 14. Automatic 'ON and OFF' lighting and air-conditional

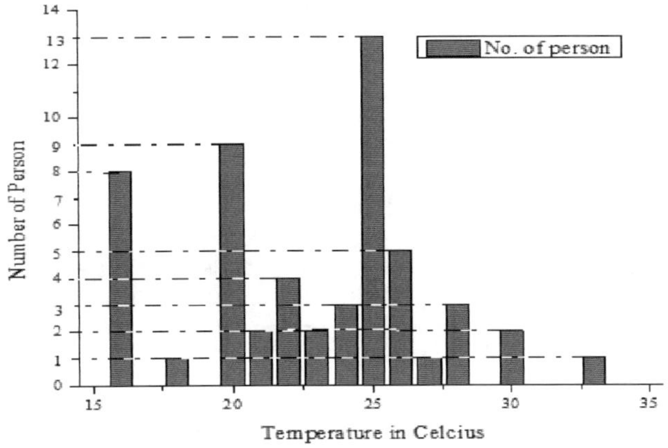

Figure 15. The temperature in a room

The proposed data from survey form will be used as input information to energy monitoring system based on human behaviour design at our Research Lab at Faculty of Electronic & Computer Engineering, Universiti Teknikal Malaysia Melaka (UTeM). This survey is very important in order to verify the reliability and accountability of the design system. Moreover, the human behaviour has been categorized as unique information and required continuous monitoring. The efficiency of the energy usage will help the administration to reduce the operational cost and improved the productivity.

4. Conclusion

Human activity cannot be control in most of the time and places. However, the information on human behaviour can be uses to create the algorithm that very useful in energy monitoring system. The survey data are combined to identify the human behaviors. The used of low price devices will not ensure the energy efficiency and power saving. Thus, the information of human behaviour in their working place are very important to plan the energy saving and higher efficiency on power consumption. However, the continuous recorded system on human behaviour is still needed to improve the system reliability. So, this survey is an important step towards green environment and better life in the future. By capturing human movement based on ambulatory technology is a relatively new field of research since it is directly related to the recent advances in miniature devices and sensors, new technologies for powerful microcontroller, high capacity memory and small power sources.

5. Acknowledgement

The authors wish to extend their thanks for the support provided by the Universiti Teknikal Malaysia Melaka (UTeM). This work was supported by the Malaysia Ministry of Education (MOE) through MTUN CoE Research Grant: MTUN/2012/UNIMAP-FKEKK/1 M00008.

6. References

[1] I. I. Attia, and H. Ashour, “Energy Saving Through Smart Home”, *The Online Journal on Power and Energy Engineering (OJPEE)*, Vol. 2, No. 3,pp. 223-227

[2] A. R. Devidas, T. S. Subeesh, and M. V. Ramesh, “Design and Implementation of User Interactive Wireless Smart Home Energy Management System”, pp. 1-6, 2012

[3] Tenaga Nasional Berhad, Malaysia, 2013, http://www.tnb.com.my/business/energy-savings-at-work.html#powerfactor

[4] A. Mishra, D. Irwin, P. Shenoy, J. Kurose and T. Zhu, “SmartCharge: Cutting the Electricity Bill in Smart Homes with Energy Storage”, *e-Energy 2012*,pp. 1-11, May 2012

[5] eGauge Monitoring. http://www.egauge.net/,2012

[6] B.H. Ng and Z. A. Akasah, "An Overview of Malaysia Green Technology Corporation Office Building: A Showcase Energy-Efficient Building Project in Malaysia", *Journal of Sustainable Development*, vol. 4, no. 5, pp. 212–228, Sep. 2011

[7] K. Aminian, " Monitoring Human Movement with Body-Fixed Sensors and its Clinical Applications- Chapter III", *Computational Intelligence for Movement Sciences: Neural Networks and Other Emerging Techniques, IDEA Group Publishing ITB12285*, 2006

[8] C. V. C. Bouten, A.A. H. J. Sauren, M.Verduin and J. D. Janssen, " Effects of Placement and Orientation of body-fixed Accelerometers on the Assessment of Energy Expenditure during Walking", *Medical & Biological Engineering & Computing*, 1997

[9] W. Wang, N. Wang, E. Jafer, M. Hayes, B. O'Flynn, and C. O'Mathuna, "Autonomous wireless sensor network based Building Energy and Environment Monitoring system design", *2010 The 2nd Conference on Environmental Science and Information Application Technology*, pp. 367–372, Jul. 2010

[10] W. Xu, M. Dong, "Tracking energy consumptions of home appliances using electrical signature data", *2012 IEEE Conference on Power and Energy Society General Meeting*, pp. 1-5, 2012

[11] M. R. Alam, "SPEED: An Inhabitant Activity Prediction Algorithm for Smart Homes",*IEEE Transaction on Systems, Man, and Cybernetics-Part A:Systems and Human*, Vol. 42, No. 4, pp. 985-990, July 2012

Journal of Physics: Conference Series **495** (2014) 012031 doi:10.1088/1742-6596/495/1/012031

Application of Genetic Algorithm to the Design Optimization of Complex Energy Saving Glass Coating Structure

F M Johar[1] F A Azmin[1] A S Shibghatullah[2] M K Suaidi[1] B H Ahmad[1] M Z A Abd Aziz[1] S N Salleh[1] M Md Shukor[1]

[1] Faculty of Electronic and Computer Engineering, Universiti Teknikal Malaysia Melaka (UTeM), Melaka, Malaysia

[2] Faculty of Information and Communication Technology[2], Universiti Teknikal Malaysia Melaka (UTeM), Melaka, Malaysia

E-mail: fauzi@utem.edu.my[1], inuyaharaf@gmail.com[1], samad@utem.edu.my[2]

Abstract. Attenuation of GSM, GPS and personal communication signal leads to poor communication inside the building using regular shapes of energy saving glass coating. Thus, the transmission is very low. A brand new type of band pass frequency selective surface (FSS) for energy saving glass application is presented in this paper for one unit cell. Numerical Periodic Method of Moment approach according to a previous study has been applied to determine the new optimum design of one unit cell energy saving glass coating structure. Optimization technique based on the Genetic Algorithm (GA) is used to obtain an improved in return loss and transmission signal. The unit cell of FSS is designed and simulated using the CST Microwave Studio software at based on industrial, scientific and medical bands (ISM). A unique and irregular shape of an energy saving glass coating structure is obtained with lower return loss and improved transmission coefficient.

1. Introduction

Nowadays, energy saving glass has become very popular in modern building and vehicles. It was proven to be very useful to maintain the temperature level inside a building or vehicle using energy saving windows. This technology falls under the Green Technology category for it helps contribute saving the energy in a building or vehicles. It is in the late 1980s when low emissivity (low-e) glass first began to contribute to energy saving through window of a building with low-e properties. It works by applying a very thin metallic oxide (e.g. silver oxide or tin oxide) on one side of float glass.

This coating blocks infrared radiation due to heat from outside of the building and almost transparent to the visible light [1]. It provides a good thermal isolation inside the building, especially during winter. Even though energy saving glass plays the role of saving the energy, useful signals in wireless communication systems such as global positioning system (GPS), mobile communication system (GSM, UMTS, 3G), wireless network (Wi-Fi) and wireless broadband (WiMax, LTE) are attenuated due to metal-oxide coating on the energy saving glass [2]. These weaknesses limit the efficiency of the energy saving glass function in wireless communication of microwave signals. A significant amount of signal attenuation can lead to received signal level (RSL) or received power exceeding the fade margin, resulting in radio link breakdown between base stations and mobile devices. In addition, the functional aspect of an energy saving glass is limited when most wireless communication signals are attenuated through the use of low-e glass. One of the most common practices to overcome the signal attenuation is by employing numerous repeaters to enhance the useful signals. However, it is not a cost effective solution and consuming more electricity for installing and operating these additional repeaters. The current study [2] ascertains that attenuation of useful signals such as GSM, GPS and personal communication signal leads to poor communication inside the building.

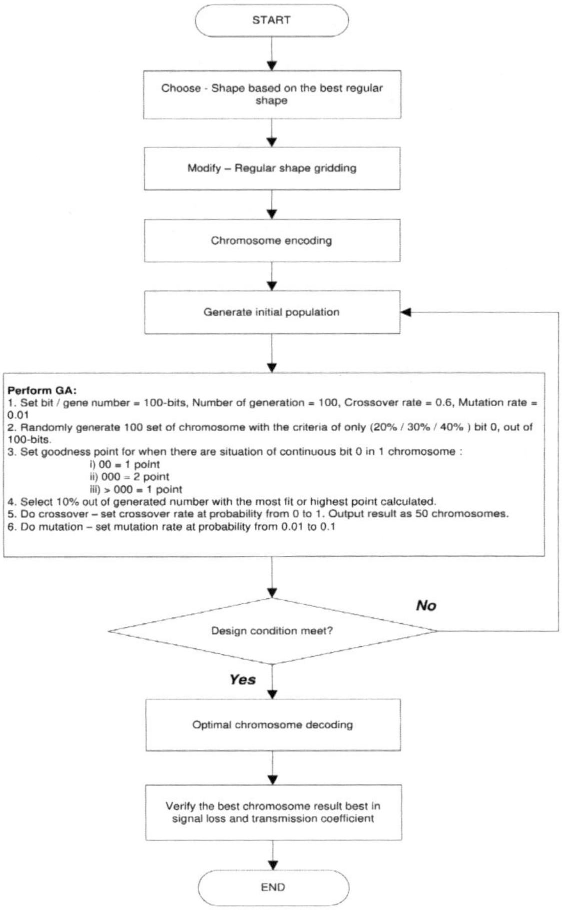

Figure 1. Flowchart of design cycle utilizing the Genetic Algorithm approach

According to [3], metamaterials have the properties that cannot be found in nature. Its properties usually gained from their designed structures which consist of periodic structure and subwavelength characteristic with particle smaller than the light wavelength with which it interacts. Known as Frequency Selective Surface (FSS), it is a planar periodic structure of identical array of patches or aperture type elements arranged in one or two dimensional plane to electromagnetic (EM) waves. The frequency characteristics such as bandwidth, transfer function, phase feature and angle stability are determined by many critical parameters such as array periodicity, unit cell's pattern, element geometry size, property of dielectric substrate and others to form an FSS. There are various methods to analyze the periodic structures of the FSS such as the mutual impedance method, the method of moments (MoM), the finite element method (FEM), the finite-difference time-domain (FDTD) method and equivalent circuit method (EC) [6-10].

An optimization method such as Genetic Algorithm (GA) is a method that helps solve problems that is unsolvable using conventional techniques. Inspired by evolutionary mechanisms, it is a popular strategy used to optimize a large number of solutions or variables in a non-linear system. GA is one of the categories of the large field of optimization algorithms. As mentioned above, it is the natural selection of the fittest individuals in a population over generations to solve a problem. In general, generations represent populations that consist of individuals that will go through an evolution process. The evolution process works using operators such as selection, crossover and mutation to name a few. A population is created from chromosomes before operators are applied to each of the individual of the population. Then, a new individual called an offspring is generated. Depending on the feature of the problem, separate positions of each chromosome are encoded as bits, characters or number. In order to use a genetic algorithm, solutions to a problem must be encode into a structure that can be stored in the computer. This is called a genome (or chromosome). In general, GA first makes use of fitness function to evaluate a population of individuals to create a new population.

Implementation of numerical modeling can help improve the design process of a complex shape. Characterization of one design structure can be more accurate and more details by employing a numerical analysis such as Finite Difference Time Domain (FDTD). Issues on modeling, frequency range, material properties or dimensions that is none of the capability of optimization can be covered by a numerical modeling. Optimization problems in electromagnetic has been solve by using Genetic Algorithms (GA) optimization for a quite a long time. There are many advantages when applying GA to any optimization problems. Unlike other optimization method, GA does not have many mathematical requirements about the particular optimization problems. Employing its nature on evolutionary, GA will search for solutions without regard to the specific inner workings of the problem [4]. The evolution operators such as selection, crossover, mutation and recombination make GA effective at performing global search. Today, GA has been proven extremely useful in designing FSS with optimal propagation properties. The exploration ability of GA as an optimizer is indeed very useful in solving complex joining problems especially when numerous parameters to be optimized at once.

In this paper, we proposed GA approach with the application of improved return loss and transmission coefficient. Pre-existing FSS designed by numerical method, Periodic Method of Moment (PMM) is optimized to be able to apply to an application such as the shape of a coating structure on energy saving glass [4]. A unique and irregular shape of an energy saving glass coating structure is obtained with lower return loss and improved transmission coefficient. The flowchart of this design method with the approach of FSS and GA is shown in Figure 1. This topic uses the approach of modeling, synthesizing that includes process of a problem, and fabrication can be planned in the future, when a new design is finalized. The topic chosen will contribute to a new and improved design of a coating structure of energy saving glass as well as green technology.

2. Problem related to GA-based FSS application in energy saving glass

A study [2] proves that attenuation of GSM, GPS and personal communication signal leads to poor communication inside a building using energy saving glass. Thus, the transmission is very low. The coating on energy saving glass is usually etched in regular shapes. Cross dipole, circular loop, hexagon, square and tripoles are some of the popular geometries of shape coated on energy saving glass [1]. This coating works like a filter where the coating part is reflected part, while the etched part left, is where transmissions are allowed to pass through. Hence, the bigger size of the coating will reflect more signals and heat.

According to this situation, it can be assumed that attenuation of signals with energy saving glass involves the use of low-e glass. Recently, a study also attests that transmission of signals improved when at least 10% of metallic coating being removed, and transmission loss of cross dipole for instance results around 25dB [2]. As the bigger size of coating etched part the better the efficiency of the transmission signals. The smartest way to deal with this problem is by introducing a complex or irregular shape of coating as well as lessens the percentage of coating area.

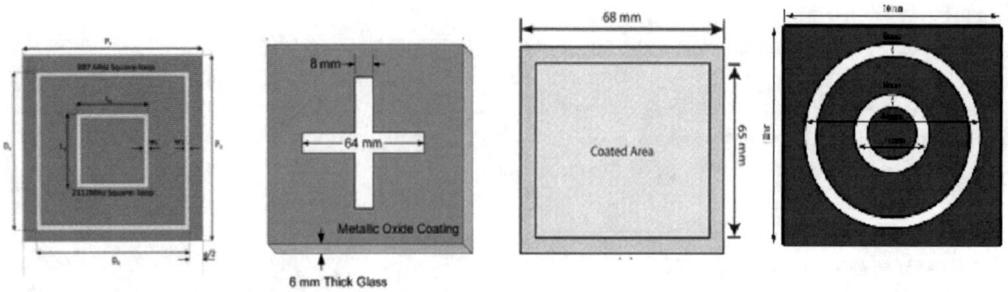

Figure 2. Regular shapes of existing coating

Current studies using regular shapes of cross-dipole for example, shows transmission loss at 25-30dB [2] of the coated area of an energy saving glass which contributes to the attenuation of useful signals. Until today, the coating of an energy saving glass still uses regular shapes as the structure of the low-e glass coating. Based on the study above, introducing a new complex shape of coating with less percentage of coating area is optimistic to improve the efficiency of microwave signal transmission. An aperture type of design is the best proposed method of previous research [5]. Aperture refers to the slot made on the coating to represent an air or free space. This particular design is closely related to a structure design named Frequency Selective Surface (FSS). It is a periodic structure of identical cells in one or two directions. Aperture and patch type element are the two categories of FSS. An irregular or complex shape can be worked out through several methods for example mathematical analysis, software, optimization or any other process.

Currently there are two issues when proposing a complex shape. First, it is a difficult thing to do since a complex shape involves small pixels and it's complicated to optimize every single pixel. Existing simulation software such as Computer Simulation Technology (CST) can be implemented to design this complex shape, but will not assure it is the best shape optimized. There would be a trouble in designing a complex shape since regarding the connectivity condition of critical point. What will happens is during fabrication process, to fabricate critical point tend to break the material used for example, glass.

3. Design Methodologies

Designing a complex shape can be a difficult task and a specific method to get an optimum result is needed. This paper demonstrate the design method a complex shape of coating on an energy saving glass by modifying the best result of regular shape design based on previous study and experiment. In this work, optimization method is introduced to help in the design process. GA is implemented to randomly generate a set of 100-bits binary and selecting the best chromosome. In order to verify that the new tool is effective in selecting the best complex shape, the result is tested by measuring the transmission losses and return loss improved from the original. The GA starts with a random population and then it generates modification bits to form an FSS through selection, crossover and mutation. A GA-based optimization flowchart is presented in Figure 1.

First of all, the best design of existing method is chosen based on the numerical or analysis characteristics. As for this particular, an initial population generated by PMM was chosen. Then, the shape is modifying into grids of pixel of 400 bits in a dimension on 20 by 20. This is called a GA initialization when involve with the chromosome encoding process, followed by generating initial population. Figure 3 shows the original design of PMM generated geometry.

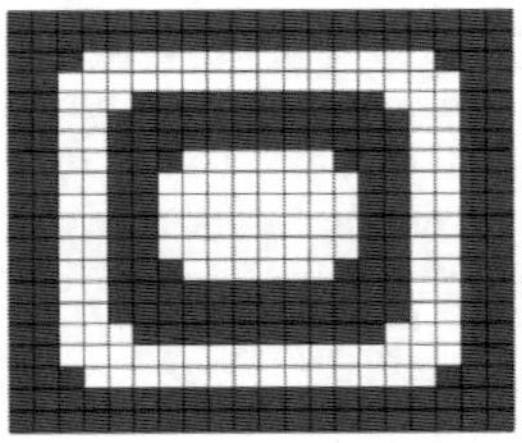

Figure 3. Pre-existing FSS design generated by Periodic Method of Moment

The next step is Encoded into pixels of 400 bits binary number it represent as coated in binary 1, and uncoated as binary 0. Here, the operation of GA applies with the constraints stated as 5% , 10% and 15% of coated side 156 bits, and to uncoated side 244 bits. Table 1 shows pseudo code to code GA.

Table 1. Pseudo code of GA with rules to crossover and mutation

```
START
        Generate 1000 chromosome according to the number of bit-0 required.
DEFINE
            Define 'goodness' point Thus, goodness point will be given as:
              i.   1 point will be given if there is two bit-0, for example: 00
              ii.  2 points will be given if there is three bit-0, for example: 000
              iii. And 1 points is given to situation when there are more than four
                   bit-0, for example: 0000, 00000, 000000++

DO
        Do selection for 10% of the best/fittest chromosome using the 'goodness' criteria. e.g :
        10% = 100 chromosomes.

        Do crossover to this 100 chromosomes. Crossover rate can be set from 0 to 1. Output
        results of 50 new chromosomes.
        Do mutation (bit complement) to these 50 chromosomes. Set mutation rate at 0.01.

SELECT
        Select best chromosome based on the highest goodness point, crossover result and
        mutation result.
REPEAT
        Repeat generation of random number IF no result.
OUTPUT
        Output results in text file for the random number generated, crossover result, mutation
        result, and goodness point result of every chromosomes.
```

The characteristics of each chromosome that carries 100bits each are determined by a measure of goodness. Since it is almost impossible and time consuming to determine the losses of each chromosome, selection method used is by giving a 'goodness' value to chromosome that has the following criteria. Logically, uncoated part must be continued side to side to produce good result. However, when the continued bit-0 is too long, it is not good either. Thus, goodness point will be given as 1 point will be given if there is two bit-0, for example: 00. Two points will be given if there is three bit-0, for example: 000. Finally, one point is given to situation when there is more than four bit 0, for example: 0000, 00000, 000000 and so on. Each chromosome will be determined by its goodness

Journal of Physics: Conference Series **495** (2014) 012031 doi:10.1088/1742-6596/495/1/012031

point. The fitness function stated must have to maximize good point. A chromosome that has maximum goodness point will be chosen for the next process which is crossover and mutation. Operators such as selection, crossover and mutation will determine the best and optimum result.

The fitness chromosome that has the maximum value of goodness is recognized as an optimum element shape. Otherwise, the process cycle to next generation is repeated until an FSS with desired characteristics is obtained. It is aimed to find the best solution for a complex design shape replacing the regular shape as well as to improve the transmission of microwave signal received inside a building using energy saving glass. Once the condition predefined is agreed, the optimum chromosome is decoded into design. Finally, using software for testing s-parameter the efficiency of the design is tested to verify the design.

4. Results and Discussion

In this part, some illustrative experimental result is shown to demonstrate the efficiency of the method proposed. GA is used to generate random binary at coated side and uncoated side at the size of a total of 20 by 20, which consists of 400 bits. The coated side is marked as coloured box, while the uncoated side is left uncoloured. As discussed in the flowchart (Figure 1) above, initial population generated by PMM is used as the original shape to be optimized.

The best result chosen is preceded to the verifying process using CST Microwave Studio. A shape is designed according to binary result obtained from the process involving GA. Return loss and transmission coefficient are recorded and compared to the original PMM design. In Figure 4, the result of the simulation of chosen design in PMM is presented. The initial return loss is at -29.57dB using this design. While transmission coefficient result is at -0.00047dB.

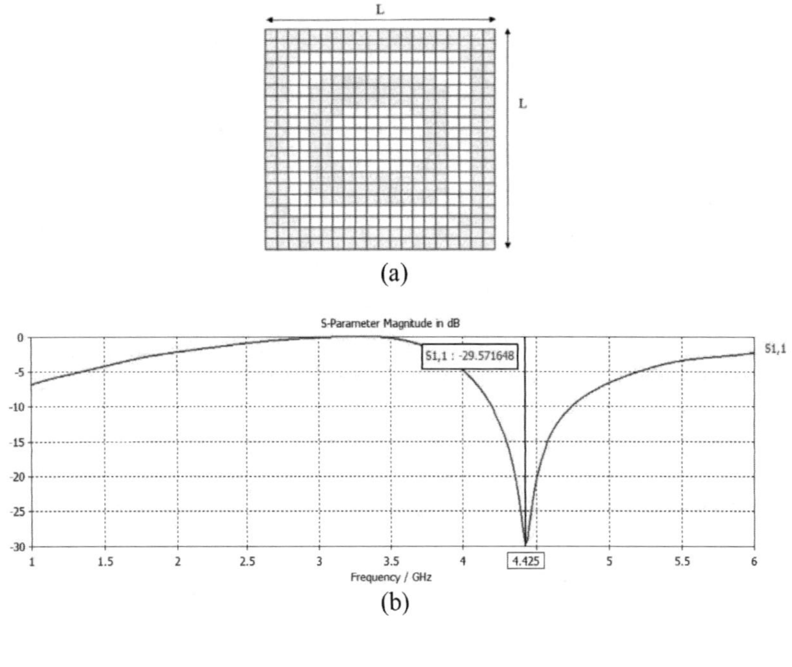

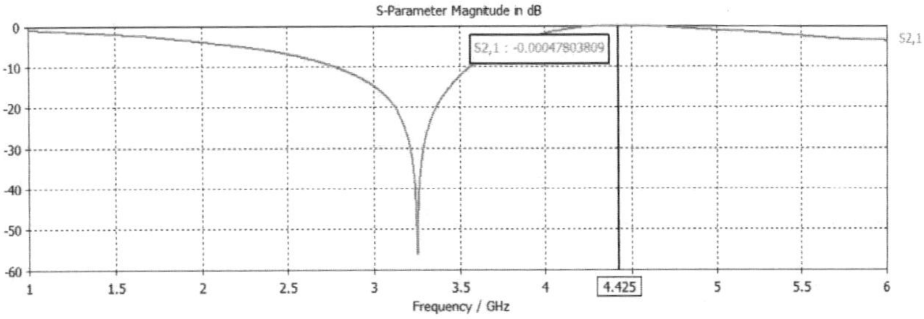

(c)

Figure 4. (a)Original initial population generated by PMM. (b) Result of return loss at 4.425GHz shows – 29.57dB. (c) Result of transmission coefficient shows -0.00047dB.

Shown in Figure 4 is the optimum result of GA being optimized at its optimum. After being applied with GA at the coated and uncoated side, result shows that optimization made at the uncoated side of 15% out of overall bits gives highest result.

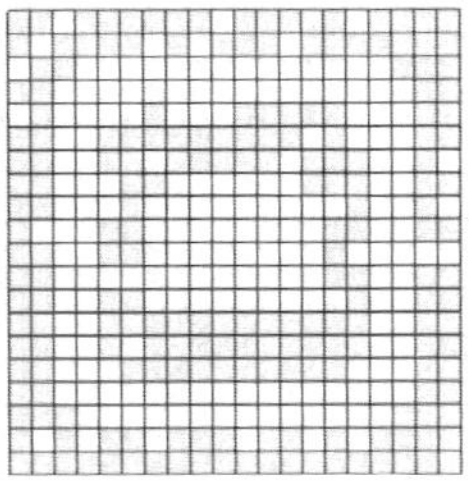

(a)

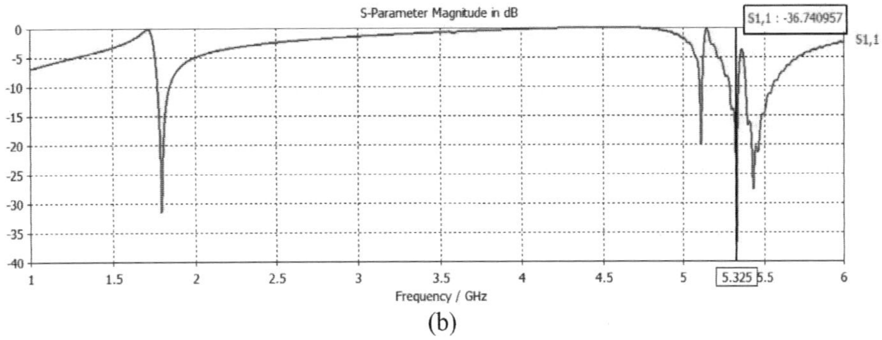

(b)

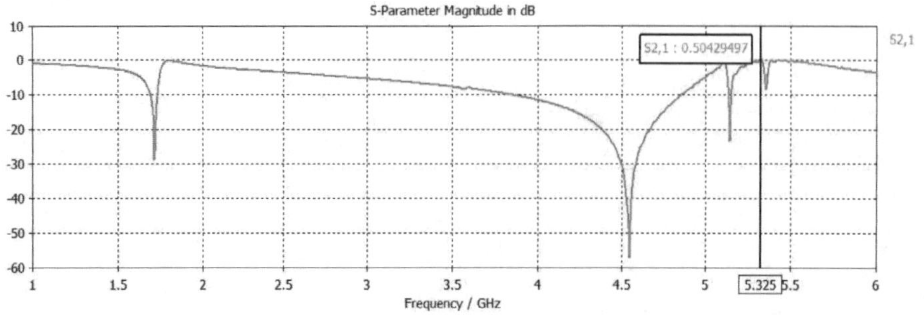

(c)

Figure 5. (a) Design of best result at uncoated side being randomized using GA. (b) Return loss, S11 at optimum frequency of 5.325GHz shows -36.7409dB. (c) Transmission coefficient shows optimum result of 0.504dB.

The other result optimized at 5%, 10% and 15% is presented in Table 2. At 15% uncoated side being optimized, it shows that highest optimum result at 5.325GHz. Results at 5% to 10% show an improvement after the uncoated part is altered.

Table 2. Result of coated and uncoated side

Uncoated side – 156 bits		
Percentage	S11 (dB)	S21(dB)
5%	-29.2233	-0.0029
10%	-29.8717	-0.0381
15%	-36.7409	0.5043
Coated side – 244 bits		
Percentage	S11(dB)	S21(dB)
5%	-30.3535	-0.0021
10%	-40.0188	-0.0188
15%	-30.0288	0.0238

While at the coated side, average result is obtained and most optimum result shows at 10% in the return loss, S11. For coated side, transmission coefficient is best at 15% GA optimization, which is 0.0238dB. It can be seen, that there are isolated pixels with or without the point of contacts in this design. In order to get a better design than a scattershot design and without having to deal with unnecessary fabrication complexity, these isolated pixels can be removed [11].

5. Conclusion

In this paper, a band pass FSS filter with sufficient performance is presented. For this process, a strong GA based optimization approach is demonstrated in detail to optimize the unit cell topology. Using GA optimization procedure, FSS can be designed to meet wider goals, including the frequency, s-parameter, return loss (S11) and transmission loss (S21). Optimum result of signals shows improvement in signal losses and transmission coefficient. It is shown in this paper that the present method is more robust and taking account of points on existing design that contributes to good result. Optimum result of signals shows improvement in signal losses and transmission coefficient. Compared to traditional design on the previous study, the return loss and transmission coefficient are affected and proven improved. For future work, the operating frequency can be varied according to other suitable application using Periodic Method of Moment and GA. A lot of variation using the pre-existing numerical method can be applied and joined with GA to come out with a unique design of many purposes of application. Processing GA on a parallel processor can save a lot of time, by running greater number of individuals in parallel.

References

[1] G. I. Kiani, L. G. Olsson, A. Karlsson, and K. P. Esselle, "Transmission of infrared and visible wavelengths through energy-saving glass due to etching of frequency selective surfaces," IET Microwaves, Antennas Propag., vol. 4, no. 7, p. 995, 2010.

[2] S. I. Sohail, G. I. Kiani, and K. P. Esselle, "Enhancing RF / Microwave Efficient Transmission through Energy Saving Glass Windows using Frequency Selective Surface," pp. 2262-2263, 2011.

[3] M. Z. A. A. Aziz, M. Shukor, B. H. Ahmad, N. A. Zainuddin, M. F. Johar, F. A. Azmin, S. N. Salleh, and M. F. A. Malek, "Impedance Modeling for the Circular Loop Frequency Selective Surface (FSS)," no. December, pp. 3-4, 2013.

[4] P. Guo, X. Wang, and Y. Han, "The enhanced genetic algorithms for the optimization design," *2010 3rd Int. Conf. Biomed. Eng. Informatics, (BMEI)* pp. 2990-2994, Oct. 2010.

[5] Munk, B. A., Frequency Selective Surfaces – Theory and Design, John Wiley and Sons, Inc., New York, 2000.

[6] Y. Liu, Q. Ye, B. Xiao, and H. Yang, "A novel application of frequency selective surface in dual-band WLAN antenna," Isape2012, pp. 512–514, Oct. 20

[7] I. Ullah, D. Habibi, and G. Kiani, "Design of RF/Microwave efficient buildings using frequency selective surface," 2011 IEEE 22nd International Symposium on Personal, Indoor and Mobile Radio Communications, pp. 2070–2074, Sep. 2011.

[8] R. J. L. and E. A. P. T.K. Chang, "Frequency selective surfaces on biased ferrite substrates," vol. 30, no. 75, pp. 3–4, 1994

[9] M.Z.A. Abd. Aziz, M. Md. Shukor and M.K. Suaidi, "Phase Shift of the Transmission Coefficients For Frequency Selective Surfaces," on 6th International Conference on Information Technology (ICIT'13), 2013.

[10] M.Z.A. Abd. Aziz, M. Md. Shukor, M.K. Suaidi, B. H. Ahmad, M. F. Johar, F.A. Azmin, S.N.Salleh and M. F. Abd. Malek, "Impedance Modeling for a Unit Cell of the Square Loop Frequency Selective Surface at 2.4 GHz," on The International Conference on IEEE RFID Technical and Application (IEEE RFID 2013), 2013.K. F. Man, K. S. Tang, and S. Kwong, "Genetic Algorithms : Concepts and Applications," vol. 43, no. 5, 1996.

[11] G. Manara, A. Monorchio and R. Mittra. "Frequency Selective Surface design based on genetic algorithms." IEEE Electronics Letters Vol. 35, No. 17 August 1999.

[12] B. G. Xia, D. h. Zhang, Y. F. Zhao, and J. Meng, "Application of the Genetic Algoritm to the Design Optimization of Milimeter Wave Spaceborne Frequency Selective Surface," pp. 962-966, 2012.

[13] A. Hassani, "An Overview of Standard and Parallel Genetic Algorithms", 1975.

[14] J. M. Johnson, Y. Rahmat-samii, and H. Ave, "Genetic Algorithms in Engineering

Electromagnetics," vol. 39, no. 4, pp.7-21, 1997.
[15] T. M. Mitchell. Machine Learning. MIT Press and The McGraw-Hill Companies, Inc, 1997.
[16] R. Kumar and S. Member, "Blending Roulette Wheel Selection & Rank Selection in Genetic Algorithms," vol. 2, no. 4, pp. 365-370, 2012.
[17] L. M. Brasd D. E. Goldberg, "Genetic Algorithms, Tournament Selection and the Effects of Noise," vol. 9, pp. 193-212, 1995.
[18] W. M. Spears and V. Anand, "A Study of Crossover Operators in Genetic Programming," in 6th International Synposium, ISMIS '91 Charlotte, N. C., USA, October 16-19, 1991 Proceedings, 1991, pp. 409-418.
[19] R. L. Haupt and S. E. Haupt, Practical Genetic Algorithms. New Jersey: John Wiley & Sons, 2004, p.253

Acknowledgments
The author would like to thank Universiti Teknikal Malaysia, Melaka, Malaysia (UTeM) and UTeM Centre of Graduate Studies (CGS), for the financial support given towards completion of this paper. The project is financially supported by the Ministry of Higher Education under the grant MTUN/2012/UTeM-FKEKK/5 M00013. The authors appreciate of the efforts of the reviewers for their helpful comments.

Implicit finite difference solution for time-fractional diffusion equations using AOR method

A. Sunarto[1], J. Sulaiman[2] and A. Saudi[3]
[1,2] School of Science and Technology,
Universiti Malaysia Sabah, Malaysia.
[3]School of Engineering and Information Technology,
Universiti Malaysia Sabah, Malaysia.

E-mail: [1]andang99@gmail.com, [2]jumat@ums.edu.my, [3]azali60@gmail.com,

Abstract. In this paper, we derive an implicit finite difference approximation equation of the one-dimensional linear time fractional diffusion equations, based on the Caputo's time fractional derivative. Then this approximation equation leads the corresponding system of linear equation, which is large scale and sparse. Due to the characteristics of the coefficient matrix, we use the Accelerated Over-Relaxation (AOR) iterative method for solving the generated linear system. One example of the problem is presented to illustrate the effectiveness of AOR method. The numerical results of this study show that the proposed iterative method is superior compared with the existing one weighted parameter iterative method.

1. Introduction

Fractional partial derivatives equations (FPDEs) have been used to govern many linear problems in variety of applications field by using fractional calculus. Therefore, there are many successful mathematical models, which are based on fractional partial derivative equations (FPDEs), have been developed [1,2,3,4]. Supposing a fractional derivative replaces the first-order time partial derivative in a diffusion model, in which this matter leads to slower diffusion [1]. In order to solve one dimensional time fractional diffusion model with constant coefficient, analytical and/or numerical solutions are available by using transform method which is one method used for numerical solution of the fractional diffusion equations (FDE) [1,5,6], finite elements together with the methods of line [3], explicit and implicit finite difference methods [7,8,9]. In fact, these finite difference schemes are available in the literature [9,10]. As we know, the explicit methods are conditionally stable.

Due to the stability of finite difference discretization schemes, this paper deals with the application of the implicit discretization scheme and Caputo's fractional partial derivative of order α being used to discretize the time-fractional diffusion equation and derive a Caputo's implicit finite difference approximation equation. This implicit finite difference approximation equation will lead the tridiagonal linear system in which the properties of the coefficient matrix of the linear system are sparse and large scale. To solve any linear system, it can be observed that many iterative methods have been proposed to solve any system of linear equations. For instance, further discussions of various iterative methods have been discussed by Young [12, 13, 14], Hackbusch [15] and Saad [16]. In addition to that, Evans [17] has also proposed block iterative methods via the Explicit Group (EG) iterative methods in order to speed up the convergence rate. Again, it can be found that the concept of block iteration has been expanded by other researchers (Ibrahim and Abdullah [23], Evans and Yousif

[24, 25], Yousif and Evans [18]) to demonstrate the efficiency of its computation cost. Apart from these iterative methods, Hadjidimos [20] has initiated the Accelerated Over-Relaxation (AOR) method which is based on the point iterative method together with two weighted parameters. Actually this iterative method is one of the efficient point iterative methods. To accelerate the convergence rate of this iteration, Martins *et al* [26] have combined the concept of the EG iterative method together with the AOR method in which this combination is called as the Explicit Group AOR (EGAOR) method for solving elliptic partial differential equations. They pointed out that the 4 point-EGAOR method is superior compared to the existing point AOR method.

Due to the advantages of the AOR method for solving partial differential equations, the main objective of this paper is to construct and examine the effectiveness of the Accelerated Over-Relaxation (AOR) iterative method for solving time fractional parabolic partial differential equations (TPPDE's) based on the Caputo's implicit finite difference approximation equation. To examine the effectiveness of the AOR method, we also implement the Successive Over-Relaxation (SOR) and Gauss-Seidel (GS) iterative methods being used control methods.

To indicate the effectiveness of AOR method, let time fractional parabolic partial differential equation (TPPDE's be defined as

$$\frac{\partial^\alpha U(x,t)}{\partial^\alpha} = a(x)\frac{\partial^2 U(x,t)}{\partial x^2} + b(x)\frac{\partial U(x,t)}{\partial x} + c(x)U(x,t) \tag{1}$$

where *a(x), b(x)* and *c(x)* are known functions or constants, whereas α is a parameter which refers to the fractional order of time derivative.

2. Preliminaries

Before constructing the finite difference approximation equation of Problem (1), we need to consider the following definitions and properties of fractional derivative theory which are used in this paper.

Definition 1. [11] The Riemann-Liouville fractional integral operator , J^α of order-α is defined as

$$J^\alpha f(x) = \frac{1}{\Gamma(\alpha)}\int_0^x (x-t)^\alpha f(t)dt\,,\ \ \alpha > 0\,,\ x > 0 \tag{2}$$

Definition 2.[11] The Caputo's fractional partial derivative operator, D^α of order -α is defined as

$$D^\alpha f(x) = \frac{1}{\Gamma(m-\alpha)}\int_0^x \frac{f^{(m)}(t)}{(x-t)^{\alpha-m+1}}dt,\ \ \alpha > 0 \tag{3}$$

with $m-1 < \alpha \le m, \mathrm{m} \in \mathrm{N},\ x > 0$

To find the numerical solution of Problem (1), we derive the Caputo's implicit finite difference approximations of Problem (1) with Dirichlet boundary conditions. To do this, we need to consider the non-local fractional derivative operator. This approximation equation can be categories as unconditionally stable scheme .Based on Problem (1), the solution domain of the problem has been restricted to the finite space domain $0 \le x \le \gamma$, with $0 < \alpha < 1$, whereas the parameter α refers to the fractional order of time derivative. To solve Problem (1), let us assume the initial and boundary conditions of Problem (1) be given as

$$U(0,t) = g_0(t),\ U(\ell,t) = g_1(t),$$

and the initial condition

$$U(x,0) = f(x),$$

where $g_0(t), g_1(t)$, and $f(x)$, are given functions. A discretize approximation to the time fractional derivative in Eq. (1), we consider Caputo's fractional partial derivative of order α, defined by [11,13,14],

$$\frac{\partial^\alpha u(x,t)}{\partial t^\alpha} = \frac{1}{\Gamma(n-1)}\int_0^\infty \frac{\partial u(x-s)}{\partial t}(t-s)^{-\alpha}\, ds, \quad t>0, \quad 0<\alpha<1 \tag{4}$$

The organization of the paper is as follows: In Section 2, the formula of the Caputo's fractional derivative operator and numerical procedure for solving time fractional diffusion equation (1) by means of the implicit finite difference method are given. In Section 3, formulation of the AOR iterative method is introduced. In Section 4 shows numerical example and its results and conclusion is given in Section 5.

3. Approximation For Fractional Diffusion Equation

In this section, the first order approximation method for the computation of Caputo's fractional partial derivative is then stated as the following expression

$$D_t^\alpha U_{i,n} \cong \sigma_{\alpha,k} \sum_{j=1}^{n} \omega_j^{(\alpha)}\left(U_{i,n-j+1} - U_{i,n-j}\right) \tag{5}$$

In which we define

$$\sigma_{\alpha,k} = \frac{1}{\Gamma(1-\alpha)(1-\alpha)k^\alpha} \quad \text{and} \quad \omega_j^{(\alpha)} = j^{1-\alpha} - (j-1)^{1-\alpha}.$$

By using formula (5), we attempt to derive a discretize equation of Problem (1). Before constructing the discretize equation, let the solution domain of the problem need to be partitioned uniformly. For some positive integers *m* and *n*, the grids sizes in space and time directions for the finite difference algorithm are defined as $h = \Delta x = \frac{\gamma - 0}{m}$ and $k = \Delta t = \frac{T}{n}$ respectively. To construct the uniformly grid network of the solution domain, let the grid points in the space interval $[0,\gamma]$ be indicated as the numbers $x_i = ih,\ i = 0,1,2,\ldots,m$ and the grid points in the time interval $[0,T]$ are labeled $t_j = jk,\ j = 0,1,2,\ldots,n$. The values of the function $U(x,t)$ at the grid points are denoted as $U_{i,j} = U(x_i,t_j)$.

Using Eq. (5) and the implicit finite difference discretization scheme, the Caputo's discrete equation of Problem (1) to the grid point centered at $(x_i,t_j) = (ih,nk)$ is given as

$$\sigma_{\alpha,k}\sum_{j=1}^{n}\omega_j^{(\alpha)}\left(U_{i,n-j+1} - U_{i,n-j}\right)$$
$$= a_1\frac{1}{h^2}\left(U_{i-1,n} - 2U_{i,n} + U_{i+1,n}\right) + b_i\frac{1}{2h}\left(U_{i+1,n} - U_{i-1,n}\right) + c_i U_{i,n}, \tag{6}$$

for *i=1,2...,m-1.*

According to Eq. (6), the discrete equation is known as the fully implicit finite difference approximation equation which is consistent first order accuracy in time and second order in space. Again the approximation equation (6) can be rewritten based on the specified time level. Therefore, we have for $n \geq 2$:

$$\sigma_{\alpha,k}\sum_{j=1}^{n}\omega^{(\alpha)}\left(U_{i,n-j+1} - U_{i,n-j}\right) = \left(\frac{a_i}{h^2} - \frac{b_i}{2h}\right)U_{i-1,n} + \left(c_i - \frac{2a_i}{h^2}\right)U_{i,n} + \left(\frac{a_i}{h^2} + \frac{b_i}{2h}\right)U_{i+1,n}, \tag{7a}$$

$$\therefore \sigma_{\alpha,k}\sum_{j=1}^{n}\omega_j^{(\alpha)}\left(U_{i,n-j+1}-U_{i,n-j}\right)=p_iU_{i-1,n}+q_iU_{i,n}+r_iU_{i+1,n},$$

where

$$p_i=\frac{a_i}{h^2}-\frac{b_i}{2h},$$

$$q_i=c_i-\frac{2a_i}{h^2},$$

$$r_i=\frac{a_i}{h^2}+\frac{b_i}{2h}.$$

Also, we get for $n = 1$,

$$-p_iU_{i-1,1}+q_i^*U_{i,1}-r_iU_{i+1,1}=f_{i,1},\quad i=1,2,...,m-1 \tag{7b}$$

where

$$\omega_j^{(\alpha)}=1,$$

$$q_i^*=\sigma_{\alpha,k}-q_i,$$

$$f_{i,1}=\sigma_{\alpha,k}U_{i,1}.$$

Based on Eq. (7b), it can be seen that the tridiagonal linear system can be constructed in a matrix form as

$$\underset{\sim}{AU}=\underset{\sim}{f} \tag{8}$$

where

$$A=\begin{bmatrix} q* & -r & & & & \\ -p & q* & -r & & & \\ & -p & q* & -r & & \\ & & \ddots & \ddots & \ddots & \\ & & & -p & q* & -r \\ & & & & -p & q* \end{bmatrix}_{(m-1)x(m-1)},$$

$$\underset{\sim}{U}=\begin{bmatrix}U_{11} & U_{21} & U_{31} & \cdots & U_{m-2,1} & U_{m-1,1}\end{bmatrix}^T,$$

$$\underset{\sim}{f}=\begin{bmatrix}U_{11}+p_1U_{01} & U_{21} & U_{31} & \cdots & U_{m-2,1} & U_{m-1,1}+p_{m-1}U_{m,1}\end{bmatrix}^T.$$

4. Formulation of Accelerated Over-Relaxation Iterative Method

Based on the tridiagonal linear system in Eq. (8), it is clear that the characteristics of its coefficient matrix are large scale and sparse. As mentioned in Section 1, many researchers have discussed various iterative methods such as Young [12,13,14], Hackbusch [15], Saad [16], Evans [17], Yousif and Evans [18,19], and Othman and Abdullah [20]. To obtain numerical solutions of the tridiagonal linear system which is generated by the Caputo's implicit finite difference approximation equation, we consider the Accelerated Over-Relaxation (AOR) iterative method [20, 21], which is the most known and widely using for solving any linear systems. To formulate AOR method, let the coefficient matrix A in (8) be expressed as summation of the three matrices

$$A=D-L-V \tag{9}$$

where D, L and V are diagonal, lower triangular and upper triangular matrices respectively.

Thus, AOR iterative method can be defined generally as [21, 22]

$$\tilde{U}^{(k+1)} = (D - \omega L)^{-1}\left[\beta V + (\beta - \omega)D + (1 - \beta)D\right]\tilde{U}^{(k)} + \beta(D - \omega L)^{-1} f \tag{10}$$

where $\tilde{U}^{(k)}$ represents an unknown vector at k^{th} iteration. The implementation of the AOR iterative method can be described in Algorithm 1.

Algorithm 1: AOR method

i. Initialize $\tilde{U} \leftarrow 0$ and $\varepsilon \leftarrow 10^{-10}$.

ii. For $j = 1,2,\ldots,n$ Implement

 a. For $i = 1,2,\ldots,m-1$ calculate

$$\tilde{U}^{(k+1)} = (D - \omega L)^{-1}\left[\beta V + (\beta - \omega)D + (1 - \beta)D\right]\tilde{U}^{(k)} + \beta(D - \omega L)^{-1} f$$

 b. Convergence test. If the convergence criterion i.e $\left\|\tilde{U}^{(k+1)} - \tilde{U}^{(k)}\right\| \le \varepsilon = 10^{-10}$ is satisfied, go to Step (iii). Otherwise go back to Step (a).

iii Display approximate solutions.

5. Numerical Example

In this section, we apply the approximation equation (7), in order in getting numerical solution of time fractional diffusion equation. One example of the time fractional diffusion equation was tested for verify the effectiveness of the Gauss-Seidel (GS), Successive Over-Relaxation (SOR) and Accelerated Over-Relaxation (AOR) iterative methods. For comparison purpose in indicating the effectiveness of these three proposed iterative methods, three criteria will be considered such as number of iterations, execution time (in seconds) and maximum absolute error at three different values of $\alpha = 0.25$, $\alpha = 0.50$ and $\alpha = 0.75$. For implementation of point iterations, the convergence test considered the tolerance error, which is fixed as $\varepsilon = 10^{-10}$.

Consider the time fractional initial boundary value problem be given as [22]

$$\frac{\partial^\alpha U(x,t)}{\partial t^\alpha} = \frac{\partial^2 U(x,t)}{\partial x^2}, \quad 0 < \alpha \le 1, 0 \le x \le \gamma, \quad t > 0, \tag{11}$$

where the boundary conditions are stated in fractional terms

$$U(0,t) = \frac{2kt^\alpha}{\Gamma(\alpha+1)}, \quad U(\ell,t) = \ell^2 + \frac{2kt^\alpha}{\Gamma(\alpha+1)}, \tag{12}$$

and the initial condition

$$U(x,0) = x^2. \tag{13}$$

From Problem (11), as taking $\alpha = 1$, it can be seen that Eq. (11) can be reduced to the standard diffusion equation

$$\frac{\partial U(x,t)}{\partial t} = \frac{\partial^2 U(x,t)}{\partial x^2}, \quad 0 \le x \le \gamma, \quad t > 0, \tag{14}$$

subjected to the initial condition

$$U(x,0) = x^2,$$

and boundary conditions

$$U(0,t) = 2kt, \qquad U(\ell,t) = \ell^2 + 2kt,$$

Then the analytical solution of Problem (14) is obtained as follows

$$U(x,t) = x^2 + 2kt.$$

Now by applying the series

$$U(x,t) = \sum_{n=0}^{m-1} \frac{\partial^n U(x,0)}{\partial t^n} \frac{t^n}{n!} + \sum_{n=1}^{\infty} \sum_{i=0}^{m-1} \frac{\partial^{mn+i} U(x,0)}{\partial t^{mn+i}} \frac{t^{n\alpha+i}}{\Gamma(n\alpha+i+1)}$$

to $U(x,t)$ for $0 < \alpha \le 1$, it can be shown that the analytical solution of Problem (11) is given as

$$U(x,t) = x^2 + 2k \frac{t^\alpha}{\Gamma(\alpha+1)}.$$

All results of numerical experiments for Problem (11), obtained from implementation of GS, SOR and AOR iterative methods are recorded in Table 1 at different values of mesh sizes, m = 128, 256, 512, 1024, and 2048.

TABLE 1. Comparison of number iterations, the execution time (seconds) and maximum errors for the iterative methods using example at $\alpha = 0.25, 0.50, 0.75$

M	Method	α = 0.25			α = 0.50			α = 0.75		
		K	Time	Max Error	K	Time	Max Error	K	Time	Max Error
128	GS	21017	37.73	9.97e-5	13601	5.92	9.86e-5	6695	2.94	1.30e-4
	SOR	465	1.54	9.95e-5	385	1.44	9.84e-5	411	1.51	1.30e-4
	AOR	431	0.33	9.95e-5	185	0.33	9.84e-5	411	0.69	1.29e-4
256	GS	77231	343.63	1.00e-4	50095	42.17	9.90e-5	24732	20.70	1.30e-4
	SOR	939	4.73	9.96e-5	769	3.99	9.84e-5	513	2.92	1.30e-4
	AOR	870	1.25	9.95e-5	709	2.27	9.84e-5	513	1.48	1.29e-4
512	GS	281598	2747.34	1.02e-4	183181	339.85	1.01e-4	90783	166.75	1.32e-4
	SOR	1886	17.51	9.96e-5	1537	14.39	9.84e-5	1025	10.33	1.30e-4
	AOR	1607	8.16	9.95e-5	1481	9.91	9.82e-5	1025	5.79	1.29e-4
1024	GS	1017140	68285.36	1.09e-4	663971	2454.53	1.08e-5	330622	1209.39	1.40e-4
	SOR	3805	68.91	9.96e-5	3073	55.68	9.84e-5	2049	37.33	1.30e-4
	AOR	3100	28.30	9.95e-5	2918	14.39	9.84e-5	2049	19.36	1.29e-4
2048	GS	3631638	58914.30	1.38e-4	2380946	17795.25	1.38e-4	1192528	8794.26	1.71e-4
	SOR	8193	430.17	9.96e-5	6145	239.84	9.84e-5	7849	303.50	1.30e-4
	AOR	8139	144.51	9.95e-5	2819	27.23	9.83e-5	7849	135.97	1.29e-4

6. Conclusion

For the time fractional diffusion problems, the paper presents the formulation of the Caputo's implicit finite difference equations to generate a linear system. Then to solve the generated linear system, the formulation and implementation of these three proposed iterative methods such as GS, SOR and AOR have been presented based on the Caputo's implicit finite difference approximation equation. From observation of all experimental results by imposing the GS, SOR and AOR iterative methods, it is obvious at $\alpha = 0.25$ that that number of iterations declined approximately by 0.66-18.53% corresponds to the AOR iterative method compared with the SOR method. Again in terms of execution time, implementations of AOR method are much faster about 53.39-78.57% than the SOR method. It means that the AOR method requires the least amount for number of iterations and computational time at $\alpha = 0.25$ as compared with GS and SOR iterative methods. This is due to the implementations of AOR iterative method have been accelerated by using the optimal value of the two weighted parameters, ω and β. In fact, these conclusions are inline with the results of Othman and

Abdullah [19]. Based on the accuracy of there three iterative methods, it can be concluded that the numerical solutions for AOR method are in good agreement.

References

[1] Mainardi, F. 1997. Fractals and Fractional Calculus Continuum Mechanics. Heidelberg: Springer-Verlag, 291-348.

[2] Diethelm, K. and Freed, A.D. 1999. On the solution of nonlinear fractional order differential equation used in the modeling of viscoelasticity, in Scientific Computing in Chemical engineering II Computational Fluid Dynamic, Reaction Engineering and Molecular Properties, Heidelberg: Springer-Verlag, 217-224.

[3] Liu, F, Anh,V, and Turner, I. 2004. Numerical solution of the space fractional Fokker-Planck equation, *JCAM*, **166**: 209-219.

[4] Meerschaert, M.M. and Tadjeran, C.. 2004. Finite difference approximation for fractional advection-dispersion flow equations, *JCAM*, **172**:145-155.

[5] Chaves, A. 1998. Fractional diffusion equation to describe Levy flight, *Phys.Lett, A* **239**:13-16.

[6] Agrawal, O.P. 2002. Solution for a fractional diffusion-wave equation defined in a bounded domain, *Nonlinear Dynamic*, **29:**145-155

[7] Liu, F, Zhuang, P, Anh, V, and Turner, I.. 2006. A fractional-order implicit difference approximation for the space-time fractional diffusion equation, *ANZIAM J*, **47**:C871-C887.

[8] Shen, S. and Liu, F. 2005. Error analysis of an explicit finite difference approximation for the space fractional diffusion equation with insulated ends, *ANZIAM* J.**46 (E)**:C871-C887

[9] Yuste, S.B. and Acedo, L. 2005. An explicit finite difference method and a new von Neumann-type stability analysis for fractional diffusion equations, *SIAM J. Numer. Anal.***42 (5)**:1862-1874

[10] Yuste, S.B. 2006. Weighted average finite difference method for fractional diffusion equations, *J.Comp.Phys.***216**:264-274

[11] Zhang, Y. 2009. A Finite Difference Method For Fractional Partial Differential Equation, Applied Mathematics And Computation **215**:524-529.

[12] Young, D.M. 1954. Iterative Methods for solving Partial Difference Equations of Elliptic Type, *Trans. Amer. Math. Soc.*,**76**:92-111.

[13] Young, D.M. 1971. Iterative solution of large linear systems. London: Academic Press.

[14] Young, D.M. 1972. Second-degree iterative methods for the solution of large linear systems. *Journal of Approximation Theory*. **5**:37-148.

[15] Hackbusch, W. 1995. Iterative solution of large sparse systems of equations. New York: Springer-Verlag.

[16] Saad, Y. 1996. Iterative method for sparse linear systems. Boston: International Thomas Publishing.

[17] Evans, D.J. 1985. Group Explicit Iterative methods for solving large linear systems, *Int. J. Computer Maths.,* **17**:81-108.

[18] Yousif, W.S. and Evans, D. J. 1995. Explicit de-coupled group iterative methods and their implementations. *Parallel Algorithms and Applications*. **7**:53-71.

[19] Othman, M, and Abdullah, A.R. 2000. An Efficient Four Points Modified Explicit Group Poisson Solver. *International Journal Computer Mathematics*. **76**:203-217.

[20] Hadjidimos, A. 1978. Accelerated OverRelaxation Method. *Mathematics of Computation.*, **32,** 149-157.

[21] Hallett, A.J.H. 1986. The Convergence Of Accelerated OverRelaxation Iterations, *Mathematics of Computation.* **47** (125): 219:223.

[22] Jun, J.Y, and Tang, M.J. 2013. Moving Finite Element Methods For Time Fractional Partial Differential equations, *Science China Mathematics*. **56** (6): 1287-1300.

[23] Ibrahim, A. and Abdullah, A.R. 1995. Solving the two dimensional diffusion equation by the Four Point Explicit Decoupled Group (EDG) iterative method. *International Journal Computer Mathematics*. **58**:253-256.

[24] Evans, D.J and Yousif, W. S. 1986. Explicit Group Iterative Methods for solving elliptic partial differential equations in 3-space dimensions. *Int. J. Computer Maths.,* **18**:323-340.

[25] Evans, D. J and Yousif, W. S. 1990. The Explicit Block Relaxation method as a grid smoother in the Multigrid V-cycle scheme. *Int. J. Computer Maths.,* **34**:71-78.

[26] Martins, M.M., Yousif, W. S., and Evans, D. J. 2002. Explicit Group AOR method for solving elliptic partial differential equations. *Neura Parallel, and Science Computation*. **10** (4): 411-422.

On the Marangoni Instability in Gas Absorption

T A Poroyko[1] and E F Skurygin[1,2]
[1]Yaroslavl State Technical University, Yaroslavl, Russia

E-mail: skouryguine@rambler.ru

Abstract. The non-steady-state process of a gas absorption by initially motionless liquid is investigated. Nonlinear analysis of the thermo-capillary instability showed that the transition of the process to unstable convective regime is determined by two sharply different characteristic times. First one is the time of neutral perturbations, when the fluctuations of the concentration of absorbable substance, temperature and the liquid velocity stop decrease and begin to grow together, but their intensity is too small to influence on the absorption rate. The second one is the time much greater than the first, when the perturbations reach sufficient values, and the absorption process turns to the convective regime. The results obtained explain the well-known experimental data on delaying of the transition of the absorption process to the unstable convective regime.

1. Introduction

It is well known that the process of a gas absorption by a liquid goes in diffusion regime only up to some critical time, after which it turns to unstable convective regime, accompanied by a higher rate of mass transfer [1].The theoretical analysis performed in [2,3] showed that the origin of diffusion-convective instability is the temperature dependence of the surface tension of liquid. Thermo-capillary effect or Marangoni effect is widely used in industry as an opportunity to intensification of chemical processes at the liquid – gas interface [4, 5]. Mathematical modeling of non-steady-state absorption involves the problem of determining the critical time of transition of the process to unstable convective regime and the calculation of mass transfer rate under the conditions of surface convection. One unsolved problem is the following: the experimentally observed critical time of the absorption of carbon dioxide by water exceeds theoretical estimations about 1,000 times [2, 3].

The aim of our work is the studying of the mechanism of the transition of absorption process from the diffusion to unstable convective regime based on the nonlinear analysis of the development of small perturbations.

In Section 2, the problem is formulated and the basic equations are determined. In Section 3, a linear analysis of small perturbations in the absorption process is performed. Preliminary evaluations are obtained. Section 4 presents the results of a nonlinear analysis of the development of small perturbations. In Conclusions, Section 5, the mechanism of the transition of the absorption process from diffusion to unstable convective regime is discussed.

---

[2]Author to whom any correspondence should be addressed

2. The problem statement. The basic equations

Non-steady-state process of the absorption of a component *A* from a gaseous phase by initially motionless liquid is studied. In the beginning of the process the concentration of component in the liquid phase is equal to zero, the liquid temperature is constant and equal to T_0. The main resistance to heat and mass transfer is concentrated in the liquid phase. Liquid layer thickness is considered infinite.

In the diffusion regime the concentration *c* of a component *A* is described by the following well-known equation [6]

$$\frac{\partial c^{(0)}}{\partial t} - D\frac{\partial^2 c^{(0)}}{\partial z^2} = 0, \tag{1}$$

Where *D* is molecular diffusivity, *t* is a time; the superscript *(0)* corresponds to the diffusion regime. The *z* coordinate is directed deep into the liquid by the normal to the liquid surface, the value $z = 0$ corresponds to the gas-liquid interface.

At the interface the concentration of the component *A* take the equilibrium value c_* corresponding to its partial pressure in the gaseous phase [7]. In the liquid depth, the concentration is equal to zero.

The solution of non-steady-state diffusion in a motionless liquid is well known [8]. Let us give an expression for the flux density of absorbable material at the interface

$$j^{(0)} = -D\frac{\partial c^{(0)}}{\partial z} = c_*\left(\frac{D}{\pi t}\right)^{1/2}. \tag{2}$$

The absorption process is accompanied by heat generation, which leads to a change in temperature and the surface tension of liquid. For a small temperature changes the surface tension σ varies linearly

$$\sigma = \sigma_0 - \varepsilon(T - T_0), \tag{3}$$

where ε is negative coefficient of surface tension. Inhomogeneity of the surface tension creates frictional stress that causes liquid motion and may cause convective instability of the process [1,4]. Under the conditions of surface convection the absorption process is described by joint solution of the equations of convective diffusion, thermal conductivity, and the equations of liquid motion.

$$\partial c / \partial t + \mathbf{v}\nabla c - D\Delta c = 0, \tag{4}$$

$$\partial T / \partial t + \mathbf{v}\nabla T - \chi\Delta T = 0, \tag{5}$$

$$\frac{\partial}{\partial t} rot\ \mathbf{v} + (\mathbf{v}\nabla) rot\ \mathrm{v} - (rot\ \mathbf{v}\nabla) - \nu\Delta rot\ \mathbf{v} = 0, \tag{6}$$

where χ is thermal conductivity, $\mathbf{v} = (\mathrm{v_x}, \mathrm{v_y}, \mathrm{v_z})$ is a vector of a liquid velocity, ν is kinematic viscosity, coordinates x, y are directed parallel to the gas-liquid interface.

Boundary conditions are as follows. At the interface we have

$$c = c_*, \quad D\frac{\partial c}{\partial z} = \Delta H \lambda \frac{\partial T}{\partial z}, \quad v_z = 0, \quad \rho\nu\frac{\partial v_x}{\partial z} = -\frac{\partial \sigma}{\partial x}, \quad \rho\nu\frac{\partial v_y}{\partial z} = -\frac{\partial \sigma}{\partial y} \tag{7}$$

for $z = 0$,

where $\lambda = \rho c_p \chi$ is specific heat conductivity of liquid, ρ is its density, c_p is its specific heat capacity at constant pressure. In the depth of the liquid the conditions are

$$c -> 0, \quad T -> T_0, \quad v_x -> 0, \quad v_y -> 0, \quad v_z -> 0 \quad \text{for } z- >\infty. \tag{8}$$

For further analysis, let us turn to dimensionless variables

$$C = c / c_*, \quad \theta = (T - T_0)/T_*, \quad (V_x, V_y, V_z) = (v_x / v_*, v_y / v_*, v_z / v_*),$$
$$\tau = t/t_*, \quad (X, Y, Z) = (x / z_*, y / z_*, z / z_*), \tag{9}$$

where the characteristic scales of the length z_*, the time t_*, the velocity v_* and the temperature T_* are equal to:

$$z_* = \frac{\lambda\mu}{\varepsilon \Delta H c_*}\left(\frac{\chi}{D}\right)^{1/2}; \quad t_* = \frac{1}{D}\left(\frac{\lambda\mu}{\varepsilon \Delta H c_*}\right)^2; \quad v_* = \frac{z_*}{t_*}; \quad T_* = (D\chi)^{1/2}\frac{\Delta H c_*}{\lambda} \tag{10}$$

$\mu = \rho\lambda$ is dynamic viscosity.

In dimensionless variables, equations (4) - (6) take the form

$$\partial C / \partial \tau + V\nabla C - Le^{-1}\Delta C = 0, \tag{11}$$
$$\partial \theta / \partial \tau + V\nabla \theta - \Delta \theta = 0, \tag{12}$$
$$\frac{\partial}{\partial t} rot\ \mathbf{V} + (\mathbf{V}\nabla) rot\ \mathbf{V} - (rot\ \mathbf{V}\nabla) - \mathrm{Pr}^{-1}\Delta rot\ \mathbf{V} = 0, \tag{13}$$

where $Le = \chi / D$ is the Lewis number, $\mathrm{Pr} = \nu / D$ is the Prandtl number.

3. The linear instability analysis

3.1. The linearized equations

Let us represent the values of concentration and temperature as the sum of the unperturbed solutions corresponding to the diffusion in motionless liquid, and sinusoidal varying small perturbations

$$C(\mathbf{R}, \tau) = C^{(0)}(Z, \tau) + c^{(1)}(Z, \tau)\cos KX; \quad \theta(\mathbf{R}, \tau) = \theta^{(0)}(Z, \tau) + \theta^{(1)}(Z, \tau)\cos KX. \tag{14}$$

Superscript (1) corresponds to a small perturbation; ***R*** is the radius vector, *K* is dimensionless wave number. In the linear approximation only normal to the surface velocity component is significant. It is represented in the form

$$V_z(\mathbf{R}, \tau) = V^{(1)}(Z, \tau)\cos KX. \tag{15}$$

The following equations are valid for undisturbed concentration and temperature

$$\frac{\partial C^{(0)}}{\partial \tau} - Le^{-1}\frac{\partial^2 C^{(0)}}{\partial Z^2} = 0: \qquad \frac{\partial \theta^{(0)}}{\partial \tau} - \frac{\partial^2 \theta^{(0)}}{\partial Z^2} = 0. \tag{16}$$

Initial and boundary conditions for (16) are given in the following form

$$C^{(0)} = 1, \ \sqrt{\mathrm{Le}}\,\frac{\partial T^{(0)}}{\partial Z} = \frac{\partial C^{(0)}}{\partial Z} \quad \text{for Z} = 0; \quad C^{(0)} -> 0, \ \ \theta^{(0)} -> 0 \ \text{ for Z} -> \infty; \tag{17}$$
$$C^{(0)} = 0, \quad \theta^{(0)} = 0; \quad for \ \ \tau = 0.$$

Well-known solution of the problem (16), (17) gives the following expressionы for the mass flux density $J^{(0)}$ and the heat flux density $Q^{(0)}$

$$J^{(0)} = (Le \cdot \pi \cdot \tau)^{-1/2}, \quad Q^{(0)} = (\pi \cdot \tau)^{-1/2}. \tag{18}$$

The linearized form of small perturbation equations is as follow

$$\left(\frac{\partial}{\partial \tau} - Le^{-1}\left(\frac{\partial^2}{\partial Z^2} + K^2\right)\right)C^{(1)} = -V^{(1)}\frac{\partial C^{(0)}}{\partial Z}, \tag{19}$$

$$\left(\frac{\partial}{\partial \tau} - \frac{\partial^2}{\partial Z^2} + K^2\right)\theta^{(1)} = -V^{(1)}\frac{\partial \theta^{(0)}}{\partial Z}, \tag{20}$$

$$\left(\frac{\partial}{\partial \tau} - \Pr^{-1}\left(\frac{\partial^2}{\partial Z^2} + K^2\right)\right)\left(-\frac{\partial^2}{\partial Z^2} + K^2\right)V^{(1)} = 0. \tag{21}$$

Let us write the boundary conditions for small perturbations. On the liquid – surface interface we have

$$C^{(1)} = 0, \quad \sqrt{Le}\,\frac{\partial \theta^{(1)}}{\partial Z} = \frac{\partial C^{(1)}}{\partial Z}, \quad V^{(1)} = 0, \quad \frac{\partial^2}{\partial Z^2}V^{(1)} = K^2\theta^{(1)} \quad for \ Z = 0. \tag{22}$$

In the depth of the liquid

$$C^{(1)} -> 0, \quad \theta^{(1)} -> 0, \quad V^{(1)} -> 0, \quad \frac{\partial V^{(1)}}{\partial Z} -> 0 \quad \text{for Z} -> \infty. \tag{23}$$

3.2. The neutral perturbations

Critical time of the transition of the process from diffusion to convective regime is determined from the conditions for the existence of neutral perturbations [2,3]. Perturbations are considered neutral if their partial derivatives with respect to time are zero. The time of neutral perturbations for a given wave number $\tau(K)$ may be found by solving the following equation [3]

$$2K\tau[\exp(4K^2\tau)erfc(2K\sqrt{\tau}) - \frac{1}{\sqrt{Le}}\exp(\frac{4K^2\tau}{Le})erfc(2K\sqrt{\frac{\tau}{\sqrt{Le}}})] = 1. \tag{24}$$

Critical time τ_{cr} is equal to the minimum on the wave numbers of the time of neutral disturbances.

$$\tau_{cr} = \min_K \tau(K) = \tau(K_{cr}), \tag{25}$$

Where K_{cr} is a critical wave number. For an infinite Lewis number the critical time is equal to π [2,3]. The following asymptotic interpolation formulae of the critical time and the critical wave number versus the Lewis number for *Le> 50* are written down in [9]

$$\tau_{cr} = \pi(1+3\pi^{1/3}Le^{-1/3}+12Le^{-2/3}), \tag{26}$$

$$K_{cr} = (2^{-1}\pi^{-2/3}Le^{1/6}-0.7Le^{-1/6}+0.9Le^{-1/2}), \tag{27}$$

For the case of Le = *79*, which corresponds to the absorption of carbon dioxide by the water, we have K_{cr}= 0.25, τ_{cr} = 8.57.

Here are profiles of neutral disturbances of concentration of absorbable substance, temperature and liquid velocity

$$C_n(Z) = B[-2\tau * \exp[\frac{4K^2\tau}{Le} + KZ] * erfc(\frac{4K\tau + Z * Le}{2\sqrt{Le * \tau}}) + 2\tau * \exp[\frac{4K^2\tau}{Le} - KZ] * erfc(2k\sqrt{\tau / Le})], \tag{28}$$

$$\theta_n(Z) = B[4\tau * \exp[\frac{4K^2\tau}{Le} - KX]\frac{erfc(2K\sqrt{\tau / Le})}{\sqrt{Le}} - 2\tau * \exp[4K^2\tau - KX] * erfc(2K\sqrt{\tau}) - 2\tau * \exp[4K^2\tau + KX] * erfc(\frac{4K\tau + X}{2\sqrt{\tau}})], \tag{29}$$

$$V_n(Z) = BZ * e^{-KZ}. \tag{30}$$

Subscript n corresponds to the neutral disturbance, B is arbitrary constant. Time τ is a function of the wave number K and satisfies (24).

3.3. *Analysis of the most rapidly growing perturbations*

The transition from diffusion to convective mass transfer regime is determined the fastest growing perturbations. Solutions (28) - (30) at the critical value of the wave number represent the perturbations of concentration, temperature and velocity, which are the first to stop decrease and begin to grow together. Let as investigate the development of these disturbances during the whole process of absorption. Non-stationary perturbations satisfy the equations (19) - (21) with boundary conditions (22) - (23). As additional conditions require their coincidence with neutral perturbations at a critical time

$$C^{(1)}(Z,\tau)=C_n(Z), \quad \theta^{(1)}(Z,\tau)=\theta_n(Z), \quad V^{(1)}(Z,\tau)=V_n(Z) \quad \text{for } \tau=\tau_{cr}, \quad \mathrm{K}=K_{cr} \tag{31}$$

Solution of the transport equations in $\tau < \tau_{cr}$ belongs to the class of ill-posed problems [10]. The additional assumptions are required to solve it. Let us define the initial conditions in the following form

$$C^{(1)}(Z,\tau)=B\sum_{i=0}^{p}M_{1,i}L_i^0(2KZ)e^{-KZ}, \quad \theta^{(1)}(Z,\tau)=B\sum_{i=0}^{p}M_{2,i}L_i^0(2KZ)e^{-KZ},$$
$$V^{(1)}(Z,\tau)=2BKZ\sum_{i=0}^{p}M_{3,i}L_i^2(2KZ)e^{-KZ} \quad \text{for } \tau=0 \tag{32}$$

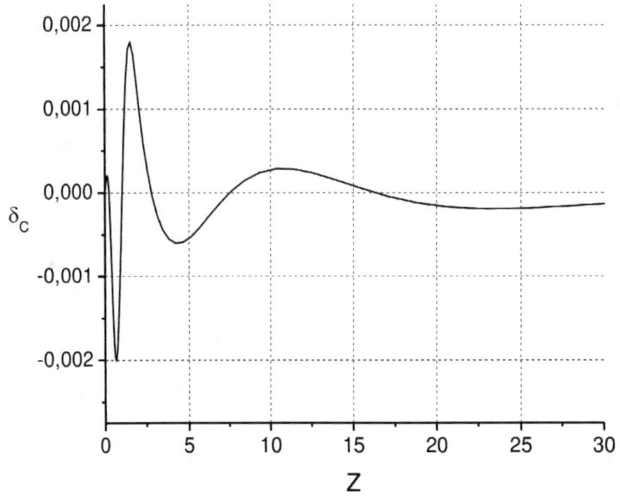

Figure 1. The difference between the numerical solution of non-stationary problem and the neutral perturbation of concentration

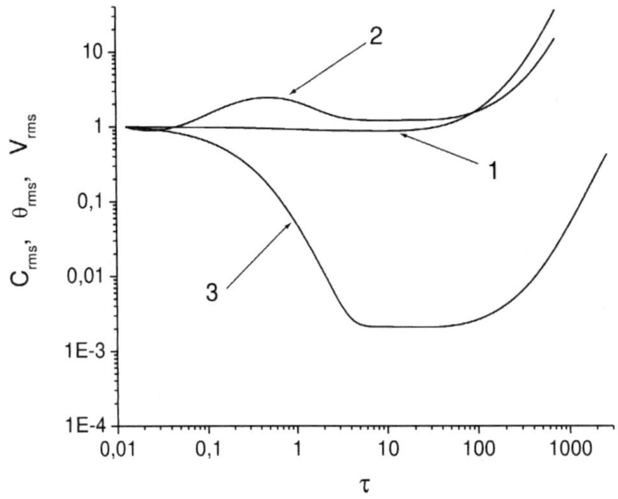

Figure 2. The intensities of disturbances versus time.
1 – C_{rms}, 2 - θ_{rms}, 3 - V_{rms}

where the constant B is defined in formulae (28) - (30), L^{β}_{i} are the Laguerre polynomials, M is a matrix $3\times(p+1)$ of undetermined coefficients, the latter are found by solving the following equations

$$\int_0^\infty Z^m C^{(1)}(Z,\tau_{cr})\exp(-KZ)dZ = \int_0^\infty Z^m C_n(Z)\exp(-KZ)dZ,$$
$$\int_0^\infty Z^m \theta^{(1)}(Z,\tau_{cr})\exp(-KZ)dZ = \int_0^\infty Z^m \theta_n(Z)\exp(-KZ)dZ, \quad (33)$$
$$\int_0^\infty Z^m V^{(1)}(Z,\tau_{cr})\exp(-KZ)dZ = \int_0^\infty Z^m V_n(Z)\exp(-KZ)dZ, \quad \text{m=0,1..p.}$$

Formulae (33) correspond to the approximate implementation of conditions (31).

Results of computations of the evolution of the fastest growing perturbation are shown in figure 1 and figure 2. Computations were performed using the following parameters: *Le*= 79, *Pr*= 7, *p* = 3. Concordance of the numerical solution of the non-stationary problem with the neutral perturbations at the critical time is shown in figure 1. Their difference being divided by the maximum amplitude of the perturbation

$$\delta_C = [C^{(1)}(Z,\tau_{cr}) - C_n(Z)] / C_{n\max}, \quad C_{n\max} = \max_Z C_n(Z) \quad (33)$$

does not exceed 0.2%.

Figure 2 presents the time dependent quadratic means for perturbations of concentration, temperature and velocity normalized to their initial values

$$C_{rms}(\tau) = \left[\int_0^\infty \left(C^{(1)}(Z,\tau)\right)^2 dZ\right]^{1/2} \left[\int_0^\infty \left(C^{(1)}(Z,0)\right)^2 dZ\right]^{-1/2},$$
$$\theta_{rms}(\tau) = \left[\int_0^\infty \left(\theta^{(1)}(Z,\tau)\right)^2 dZ\right]^{1/2} \left[\int_0^\infty \left(\theta^{(1)}(Z,0)\right)^2 dZ\right]^{-1/2}, \quad (34)$$
$$V_{rms}(\tau) = \left[\int_0^\infty \left(V^{(1)}(Z,\tau)\right)^2 dZ\right]^{1/2} \left[\int_0^\infty \left(V^{(1)}(Z,0)\right)^2 dZ\right]^{-1/2}.$$

On the time interval $\tau < \tau_{cr}$, $(\tau_{cr} = 8.57)$ the intensities of concentration and temperature perturbations are virtually unchanged, but the intensity of the velocity perturbations sharply diminishes, its root mean square decreases by approximately 500 times. The initial intensity restores at the time greater than 3000; that exceeds the critical time more than 300 times. Transition from diffusion to convective mass transfer regime take place when the mass flux, caused by the velocity and concentration, being equal to $\frac{1}{2}V^{(1)}C^{(1)}$, in the main part of the diffusion layer becomes

of the same order as the diffusion flux in a motionless liquid $J^{(0)}$, i.e. when the perturbations of the velocity and the concentration are sufficiently large. The analysis has shown that the most rapidly growing perturbations decrease sharply before starting to rise. Critical time, being defined as the time of neutral perturbations, corresponds to the prominent minimum of the perturbations of velocity and concentration and cannot be regarded as a time of the transition from diffusion to convective regime. Such a transition can only take place much later when the disturbances reach the required intensity. Next, let us consider the mechanism of the transition from diffusion to convective mass transfer on the basis of nonlinear analysis of perturbations.

4. Nonlinear analysis.

The values $C^{(0)}$ and $C^{(1)}$ in formula (14) can be regarded as the zero and the first Fourier components of the concentration. In the linear analysis the zero Fourier component is unperturbed concentration corresponding to diffusion in a motionless liquid. Let us perform nonlinear analysis, taking into account the influence of disturbances on the zero Fourier components of concentration. Let us replace the linearized equation (15) to the following

$$\frac{\partial C^{(0)}}{\partial \tau} - Le^{-1}\frac{\partial^2 C^{(0)}}{\partial Z^2} = -\frac{1}{2}\frac{\partial}{\partial Z}(V^{(1)}C^{(1)}). \tag{35}$$

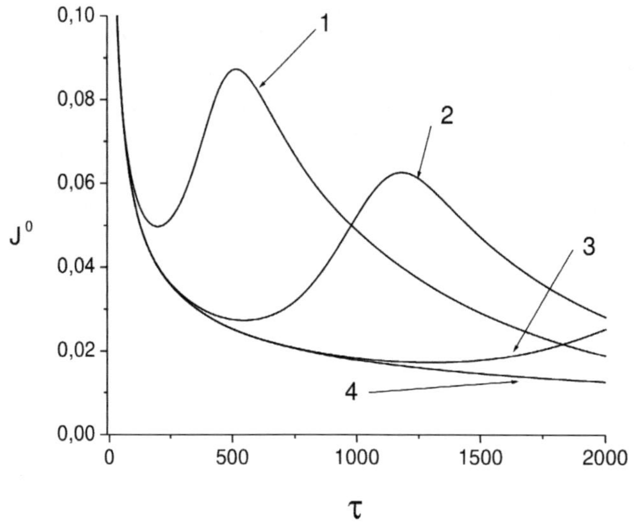

Figure 3. Mass flux density at different initial intensity of concentration disturbances. 1 - ε =0.1, 2 - ε =0.01, 3 - ε =0.001, 4 – flux density in motionless liquid

We calculated the perturbations of concentration, temperature and velocity for the initial conditions (31) and different values of the parameter B, which determines the initial intensity ε of concentration perturbations. Their influence on the zero Fourier component of the flux density of absorbed substance

$$J^0(\tau) = -Le^{-1}\frac{\partial}{\partial Z}C^{(0)}(Z,\tau) \quad \text{for } Z = 0 \tag{36}$$

is shown in Figure 3. The time when the flux under the condition of surface convection begins to exceed the flux in motionless liquid is quite clearly visible on the chart for all initial intensities of concentration disturbances. It allows us to conclude the presence of a second critical time when the developed the velocity and concentration disturbances causes the transition from diffusion to convective mass transfer regime.

5. Conclusions

The mechanism of the transition of absorption process from diffusion to unstable convective regime is as follows. Upon reaching the first critical time defined as the time of neutral disturbances, the fluctuations of concentration, temperature and velocity are no longer decrease and begin to grow together. However, their intensity is too small; they cannot influence the interfacial mass transfer rate. It takes time, much greater the first critical for these disturbances to grow and reach the value necessary to change the flux of absorbed substance. At this time, much exceeding the first critical, concentration and velocity perturbations grow much faster than zero Fourier component of the concentration; that creates clearly defined second critical time, when the flux under the conditions of the surface convection begins to exceed the flux to motionless liquid. Executed simulations of the evolution of small perturbations of concentration, velocity and temperature do not allow to accurately calculating the second critical time, however, our analysis shows that this value is much larger than the first critical time.

References

[1] Muenz K and, Marchello J M 1965 Surface motion and gas absorption*AICh J.* **12** 249

[2] V.V.Dilman,V.I.Naydenov,V.V.Olevsky 1992 Marangonidiffusive_thermal instability in absorption *Russian Chemical Industry* **69** 8 25

[3] Dil'man V V,Kulov N N,Lotkhov V A,Kaminskii V A, and Naidenov V I 1998 On the differencein rates of absorption and desorption of gases *Theor. Found. Chem. Eng.***32** 337,

[4] Scriven L E and Sterling C V 1960 The Marangoni Effects *Nature***187**, 186

[5] NepomnyashchyA,Simanovskii I, Legros J C 2006:*Interfacial Convection in Multilayer Systems* (New York: Springer Science+Business Media LLC)

[6] Vladimirov V S 1983 *Equations of Mathematical Physics* (Moscow: Mir Publishers)

[7] Bird R,Stewart W and Lightfoot E 2007*Transport Phenomena.* (New York : Wiley)

[8] Daniels F andR.Alberty R 1975*Physical chemistry* (New York: Wiley)

[9] PoroykoT A and Skurygin E F 2013Effect of Thickness of Liquid Layer on Instability of Absorption *Theor. Found. Chem. Eng.***47** 682

[10] Tikhonov A N,Goncharsky A V, Stepanov V V and Yagola A G 1995*Numerical Methods for the Solution of Ill-Posed Problems* (Dordrecht/Boston/London:Kluwer academic publisher)

Journal of Physics: Conference Series **495** (2014) 012034 doi:10.1088/1742-6596/495/1/012034

Selective fermentation of pitted dates by *S. cerevisiae* for the production of concentrated fructose syrups and ethanol

Meilana Dharma Putra, Ahmed E. Abasaeed*, Mohamed A. Zeinelabdeen, Mohamed H. Gaily, Ashraf K. Sulieman

Chemical Engineering Department, King Saud University, PO Box 800, Riyadh 11421 Saudi Arabia

E-mail: abasaeed@ksu.edu.sa

Abstract. About half of worldwide production of dates is unconsumed. Dates contain over 75 % reduced sugars (mostly glucose and fructose with nearly equal amount). Compared to the commercial *Saccharomyces cerevisiae* wild strain, the strains ATCC 36858 and 36859 could produce high concentration fructose syrups. The fructose fractions obtained were 95.9 and 97.4% for ATCC 36858 and 86.5 and 91.4% for ATCC 36859 at 30 and 33°C, respectively. Fructose yields higher than 90% were obtained using ATCC 36858 compared to those obtained using ATCC 36859 which were 87.3 and 66.1% at 30 and 33°C, respectively. The ethanol yield using ATCC 36858 was higher than that using ATCC 36859 by 16 and 9% at 30 and 33°C, respectively. Through this finding, the production of fructose and ethanol from date extract is a promising process. Moreover, the fructose fractions obtained here (about 90%) are much higher than those obtained with the commercial process, i.e. 55 % fructose syrups.

1. Introduction

Date palm trees are one of the most important agricultures grown in many regions of the world, i.e., Arabian Peninsula, North Africa and Middle East. Countries such as Australia, India, Pakistan, Southern Africa, Mexico, South America, and United States have started growing date trees [1]. The Kingdom of Saudi Arabia produced over million tons of dates in 2010 [2] compared to about 8 million tons of dates was worldwide produced in 2010 [3]. Though date is one of the staple foods, about half of dates are still unconsumed [4]. Date syrups are very rich in sugars (fructose and glucose), a small amount of sucrose, minerals and vitamin [5]. Pitted date contains over 75 % sugars of its dry weight. Half of date sugar is fructose. Fructose is the sweetest natural sugar that is 30 % sweeter than sucrose [6]; thus it is used as sweetener (nowadays marketed as high fructose corn syrup) and also used in the food, confectionery and beverage industry [6].

Commercial fructose is produced via enzymatic saccharification of starch to glucose and a subsequent enzymatic isomerization of the glucose. About 42 % fructose syrup is obtained due to equilibrium limitations. Because glucose and fructose are isomer, the separation of them is difficult. To concentrate the fructose fraction in the syrup, costly multistage chromatograph separation is employed to achieve 90 % fructose fraction [7]. The commonly marketed 55 % fructose syrups are obtained by blending 42 and 90% fructose syrups. Recently, a method using ionic liquids to separate fructose and glucose has been invented [8]. On the other hand, the spoilage dates have great industrial potential when used to produce fructose. Moreover, the ethanol produced from agricultural waste has been suggested to be a promising alternative in the fuel market [9,10]; it will therefore be a green solution in the sustainable energy [11].

A selective fermentation process has been suggested to produce fructose and ethanol. These productions has been also studied either using sucrose media or artichoke juice by *S. cerevisiae* ATCC 36858 or ATCC 36859 [12,13]. More recently, *S. cerevisiae* ATCC 36858 has been utilized for the production of fructose and ethanol from dates extract [14,15].

Underutilized dates provide a suitable opportunity for the production of fructose and ethanol. In this paper, the performance of *S. cerevisiae* ATCC 36858 and 36859 on date extract media will be studied. The commercial wild strain of *S. cerevisiae* to produce ethanol will be also compared to the ATCC strains. The purpose of this study is to produce the fructose syrups from date extract with high fructose fraction without significant fructose losses.

2. Materials and methods

2.1. *Substrate extraction and preparation*

A low quality variety of dates, locally called Ruzaiz, were used in this investigation. The substrate was prepared by using 1 L of deionized water to extract the sugar content from 400 grams of pitted date in a temperature controlled shaker water bath at 40°C for 2 hours according to the procedure developed by Gaily et al. [16]. The extract was twice-filtered to remove fibers. Sugar content of dates was in the range 75-80% (w/w) and was composed mainly of equal amounts of fructose and glucoses. The initial total sugar concentration was adjusted as required by adding deionized water. The substrate was sterilized in an autoclave sterilizer at 121°C for 15 minutes.

2.2. *Yeast propagation and media*

In this contribution, three strains of *Saccharomyces cerevisiae* yeast were used: a commercial over-the-shelf strain that is used in bakeries (hereafter called Wild strain) and two glucose selective strains supplied by the American Type Culture Collection (ATCC, Manassas, VA, USA), namely: ATCC 36858 and ATCC 36859. The inoculation medium for the Wild *S. cerevisiae* strain was composed of 3g yeast extract, 3g malt extract, 5g peptone and 10g dextrose dissolved in 1L of deionized water. 200 ml of the inoculum medium were transferred to a 500 ml conical flask and autoclaved at 121°C for 15 minutes before adding 2g of the Wild yeast to it. The content was allowed to propagate for two days in a temperature controlled water bath shaker at 120 rpm and 30°C. The two selective strains (ATCC 36858 and ATCC 36859) were removed from their ampoules and revived according to ATCC procedure and were allowed to grow in a suitable liquid medium at 30°C. The composition of the inoculum medium for ATCC 36858 was similar to that of the Wild strain. The composition of the medium for ATCC 36859 was composed of 10g glucose, 30g yeast extract, 3.5g peptone, 2g KH_2PO_4, 1 g $MgSO_4.7H_2O$ and 1g $(NH_4)_2SO_4$ dissolved in 1L of deionized water. The inoculum media were then transferred to 500 ml conical flasks and autoclaved at 121 °C for 15 minutes. The selective strain was then added to its sterilized inoculum medium. The mixture was propagated in a temperature controlled shaker water bath at 120 rpm and 30°C for two days to be ready for fermentation experiments.

2.3. *Fermentation experiment*

The fermentation experiments were conducted in a 1 liter batch fermentor (Lambda Minifor Laboratory Fermentor, Sihlbruggstrasse, Switzerland) using the three strains (Wild, ATCC 36858 and ATCC 36859). The fermentations were performed at a constant agitation of 120 rpm and two temperatures i.e., 30 and 33°C. In all fermentation experiments, 400 ml of the mixture of the substrate (340 ml) and the inoculation medium containing yeast (60 ml) was fed to the fermentor. The substrate was fed initially into the fermentor vessel and the whole system was sterilized for 15 minutes at 121°C before adding medium containing yeast. The fermentor was then turned on and the temperature, agitation and pH were monitored during fermentation. Samples were aseptically collected for analysis.

2.4. Sample analysis
Samples were analyzed using High Performance Liquid Chromatography (HPLC). Samples were micro-centrifuged before injection to the HPLC to separate cells and fibers. The HPLC, (Agilent 1200 Infinitely series) was equipped with RID detector. Aminex® column (Cat. #125-0115; 150 x 7.8mm from BIO-RAD) was used for detection of sugars (i.e., fructose, glucose and sucrose) and ethanol. The column was maintained at 40°C. 1 mM sulfuric acid solution was used as a mobile phase at a flow rate of 0.8 ml/min.

3. Results and discussion

Figure 1 show comparison between selective fermentation and non-selective fermentation of date extract at 30°C as presented in section 3.1. The comparison between *S. cerevisiae* strain ATCC 36858 and ATCC 36859 are presented in Figures 2-4. The fructose yield, fructose fraction in sugar and ethanol yield are shown in Figure 5. The fructose yield is calculated based on the amount of fructose present (g) at the end of fermentation to its initial amount (g). Fructose fraction denotes the amount of fructose (g) to the amount of total sugars (g) at the end of process. The theoretical ethanol yield is defined as the produced ethanol (g) to the consumed total sugars (g) at the end of fermentation. These parameters are given as percentage (%) in the figures.

3.1. Selective and non-selective fermentations of date extract
The kinetic profiles resulting from the non-selective fermentation of the date syrup using a commercial *S. cerevisiae* strain, i.e. wild strain at 30°C are shown in Figure 1a. It is clear from the figure that the wild strain fermented both glucose and fructose, with a slightly lower initial rate of fructose fermentation. The two sugars were consumed after a fermentation time of about 24 h. The concentration of ethanol is high due to the consumption of the two sugars. These results reveal the nonselective nature of this strain.

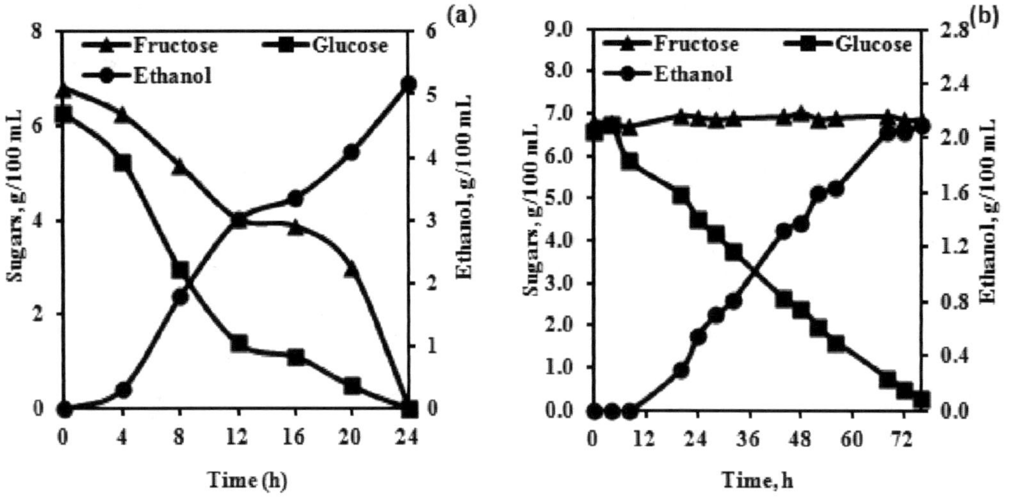

Figure 1. Kinetic profiles at 30°C in dates extract: (a) non-selective fermentation by wild *S. cerevisiae*; (b) selective fermentation by *S. cerevisiae* ATCC 36858

On the other hand, at the same temperature compared to the wild strain, the *S. cerevisiae* strain ATCC 36858 was highly selective towards fermentation of glucose (Figure 1b). When glucose has been completely fermented, the fructose component was almost unaffected. The ethanol yield was less than that obtained with the wild strain as shown in Figure 1b due to probably the higher total sugar concentration (unconsumed fructose) during fermentation that relates to osmotic phenomena [17]. This is also reasonable since the ethanol started to be produced after 20 h for selective fermentation, while it was produced at 4 h for non-selective one. Thus, a slower fermentation occurred for the selective process that continued to 72 h.

3.2. Selective fermentation by S. cerevisiae ATCC 36858 and 36859

Figure 2 presents the kinetic profiles of glucose for both ATCC 36858 and 36859 at 30 and 33°C. At the beginning of process, the glucose was consumed slowly, then after 12 h it dropped at higher rates. There are no major differences of trend between 30 and 33°C for ATCC 36858; in contrast, for ATCC 36859 at 33°C, the glucose quickly decreased even it is completely consumed before 72 h compared to at 30°C.

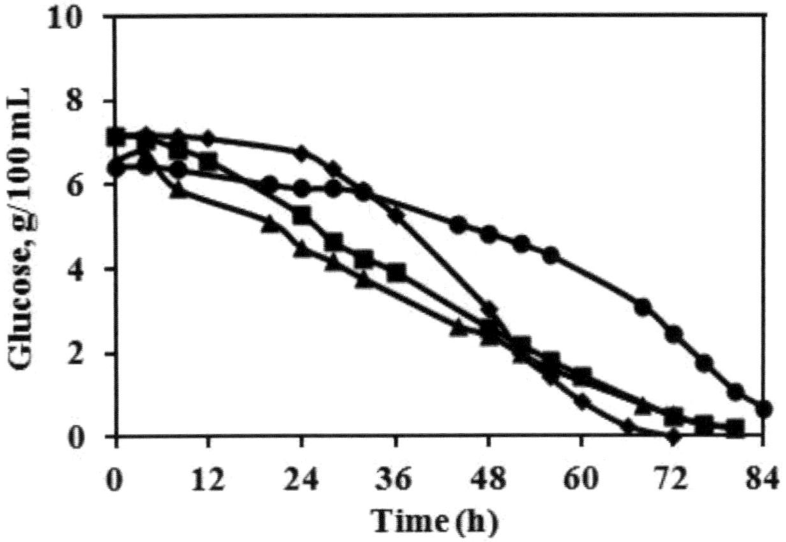

Figure 2. Kinetic profiles of glucose: (ATCC 36858-30°C (closed triangles)); (ATCC 36858-33°C (closed squares)); (ATCC 36859-30°C (closed circles)); (ATCC 36859-33°C (closed rhombuses)).

The kinetic profiles of fructose for both ATCC 36858 and 36859 at 30 and 33°C are shown in Figure 3. The fructose was slightly consumed by the both strain; the exception was for ATCC 36859 at 33°C which showed significant fructose consumption. The fructose dropped up to 4 g/100 ml at 72 h which is corresponding to 42% loss of fructose. It seems that the selectivity of the yeast ATCC 36859 significantly declined at 33°C as indicated by fast consumption of fructose.

Figure 4 shows the kinetic profiles of ethanol for both ATCC 36858 and 36859 at 30 and 33°C. It is clear from the figure that increasing the temperature increased the ethanol production. The production of ethanol for ATCC 36858 was higher that ATCC 36859 at 30°C. The significant production for ATCC 36859 at 33°C was related to the higher consumption of fructose. However, the earlier (before 24 h) ethanol produced for ATCC 36858 indicates that the yeast has short lag phase of growth [14] as manifested by earlier glucose consumption (before 12 h) compared to ATCC 36859.

The fermentation performances, measured in terms of fructose yield, fructose fraction and ethanol yield, can be observed from Figure 5. It is clear that the higher temperature led to the lower fructose yield. For all temperatures, the fructose yield for ATCC 36858 (> 90%) were much higher than ATCC 36859 which are less than 88%. 100.6% fructose yield for ATCC 36858 at 30°C was due to the sucrose hydrolysis to glucose and fructose; this finding shows the ability of this strain to hydrolyze the sucrose [12]. The fructose fraction in sugar and the ethanol yield for ATCC 36858 were also higher than those for ATCC 36859.

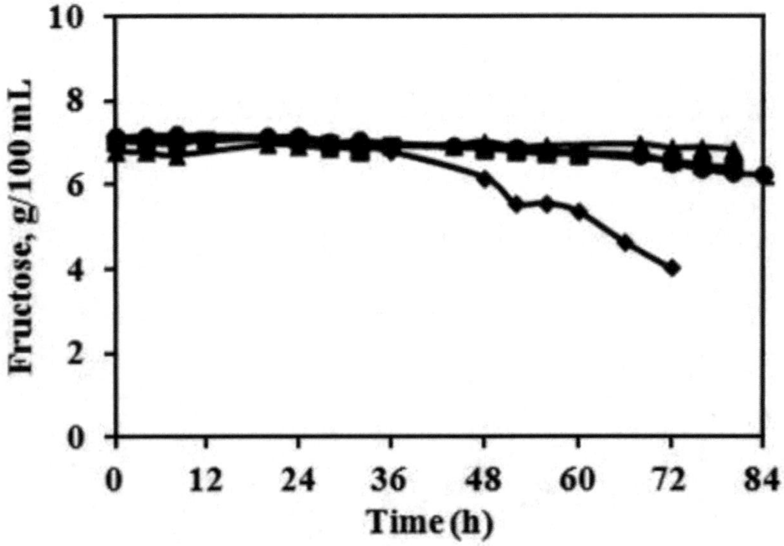

Figure 3. Kinetic profiles of fructose: (ATCC 36858-30°C (closed triangles)); (ATCC 36858-33°C (closed squares)); (ATCC 36859-30°C (closed circles)); (ATCC 36859-33°C (closed rhombuses)).

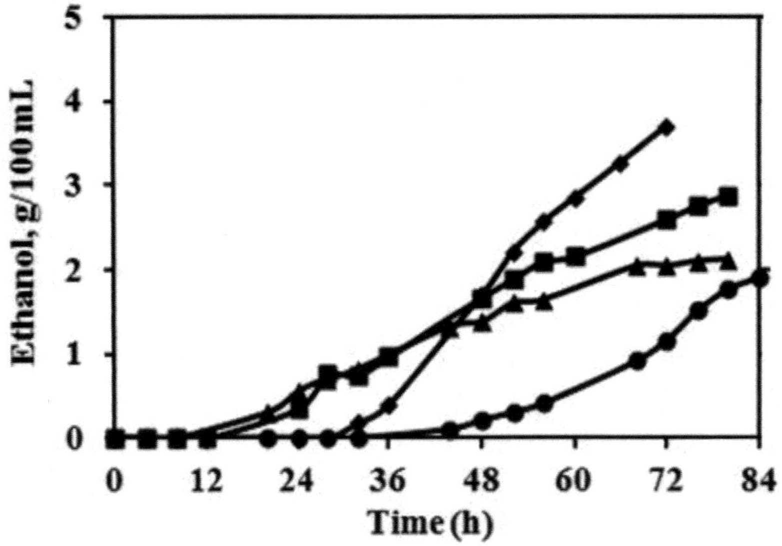

Figure 4. Kinetic profiles of ethanol: (ATCC 36858-30°C (closed triangles)); (ATCC 36858-33°C (closed squares)); (ATCC 36859-30°C (closed circles)); (ATCC 36859-33°C (closed rhombuses)).

Furthermore, increasing temperature 36858 increased ethanol yield for ATCC and ATCC 36859 by 12% and 23%, respectively. It has been reported that the optimum growth of yeast at around 30°C [18]; here the highest fructose yield was obtained at 30°C. On the other hand, the optimum ethanol was achieved at some temperatures slightly above those required for optimum yeast growth [19].

These findings also show that the wild yeast is favored when production of ethanol is the target (i.e., highest ethanol yield), while the selective strains ATCC 36858 and 36859 are favored for the simultaneous production of fructose and ethanol.

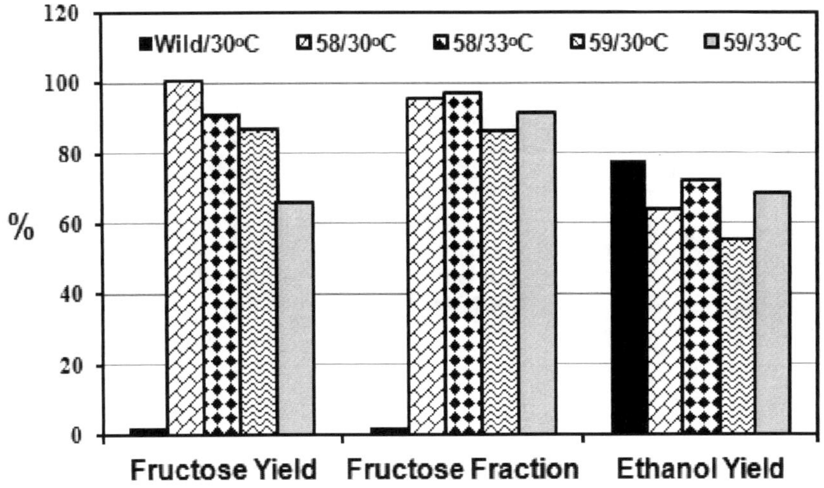

Figure 5. Performance of fermentation with various yeast strains

4. Conclusion

Simultaneous production of fructose and ethanol from date extract was achieved by using two glucose-selective strains of *S. cerevisiae*, namely: ATCC 36858 and 36859 from date extract. On the other hand, the production of ethanol only was achieved by commercial yeast which fermented fructose. Fructose syrups with about 90% fructose fraction in sugars were obtained using the glucose-selective strains. The higher fermentation temperature led to the higher fructose yield. The process using *S. cerevisiae* ATCC 36858 showed better performances, i.e., higher fructose yield, fructose fraction and ethanol yield than using *S. cerevisiae* ATCC 36859. The higher ethanol yield was obtained in higher temperature. Therefore, the selective fermentation of glucose in the date extract is a very promising way for the production of high concentration fructose syrup.

Acknowledgments
The authors extend their appreciation to the National Science, Technology and Innovation Plan (NSTIP) at King Saud University for their generous support and funding of this study as a part of project # 08-ADV391-02.

References
[1] Chao C T and Krueger R R 2007 *HortScience* **42** 1077
[2] Al-Shreed F, Al-Jamal M, Al-Abbad A, Al-Elaiw Z, Abdallah A B and Belaifa H 2012 *J. Dev. Agric. Econ.* **4** 268
[3] Jain S M 2012 *Emir. J. Food Agric.* **24** 400
[4] Moshaf S, Hamidi-Esfahani Z and Azizi M H 2011 *World Acad. Sci. Eng. Technol.* **57** 521
[5] Baliga M S, Baliga B R V, Kandathil S M, Bhat H P and Vayalil P K 2011 *Food Res. Int.* **44** 1812
[6] Khosravanipour Mostafazadeh A, Sarshar M, Javadian S, Zarefard M R and Amirifard Haghighi Z 2011 *Sep. Purif. Technol.* **79** 72
[7] Hanover L M and White J S 1993 *Am. J. Clin. Nutr.* **58** 724S
[8] AlNashef I M, Gaily M H, Al-Zahrani S M and Abasaeed A E 2009 *US Patent* **7942972**
[9] Lorenzo F, Marina B, Maryna S, Shaunita R H, van Zyl W H and Sergio C 2010 *Chem. Eng. Trans.* **20** 97
[10] Petridis L and Smith J C 2008 *J. Phys. Conf. Ser.* **125** 012055

[11] Anggono W, Wardana I N G, Lawes M, Hughes K J, Wahyudi S, Hamidi N and Hayakawa A 2013 *J. Phys. Conf. Ser.* **423** 012015
[12] Atiyeh H and Duvnjak Z 2001 *J. Chem.Technol. Biotechnol.* **76** 1017
[13] Koren D W and Duvnjak Z 1991 *J. Ind. Microbiol.* **7** 131
[14] Putra M D, Abasaeed A E, Al-Zahrani S M, Gaily M H, Sulieman A K, Zeinelabdeen M A and Atiyeh H K 2013 *Biotechnol. Lett.* doi:10.1007/s10529-013-1388-y
[15] Gaily M H, Sulieman A K, Zeinelabdeen M A, Al-Zahrani S M, Atiyeh H K and Abasaeed A E 2012 *Afr. J. Biotechnol.* **11** 8212
[16] Gaily M H, Elhassan B M, Abasaeed A E and Al-Zahrani S M 2010 *Int. J. Food Eng.* **6** 12
[17] Jones R P, Pamment N and Greenfield P F 1981 *Process Biochem.* **16** 42
[18] Salvadó Z, Arroyo-López F N, Guillamón J M, Salazar G, Querol A and Barrio E 2011 *Appl. Environ. Microbiol.* **77** 2292
[19] Brown S and Oliver S 1982 *Biotechnol. Lett.* **4** 269

Journal of Physics: Conference Series **495** (2014) 012035 doi:10.1088/1742-6596/495/1/012035

Phenomena of Foamed Concrete under Rolling of Aircraft Wheels

Chun-shui JIANG, Hong-yu YAO, Xian-bo XIAO[1], Xiang-jun KONG, Ya-jie SHI
China Academy of Civil Aviation Science and Technology,
Xibahe Beili #24, Chaoyang District, Beijing, 100028, China

Abstract. Engineered Material Arresting System (EMAS) is an effective technique to reduce hazards associated with aircraft overrunning runway. In order to ascertain phenomena of the foamed concrete used for EMAS under rolling of aircraft wheel, a specially designed experimental setup was built which employed Boeing 737 aircraft wheels bearing actual vertical loads to roll through the foamed concrete. A number of experiments were conducted upon this setup. It is discovered that the wheel rolls the concrete in a pure rolling manner and crushes the concrete downwards, instead of crushing it forward, as long as the concrete is not higher than the wheel axle. The concrete is compressed into powder in-situ by the wheel and then is brought to bottom of the wheel. The powder under the wheel is loose and thus is not able to sustain wheel braking. It is also found that after being rolled by the wheel the concrete exhibits either of two states, i.e. either 'crushed through' whole thickness of the concrete or 'crushed halfway', depending on combination of strength of the concrete, thickness of the concrete, vertical load the wheel carries, tire dimension and tire pressure. A new EMAS design concept is developed that if an EMAS design results in the 'crushed through' state for the main gears while the 'crushed halfway' state for the nose gear, the arresting bed would be optimal to accommodate the large difference in strength between the nose gear and the main gear of an aircraft.

1. Introduction

Aircraft overrunning runway is one of the major threats to aviation safety. Engineered Material Arresting System (hereinafter abbreviated as EMAS) has been proven to be an effective technique to address this issue [1]. EMAS comprises foamed concrete blocks with thicknesses of hundreds of millimeters that are arranged on the ground beyond runway end. Once an aircraft overruns runway and then runs into EMAS, the foamed concrete will get crushed as rolled by the aircraft wheels. In this way, drag forces are generated to the aircraft landing gears, and the aircraft is therefore slowed down gradually and eventually is stopped in the EMAS bed for safety. Studies on EMAS mainly focus on the analysis of the drag forces imposed on the aircraft landing gears as well as the dynamic simulation for arresting process. Wang et al. [2] and Cook et al. [3,4] assumed that the drag force landing gear subjected to is determined by the product of forward project area of the wheel portion immersed in the foam material and crush strength of the foam material, which is called 'horizontal push model' in this paper. Li et al. [5] assumed that the drag force suffered by wheel is similar to that when the wheel moves in dry snow. DeLoach et al. [6] have drawn a conclusion that the drag force is related to the

1 Xianbo.xiao@gmail.com

speed of aircraft or the second power of the speed on assumption that the arresting process is similar to the movement of an object in a fluid. Heymsfield et al. [7] believed that after an aircraft's bogie landing gear runs into the arresting bed, the rear wheel advances in the rut created by the front wheel, thus no drag force is created on the rear wheel. Physical basis of these analysis and simulations mostly rely on the experimental results in terms of material mechanics, without experimental results under actual conditions, i.e. under rolling of aircraft wheels. In seeking for the experimental results under actual conditions, the greatest fidelity could be undoubtedly achieved by conducting a test of taxiing a real aircraft into an arresting bed, but the cost and the safety risks are great and few test parameters are adjustable. Simulated experiments in laboratory are of fidelity between the real aircraft test and the material mechanics experiment, while design and construction of the experimental device are still problems to be solved. A device for simulation was described [8]. To be exact, a small sized aluminum cylinder simulating aircraft wheel of reduced scales was hung up and then ploughed through foamed concrete in the way of pendulum. In this paper, this process is regarded distinct from the situation that an elastic pneumatic rubber tire rolls (might accompanied with sliding) in the foam. DeLoach et al [6] also launched an interesting EMAS experiment by using a runway friction coefficient measurement vehicle with a speed of up to tens of kilometers per hour, but the wheel in the experiment was not a real aircraft wheel. Furthermore, because of small size of the wheel, the wheel was managed to penetrate the foam by only about 100 mm so as to avoid the interference between the wheel fixture and the foam material to assure correct drag measurement. So, the experiment configuration made a considerable difference from the actual conditions. In order to clarify phenomena of the material under rolling of aircraft wheels and to provide a physical basis for mathematical simulation of the arresting process, a special experimental device was designed and built. A number of experiments were carried out upon the device and some interesting phenomena were observed.

2. Experimental Device and Methods

Aircraft Experimental device is shown in Figure 1 [9]. Platform (2) was free to move in its longitudinal direction on the support of rollers (1) and was constrained by lateral rollers from moving in its lateral direction. Foamed concrete blocks (3) were arranged in line on the platform and constrained with baffles on both sides in the rear end. Front of the concrete had been sawed into ramp to eliminate the impact when the wheel rushes into the material. Real aircraft wheel (4) can rotate freely on wheel fixture (5). Load (6) was stacked on top of wheel fixture to subject the wheel to a practical shared weight of an aircraft. The wheel fixture was free to move vertically while held laterally by lateral rollers.

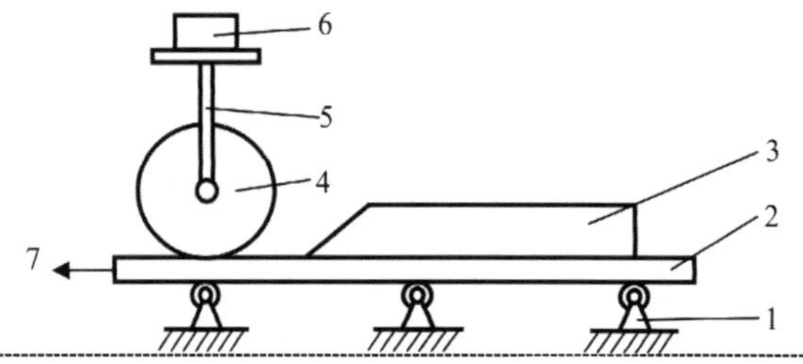

Figure 1. Schematic diagram of the experiment device

1: Roller, 2: Platform, 3: Foamed concrete, 4: Aircraft wheel, 5: Wheel fixture, 6: Load, 7: Pull

Experiments started with pulling (7) the platform at a constant speed, which makes the wheel roll into the foam via the ramp. Sensors recorded with time forces in three-direction exerted on the wheel fixture, vertical displacement of the wheel fixture, rotation speed of the wheel, speed of the platform, pulling force to the platform and other parameters involved. The device employed main wheel and nose wheel respectively, of a Boeing 737-300 aircraft and adopted static loads on each wheel of this type of aircraft under various payload conditions. Notablely, deviation of the experiment conditions from the actual situation lies in lower wheel speed (250 mm/s), wheel fixture lacking damping mechanism of aircraft landing gear and the interference of the wheel fixture with foam material when the wheel penetrated the material deeply.

In order to clarify phenomena around rear wheel of a bogie landing gear rolling in arresting bed, rolling-twice experiment was designed. As the name implies, after wheel rolled arresting bed for the first time, the wheel was driven to roll the as-rolled bed once again. The same aircraft wheel advanced in the rut produced in the first rolling, which is equivalent to the effect that the rear wheel of bogie landing gear rolls the arresting bed.

Foamed concrete used in the experiments was made of mixture of cement, fly ash, fiber, vesicant and water. Thicknesses of the concrete ranged from 240 to 600mm. Most pores in the material are enclosed with diameters of 1 ~ 2 mm. Porosity of the material is around 75% and density 200 Kg/m3. The microstructures and mechanical properties of the material have been reported [10].

3. Experimental Results

3.1. Rolling or sliding

Under a variety of experimental conditions, it is recorded by the sensors that linear speed of the platform accords with rotational speed of the wheel, indicating that the wheel rolls material in a pure rolling manner.

3.2. Movement of the foamed material

Figure 2 presents appearance of the foamed material after the experiment. Almost all affected foam material was compressed in-situ into the rut with little scattered, at least under conditions of this experiment (aircraft wheel speed of 250 mm/s and material thickness of 310 mm). As the material thickness increases, the pushing-forward effect of wheel on the material increases, so a small amount of material was pushed aside, even though the proportion was still small (see below).

Figure 2. Foamed concrete appearance after the experiment

3.3. Pushing forward to the foam material by aircraft wheel

Figure 3 shows the appearance of foam material in front of wheel in experiment. The material was not pushed away, but instead, was compressed into the bottom of the wheel. Cracking of material in front of wheel and deformation of the wood board used as stopper indicate the presence of a forward pushing exerted by the wheel. Longitudinal section of this part of material is shown in Figure 4. The left part is uncrushed foam material while the right part presents indentation contour of wheel, and the middle is material powder made by compression. Crushing of material is characterized by severe localization. The material powder in front of wheel gets thicker and thicker from top down, and basically non-existence at the top. This feature does not correspond to the 'horizontal push model' [2-4] where the crushed material amount in front of wheel (i.e. the thickness of powder) should be independent of the height. This indicates that the material crushing is mainly attributed to vertical or diagonal compression. Since the wheel front is in form of arc, the reaction force of the material compression could be decomposed into a component of horizontal drag force.

The increased material thickness will strengthen the pushing forward effect of aircraft wheel to the material. When the material thickness (600 mm) slightly surpasses the height of wheel axle (the test main wheel radius is 515 mm), surface layer of material was pushed away in the rolling (Figure 5). After the experiment, longitudinal section of the material in front of wheel (Figure 6) showed that the bulk of material was still compressed into powder by the wheel. Hence, under conditions that material is slightly higher than the wheel axle, the compression process is essentially that materials get crushed into powders by rotating wheel, and only material portion above the wheel axle is appropriate for the horizontal push model. Additionally, appearance of rut after wheel rolling shows that, for the material slightly higher than the wheel axle, two shallow grooves beside the deep wheel rut, on the top surface of the material, were created by the wheel fixture's scratching across the material, and the cross section of the shallow grooves corresponds to the projection of the wheel fixture.

Figure 3. Foamed concrete in front of the wheel

Figure 4. Longitudinal section of foamed concrete in Fig. 3

Figure 5. Surface material portion was pushed away when the material is higher than wheel axle

Figure 6. Longitudinal section of foamed concrete in Fig. 5

3.4. 'Crushed through' and 'crushed halfway'

It was found that under certain combinations of material strength, material thickness, wheel size, tire pressure and vertical load, either 'crushed through' whole thickness of the concrete or just 'crushed halfway' may occur in the rolled foam material. Figure 7 shows cross-sections of two ruts, corresponding to 'crushed through' (left) and 'crushed halfway' (right) respectively, after rolling. The latter state is correlated to a small drag force and low material utilization efficiency. Drag force can be enhanced by increasing vertical load rather than thickening the material. Whereas the 'crushed through' state is just in the opposite rules.

'crushed through' 'crushed halfway'

Figure 7. Cross-sections of ruts after rolling

3.5. Phenomena in the rolling-twice experiments

Compared with the first rolling, sinkage of wheel axle in the second rolling was increased by 26 mm, and the rut depth was increased by 18 mm, indicating that the material powder at bottom of the rut had been further compressed.

4. Discussions

The experiments showed that the wheels move in the foam materials in the form of pure rolling. Deposit of crushed material powder was observed beneath the wheel after experiment. It is impossible for the powder to bear the shear force generated from wheel braking due to its extremely low shear strength. Thus, although the standard design conditions for EMAS regulated by the Federal Aviation Administration (FAA) of the United States of America describe that friction coefficient equals 0.25 [11] when aircraft is braked, braking effectiveness can be achieved only in the setback section in front of arresting bed rather than in the bed. It has been verified with real aircraft tests that braking in arresting bed has no effect on drag force [12].

Some researchers [4, 6] reckon that during the arresting process, foamed concrete material is driven by wheels at high speeds in motion, and thus drag force expression should comprise the contribution of material kinetic energy 1/2 ρv2, where ρ is specific weight of the material and v the aircraft speed. Experimental results presented this paper show that almost all the affected foam material was compressed in-situ into the rut, so the kinetic energy of the material could be ignored. Speed effect on the drag force should be relevant to the loading speed effect on compression resistance of the material. The rotation of wheel results in different speeds at which each point at tire front compresses the material, which complicates the analysis of speed effect on drag force.

Crushing of foamed concrete material is characterized by severe localization and discontinuity, corresponding to the experimental results of material mechanics [10]. Therefore, the application of continuum mechanics to the analysis of drag force will meet great challenge, and in deed some finite element analysises obtained unsatisfactory results [13, 14]. Some researchers [7] discretized a tire into radial springs, and discretized the foam material into a series of sliding friction pairs.

As limited by the clearance above ground of engine that is hung below wing, the typical thickness of material for EMAS system is usually about 600 mm. Together with blocking up effect of crushed powder on the wheel and the possible 'crushed halfway' state, the material will not be substantially higher than the axle of aircraft main wheel (for example, radius of Boeing 737-300 main wheel is 508 mm). In summary, it can be concluded that the producing mechanism of drag force is that the material gets crushed into powder in situ vertically or diagonally, instead of the horizontal push model reported [2-4]. And the material strength is relevant to the strength when it is further compressed into powder, rather than the strength when the material commences cracking. Meanwhile, drag force produced by surface material portion higher than the wheel axle or interfered by structures beside wheel (such as axle, landing gear piston, etc.), can be calculated following horizontal push model.

On the other hand, Boeing 737-300 aircraft has nose wheels in radius of 340 mm, much smaller than the typical material thickness. So, the wheel axle and gear piston of the nose gear will definitely interfere with the material severely. No matter whether horizontal push model or other models is used, it would be difficult to accurately calculate the drag force nose gear subjected to. Because the size and amount of nose wheels are both smaller than those of main wheels, the inaccuracy of the drag force of nose wheels would not affect the stopping distance substantially, but is a challenge to the structure safety evaluation of nose gear of low load capacity. One of the solutions to address this issue is to reduce the material thickness, but this is always impeded by the demands for large drag force on main gear. Fortunately, it is worthy of trying to apply combination of the two states of 'crushed through' and 'crushed halfway' novelly discovered in this paper to solve the problem. If the 'crushed halfway' state occurs with nose gear while the 'crushed through' state occurs with main gear respectively through deliberate design, the structure safety and accuracy of drag force calculation of nose gear will be ensured due to the small immersion depth and resultant small drag force. Moreover, the main gear with a large load capacity will exert a large drag force adequately with the help of great immersion depth, and drag force calculation accuracy will be improved because of the high axle of aircraft wheel. In this way, an arresting bed can be optimized to accommodate different characteristics of nose and main gear.

The rolling-twice experiment results show that the rear wheel of bogie gear still compressed the material powder effectively to a small extent, which produces a small amount of drag force. Because rear wheel in the rut lacked supporting material at its front edge, and then entire vertical load was transferred to the material powder under bottom of wheel, the powder would be further compressed. It can be speculated that if the 'crushed halfway' state is with front wheel, the drag force rear wheel subjected to will increase.

5. Conclusions

Experiments of real aircraft wheel rolling through the foamed concrete were conducted upon a specially designed and built device. Some interesting phenomena of the material under rolling of the wheels were discovered, which can lend a physical basis for drag force analysis and mathematical simulation of the arresting process. The main conclusions can be summarized as follows:

- When the foam material is not higher than wheel axle, the wheel compresses the material in a pure rolling manner rather than compresses horizontally. The foam is compressed in-situ into powder and brought down to the bottom of wheel, with no kinetic energy endowed. Moreover, the shear strength of the powder pile is not enough to sustain wheel braking.
- Crushing of the foam material exhibits severe localization and discontinuity under compression of aircraft wheels.

- The material portion above the wheel axle and that interfered with structures besides the tire is pushed forward, appropriate for horizontal push model.
- Under certain combinations of material strength, material thickness, tire size, tire pressure and vertical load, either 'crushed through' whole thickness of the concrete or 'crushed halfway' may occur in the foam material after rolling. An optimized arresting bed could be designed by satisfying at the same time the distinct characteristics of both nose gear and main gear of an aircraft, with a reasonable exploitation of these two states.
- A small amount of drag force is generated on rear wheel of a bogie gear. And if 'crushed halfway' state is with front wheel of a bogie gear, the drag force exerted on the rear wheel would be further increased.

6. Acknowledgements

This research was financially supported by the National Science and Technology Support Program of China (2012BAG04B02), the Civil Aviation Joint Research Fund Key Project of National Natural Science Foundation of China (U1233203), and the Science and Technology Project of Civil Aviation Administration of China (MHRDZ201123).

7. References

[1] Zeng L, Kong X J, Shi Y J and Xiao X B 2012 *Aviation Safety* **3** 13-16
[2] Wang W, Chang H 2009 *Journal of Civil Aviation University of China* **27** 10-13
[3] Cook B et al 1995 DOT/FAA/CT-95 *FAA Technical Center*
[4] Cook R F 1987 DOT/FAA/PM-87-27 *Federal Aviation Administration*
[5] Li F Y, Jiao Z X, Gui Y Q and Wang L 2010 *Journal of Beijing University of Aeronautics and Astronautics* **36** 1-4
[6] DeLoach R et al 2009 *the 47th AIAA Aerospace Sciences Meeting including the New Horizons Forum and Aerospace Exposition, Orlando* FL(USA) 1156.
[7] Heymsfield E et al 2012 *Journal of Transportation Engineering* **138** 284-292.
[8] Barsotti M A et al 2009 Airport Cooperative Research Program, Report 29, *Washington D.C.: Transportation Research Board of the National Academies*
[9] Kong X J, Shi Y J, Xiao X B et al Development of an Experimental Device for Engineered Arresting Materials, Experimental Mechanics (to be published)
[10] Zhao R, Guo W G, Shi Y J and Zeng L 2011 Experimental study on anti-penetration behavior of foamed composite material, *The International Symposium on Impact Dynamics*
[11] 2005 Engineered Materials Arresting Systems (EMAS) for aircraft overruns, Advisory Circular No. 150/5220-22A, *Washington D.C.: Federal Aviation Administration.*
[12] Cook R 1993 DOT/FAA/CT-TN93/4 *USA: FAA Technical Center*
[13] Shi Y J 2010 *Proceedings of the 11th International LS-DYNA Users Conference, Detroit* 16-21
[14] Barsotti M 2012 *FEA Information Engineering Journal* **1** 16-37

ScieTech 2014 IOP Publishing
Journal of Physics: Conference Series **495** (2014) 012036 doi:10.1088/1742-6596/495/1/012036

Kazakh Traditional Dance Gesture Recognition

A K Nussipbekov [1], E N Amirgaliyev [1] and Minsoo Hahn [2]
[1] Faculty of Mechanics and Mathematics, Al-Farabi Kazakh National University, Almaty, Kazakhstan
[2] Digital Media Lab, Korea Advanced Institute of Science and Technology, Daejeon, South Korea

E-mail: abai.nk@gmail.com

Abstract. Full body gesture recognition is an important and interdisciplinary research field which is widely used in many application spheres including dance gesture recognition. The rapid growth of technology in recent years brought a lot of contribution in this domain. However it is still challenging task. In this paper we implement Kazakh traditional dance gesture recognition. We use Microsoft Kinect camera to obtain human skeleton and depth information. Then we apply tree-structured Bayesian network and Expectation Maximization algorithm with K-means clustering to calculate conditional linear Gaussians for classifying poses. And finally we use Hidden Markov Model to detect dance gestures. Our main contribution is that we extend Kinect skeleton by adding headwear as a new skeleton joint which is calculated from depth image. This novelty allows us to significantly improve the accuracy of head gesture recognition of a dancer which in turn plays considerable role in whole body gesture recognition. Experimental results show the efficiency of the proposed method and that its performance is comparable to the state-of-the-art system performances.

1. Introduction

Gesture recognition is an important, interdisciplinary area which plays significant role in HCI (Human Computer Interaction) and it is used in a wide range of applications including robotics [1], medicine [2] computer games [3] education [4], sport [5] or dance as considered herein. Dance gesture recognition and classification is related to full body recognition task. Nowadays, the systems able to recognize and classify dance gestures, have a big impact on our day life. Recent researches like [6], [7] show that interactive gaming systems help to facilitate exercise and rehabilitation and even affect on academic performance. Interactive dance systems help to evaluate dancers performance [8] and can be used in online teaching [9].

There has been considerable progress on this topic, however the problem is still quite difficult and challenging. Correct pose estimation is one of the key components. One should take into account that pose estimation should be calculated in 3D space for better performance and robustness. Kazakh traditional dance recognition may differ from other types of dance recognition tasks because of the specificity of Kazakh traditional clothes (see figure 1). Some of the clothe parts can make it complicate to use some well known pose estimation techniques, while the other parts may conversely help to increase accuracy. Another point is that in Kazakh traditional dances the head movements are important, because there are a lot of poses which may differ only in head orientation and that is why calculating correct head pose orientation is vital part for the accuracy of whole body recognition task.

Kazakh traditional dances date back to the old days of nomads and are historically and culturally valuable heritage. To the best of our knowledge, this problem has not been addressed anywhere yet. Figure 2 demonstrates the overview of our proposed system.

Figure 1. The specificities of Kazakh traditional women clothes are long and wide skirt and a high headwear called "Saukele".

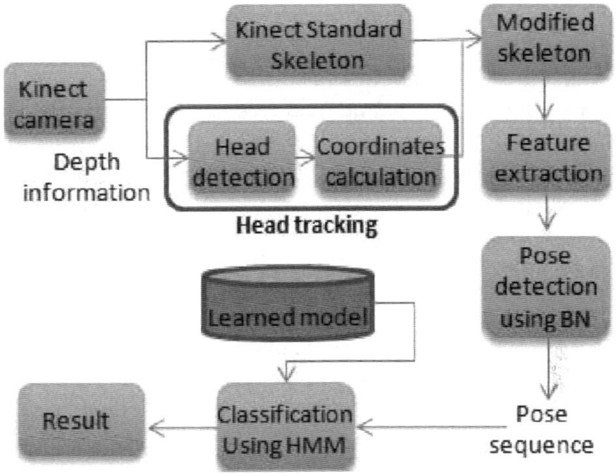

Figure 2. System Overview

In our work we extended the human skeleton, provided by Microsoft Kinect SDK, by applying headwear tracking. The modified skeleton (see figure 8a) is then used for pose recognition by learning tree-structured Bayesian network model using EM and k-means algorithm and finally gesture recognition is performed by applying Hidden Markov Model. As experiment results show, the proposed modified skeleton gives us more precise head movement information which, in turn, increases the recognition accuracy of the whole body for the Kazakh traditional dances.

2. Related Work

As dance gesture recognition is a special case of general gesture recognition task, many of the traditional approaches are used here. They have been addressed in many researches. Generally, these works can be divided according to the technology they use as well as the algorithm and techniques they apply for retrieving features and performing recognition.

There are many works which use different type of devices like electro-magnetic and optical sensors, accelerometers that are fixed on a human body or embedded in to the environment. For example [10] used body sensor network in their work, they applied hierarchical model in order to recognize actions on node (low) level and portable devices (high) level. Similarly, H. Junker *et al.* [11] used body-worn inertial sensors and used a kind of two stage approach, firstly using similarity search to select specific motion events and then classifying them by hidden Markov model. In contrast, [12] proposed to use smart clothes based on conductive elastomer sensors which author say, is a new kind of strain sensors, compared to previous works, they were able to reconstruct shoulder, wrist, elbow and hands movements. There are some studies which are related directly to the dance gesture recognition problem using sensor approach. For instance, [13] created a real-time gesture driven interactive system with multimodal feedback for performing dance. In their study they operated with 41 markers attached to a dancer to obtain 3D coordinates of 10 body parts, however the latest Kinect SDK, which we use in our work, provides us with more body parts, in particular it includes hands and feet.

Another type of approaches are based on cameras. Camera based approaches, in turn, may also be divided into simple, stereo and depth cameras. Nam Vo *et al.* [14] have proposed to use skin color information taken from the web camera. They detect skin by recognizing face using boosted Haar wavelet classifier, determine skin with the help of thresholds, after that they use skin information to detect other parts and finally apply several recognition techniques. Additionally some other colorful objects may be used like colorful markers, gloves and etc. [15], [16]. An alternative approach is to use silhouettes as main feature. In [17], for example, they used silhouette energy image and variability models, and in [18], they proposed to adapt silhouette directionality-based feature vectors from silhouette contours.

Stereo cameras are yet another way not only to just recognize gestures but perform it in 3D space. Thus, [19] got high recognition results by using a novel approach based on multilinear analysis, they extracted low dimensional pose orientation coefficient vectors by performing tensor decomposition and projection.

Although all of the above mentioned works got promising results, they still have significant disadvantages. Sensors, for instance, are often cumbersome and movement-restraint, especially when there are too many wires. Another problem is that some of them are very expensive. Cameras are very sensitive to illumination and have certain complexities with occlusion, clutter and also often angle view dependent.

In terms of algorithms generally researchers can use Support Vector Machine (SVM) [17], [19], Neural Networks [20], K-means clustering [18], decision trees [15], Hidden Markov Model [1], [21], [22] and many other approaches are used today. Choosing appropriate algorithm depends on the purpose of work.

In our own method we use a Kinect camera which was launched in 2010 by Microsoft (see figure 3). It has been used in a wide range of researches like [4], [9]. The advantage of this camera over other traditional ones is that beside its RGB camera and multi-array microphone, it includes an infrared laser projector combined with a monochrome sensor, that allows it to capture 3D scenes in any available lighting conditions. Latest Kinect SDK provides developers with 640x480p depth image and skeleton consisting of 20 joints. Kinect has many benefits unlike previously explained methods: people do not need to wear special sensors which make it natural to use, it does not suffer from the shortage of illumination as regular cameras do, it gives more precise 3D coordinates and at the same time it's angle view independent compare to monocular cameras and some other advantages that make it better to use in our work. In our project we used Bayesian network for skeleton to robustly detect poses, and by using Kinect we didn't have to use additional skeletonization procedures to get skeleton information as it was done by K. Srijeyanthan *et al.* [23], [22] and then used in their another study for their framework. In terms of classification algorithms, we used Expectation Maximization and standard k-means algorithm to detect poses and Hidden Markov Model to determine the sequential nature of movements.

3. Feature extraction and preprocessing

3.1 Skeleton features extraction

As it was mentioned above, Kinect provides us with skeleton consisting of 20 joints each having its 3D coordinates (X, Y, Z). Using human skeleton instead of other approaches explained before gives us several advantages. First of all, we do not rely on dancer's visual appearance like color of clothes. Moreover, despite the fact that people skeleton can be different in size, such as they may have different bone lengths, by using joint angle values as a feature set instead of joint coordinates, we make our system person independent. In our work we adopted similar features proposed in [24]. We use inclination and azimuth angle values (see figure 4) of particular joint vectors. These joints are: Right Hand (RH), Left Hand (LH), Right Elbow (RE), Left Elbow (LE) and Head.

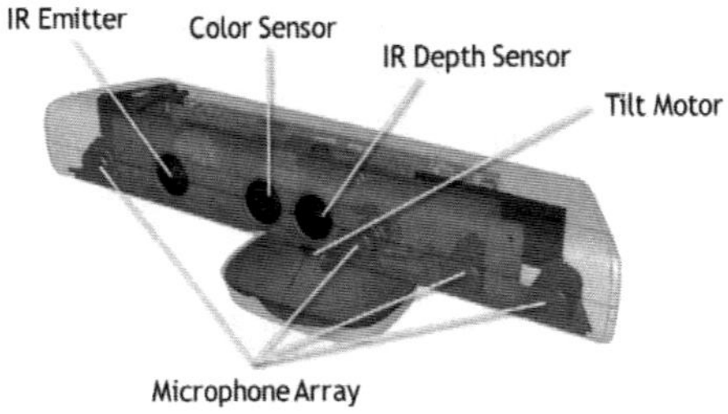

Figure 3. Kinect camera

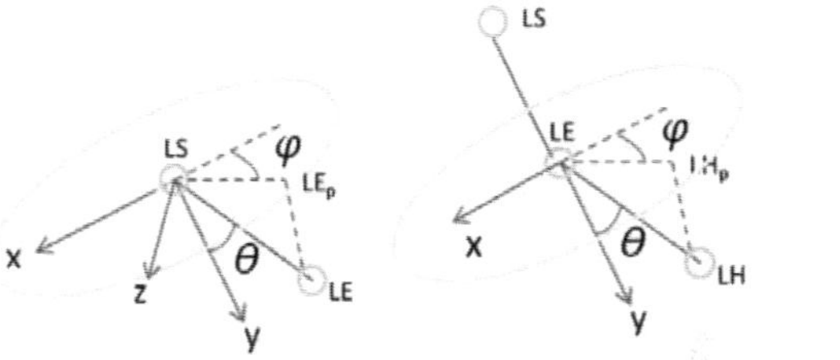

Figure 4. Spherical coordinate system for first (left) and second-degree (right) joints (LH-left hand, LS-left shoulder, LE-left elbow, LEp and LHp – their projections, θ-inclination angle, φ-azimuth angle)

Joints were divided into first-degree, like head and elbows which are calculated relative to torso, and second-degree joints, like hands that are calculated relative to its parent joint (see figure 4). Because of specificity of Kazakh traditional clothes (see figure 1), legs can't be tracked directly from Kinect skeleton. Inferred joints that supposed to help in such situations often make unrealistic motion. It could be possible to use depth silhouette or color information for this problem but it will degrade time performance of the overall system. On the other hand, we observe that leg movements are not so important in recognition of Kazakh traditional dances.

3.2 Skeleton modification

Head movements are important in Kazakh traditional dances and their precise angle calculation is essential. We observe that standard Kinect skeleton can not give us a favorable head tracking accuracy, so we decided to use our own method that is based on dancer's headwear tracking.

In our proposed method, firstly, we isolate a human performing dance from the background. We do it by performing background subtraction based on auto thresholding principle proposed by [25] after obtaining histogram of depth image (see figure 5). Next, we calculate a region of interest (ROI). It is an area which is just over the head (se figure 6). The location of ROI is calculated by using Kinect skeleton head location information and its size calculated depending on how far a performer stands from the camera. After, we search for headwear in that region by observing "blobs", there are a lot of ready solutions for such kind of problems and in our project we used EmguCV library for that purpose. Then we calculate the center of mass of our object by calculating image moments. We find X and Y coordinate on the image, and after, we calculate a depth value at that point. Finally we have to transform these coordinates to the real world ones. We map them to Kinect skeleton and thus get X, Y, Z world coordinates of our headwear.

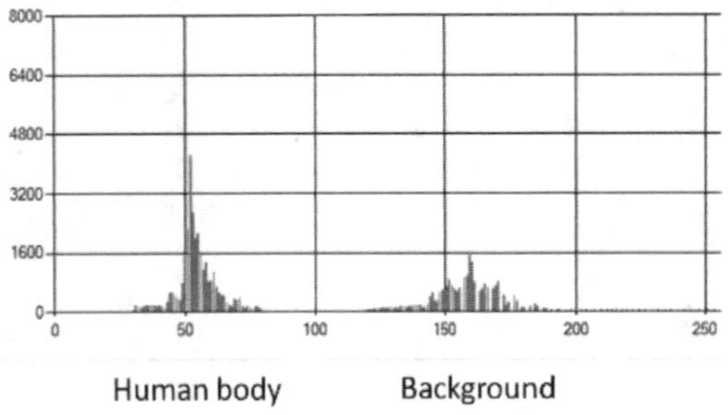

Figure 5. Depth histogram.

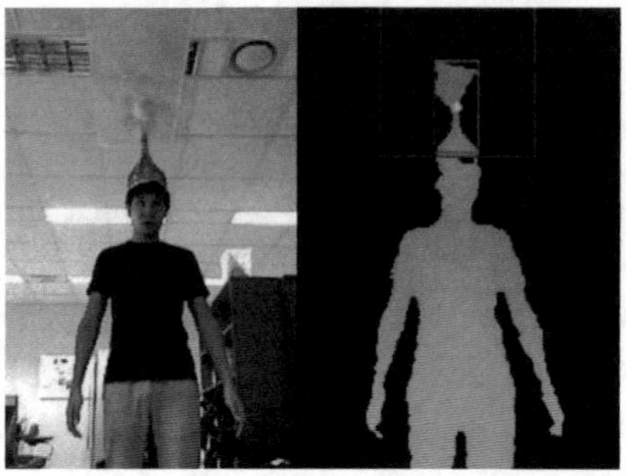

Figure 6. ROI and headwear calculation from depth image.

In order to test our proposed method we fixed laser on the top of the headwear and a linear grid, marked with distance values, to the ceiling of a room and then observed the displacement (w) of the laser point on that grid (see figure 7). Knowing the distance between a person's headwear and ceiling (h), we calculated real head inclination angle values, which is $\arctan(\theta)$, and compared them with our proposed method and Kinect skeleton's head inclination values (see figure 9). Experiment results demonstrate that our proposed new calculated skeleton "bone" (see figure 8a) gives us more precise and accurate head inclination values (see figure 9a, 9c) compare to standard skeleton's head (see figure 9b, 9d). We made this experiment in XY and YZ plane and we noticed that this is especially noticeable for small head inclinations and inclinations in YZ plane.

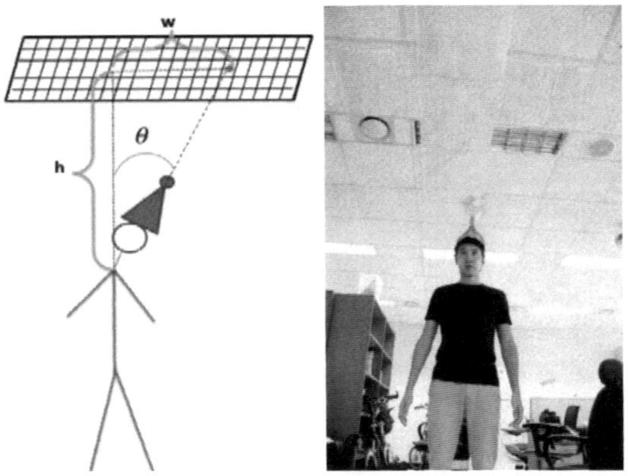

Figure 7. Measurement of real inclination angle values of head.

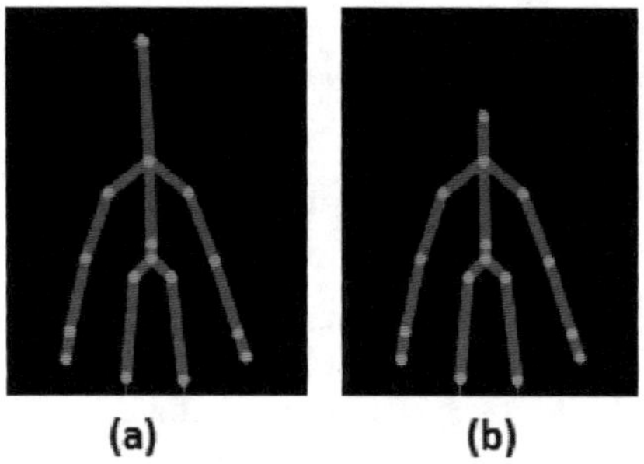

Figure 8. a) extended skeleton. b) standard skeleton

Another thing that we found is that our proposed method produces more clear and expressed graph (see figure 10 a) of head movements from which we can easily extract features we need compared to standard skeleton. Standard skeleton's head, in turn, experiences jitters while moving which affects on the diagram (see figure 10 b) of periodical head movements.

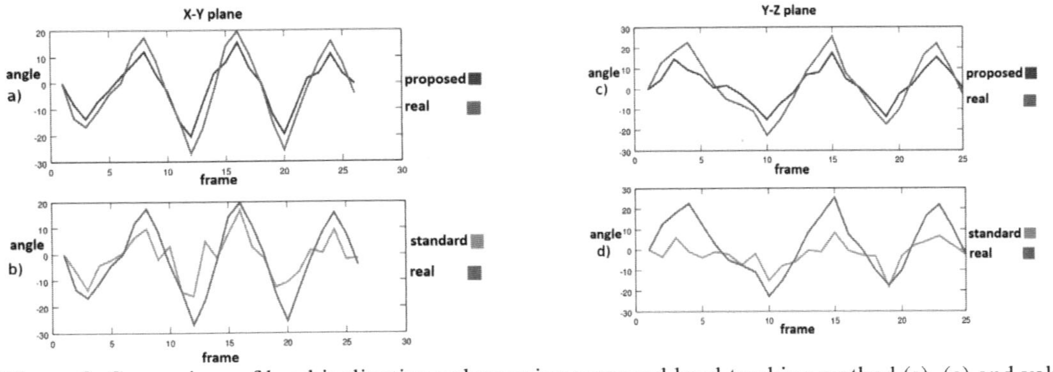

Figure 9. Comparison of head inclination values using proposed head tracking method (a), (c) and values taken from Kinect skeleton's head (b), (d) with real values

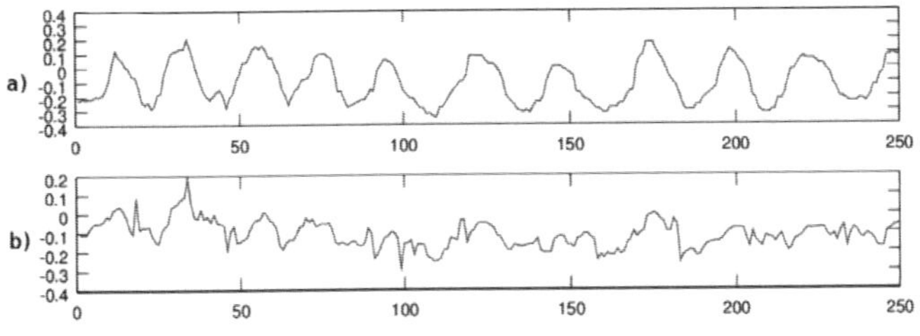

Figure 10. Proposed method (a) gives more clear and expressed diagram compare to Kinect skeleton's head (b) gives.

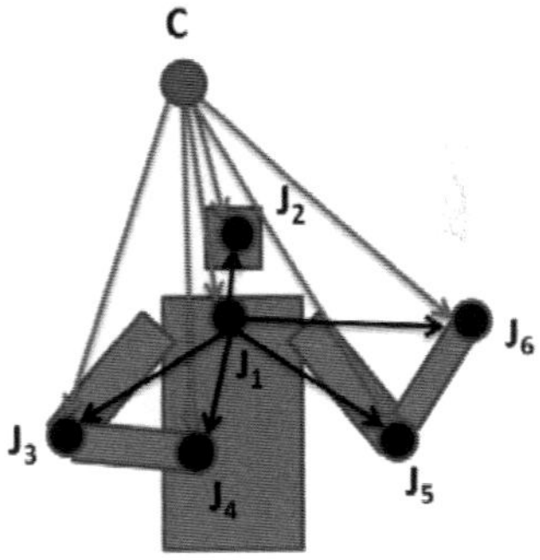

Figure 11. Skeleton. Joint (J) and class (C) dependencies.

4. Recognition

4.1 Pose recognition

Tree structured models and Bayesian network has been widely used in human pose estimation [26], [27]. The recently opened Stanford PGM online course [30] also adopted it in their class. We used an improved version of their proposed model for our own purposes.

The good thing about them is that it allows us not only to represent a pose but to take into consideration the kinematic constraints of human body structure. In other words it takes into account different dependencies among body parts which makes the pose estimation more realistic and efficient.

The structure is shown in figure 11. Each node except torso (J_1) has its own physical parent node and class parent node whom it belongs to. Class node represents a posture class. A torso, being a root node, has only one parent node which is a class node. We parameterize CPD values of each node by using Gaussian and Conditional Linear Gaussian (CLG) distribution.

A torso is parameterized by following simple Gaussian distribution:

$$\theta \mid C = k \approx N(\mu_k^{\theta}, \sigma_k^{\theta 2}), \qquad (1)$$

$$\varphi \mid C = k \approx N(\mu_k^{\varphi}, \sigma_k^{\varphi 2}), \qquad (2)$$

where k are indexes of pose classes. Other body parts have two parents, thus they will be parameterized by following linear Gaussian (CLG):

$$\theta_i \mid \mathrm{J}_{p(i)}, C = k \approx N(\beta_{ik}^{(1)} + \beta_{ik}^{(2)}\theta_{p(i)} + \beta_{ik}^{(3)}\varphi_{p(i)}, \sigma_{ik}^{\theta 2}), \qquad (3)$$

$$\varphi_i \mid \mathrm{J}_{p(i)}, C = k \approx N(\beta_{ik}^{(4)} + \beta_{ik}^{(5)}\theta_{p(i)} + \beta_{ik}^{(6)}\varphi_{p(i)}, \sigma_{ik}^{\varphi 2}), \qquad (4)$$

where $i = 1, 2, ..., 5$ are body part indexes, k is the pose class number, $p(i)$ is the parent of $i'th$ node.

Each class of pose is related to its own set of conditional linear Gaussian parameters. And we learn these parameters performing Expectation Maximization (EM) algorithm by iterating between calculating probabilities of pose classes $P(C = k \mid \mathrm{J}_1, ..., \mathrm{J}_6)$ which, in turn, calculated from the joint probability of pose and class, and estimating CLG values using these probability values.

For better performance we added standard *k*-means clustering algorithm before running EM algorithm which classifies pose classes in order to use them as initial class probability values in EM algorithm.

In our study each posture class is represented as a unique symbol which is used in our following part.

4.2 Gesture recognition

Now, having obtained the set of poses, we would like to determine the gesture to which this sequence belongs to. We use discrete Hidden Markov Model (HMM) in this problem because they handle the sequential nature of the gestures consisting of these poses.

HMM was explained in details in a Rabiner's tutorial [28]. It is characterized by N hidden states $S = \{\mathrm{S}_1, S_2, ..., S_N\}$ and M observation symbols (alphabet) $V = \{v_1, v_2, ..., v_M\}$. The transition probability between states is in $N \times N$ matrix $A = \{a_{ij}\}$ where $a_{ij} = P[\mathrm{q}_{t+1} = \mathrm{S}_j \mid \mathrm{q}_t = \mathrm{S}_i]$, q_t is a state at time *t*. The observation probability matrix $B = \{\mathrm{b}_j(\mathrm{k})\}$, where $b_j(k) = P[v_k \mid q_t = S_j]$, where $1 \le k \le M$ and $1 \le i, j \le N$. And the initial state distribution probability $\pi_i = P[q_1 = S_i]$. HMM is used for different kind of problems that can be divided into two types: inference and learning.

In our work we use HMM to firstly learn models of different dance gestures and then use them for calculating the probability of some specified sequence of poses (observed sequence). Learning a model is the problem of adjusting the model parameters $\lambda = (A, B, \pi)$ to maximize the probability of output symbols given the model. It is solved by using Baum-Welch algorithm.

Every taken gesture from camera is, as it was mentioned above, represented as a sequence of symbols which we recognize by calculating $P(O|\lambda)$, the probability of observation sequence $O = O_1O_2...O_T$, given the model λ and selecting the class which gives the highest probability for a given observation

$$\arg\max\{P(O|M_1), P(O|M_2), ..., P(O|M_n)\}, \quad (5)$$

where M_i is HMM model for $i'th$ gesture and n is number of gestures.

5. Experiment results

We analyzed many Kazakh traditional dances and noticed that many of them share same or almost same gestures. Therefore we decided to classify gestures rather than whole dances. There are no specific names for these gestures so we named them according to their main poses [31]. In our experiment we used 12 dancers each performing 10 dance gestures from different dance types 5 - 8 times. The dataset was manually labeled. In total we collected about 800 dance gestures, 70% of which was used for training and the remaining 30% for testing. We evaluated our and the following table 1 demonstrates results.

Table 1. Gesture recognition accuracy results. Accuracy (a) - initial recognition result, accuracy (b) - after adding headwear tracking.

Dance gesture	Accuracy (a)	Accuracy (b)
Salem	84.0%	84.2%
Qus qanat	97.0%	98.7%
Qos muyiz	92.1%	97.3%
Saukele	85.0%	92.0%
Sanyr muyiz	83.3%	87.6%
Belbeu	90.8%	90.0%
Kamshy	86.0%	89.1%
Qol zhaiyu	86.7%	89.4%
Shashyn oru	79.4%	85.8%
Aq qu	88.0%	94.1%
		90.82%

$$Accuracy = (N_c \times N_T^{-1}) \times 100, \quad (6)$$

where, N_c - number of correctly classified gestures, N_T - total number of gestures.

It can be seen from the table that the lowest final performances among our selected gestures is 84%. This is because of noise persisting in Kinect depth image which affects on skeleton accuracy for low expressed movements e.g. Qol Zhaiyu and Shashyn Oru gestures, another reason is that joints can overlap each other which affects on azimuth angle calculation accuracy e.g. Salem gesture.

However by using our new head tracking approach we got higher accuracy (see table1 Accuracy (b)) for such gestures like Aq Qu (94.1% compare to 88%), Saukele (92% compare to 85%), this is because head movements are one of the key components in these gestures. Other gestures, where head movements are not so expressed, remained almost the same. Kazakh traditional dances are different from many other dances. And as it was mentioned above, to the best of our knowledge, this problem has not been addressed anywhere yet so we cannot compare it with other works. But generally 90.82% is a very good accuracy on the dance recognition task. For example [29] has got 92% on average for the recognition of Bali traditional dances.

6. Conclusion

"Kazakh traditional dance gesture recognition" is interesting and challenging task. We used Microsoft Kinect camera in our work in order our system to be easy and natural to use. We adopted some well known statistical models like Bayesian network, hidden Markov models and used appropriate algorithms like EM algorithm, k-means clustering, Gaussian distribution to parameterize and learn these models. Our main contribution however was improvement of head gesture recognition by our own proposed head tracking method. Finally we got high recognition results which are demonstrated in last section. The proposed work carries a cultural significance for Kazakh traditions. It may be used for dance evaluation systems as well as in computer games.

References:

[1] H. Yang, A. Park, S. Lee, and S. Member, "Gesture Spotting and Recognition for Human – Robot Interaction," vol. 23, no. 2, pp. 256–270, 2007.

[2] K. Englmeier, C. Krapichler, M. Haubner, M. Seemand, and M. Reiser, "Virtual reality and multimedia human-computer interaction in medicine."

[3] M. Labs and M. Corporation, "USING HUMAN BODY GESTURES AS INPUTS FOR GAMING VIA DEPTH ANALYSIS," pp. 993–996, 2008.

[4] V. Thakkar, A. Shah, M. Thakkar, A. Joshi, and N. Mendjoge, "Learning Math Using Gesture," *International Conference on Education and e-Learning Innovations*, pp. 1–3, Jul. 2012.

[5] K. Takano and K. F. Li, "Classifying Sports Gesture Using Event-Based Matching in a Multimedia E-learning System," *2012 26th International Conference on Advanced Information Networking and Applications Workshops*, pp. 833–838, Mar. 2012.

[6] Z. Gao, P. Hannan, P. Xiang, D. F. Stodden, and V. E. Valdez, "Video game-based exercise, Latino children's physical health, and academic achievement.," *American journal of preventive medicine*, vol. 44, no. 3 Suppl 3, pp. S240–6, Mar. 2013.

[7] M. J. D. Taylor, D. McCormick, T. Shawis, R. Impson, and M. Griffin, "Activity-promoting gaming systems in exercise and rehabilitation," *The Journal of Rehabilitation Research and Development*, vol. 48, no. 10, p. 1171, 2011.

[8] J. James, T. Ingalls, G. Qian, L. Olsen, D. Whiteley, S. Wong, and T. Rikakis, "Movement-based interactive dance performance," *Proceedings of the 14th annual ACM international conference on Multimedia - MULTIMEDIA '06*, p. 470, 2006.

[9] S. Essid, D. Alexiadis, R. Tournemenne, M. Gowing, P. Kelly, D. Monaghan, P. Daras, A. Dremeau, and N. E. O'Connor, "An advanced virtual dance performance evaluator," *2012 IEEE International Conference on Acoustics, Speech and Signal Processing (ICASSP)*, pp. 2269–2272, Mar. 2012.

[10] L. Wang, T. Gu, X. Tao, and J. Lu, "A hierarchical approach to real-time activity recognition in body sensor networks," *Pervasive and Mobile Computing*, vol. 8, no. 1, pp. 115–130, Feb. 2012.

[11] H. Junker, O. Amft, P. Lukowicz, and G. Tröster, "Gesture spotting with body-worn inertial sensors to detect user activities," *Pattern Recognition*, vol. 41, no. 6, pp. 2010–2024, Jun. 2008.

[12] D. De Rossi, R. Bartalesi, F. Lorussi, a. Tognetti, and G. Zupone, "Body Gesture and Posture Classification by Smart Clothes," *The First IEEE/RAS-EMBS International Conference on Biomedical Robotics and Biomechatronics, 2006. BioRob 2006.*, pp. 1189–1193.

[13] T. Ingalls, L. Olson, J. James, and T. Rikakis, "A gesture-driven multimodal interactive dance system," *2004 IEEE International Conference on Multimedia and Expo (ICME) (IEEE Cat. No.04TH8763)*, pp. 1579–1582, 2004.

[14] N. Vo, Q. Tran, T. B. Dinh, T. B. Dinh, and Q. M. Nguyen, "An Efficient Human-Computer Interaction Framework Using Skin Color Tracking and Gesture Recognition," *2010 IEEE RIVF International Conference on Computing & Communication Technologies, Research, Innovation, and Vision for the Future (RIVF)*, pp. 1–6, Nov. 2010.

[15] Y. Iwai, K. Watanabe, Y. Yagi, and M. Yachida, "Using Colored Gloves," pp. 662–666, 1996.

[16] H. Ukida, S. Kaji, Y. Tanimoto, and H. Yamamoto, "Human Motion Capture System Using Color Markers and Silhouette," no. April, pp. 151–156, 2006.

[17] M. Ahmad and S.-W. Lee, "Recognizing human actions based on silhouette energy image and global motion description," *2008 8th IEEE International Conference on Automatic Face & Gesture Recognition*, pp. 1–6, Sep. 2008.

[18] M. Singh, S. Member, A. Basu, and S. Member, "Human Activity Recognition Based on Silhouette Directionality," vol. 18, no. 9, pp. 1280–1292, 2008.

[19] B. Peng and G. Qian, "Binocular dance pose recognition and body orientation estimation via multilinear analysis," *2008 IEEE Computer Society Conference on Computer Vision and Pattern Recognition Workshops*, pp. 1–8, Jun. 2008.

[20] N. Kim, Y. An, and B. Cha, "Gesture Recognition Based on Neural Networks for Dance Game Contents," *2009 International Conference on New Trends in Information and Service Science*, pp. 1134–1139, Jun. 2009.

[21] L. Xia, C. Chen, and J. K. Aggarwal, "View Invariant Human Action Recognition Using Histograms of 3D Joints The University of Texas at Austin," pp. 20–27, 2012.

[22] C. N. Joseph, S. Kokulakumaran, K. Srijeyanthan, a. Thusyanthan, C. Gunasekara, and C. D. Gamage, "A framework for whole-body gesture recognition from video feeds," *2010 5th International Conference on Industrial and Information Systems*, pp. 430–435, Jul. 2010.

[23] K. Srijeyanthan, a Thusyanthan, C. N. Joseph, S. Kokulakumaran, C. Gunasekara, and C. Gamage, "Skeletonization in a real-time gesture recognition system," *2010 Fifth International Conference on Information and Automation for Sustainability*, pp. 213–218, Dec. 2010.

[24] M. Raptis, D. Kirovski, and H. Hoppe, "Real-time classification of dance gestures from skeleton animation," *Proceedings of the 2011 ACM SIGGRAPH/Eurographics Symposium on Computer Animation - SCA '11*, p. 147, 2011.

[25] P. T. Using and I. S. Method, "Picture Thresholding Using," no. 8, pp. 630–632, 1978.

[26] L. Zhang, J. Chen, Z. Zeng, and Q. Ji, "2D and 3D upper body tracking with one framework," *2008 19th International Conference on Pattern Recognition*, pp. 1–4, Dec. 2008.

[27] D. P. Huttenlocher, "Beyond trees: common-factor models for 2D human pose recovery," *Tenth IEEE International Conference on Computer Vision (ICCV'05) Volume 1*, pp. 470–477 Vol. 1, 2005.

[28] L. R. Rabiner, "A tutorial on hidden Markov models and selected applications in speech recognition," *Proceedings of the IEEE*, vol. 77, no. 2, pp. 257–286, 1989.

[29] Y. Heryadi, M. I. Fanany, A. M. Arymurthy, F. I. Komputer, and U. Indonesia, "A Syntactical Modeling and Classification for Performance Evaluation of Bali Traditional Dance," pp. 978–979, 2012.

[30] Probabilistic Graphical Models by Daphne Koller, available at https://www.coursera.org/course/pgm, December 8, 2013

[31] Pazilova M. K., Shiskkova N.V., "Kazakh Dances", Omsk Regional College of Culture and Arts, 2010.

Journal of Physics: Conference Series **495** (2014) 012037 doi:10.1088/1742-6596/495/1/012037

The development of GPS *TroWav* tool for atmospheric – terrestrial studies

W Suparta[1,*]

[1]Institute of Space Science (ANGKASA), Universiti Kebangsaan Malaysia, 43600 Bangi, Selangor Darul Ehsan, Malaysia

Email: wayan@ukm.my

Abstract. We have developed an efficient tool to process dual-frequency Global Positioning System (GPS) signals and the surface meteorological data, called the Tropospheric Water Vapor (*TroWav*) program. *TroWav* is a stand-alone program to compute atmospheric precipitable water vapor (PWV). The source of the program is developed using Matlab™ and the graphical user interface for the system was developed using a Visual Basic. The algorithms of the program capable to compute satellite elevation angle, Zenith Tropospheric Delay (ZTD), Zenith Hydrostatic Delay (ZHD), Zenith Wet Delay (ZWD) and mapping function. The tool is very practical and useful for sustainable atmospheric management.

1. Introduction

Precipitable water vapor (PWV) is one of the paramount importance parameters for atmospheric studies that can improve numerical weather prediction (NWP), nowcasting and forecasting, and viable use to study the thunderstorms, flooding, natural climate variability (global warming, climate change and El Niño–Southern Oscillation), precipitation budget, teleconnections, terrestrial coupling, etc. Global Positioning System (GPS) has an established reputation as a reliable technique of positioning equipment and has a complete turnkey capability in all areas of process equipment and space supply from integrated design through to engineering, fabrication, commissioning and testing. In contrast to atmospheric science applications, improvement of the system should be carried out to solve the data complexity with good practices and useful to support sustainable development.

GPS Meteorology technique has been developed since 1990s to measure PWV with superior in temporal and spatial resolution [1],[2]. Suparta and Alhasa [3] has proposed to estimate the PWV using an adaptive neuro-fuzzy inference system (ANFIS) due to the GPS data are not always available for a full 24-hour. However, as for a comparison of PWV from ANFIS, PWV derived from GPS data is necessary. Therefore, a suitable tool at low-cost and with high accuracy, named as the Tropospheric water vapor (*TroWav*) has been initiated since 2003 and the study was first started from the polar region. The tool is to compute PWV using GPS signals and the surface meteorological data [4]. The result of the PWV has been validated with radiosonde data [5]. Since then, the program has been updated that suitable for ENSO studies over the Borneo region [6]. In this paper, we updated the development of *TroWav* program for atmospheric-terrestrial studies in the equatorial region, so-called the *TroWav* 2.0.

2. Concept and System Design

In this state-of-art, we first develop a concept of determination of PWV from GPS signals perspective. Setup of measurement system and data processing are then constructed. Implementation of the designing system will be realized in the Matlab program. Testing and analysis result will be conducted to do the best performance to the system. In the following is describing the first step.

2.1. Deriving PWV from Ground-Based GPS Meteorology

When the radio signals traveling through the Earth's atmosphere to a receiver on the ground, it is retarded and affected significantly by the presence of free electrons in the ionosphere and the refraction (see Figure 1). The refraction of the GPS signal is induced by refractivity of gases, hydrometeors, and other particulates in the neutral atmosphere (the troposphere and stratosphere). Suparta et al. [7] described in detail propagation delays in the troposphere, where total delays can be expressed as the sum of the hydrostatic and wet components. In general, the total tropospheric delay following Davis et al. [8] and modified by Suparta [9] as

$$ZTD = m_{hyd}(\theta) \cdot SHD + m_{wet}(\theta) \cdot SWD \tag{1}$$

where ZTD is zenith tropospheric delay or equal to total tropospheric delay (D_s^{Trop}), *SHD* is the slant hydrostatic delay, *SWD* is the slant wet delay, θ is the satellite elevation angle that can be extracted from the GPS signals, and $m_{hyd}(\theta)$ and $m_{wet}(\theta)$, hydrostatic and wet mapping functions, respectively. On the other hand in equilibrium state, the ZTD can be expressed as [9]

$$ZTD = ZHD + ZWD \tag{2}$$

In practices, ZTD is computed using Modified Hopfield model [4,10]:

$$ZTD = 10^{-12} N_{j,0}^{Trop} \begin{bmatrix} 1 + 4a_j \frac{r_j^2}{2} + \left(6a_j^2 + 4b_j\right)\frac{r_j^3}{3} + 4a_j\left(a_j^2 + 3b_j\right)\frac{r_j^4}{4} + \left(a_j^4 + 12a_j^2 b_j + 6b_j^2\right)\frac{r_j^5}{5} + \ldots \\ 4a_j b_j\left(a_j^2 + 3b_j\right)\frac{r_j^6}{6} + b_j^2\left(6a_j^2 + 4b_j\right)\frac{r_j^7}{7} + 4a_j b_j^3 \frac{r_j^8}{8} + b_j^4 \frac{r_j^9}{9} \end{bmatrix} \tag{3}$$

$$a_j = -\frac{\sin\theta}{h_j} \text{ and } b_j = -\frac{\cos^2\theta}{2h_j R_E} \tag{4}$$

where for the hydrostatic component subscript j is replaced by h and for the wet component subscript j is replaced by w respectively. In (3) and (4), h_j (in meters) represent h_h ($h_h = 40136 + 148.72T$) and h_w (set to 11 km) is the effective height for the hydrostatic and wet components respectively. R_E is radius of the Earth and taken as 6378137 meters. $N_{j,0}^{Trop}$ is the total refractivity at the surface of the Earth.

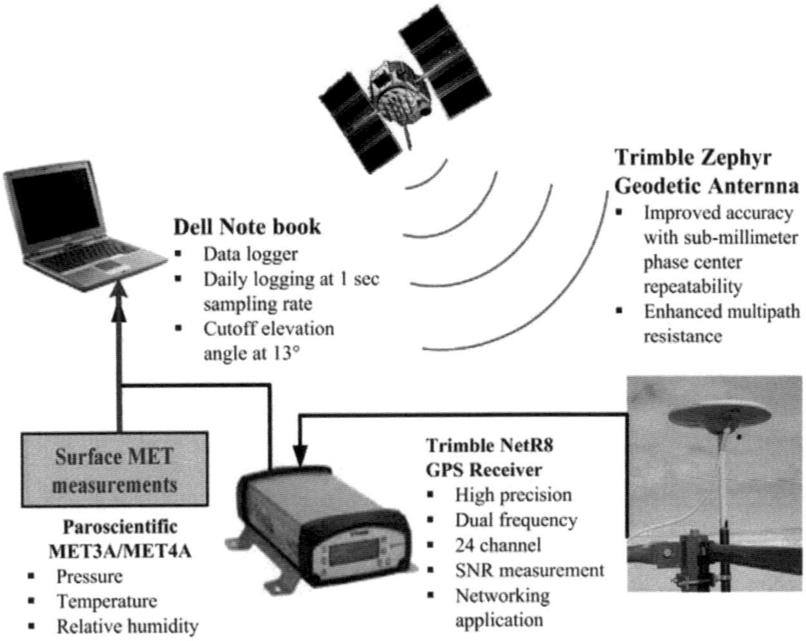

Figure 1. Ground-based GPS measurements system for PWV

The ZHD in (2) is calculated using the Saastamoinen model [11]. The model uses the surface pressure (P_s) and a correction factor to correct the local gravitational acceleration at the center mass of atmospheric column, which can be expressed as,

$$ZHD_{SAAS}(P,\lambda,h) = \frac{(2.2768 \pm 0.0024)P_s}{(1 - 0.00266\cos(2\lambda) - 0.00028\,h)} \tag{5}$$

where λ is the latitude of site (in degree), P_s is the total surface pressure (hPa) and h is the height of the site above the ellipsoid (in km) measured by GPS receiver.

The ZWD is calculated from GPS measurements by separating the ZTD from ZHD. Note that the ZTD was mapped to satellite elevation angle. PWV now can be calculated using Bevis et al. [1] equation:

$$PWV = \pi(T_m)ZWD_{GPS} \tag{6}$$

where the dimensionless $\pi(T_m)$ is a conversion factor that varies with the summation on the local climate (e.g. location, elevation, season and weather) and depend on a weighted mean temperature (T_m) as given by

$$\pi(T_m) = \left[\rho_{lw} R_v \left(k_2' + k_3 / T_m\right)\right]^{-1} 10^6 \tag{7}$$

where ρ_{lw} and R_v are the density of the liquid water and specific gas constant for water vapor, respectively. From (13), k_2' and k_3 are the refraction constants, k_2'= (22.1 ± 2.2) K mbar^{-1} and k_3= (3.739 ± 0.012) x 10^5 K^2 mbar^{-1}, respectively. The mean temperature T_m is estimated linearly by [12],

$$T_m = 70.2 + 0.72\, T_K \tag{8}$$

where T_K is surface temperature (Kelvin) measured at the site of GPS receiver. Measuring of PWV from ground-based GPS receiver is depicted in Figure 1. GPS data is recorded every 30s.

2.2. System Design

The system is constructed to properly work with faster process and accurate quantification. Data cleaning in terms of time series missing and errors were solved properly by using MATLAB. Matlab is used as a versatile tool to efficiently perform tasks, simply by providing a toolbox and can simulate different signals, and the processing involved. Figure 2 shows the flowchart for designing of PWV obtained from ground-based GPS and the surface meteorological data. GPS receiver from Trimble product will produce the raw data in *.dat file. This file can be extracted to RINEX (Receiver Independent Exchange) format (*.*obs and *.*nav) by using a Translate/Edit/Quality Check (TEQC) routine developed by UNAVCO (http://www.unavco.org). In case the GPS data is obtained from another source like SOPAC (Scripps Orbit and Permanent Array Center), we need to convert the GPS data from Hatanaka format (*.d) to RINEX using software that has been provided at the http://terras.gsi.go.jp/ja/crx2rnx.html.

As seen in Figure 2, we required the surface meteorological data (pressure (P), temperature (T) and relative humidity (H)) near or co-located with the GPS receiver to process PWV. Note that propagation signal in the ionosphere is complex than in the troposphere. On the other hand, neutral atmosphere is independent on the GPS frequencies. Before PWV is computed, the position of the reference station, which using the World Geodetic System 1984 (WGS84) frame is determined. The next task is to compute the satellite position by using Keplerian equation. Satellite elevation angle is strongly important in the processing of GPS data. This parameter is used to map the delay into zenith direction, known as the mapping function. In (1), $m_{hyd}(\theta)$ and $m_{wet}(\theta)$ are computed using the Vienna Mapping Function (VMF1) [13]. Updated Niell mapping function with VMF1 and use the same T_m equation in (8), the *TroWav* program for this update is called the *TroWav* 1.0. T_m in (8) is very suitable for mid-latitude regions. The latest tool is *TroWav* 2.0, which updating the following T_m equation that capable use for equatorial region [14]:

$$T_m = 0.83663\, T_s + 48.103 \tag{15}$$

where T_s is similar to T_K. Finally, the surface meteorology data (P, T and H) should be the same size matrix with the GPS data to produce PWV.

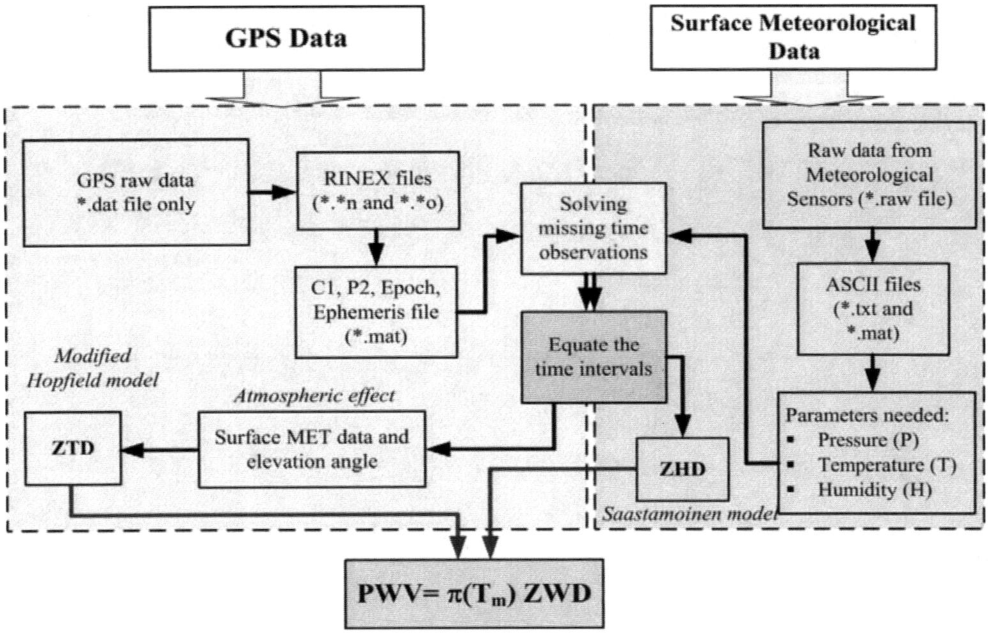

Figure 2. Design system of GPS and meteorology data processing

3. System Testing and Analysis Result

The graphical user interface (GUI) of the *TroWav* system is developed using Microsoft Visual Basic 7.0. The program is currently running in a 32-bit version of Windows. There are six menus in the top left side of Figure 3. The **GPS** menu will take input of GPS data or RINEX files (*.obs and *.nav), and converted into *.mat or ASCII files to easier processed in Matlab. In the **processing** menu, the satellite elevation angle will be processed. Source code for GPS data cleaning and to fix the data errors have been developed with Matlab program. Satellite elevation angle is computed based on the input of GPS position (latitude, longitude and ellipsoidal height), pseudorange (C1, P2), ephemeris and epoch. All these parameters are obtained from RINEX files. The input of surface meteorological data (P, T and H) is processed in the **surface meteorology** menu. In this menu, the system was also capable to plot the three of parameters. The hearth of this program is **GPS MET** menu, which compute the ZHD, ZTD, ZHD and PWV as demonstrated in Figure 4. The program is also designed capable to select the interval data based on the availability of meteorology data for possible computation of above four parameters: 30s (default), one-min, 10-min, one-hour and 3-hour. Explanation on how to use the *TroWav* program is compiled in the **about** menu.

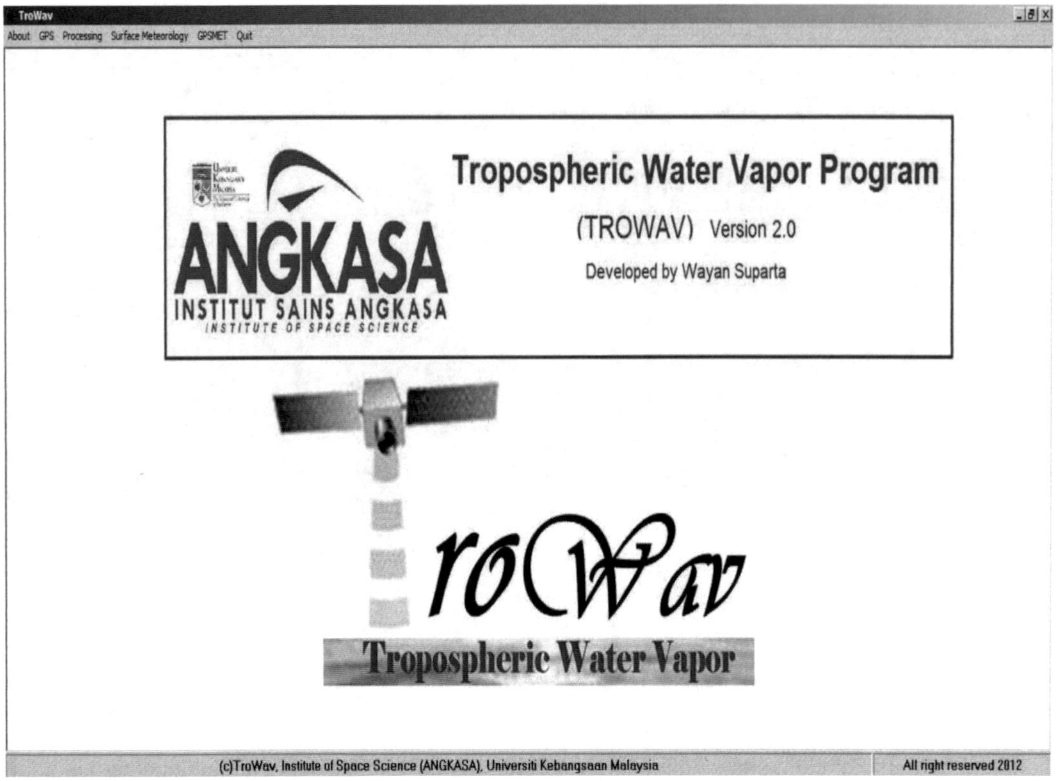

Figure3. Front menu of *TroWav* system

The testing result from the *TroWav* screen is demonstrated in Figure 4. Example data used here is collected from Casey station (CAS1) in Antarctica with period of testing is on September 7, 2013. This station is managed by Australian Antarctic Division. From the figure, the four GPS meteorology parameters (ZHD, ZTD, ZWD and PWV) showed a similar pattern. The processing results are in 30 seconds resolution (default), means that the daily size of data is 2880. The ZHD demonstrated very clear and smooth. Their pattern is clearly dependent on the surface pressure and position of GPS receiver. For the three another parameter showed noisy from 00:00 to 08:00 UT, which possibly degrading the GPS signals due to atmospheric conditions. On the other hand, the ZTD pattern closely followed the ZHD trend which clarified that the hydrostatic part is contributed about 90% in the total of ZTD component, and the rest is contribution from the wet part. Note that not all the GPS meteorology parameters will show a similar pattern or trend as in Figure 4. This depends on the atmospheric conditions of the station, geography and typical weather or current climate.

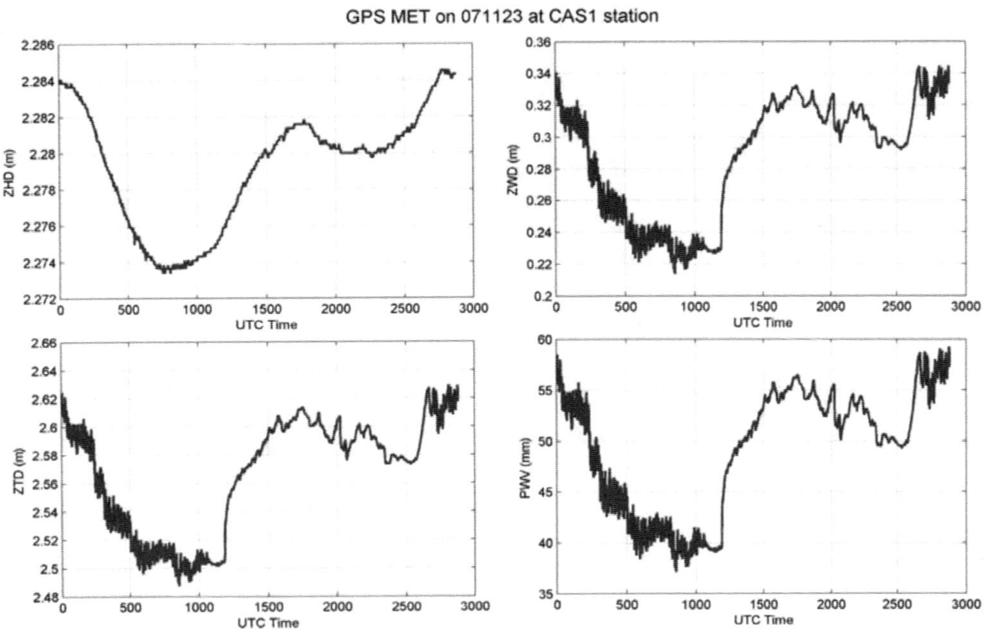

Figure 4. GPS meteorology parameters from *TroWav* 2.0 result for CAS1 station

4. Conclusion

In this paper, a promising tool with effective-cost to process GPS and the surface meteorological data to produce PWV as a climate indicator for the atmospheric and terrestrial studies was developed. The program, namely *TroWav* is capable to plot the processing data (ZHD, ZTD, ZWD and PWV) and also save the data and the figure for future use. The system is currently running under a 32-bit of Windows platform. By this tool, GPS user capable to improve the positioning accuracy via atmospheric effects quantification as well as utilize the GPS signals for sustainable development.

Acknowledgments

This work was partially supported by the Ministry of Education Malaysia (MOE) under grant FRGS/2/2013/TK03/UKM/02/3, Ministry of Science, Innovation and Technology Malaysia (MOSTI) under grant 06-01-02-SF0808 and Academy Science of Malaysia under grant PKT 1/2003. Thank to Kemal Maulana Alhasa in helping the development of *TroWav* GUI.

References

[1] Bevis M, Businger S, Herring T A, Rocken C, Anthes R A and Ware R H 1992 GPS meteorology remote sensing of atmospheric water vapor using the global positioning system *J.Geophys. Res.* **97** 15787

[2] Rocken C, Ware R, Van Hove T, Solheim F, Alber C, and Johnson J 1993 Sensing atmospheric water vapour with the global positioning system *Geophys. Res. Lett.* **20** 2631

[3] Suparta W and Alhasa K M. 2013 Estimation of precipitable water vapor using an adaptive neuro-fuzzy inference system technique *Information and Communication Technology, Lecture Notes in Computer Science* **7804** 214

[4] Suparta W, Abdul Rashid ZA, Mohd Ali MA, Yatim B and Fraser G J 2008 Observation of Antarctic precipitable water vapor and its response to the solar activity based on GPS sensing *J. Atmos. Sol. Terr. Phys.* **70** 1419

[5] Suparta W 2010 Using a global positioning system to estimate precipitable water vapor in Antarctica *Polar Geog.* **33** 63

[6] Suparta W, Jit Singh M S, Mohd Ali, M A, Yatim B, Mohd Yatim A N 2011 GPS water vapor monitoring and TroWav updated for ENSO studies *Proc. Int. Conf. on Instrumentation, Communications, Information Technology and Biomedical Engineering (ICICI-BME)* p 35

[7] Suparta W, Ali MAM, Yatim B, Misran N 2007 Remote sensing of antarctic atmospheric water vapour using ground-based gps meteorology *Proc. 5th Student Conference on Research and Development 2007 (SCOReD 2007)* p 1

[8] Davis J L, Herring T A, Shapiro I I, Rogres A E E and Elgered G 1985 Geodesy by radio interferometry: Effects of atmospheric modeling errors on estimates of baseline length *Radio Sci.* **20** 1593

[9] Suparta W 2012 Detection of a katabatic event with GPS meteorology measurements at Scott Base Antarctica *Meteorol. Atmos. Phys.* **155** 153

[10] Hofmann-Wellenhof B, Lichtenegger H, Collins J 2001 *Global Positioning System: Theory and Practice* 7th rev. ed., Springer Verlag

[11] Saastamoinen J 1972 Introduction to practical computation of astronomical refraction *Bulletin Geodesique* **106** 383

[12] Bevis M, Businger S, Chiswell S 1994 GPS meteorology: Mapping zenith wet delays onto precipitable water *J. Appl. Meteorol.* **33** 379

[13] Boehm J, Werl B, Schuh H 2006 Troposphere mapping functions for GPS and very long baseline interferometry from European Centre for Medium-Range Weather Forecasts operational analysis data *J. Geophys. Res.* **111** B02406

[14] Suparta W, Iskandar A 2013 Modeling of weighted mean temperature over the Western Pacific region to estimate GPS PWV *Proc. 2013 IEEE Inter. Conf. Space Sci. Comm. (ICONSPACE)* p 190

Journal of Physics: Conference Series **495** (2014) 012038 doi:10.1088/1742-6596/495/1/012038

Analysis distribution of galactic cosmic rays particle energy with polar orbit satellite for Geant4 application

W Suparta[1,*] **and W S Putro**[1]

[1]Institute of Space Science (ANGKASA), Universiti Kebangsaan Malaysia, 43600 Bangi, Selangor Darul Ehsan, Malaysia.

Email: wayan@ukm.my

Abstract. Galactic Cosmic Rays (GCR) are photon waves originating from astrophysical sources which traverse through the interstellar/interplanetary medium and reaching the terrestrial atmosphere. The energies of Galactic Cosmic Ray particles up to and exceeding 10^{12} eV, and this spectrum are peaked around 1 GeV. The National Aeronautics and Space Administration (NASA) provide satellite mission for monitoring the energy GCR particles in polar orbit, so-called the ACE and OMNI. In this paper, we analyze results from measurement error of GCR sensor. The error result is obtained by comparing the measurements from GCR sensor with ground-based neutron monitors at Bartol University. The measurements were taken for two periods during a Solar Particle Event (SPE) maximum on 14 July 2000 and 28 October 2003. The largest value of measurement error from GCR sensor in this study is OMNI satellites. After the error results were obtained, they were applied into Geant4 simulation. This simulation shows the shape of particle energy distribution of GCR sensors. The simulation has been tested and can be operated very well under Linux based platform.

Keywords: *Galactic Cosmic Rays, Measurement error, Geant4, Polar orbit, Simulation*

1. Introduction

The radiation of space radiation environment is composed different particles (neutron, electron, protons and heavier charged particle), with large energy range up to 10^{20} eV. Among the crucial components are ionizing charged particles, i.e. the particle trapped in the radiation belts, the solar particle event (SPE) and galactic cosmic ray (GCR) radiation [1]. GCRs are high energy charged particle that originated from the outside of solar system, which strongly causes Single Event Effect (SEE) in microelectronics and photonics [2]. The phenomena of GCR radiation will be peaked at SPE activity. GCRs dominated the effect in terms of linear energy transfer (LET) in satellite application. These effects on spacecraft system and instruments near Earth orbit are difficult quantified due to highly energetic particles in the heavy ion component compared to polar orbiting satellites [3].

Boscherini et al [4] discussed more details about the radiation damage of electronic in space, and Tajima [5] also discussed a detector for cosmic rays measurement on ground and in space, whereas in his study emphasizes error result of GCR sensor on satellite based by comparing this sensor with detector from ground-based. Our work progress is to analyze the energetic particles for developing a GCR sensor for equatorial inclination satellite that later can be operated in Low-Earth Orbit (LEO).

In this paper, we started first to study the distribution of galactic cosmic rays particle energy during a SPE maximum occurred on 14 July 2000 and 28 October 2003. This study is very important because satellite is a complex device with electronic components. If the satellite is attached by maximum radiation then the electronic component would be damage by LET from space, then the sensor accuracy will be affected seriously. In line with above purpose, Geant4 (GEometry ANd Tracking) is proposed to simulate the passage of particle through matter.

2. Methodology

2.1 Data Collection

Data used in this study were taken from ACE and OMNI satellites that can be accessed through http://cdaweb.gsfc.nasa.gov/cgi-bin/eval2.cgi, and ground-based neutron monitors at Bartol University is downloaded from http://neutronm.bartol.udel.edu/~pyle/bri_table.html. ACE satellite is placed at an altitude of 1.5 million km with an inclination of (i~29°). ACE satellite equipped with a sensor called CRIS (Cosmic Ray Isotope Spectrometer) recorded isotopes of galactic cosmic ray nuclei from He to Zn over the energy range from ~100 to 600 MeV/nucleon [6]. For OMNI satellite, it is placed at an altitude of 600 km with an inclination of (i~90°) equipped with Energy Proton Fluxes and Electron Detector sensor [6]. ACE and OMNI satellites record data every 1 hour, and recording of ground-based neutron monitors is also similar interval, which is every 1 hour. In this study, we are comparing the distribution of GCR's particle for 2 periods (14 July 2000 and 28 October 2003) from satellites and ground-based neutron monitors (see Table 1).

Table 1. Ground-based neutron monitors data from Bartol University at four selected stations

Date	McMurdo			Newark			South Pole			Thule		
	Corr	Uncorr	Press	Corr	Uncorr	Press	Corr	Uncorr	Press	Corr	Uncorr	Press
:	:	:	:	:	:	:	:	:	:	:	:	:
:	:	:	:	:	:	:	:	:	:	:	:	:
2000/1	8462	8029	735.3	3056	3033	760.8	8859	9331	504.8	504.8	3822	756.5
2000/2	8485	8043	735.4	3030	3005	760.9	8892	9358	504.9	3927	3817	756.9
2000/3	8458	8010	735.5	3038	3010	761.0	8894	9350	505.0	3938	3813	757.2
:	:	:	:	:	:	:	:	:	:	:	:	:

2.2 Error Analysis

Error analysis is one of the analysis techniques that used in statistic and mathematics. Nowadays, it has been widely used in every case study of measurement. Some errors can be divided into two classes: systematic and random. Systematic errors are errors which tend to shift all the measurements in a systematic way, so their mean value is displaced, and random errors are errors which fluctuate from one measurement to another. In this paper, we choose this method for error analysis of each sensor satellite. Some data set of repetitive measurements is often expressed as a single representative number called the mean or average. Below is the formulas that commonly used in the analysis of measurements errors, as described in detail by Hongchem [7]. The mean ($\bar{x}$) is the sum of individual measurements (x_i) divided by the number of measurements (n).

$$\bar{x} = \sum_{i=1}^{n} \frac{x_i}{n} \tag{1}$$

The average deviation, $\Delta\bar{x}$ is used when a data set contains less than 5 repetitive measurements. A small value of average deviation indicates data points clustered is closely around the mean and good precision.

$$\Delta\bar{x} = \pm \sum_{i=1}^{n} \frac{|x_i - \bar{x}\,|}{n} \quad (\Delta\bar{x}, N \leq 4) \tag{2}$$

The absolute value is taken of the deviation from the mean $|x_i - \bar{x}\,|$ and no information is gained about the direction of the error. The relative value of average deviation is the average deviation divided by the average and then expressed as a percentage:

$$|x_i - \bar{x}\,| = \frac{\Delta\bar{x}}{\bar{x}} \times 100\ \% \tag{3}$$

The estimated standard deviation (*S*) is used to express the precision of the measurements with the number of degrees of freedom (*N−1*) expressed as

$$S = \pm\sqrt{\sum \frac{(x_i - \bar{x}\ )}{N-1}} \tag{4}$$

The accuracy of a result can be quantified by calculating the percentage of error. The percentage of error can only be found if the true value is known. Although the percentage error is usually written as an absolute value, it can be expressed as a negative or positive sign to indicate the direction of error from true value.

$$\%\,Error\ =\ \frac{(x_T - x_E)}{x_T} \times 100\ \% \tag{5}$$

where x_T and x_E are true and experimental values, respectively.

3. Results and Discussion

Figures 1 and 2 show the comparison distribution particle of GCR, simulated with Geant4 on 14 July 2000 and 28 October 2003 from ground-based neutron monitors at Bartol University for the station of McMurdo, Newark, South Pole, and Thule. As shown in the figure, the distribution of particle was most effectively influenced by SPE phenomena, where GCR's energy can peak to around 1 GeV during the increased of SPE. GCR's activities on 28 October 2003 is super active than those of SPE on 14 July 2000, which produce solar eruption maximum, as indicated by a sudden burst of solar particles. Based on these phenomena, our next task is to create a suitable simulation of GCR's particles energy activity of SPE on 14 July 2000 and 28 October 2003 for ACE satellite with Geant4.

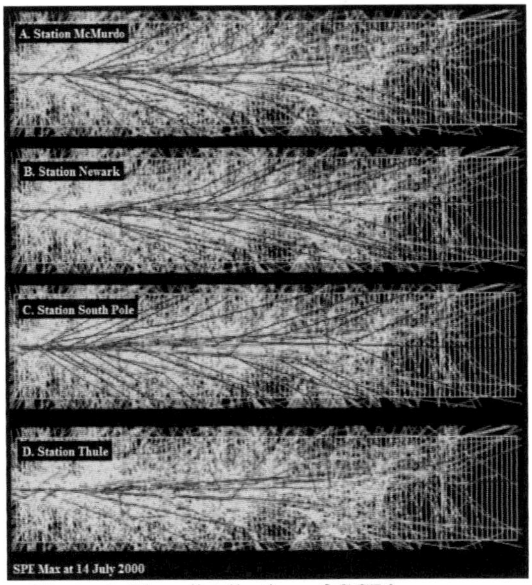

Figure 1. Flux distribution of GCR's energy particle for SPE max on 14 July 2000

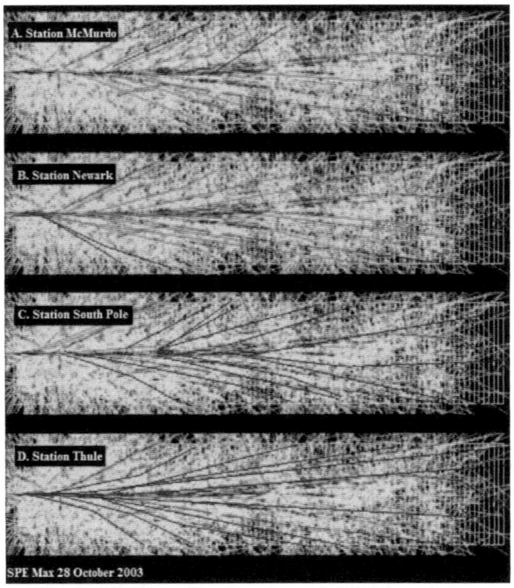

Figure 2. Flux distribution of GCR's energy article for SPE max on 28 Oct 2003

Figures 3 and 4 show the ACE satellite passing through ground-based neutron monitors, recorded at Bartol University region on 28 October 2003. Figure 4 shows the distributions of GCR's energy at Thule area of 287 MeV during SPE maximum on 28 October 2003 from ACE satellite that has a CRIS sensor. Although, ACE satellite is placed at an altitude of 1.5 million km (i~29°), it has an accurate measurement of GCR's. The sensor capable to detect distribution energy for Ni, Ne, Proton, Electron and Neutron fluxes. The error values of CRIS for Thule area are 1.39 % during SPE maximum on 28 October 2003. Table 2 and Table 3 show that OMNI satellite has bigger error compared to ACE satellite. OMNI satellites are placed at an altitude of 600 km (i~90°), which possibly limited to detect the particle distribution. The different orbit between ACE and OMNI is more than 1 million km.

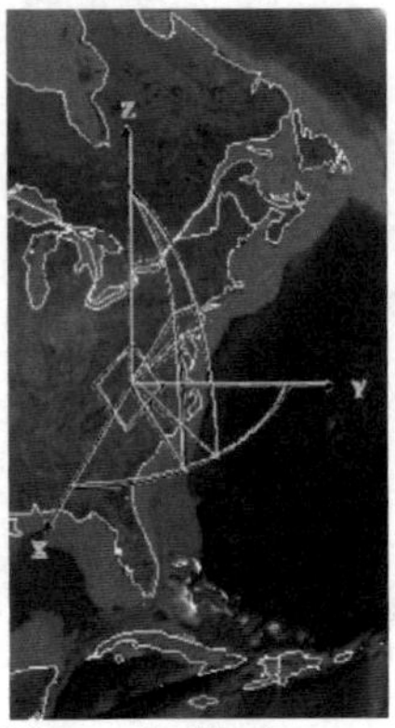

Figure 3. Thule area

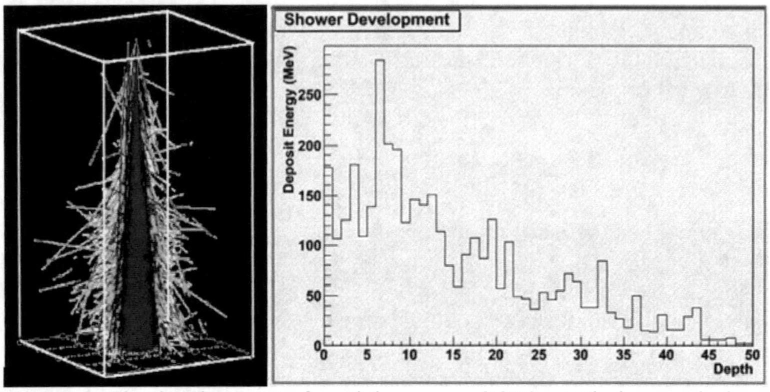

Figure 4. The development of particle energy GCR's at Thule area during SPE maximum on 28 October 2003 by ACE satellite

Table 2. Error analysis of SPE maximum for 14 July 2000.

Station	ACE Satellite			OMNI Satellite		
	deviation (S)	Mean	Error (%)	Mean	deviation (S)	Error (%)
McMurdo	$\pm$0.35	16.37	1.32%	24.42	$\pm$0.41	2.29%
Newark	$\pm$0.28	15.81	1.23%	21.53	$\pm$0.55	2.12%
South Pole	$\pm$0.67	25.43	1.63%	37.86	$\pm$0.72	2.53%
Thule	$\pm$0.34	16.34	1.31%	24.23	$\pm$0.48	2.38%

Table 3. Error analysis of SPE maximum for 28 October 2003.

Station	ACE Satellite			OMNI Satellite		
	deviation (S)	Mean	Error (%)	Mean	deviation (S)	Error (%)
McMurdo	$\pm$0.42	32.36	1.24 %	43.39	$\pm$0.56	2.38%
Newark	$\pm$0.36	25.21	1.36 %	35.32	$\pm$0.50	2.24%
South Pole	$\pm$0.77	38.48	1.67 %	45.15	$\pm$0.87	2.68%
Thule	$\pm$0.49	26.24	1.39 %	36.28	$\pm$0.58	2.33%

Table 4 and Table 5 show ACE satellite has a good resolution of area measurement compared to OMNI satellite, which indicated that orbiting area is very important to detect radiation. ACE is a complex satellite that can detect distribution energy of GCR particle; not only Proton and Electron fluxes, but all component particles of GCR. With Geant4 simulation, we can detect that the radiation area has been affected by GCR's. From the table, maximum radiation occurred on 14 July 2000 and 28 October 2003 is at South Pole area, which is located in highly latitude. This location is very dangerous area during SPE because the particle energy deposited by GCR's was more than 200 MeV, and this radiation could be a cancer risk for human [9].

Table 4. Simulation of Geant4 for radiation area of GCR's and deposit energy on 14 July 2000

Station	ACE Satellite		OMNI Satellite	
	Area (X, Y, Z) km	Energy (MeV)	Area (X, Y, Z) km	Energy (MeV)
McMurdo	15, 17, 40	270	11, 12, 35	296
Newark	16, 16, 42	258	10, 13, 32	262
South Pole	14, 16, 43	273	12, 15, 36	298
Thule	15, 13, 41	268	13, 14, 34	277

Table 5. Simulation of Geant4 for radiation area of GCR's and deposit energy on 28 October 2003

Station	ACE Satellite		OMNI Satellite	
	Area (X, Y, Z) km	Energy (MeV)	Area (X, Y, Z) km	Energy (MeV)
McMurdo	18, 17, 48	290	14, 13, 33	296
Newark	17, 18, 44	271	13, 11, 37	285
South Pole	16, 17, 50	293	12, 14, 38	307
Thule	15, 18, 49	287	11, 13, 36	291

4. Conclusion

We have successfully to do an analysis of error measurement of GCRs particle for polar orbit satellite. A simulation of GCR's energy particle during a SPE maximum was also performed with Geant4 software on major solar events occurred at 14 July 2000 and 28 October 2003. OMNI satellite found an error of 2.86% with energy particle is 307 MeV for South Pole station, and for 14 July 2000, the error values are 2.53% with energy particle is 298 MeV. During SPE maximum, a small area of Antarctica (South Pole and McMurdo) has high deposit energy from GCR of 280 MeV for 28 October 2003 and more than 260 MeV for 14 July 2000. The maximum area during distribution of GCR's energy particles are 16, 17, and 50 km on 28 October 2003 for South Pole station. Then 14, 16, and 43 km for 14 July 2000, which indicated that the event on 28 October 2003 is stronger than those of 14 July 2000. The largest value of measurement error from GCR sensor in this study is OMNI satellites.

Our next study is to do a modeling of GCR radiation for a satellite application in equatorial region like Near Equatorial Orbit (NEqO). This is in order to generate a real-time fluxes radiation of GCR prediction, as a part of an early warning system development.

5. Acknowledgments

This research is supported by Ministry of Education Malaysia (MoE) under grant FRGS/2/2013/SG02/UKM/02/3 and Ministry of Science, Innovation and Technology Malaysia (MOSTI) under grant 06-01-02-SF0808. The authors would like thank to National Aeronautics and Space Administration (NASA), Bartol University for providing data used in this study, and Geant4 forums for advising in Geant4 programming under Linux platform.

References

[1]. Friedberg W and Copeland K 2011 Ionizing radiation in earth's atmosphere and in space ear earth *Federal Aviation Administration Final Report* 32pp.

[2]. Bourdarie S and Xapsos X 2008 The near – earth space radiation environment *IEEE Transactions on Nuclear Science* **55** 1810

[3]. Benton E R and Benton E V 2001 Space radiation dosimetry in low – earth orbit and beyond *Nuclear Instruments and Methods In Physics* B **184** 255

[4]. Boscherini M et al. 2003 Radiation damage of electronic components in space environment *Nuclear Instruments and Methods in Physics Research* A **514** 112

[5]. Tajima H 2007 Detector for cosmic rays on ground and in space *Nuclear Instruments and Methods In Physics* A **581** 12
[6]. eoPortal 2013 *Satellite Missions Database* https://directory.eoportal.org/web/eoportal/satellite-missions
[7]. Hongchem 2013 *Analysis of Errors* https://eee.uci.edu/programs/hongchem/RDGerroranal.pdf
[8]. Trauth M H 2010 *MATLAB® Recipes for Earth Sciences, third edition* Springer-Verlag New York p 235
[9]. A Cucinotta F and Durante M 2006 Cancer risk from exposure to galactic cosmic rays: implications for space exploration by human beings *The Lancet Oncology* **7** 431

Journal of Physics: Conference Series **495** (2014) 012039 doi:10.1088/1742-6596/495/1/012039

Forecasting the Earth's Trapped Particle Distribution Using Hierarchical Bayesian Spatio Temporal Model

Wayan Suparta[1,*], **Gusrizal**[1]

[1]Institute of Space Sciences (ANGKASA), Universiti Kebangsaan Malaysia, 43600 Bangi, Selangor Darul Ehsan, Malaysia

E-mail: wayan@ukm.my

Abstract. We employed the Hierarchical Bayesian spatio temporal (HBST) Gaussian Process (GP) model for forecasting the distribution of the Earth's trapped particle. The model was applied in the South Atlantic Anomaly (SAA) region. Data from 1-30 January 2000 of >30 keV electron flux acquired by National Oceanic and Atmospheric (NOAA) 15 satellite was carried to model. The purpose was to forecast the flux value on 31 January 2000. Gridding process of 10x10 lot-lan was performed after cleaning and log transforming data. The HBST GP model was undertaken by implementing the Monte Carlo Markov Chain (MCMC) method. The forecasting result was interpolated by using Kriging technique to draw the distribution map of particle flux. Statistical validation represented by mean square error, root mean square error, mean absolute error, mean absolute percentage error, bias, relative bias, and mean relative separation shows good indicators. The visual validation also figured a quite similarity with NOAA's map that the model capable to forecast the particle flux.

1. Introduction

Trapped particle is one of major radiation in the space environment. It is produced by solar flares and CMEs and flows out to the Earth's atmosphere by the solar wind [1]. Due to geomagnetic field line the charged particle is trapped into two areas of Van Allen radiation belt: the inner radiation belt and the outer radiation belt. Trapped particle could cause harmful effects to the spacecraft that pass through their region especially for low earth orbit (LEO) satellite [2]. Therefore, it is important to model the distribution of Earth's trapped particle. There are several trapped particle models that have been developed, mostly are based on physics - magneto hydro dynamics (MHD) frame, such as SWMF/BATS-R-US with RCM [3], Fok Ring Current [4], Plasma sphere [5], CIMI [6]. The most used and the stable one, the AE-9/AP-9 [7] is run in a statistical modeling framework.

In this work, we choose to employ a statistical model, named hierarchical Bayesian spatio temporal model (HBST). The unique of this model because it works on a geographical coordinate rather than the (L, B) coordinate like the others. Although the approach of representing the trapped particle in magnetic coordinate is preferred in scientific application, the necessity of simple displayed model for the end user like satellite operator and designer is taking into account [8]. In addition, the model can be employed for both proton and electron in every condition of solar activity. The model can also perform a long time forecast like daily or weekly of solar trapped particle flux value. Our objective of this work is attempted to build the trapped particle distribution forecasting system for satellite operator and designer.

In this paper, the model developed is applying for the South Atlantic Anomaly (SAA) region, before applied in the whole area of Earth atmosphere. SAA is an atmosphere region centered over Brazil sky which lies from -90° to 40° of longitude and from 0° to -50° of latitude, at the altitude of ~500 km. The SAA is a main source of trapped particle in the inner radiation belt. We use the data acquired by National Oceanic and Atmospheric Administration (NOAA) 15 satellite, a polar orbit satellite with inclination 98.5° and altitude ~807 km. Based on this condition, our distribution forecasted is valid in the altitude of 300-800 km and in the shell of $L \leq 1$. We also perform the forecast on a daily basis.

2. Methodology

2.1. NOAA Data

The NOOA 15 data (http://satdat.ngdc.noaa.gov/sem/poes/data/avg/txt/) consists of 41 columns that contain time, locations, and flux values of observed particles, with the energy range from 30 keV to 200 MeV. For simplicity purpose, we chose the lowest energy level, >30 keV electron flux. The details explanation for the data can be access through NASA - National Geophysical Data Center (NGDC) (http://satdat.ngdc.noaa.gov/sem poes/docs/readme_16s_ascii.txt). In this work, we use of electron data, based on the location of the SAA. Once data sorted, then the next phase are data cleaning, performing logarithmic transformation to stabilize the data distribution, and finally gridding them.

The important step in the data process is gridding the data. NOAA 15 completes its orbit in about 1 hour and 40 minutes. Due to that fact, a point of observation could not be observed in a long period of time. We have calculated that a point in a grid of 1x1 longitude and latitude will be observed again in about two months, and this make impossible to do a forecast due to the lack of data. Therefore, gridding data is proposed to solve this problem. We use the 10x10 gridding system (see Figure 1).

Each data in a grid was averaged with its counterparts and put the average value in the center of the grid. The new value and the new location then become a representative of a grid. We employed this procedure to 65 grids of 10x10 sizes. For validation purpose, we chose 15 points randomly as the validation points, and the rest as the model input as shown in Figure 1.

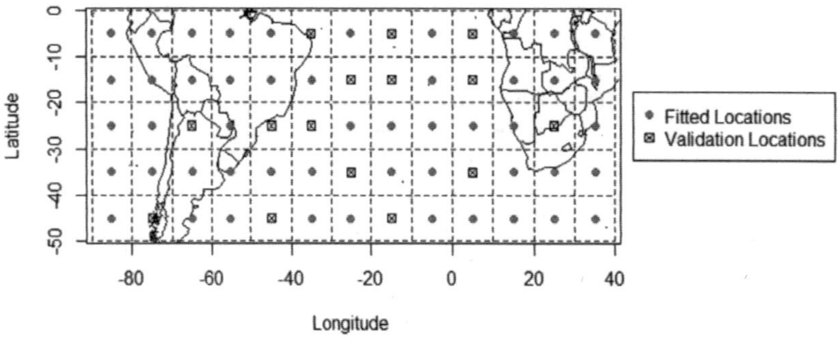

Figure 1. The validation and fit points in SAA with 10x10 grids

The validation process was conducted by comparing the values in validation points obtained by the forecasting process with the original values obtained by observation. Data from 1-30 January 2000 has chosen as the mode fitting. Our target is to forecast the flux value on 31 January 2000 with the model obtained.

2.2. Hierarchical Bayesian Spatio Temporal Model

HBST model is a statistical modeling technique that deals with space time modeling in Bayesian approach. Considering Gelfand [9], the HBST has a structure:

First stage : Data model $[Z|E, \theta]$

Second stage : Process model $[E|\theta]$

Third stage : Parameter model $[\theta]$

where the data defined by Z, the (hidden) process is defined by E , and the unknown parameters are specified by θ.

First of all, we define the generic notations used throughout this paper, and the formula related to them [10]. Let $Z(s_i, t)$ is denoted the logarithmic value of SAA's > 30 keV electron flux at site $\boldsymbol{s}_i$ and time t, i = 1,..., n and t=1,..., T. $E(s_i, t)$ is as a true value corresponding to $Z(s_i, t)$. We put T = 30 at this work to accommodate selected data from 1-30 January 2000 as the model input. We put 31 January 2000 as our forecasting target, notified as T+1. We choose to define both Z and E in vector notation as $Z_t = (Z(s_i, t), ..., Z(s_n, t))'$ and $E_t = (E(s_i, t), ..., E(s_i, t))'$. Therefore the HBST GP model [11] is expressed by:

$$Z_t = E_t + \epsilon_t \tag{1}$$

$$E_t = X_t\beta + \eta_t \tag{2}$$

with $\epsilon_t = \left(\epsilon(s_{1,}t), ..., \epsilon(s_{n,}t)\right)' \sim N(0, \sigma_\epsilon^2 I_n)$ is an error process. Included in this term, the σ_ϵ^2 is a nugget effect, and is homogeneous in space and time, whereas the I_n is the n x n identity matrix. The covariates influencing the Z value is denoted by X_t with n x p matrix size, and $\beta = (\beta_1, ..., \beta_p)'$ is the p x 1 vector of X_t, respectively. NOAA 15 data does not provide any covariate data for the trapped flux, so the intercept value is employed in Equation 2. Furthermore, the $\eta_t = \left(n(s_1, t), ..., n(s_n, t)\right)' \sim N(0, \Sigma_n)$, is a spatially correlated error. Included in, the $\Sigma_n = \sigma_\eta^2 S_n = \sigma_\eta^2 \kappa(s_i, s_j; \phi, \upsilon)$ is a variance-covariance matrix, and have dimension n x n, $i, j = 1, ..., n$. Afterward, the σ_η^2 is the site invariant common variance. The $\kappa(.; \phi, \upsilon)$ expresses the spatial correlation matrix with spatial decay ϕ, and smoothness parameter υ. The error parametersϵ_t, and η_t, are independent each other. For future reference, **z** is denoted all observed data, **x** is denoted all covariates data, and **E** represent all augmented values. Lastly, we use **θ** symbol to assert all parameters used ($\boldsymbol{\theta} = (E, \beta, \sigma_\epsilon^2, \sigma_\eta^2, \emptyset)$).

For forecasting purpose, the HBST GP model at any observed point s_i on T+1 day is expressed in terms below:

$$\mathrm{Z}(\mathrm{s_i}, \mathrm{T}+1) = \mathrm{E}(\mathrm{s_i}, \mathrm{T}+1) + \epsilon(\mathrm{s_i}, \mathrm{T}+1) \tag{3}$$

$$\mathrm{E}(\mathrm{s_i}, \mathrm{T}+1) = \mathrm{x}'(\mathrm{s_i}, \mathrm{T}+1)\beta + \eta(\mathrm{s_i}, \mathrm{T}+1) \tag{4}$$

with posterior predictive distribution of $\mathrm{Z}(\mathrm{s_i}, \mathrm{T}+1)$ given **z** is denoted by:

$$\pi(\mathrm{Z}(\mathrm{s_i}, \mathrm{T}+1)|\mathbf{z}) = \int \pi(\mathrm{Z}(\mathrm{s_i}, \mathrm{T}+1)|\boldsymbol{\theta}, \mathbf{E}, \mathrm{E}(\mathrm{s_i}, \mathrm{T}+1), \mathbf{z})\, \pi(\mathrm{E}(\mathrm{s_i}, \mathrm{T}+1)|\boldsymbol{\theta}, \mathbf{z})$$

$$\pi(\boldsymbol{\theta}, \mathbf{E}|\mathbf{z})d\mathrm{E}(\mathrm{s_i}, \mathrm{T}+1)d\mathbf{E}d\boldsymbol{\theta} \tag{5}$$

Finally, to summarize this method, we perform the following algorithm to predict the $\mathrm{Z}(\mathrm{s_i}, \mathrm{T}+1)$, by using Monte Carlo Markov Chain (MCMC)-Gibbs sampling method with j iterations:

1. Draw a sample $\boldsymbol{\theta}^{(j)}$, and $\mathbf{E}^{(j)}, j \geq 1$ from Equation 5.
2. Draw $\mathrm{E}^{(j)}(\mathrm{s_i}, \mathrm{T}+1)$ from $\mathrm{N}\left(\mathrm{x}_{\mathrm{T}+1}'\beta^{(j)}, \sigma_\eta^{2(j)}\right)$.
3. Finally draw $\mathrm{Z}^{(j)}(\mathrm{s_i}, \mathrm{T}+1)$ from $\mathrm{N}\left(\mathrm{E}^{(j)}(\mathrm{s_i}, \mathrm{T}+1), \sigma_\epsilon^{2(j)}\right)$.

Once the value of Z at T+1 day obtained, then the result is displaying in a geographic map by using Kriging interpolation technique. Detail of the Kriging interpolation technique applied on the distribution of trapped particle over SAA region can be found at Suparta et al. [12]. Figure 2 is the flow chart of this work methodology.

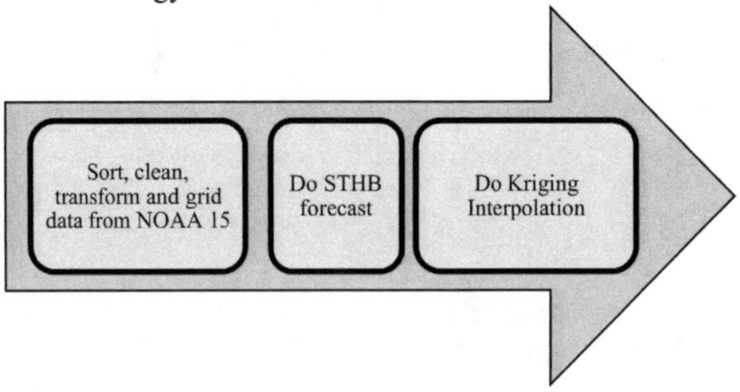

Figure 2. The flow chart of trapped particle flux distribution forecasting

3. Result and discussion

3.1. Statistical Validation

Statistical analysis was performed through seven validation parameters: mean squared error (MSE), root mean squared error (RMSE), mean absolute error (MAE), mean absolute percentage error (MAPE), bias (BIAS), relative bias (rBIAS) and the mean relative separation (rMSEP). Details formulas of the validation parameters can be found at Bakar and Sahu [13]. Table 1 shows the results of validation process.

Table 1. Validation result of GP model

MSE	RMSE	MAE	MAPE	BIAS	rBIAS	rMSEP
0.5577	0.7468	0.6691	19.3589	0.1796	0.0457	0.4969

From Table 1, we can conclude that the forecast values and the observed data have a high degree of similarity since the validation parameters produced the small numbers. This allows us to draw the conclusion that the HBST GP model can work well and give an accurate result in the term of forecasting.

3.2. Visual Analysis

The visual analysis was performed by comparing the Kriging results, both estimates and variance with several patterns of map distribution, i.e. daily data plot, NOAA's distribution map, Kriging interpolation on the raw data and Kriging interpolation on the grid data. Figures 3a and 3b present the electron >30 keV daily plot on 31 January 2000 and its distribution map drawn by NOAA, respectively. Figures 4, 5, and 6 present the distribution maps of electron estimation and variance obtained by ordinary Kriging method on STHB-GP model forecast result, empirical, and grid data, respectively. We could understand from Figure 3a that the data collected is so lack, and is a challenging work to deal with it. As for Figure 3b, nine days data are collected by NOAA to get the distribution map. On the other hand, we could produce a 1-day forecast value of SAA electron flux and display in a distribution map as in Figure 4a, while Figure 4b refers to the variance of Kriging interpolation on the forecasting result. When we compare the estimation of Kriging from Figures 4a, 5a, and 6a, we could see that the pattern of forecasting estimate (Figure 4a) is slightly different from the others and the pattern is more widened rather than concentrated. It is also reflected a decreasing value of flux maximum log in a significant rate, that is, from > 6 in the observed data to <5 in the forecast estimate. This visual analysis showed an opposite result with the statistical validation which showed very good results. Gridding process is probably as major factor to this problem as a consequence of averaging and centering.

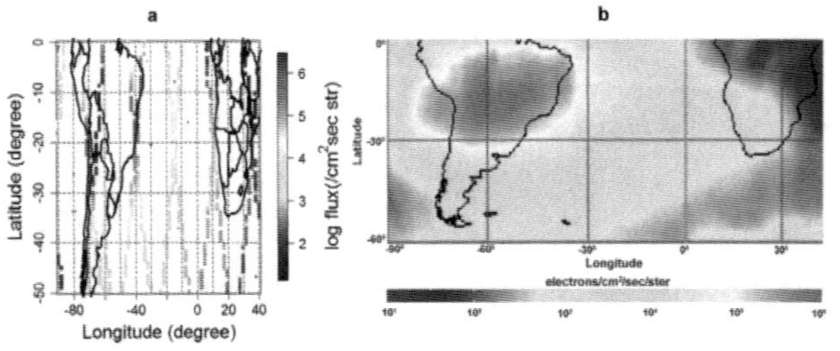

Figure 3. Mep0e1 data on 31 January 2000 for (a) raw data and (b) NOAA's SAA on 23-31 January 2000

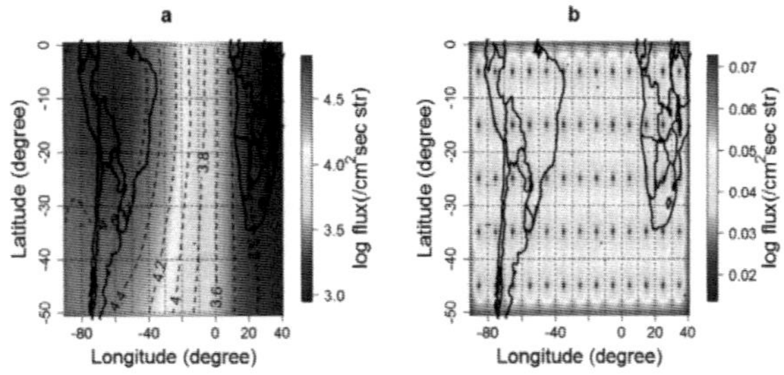

Figure 4. Flux Distribution map of >30 keV electron over SAA region on 31 January 2000, for (a) forecast estimation and (b) forecast variance

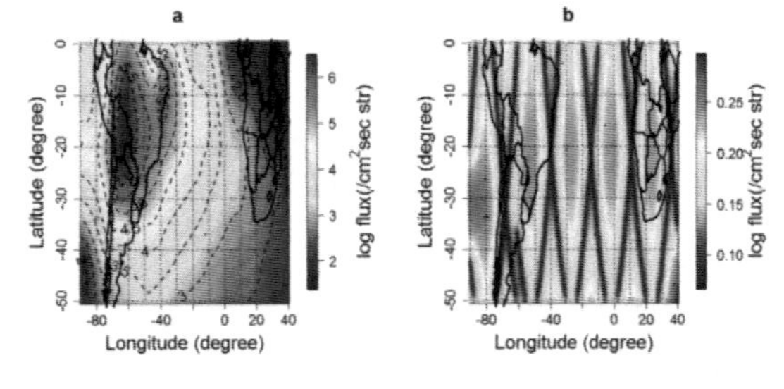

Figure 5. Flux Distribution map of >30 keV electron over SAA region on 31 January 2000, for (a) raw data estimation and (b) raw data variance

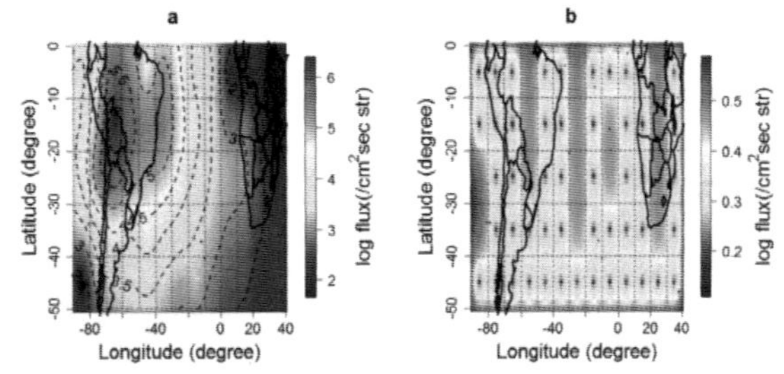

Figure 6. Flux Distribution map of >30 keV electron over SAA region on 31 January 2000 for (a) grid data estimation and (b) grid data variance

4. Summary and Future Works

HBST-GP model was successful to perform the forecasting of Earth trapped particle over the SAA region. We also achieved to deliver the convenient way to figure the >30 keV electron flux distribution map by implementing the Kriging interpolation method on a geographical map. Statistical analysis showed good results whereas the visual gave a slightly different pattern. It was also seen a significant decreasing of the flux maximum log value. To accomplish these problems, two ways are suggested to apply at the future work, i.e. shrinking the grid size and using more data from NOAA satellite series.

Acknowledgments

This work has been supported by the Ministry of Science, Technology and Innovation Malaysia (MOSTI) under Science Fund 06-01-02-SF0808 grant. Highly appreciation is also expressed to NOAA for providing the data. Thank to Mr. Muhammad Marizal for advising in R programming.

References

[1] Vainio R , Desorgher L, Heynderickx D, Storini M, Fluckiger E, Horne R B , Kovaltsov G A , Kudela K, Laurenza M, Lawlor S M, Rothkaehl H, and Usoskin I G 2009 Dynamics of the earth's particle radiation environment *Space Sci. Rev.* **147** 187

[2] Horne R B , Glaurent S A, Meredith N P, Boscher D, Maget V, Heynderickx D, and Pitchford D 2013 Space weather impact on satellites and forecasting the earth's electron radiation belts with SPACECAST *Space Weather* **11** 169

[3] Tóth G et al. 2005 Space Weather Modeling Framework: A new tool for the space science community *J. Geohys. Res.***110** A12226

[4] Fok M C, Wolf R A, Spiro R W, and Moore T E 2001 Comprehensive computational model of the earth's ring current *J. Geohys. Res.***106** 8417

[5] Pierrard V, Goldstein J, Andre N, Jordanova V K, Kotova G A, Lemaire J F, Liemohn M W, Matsui H 2009 Recent progress in physics-based models of the plasmasphere *Space Sci. Rev.* **145** 193

[6] Fok M C, Glocer A, Zheng Q, Horne R B, Meredith N P, Albert J M, and Nagai T 2011 Recent developments in the radiation belt environment model *J. Atmos.Sol-Terr Phys.* **73** 1435

[7] Ginet G P et al. 2013 AE9, AP9 and SPM: New model for specifying the trapped energetic particle and space plasma environment *Space Sci. Rev.* **179** 579

[8] Hastings D and Garreth H 2004 *Spacecraft-Environment Interactions* 2004 (Cambridge: Cambridge University Press) chapter 3 p 66

[9] Gelfand A E 2012 Hierarchical modeling for spatial data Problems *Spatial Statistics* **1** 30

[10] Sahu S K, Bakar K S, and Awang N 2013 Bayesian forecasting using hierarchical spatio-temporal models with applications to ozone levels in the eastern united states *Technical Report* School of Mathematics University of Southampton Southampton

[11] Banerjee S, Carlin B P, and Gelfand A E 2004 *Hierarchical modeling and analysis for spatial data* (Boca Raton: Chapman & Hall/CRC)

[12] Suparta W, Gusrizal, Mohd. Alauddin M A, Ahmad N 2013 Distribution of energetic particle fluxes in the vicinity of saa according to kriging interpolation *Proc. 2013 IEEE Intern. Conf. Space Sci. Comm. (ICONSPACE)*, p 17

[13] Bakar K S and Sahu S K 2013 **spTimer**: Spatio-temporal bayesian modeling using R *Technical Report* School of Mathematics University of Southampton Southampton

Spatial Analysis of Galactic Cosmic Ray Particles in Low Earth Orbit/Near Equator Orbit Using SPENVIS

W Suparta[1] and S K Zulkeple[1]
Institute of Space Science (ANGKASA), Universiti Kebangsaan Malaysia, 43600, Bangi, Selangor, Malaysia

E-mail: wayan@ukm.my

Abstract. The space environment has grown intensively harmful to spacecraft and astronauts. Galactic cosmic rays (GCRs) are one of the radiation sources that composed of high energetic particles originated from space and capable of damaging electronic systems through single event upset (SEU) process. In this paper, we analyzed GCR fluxes at different altitudes by using Space Environment Information System (SPENVIS) software and the results are compared to determine their intensities with respect to distance in the Earth's orbit. The altitudes are set at low earth orbit (400 km and 685 km), medium earth orbit (19,100 km and 20,200 km) and high earth orbit (35,793 km and 1,000,000 km). Then, within Low Earth Orbit (LEO) near the equator (NEqO), we used altitude of 685 km to compare GCRs with the intensities of solar particles and trapped particles in the radiation belt to determine the significance of GCRs in the orbit itself.

Keywords: Galactic cosmic rays, SPENVIS, Low earth orbit, Near equatorial orbit

1. Introduction

Space radiation environment is made up of three principle sources, mainly galactic cosmic rays (GCRs), trapped particles and solar particle events (SPE) [1]. GCRs are composed of 87% are protons, 12% are helium ions and 1% is heavy ions [2]. Trapped particles consisted of protons and electrons trapped in the Earth's geomagnetic field called the Van Allen's radiation belts where they moved in complex motion. SPE are energetic particles emitted by the sun during solar flares and coronal mass ejections (CMEs) where it included electrons, protons and heavier charged particles. All three sources are affected by the Earth's magnetic field [1]. However, magnetic polarity only affects GCR fluxes and not trapped particles or solar particle event fluxes [3].

Many spacecraft and artificial satellites are located at low earth orbit (LEO) because higher orbits put them at risk of higher radiation damage and exposure. However, despite being at low altitudes, they are still exposed to relatively high radiation because of variations in the solar cycle phase [3]. Due to the wide range of heavier ions, they have high LET and can be highly penetrating, hence capable of posing damaging consequences.

1.1 Galactic Cosmic Rays

GCR is dependent on solar activity whereby its intensity is at the highest during solar minimum and lowest during solar maximum conditions [4]. Charged particles had the tendency to follow the Earth's geomagnetic field lines. They tend to point towards the surface at higher inclinations, allowing deeper penetration of incident particles so GCR quite often dominate at the poles. The field lines are parallel to the surface at lower inclinations so when GCR hit Earth, most are often deflected away.

1.2 SPENVIS

It is a powerful software application developed and maintained at Belgian Institute for Space Aeronomy (BIRA-IASB) since 1996 and is an on-going European Space Agency (ESA) project that incorporated many models and tools to describe the space environment and its effects on materials [5,6]. It is accessed online and is available through its homepage http://www.spenvis.oma.be.

2. Methodology

This paper contained two aims. The first was to look into GCR fluxes at different altitudes to determine the significance level of GCR at low earth orbit (LEO) as compared to middle earth orbit (MEO) and high earth orbit (HEO). The second aim was to compare the GCR fluxes with that of trapped particles and SPE to determine significance level between the three radiation sources. We used Razaksat satellite orbit data because the focus of this research was on LEO/NEqO in which Razaksat was the best representative with nominal altitude of 685 km at 9° inclination. The software tool used was SPENVIS as it allowed users to generate orbital parameters to determine space weather conditions hence, it was a suitable tool to fulfill both aims.

2.1 Altitude Dependency

Orbital data are obtained following the availability of well-known satellites. The altitudes used to represent different orbits are shown in Table 1.

Table 1. Altitudes taken from each orbits following the orbital data of selected satellites

Low Earth Orbit (LEO)	Middle Earth Orbit (MEO)	High Earth Orbit (HEO)
400 km	19,100 km	35,793 km
685 km	20,200 km	1,000,000 km

Mrigakshi *et al* [7] described that there was a significant peak of GCR in November 2009 in which that particular month was the solar minimum occurring between solar cycles 23 and 24. The comparison of GCR intensities between solar maximum and solar minimum following Razaksat orbital data is shown in Figure 1. Solar maximum was taken during Halloween Storm 2003 which occurred from 19 October until 7 November 2003 [8]. It appeared that there was no significant difference. Therefore, we used 30 days data from November 2009 throughout this study under a quiet magnetosphere.

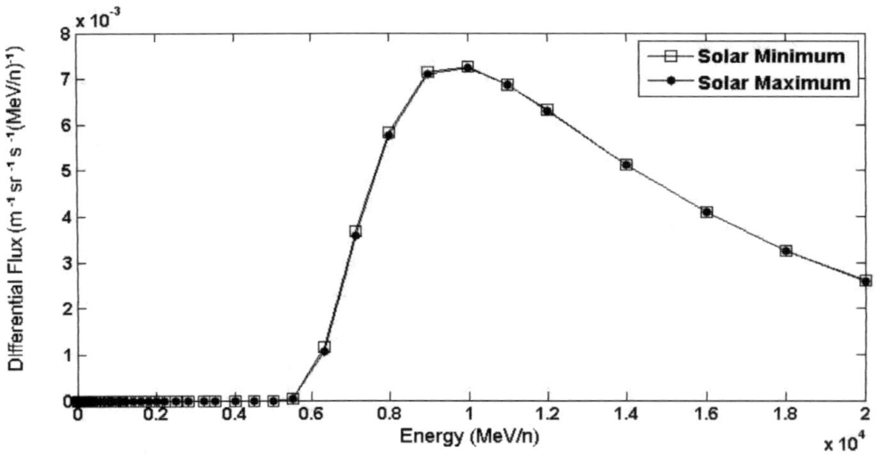

Figure 1. GCR fluxes during solar maximum and solar minimum generated from SPENVIS.

2.2 Space Radiation Environment in LEO/NEqO Orbit
Because proton made up about 87% in GCR composition, we used hydrogen particle to represent the GCR flux in this study. Then we compared it with the fluxes from trapped particles and the solar particle event at a fixed altitude of 685 km, following Malaysian satellite Razaksat orbital data in representing LEO/NEqO orbit.

3. Results and Discussion
A comparison of hydrogen particle fluxes at different altitudes is presented in Figure 2. All fluxes showed a declining trend with increasing energies but increase with respect to altitudes. However, hydrogen flux at 400 km is higher than that at 685 km due to different inclinations input in SPENVIS. At 400 km the inclination was set at 51.65° following the ISS satellite orbital data while the inclination for the altitude 685 km was set as 9° which followed the Malaysian satellite Razaksat orbital data.

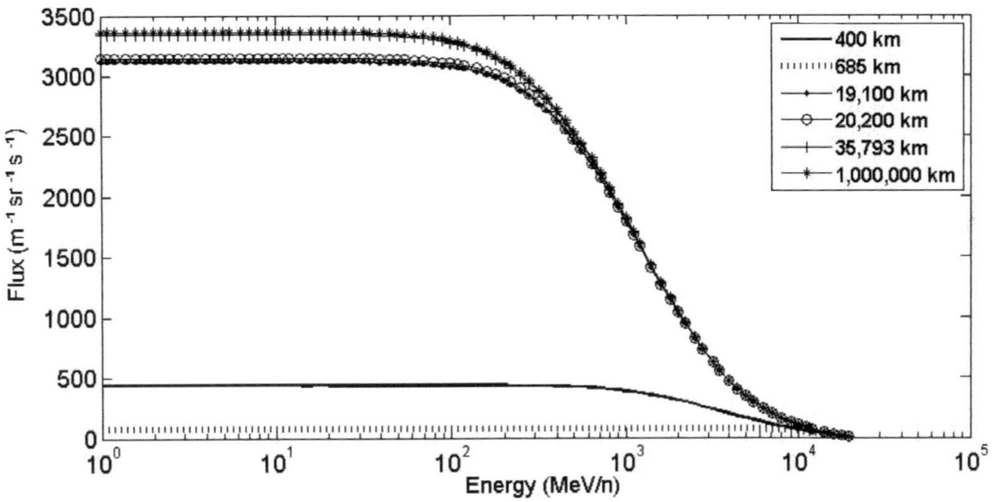

Figure 2. Hydrogen fluxes at different altitudes on November 2009 (average 30 days data)

Because radiation exposure of spacecraft in LEO is dependent on orbit inclination and altitudes [1], the following Figures 3, 4 and 5 showed how different inclinations would affect GCR intensities for three types of orbits. There was a significant difference of hydrogen fluxes at different orbital inclinations in LEO. However, they are the same for MEO and HEO cases regardless of orbital inclinations.

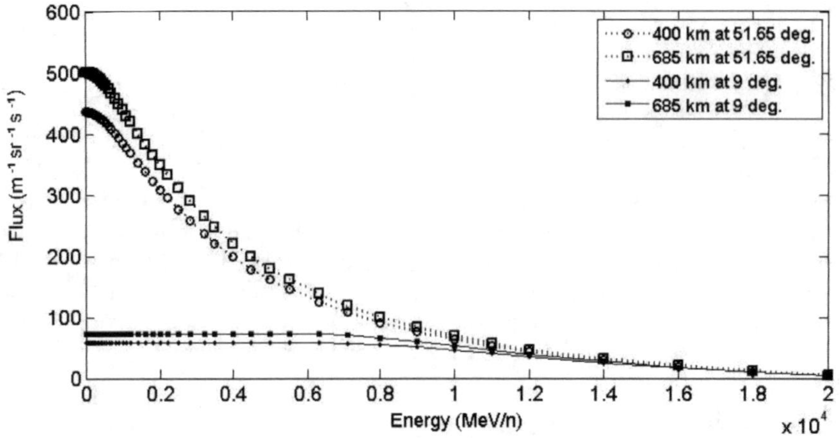

Figure 3. Hydrogen fluxes at different inclinations in low earth orbit (LEO)

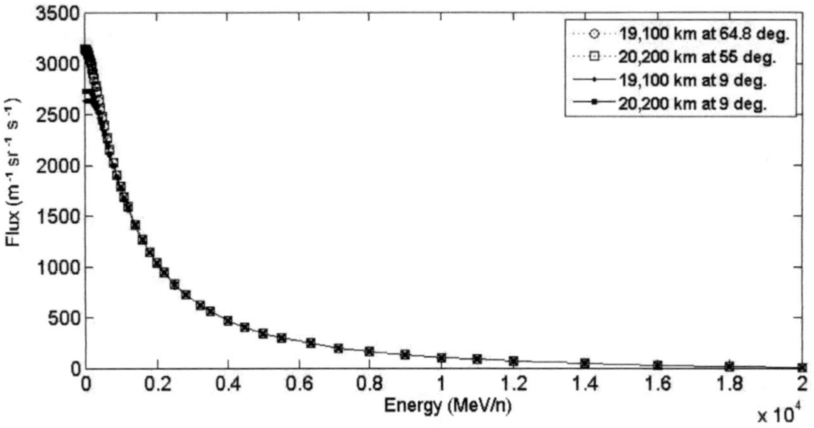

Figure 4. Hydrogen fluxes at different inclinations in middle earth orbit (MEO)

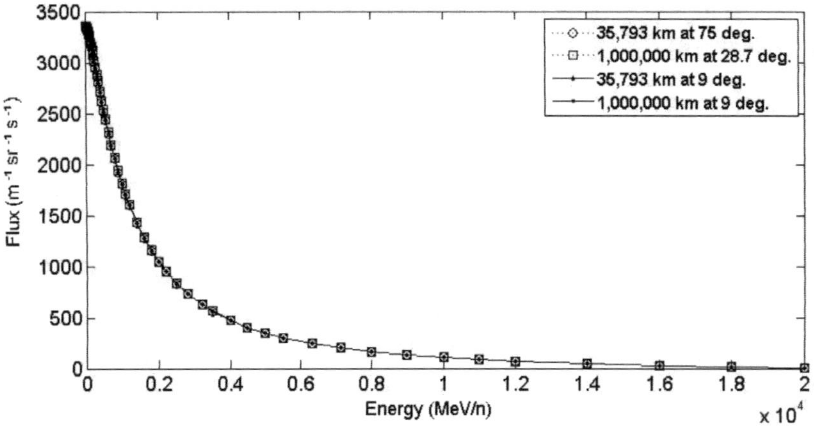

Figure 5. Hydrogen fluxes at different inclinations in high earth orbit (HEO)

Within LEO at the altitude of 685 km contained GCR with energies 10 GeV and above, trapped proton of less than 400 MeV and trapped electron of less than 10 MeV (Figures 6 and 7). This showed that in LEO/NEqO orbit, GCR and trapped particles occupied different range of energies with each being dominant at specific energy level. In other words, LEO/NEqO orbit contained high energy GCR and relatively lower energy trapped particles. The maximum fluxes for each particle type are in Table 2. The solar particle events, however, had no values obtained through SPENVIS using the Razaksat orbital data (figure not shown).

Table 2. Peak values for galactic cosmic ray and trapped particles at 685 km altitude following Razaksat satellite orbital data

Particles	Peak Values
GCR	~0.007 $m^{-1}sr^{-1}s^{-1}(MeV/n)^{-1}$
Trapped Proton	~2 $cm^{-2}s^{-1}(MeV)^{-1}$
Trapped Electron	~500 $cm^{-2}s^{-1}(MeV)^{-1}$

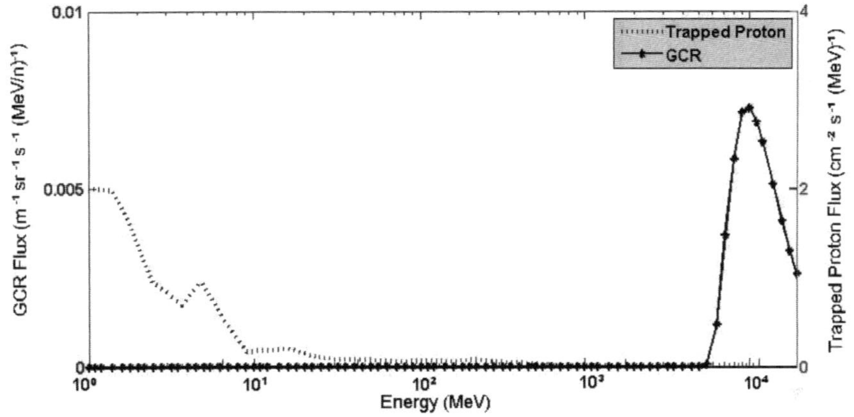

Figure 6. GCR and trapped proton fluxes at 685 km altitude

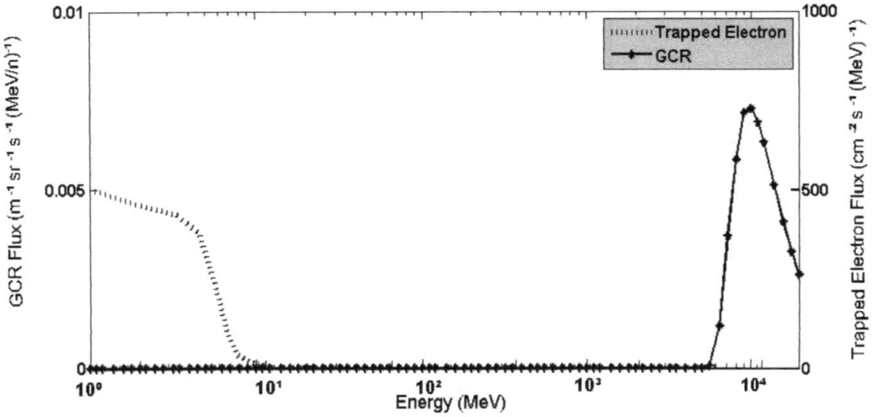

Figure 7. GCR and trapped electron fluxes at 685 km altitude

4. Conclusion and Future Work

This study aimed to describe GCR in the form of hydrogen flux in LEO/NEqO orbit by using SPENVIS. Consequently, altitudes played a role in GCR exposure to spacecraft and astronauts whereby higher altitudes gave higher GCR intensities. On the other hand, orbital inclinations have effect only towards LEO orbits but made no difference in MEO and HEO orbits in regards with GCR intensities.

Within LEO/NEqO orbit contained GCR and trapped particles but no solar particle events were detected through SPENVIS. Between the two space radiation sources, GCR dominated the higher energy range of about 10 GeV and above while trapped particles dominated the lower energy range of 400 MeV for trapped protons and 1 MeV for trapped electrons.

Space radiation consisted of three principle sources that included GCR, trapped particles and SPE, however, SPE were unable to be detected from SPENVIS. Hence, in order to better assess LEO/NEqO orbits, it would be interesting if all the three sources can be determined. Other methods such as Geant4 or International Space Station (ISS) data source can be implemented to obtain the missing SPE data. The actual level of risk posed by GCR can be further investigated through comparison with polar orbits. Moreover, materials such as gallium-arsenide or silicon that made up the satellites are protected by aluminium shielding and by considering various thicknesses can determine the level of protection it provided against the space radiation environment.

Acknowledgements

This research is supported by Ministry of Education Malaysia (MoE) under grant FRGS/2/2013/TK03/UKM/02/3 and Ministry of Science, Innovation and Technology Malaysia (MOSTI) under grant 06-01-02-SF0808. We would also like to thank the SPENVIS team for providing the software tool.

References

[1] Benton E R and Benton E V 2001 Space radiation dosimetry in low-Earth orbit and beyond *Nucl. Instr. and Meth. In Phys. Res. B* **184** 255-294

[2] Simpson J A 1983 Elemental and isotopic composition of the galactic cosmic rays *Annu. Rev. Nucl. Part. Sci.* **33** 323 382

[3] Bourdarie S and Xapsos M 2008 The space radiation environment *IEEE Trans. in Nucl. Sc. (TNS)*

[4] Badhwar G D and O'Neill P M 1996 Galactic cosmic radiation model and its application *Adv. Space Res.* **17** (2)7-(2)17

[5] Kruglanski M, Messios N, Donder E D, Gamby E, Calders S, Hetey L, Evans H and Daly E 2009 Last upgrades and development of Space Environment Information System (SPENVIS) *IEEE* 978-1-4577-0493-2

[6] Lawrence G, Reid S and Kruglanski M 2010 Space Environment Information System: Applicability for mission design and operations *American Institute of Aeronautics and Astronautics*

[7] Mrigakshi A I, Matthia D, Berger T, Reitz G and Wimmer-Schweingruber W 2013 Estimation of galactic cosmic ray exposure inside and outside the Earth's magnetosphere during the recent solar minimum between solar cycles 23 and 24 *Adv. Space Res.* **52** 979-987

[8] http://www.nasa.gov/topics/solarsystem/features/halloween_storms.html

Journal of Physics: Conference Series **495** (2014) 012041 doi:10.1088/1742-6596/495/1/012041

Ionic Wind Phenomenon and Charge Carrier Mobility in Very High Density Argon Corona Discharge Plasma

M.Nur[1], N. Bonifaci[2] and A. Denat[2]

[1] Physics Department, Faculty of Sciences and Mathematic, Diponegoro University, Semarang 50275 Indonesia
[2] G2E.Laboratory, CNRS and Joseph Fourier University 25 rue des Martyrs, 38042 Grenoble, France
E-mail: m.nur@undip.ac.id

Abstract. Wind ions phenomenon has been observed in the high density argon corona discharge plasma. Corona discharge plasma was produced by point to plane electrodes and high voltage DC. Light emission from the recombination process was observed visually. The light emission proper follow the electric field lines that occur between point and plane electrodes. By using saturation current, the mobilities of non-thermal electrons and ions have been obtained in argon gas and liquid with variation of density from 2,5 10^{21} to 2 10^{22} cm^{-3}. In the case of ions, we found that the behaviour of the apparent mobility inversely proportional to the density or follow the Langevin variation law. For non-thermal electron, mobility decreases and approximately follows a variation of Langevin type until the density ≤ 0,25 the critical density of argon.

1. Introduction

Research on the plasma corona have been carried out both in theoretical , experimental and applicative purposes [1.2.3]. Corona plasma applications has entered every aspect of the field of food technology [4]. Effect of ions in the plasma corona wind generated by electrodes on the field of multi - point heat transfer technology that enables such as drying technology has been done for example by Kim [5]. Technology which produces winds corona corona is often also called "ionic wind" has been used to reduce dust and oil dirt on insulators, and it has been patented in the U.S. Patent by Gelfan [6]. In the case of plasma corona, the mobility of ions are very interesting to discuss. Theoretical approach has been publish by **Robinson** [7] and Choelo [8]. Plasma corona with point to plane electrodes system has been used to determine the non- thermal electron mobility in very pure and high density of argon and nitrogen gaseous [9]. In 1971 Choelo [8] performed an analysis of the relationship corona current, voltage, distance between electrodes of an electrode system which generates a field is not symmetric. Under conditions of corona discharge ion flow is one type of ion (unipolar). Unlike the arc discharge conditions can charge flowing positive ions, negative ions and electrons (bipolar). Unipolar ion current flowing to the mobility μ by charge density ρ (r,t) and current density j = ρv without experiencing diffusion in an electric field E (r,t) , changes in the charge density (ρ) along the flow is:

$$\frac{d\rho}{dt} = \frac{\partial\rho}{\partial t} + v.\nabla\rho \qquad (1)$$

Obtained from the continuity equation

$$\frac{\partial\rho}{\partial t}+\nabla.(v\rho)=\frac{\partial\rho}{\partial t}+v.\nabla\rho+\rho\nabla.v=0 \tag{2}$$

Then equation 1 can be written

$$\frac{d\rho}{dt}=\rho\nabla.v=-\mu\rho\nabla.E=\frac{-\mu\rho^2}{\varepsilon_0} \tag{3}$$

Equation (3) can be integrated to obtain the
(4)

$$\frac{1}{\rho(t)}-\frac{1}{\rho_0}=\frac{\mu}{\varepsilon_0}(t-t_0) \tag{4}$$

Assuming the ions are no space charge effect of the ion flow time (T) , then the equation (9) can be minimized

$$\frac{1}{\rho}-\frac{1}{\rho_0}=\frac{\rho-\rho_0}{\rho.\rho_0}=\frac{\mu}{\varepsilon_0}(t-t_0)=\mu T$$

$$\frac{\Delta\rho}{\rho}=-\frac{\rho_0\mu T}{\varepsilon_0}=\frac{\rho_0}{\rho_s}$$

$$\rho_s\equiv\frac{\varepsilon_0}{\mu T} \tag{5}$$

and the meeting called unipolar ion saturation.

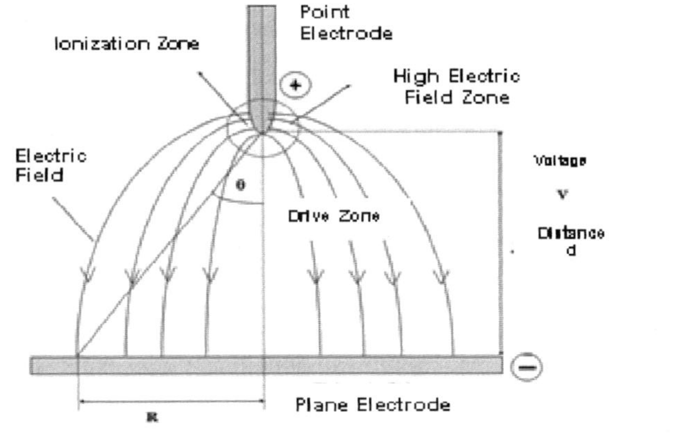

Figure 1. Sketch of a positive corona discharge electrode configuration hyperbolic (dot) - plane field .

If the ion flow distance (L) in the electric field (E) can be obtained flow time T = L / μE , it is known that the flow rate of the average ion v = μE = μV / L and ρs can be restated as ρs = ε0E / L in order to obtain the saturation current density

$$J_s = \mu\varepsilon_0 V^2 / L^3$$

$$j_s = \rho_s . \mu_s = \frac{\mu\varepsilon_0 E^2}{L} \qquad (6)$$

Equation (6) is called the ion saturation current density .
If given voltage corona (V) the average speed along the field is v = μE = μV / L and flow time L / v minimum , then

$$\rho_L = \varepsilon_0 \overline{E} / L$$

$$j_L \approx \mu\overline{E}\sigma_L = \mu\varepsilon_0 \overline{E}^2 / L = j_s$$

or

$$J_s = \mu\varepsilon_0 V^2 / L^3 \qquad (7)$$

Equation (7) is called the saturation current density corona . By taking hiperbolid - field geometry simplification , distribution and utilization of Warburg formulated , and the flow of ions along the distance between the point and the plane Choelo gain saturation current of

$$I_s \approx 2\mu\varepsilon_0 V^2 / d \qquad (8)$$

In experiments, saturation current (I) can be measured for the operating voltage (V) of a certain corona discharge formed by the configuration of the field point distance between the point and the field of d . If the calculated threshold voltage corona Vi then the saturation current to voltage relationship is shown by the equation :

$$I_s = \frac{2\mu\varepsilon_0}{d}(V - V_i)^2 \qquad (9\,)$$

Robinson approach [7] through a symmetrical field and Choelo [8] specifically for the hyperbolic plane electrodes (generally also produce a symmetrical field) produces the same equation. By using the experimental data the relationship between current and voltage, and using the formulation Robinson and Choelo the rate of charge carrier mobility (ion mobility) as well as velosity of ions can be determined.

2. **Experimental Set up**

The electrical measurements (current, pulse pause time) are performed by a device developed in the laboratory. The high voltage generator was provided by a stabilized DC voltage 0, ± 20 kV (Spellman RHSR/20PN60) connected to the point (tip) while the electrode plane was grounded. The intensity of the average current has been measured either by means of a galvanometer (Sefram Vérispot) a Keithley electrometer (Model 610C). The current pulses were visualized using a set preamplifier followed by a Tektronix oscilloscope (model 7633). In the field of pulse current, the pulses of light emitted were simultaneously detected using a photomultiplier tube (model DARIO 56AVP) with a spectral response of between 300 and 650 nm (maximum sensitivity at 400 nm) and the pulse current.. Our experimental set up is shown in the figure 2.

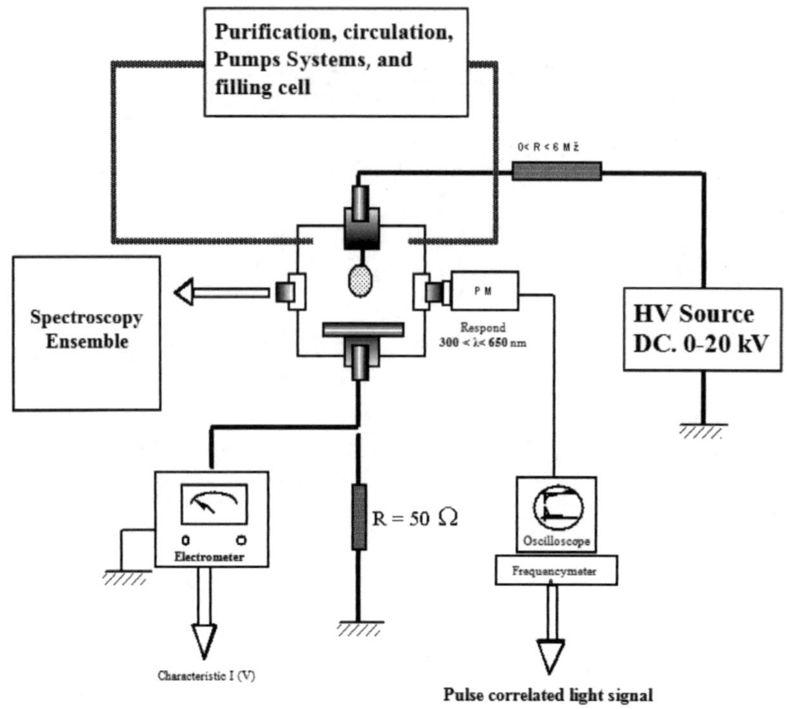

Figure 2. Sketch of experimental set up

3. **Result and Discussion**

The study of transport of charge carriers in a corona discharge in argon has been curried out from 0.5 MPa to 9,7 MPa. Before presenting the experimental results, we recall that the corona discharge, in our experimental conditions (ie very small radius of curvature of the tip and high density of the fluid) can be decomposed into two distinct regions: a region of ionization confined near the tip and a transport region. These two regions have the following characteristics. Region of ionization is very bright, its maximum length measured by visualization is about ten times smaller than the distance point-plan and the calculation shows that the entire charge is created in this region. The temperature Tp of the region similar to a plasma is unknown. By cons its pressure equals the pressure P_∞ gas or, in the case of a liquid, the hydrostatic pressure applied on it (for a stationary corona discharge, there is no pressure difference between the area ionization and the surrounding environment). The plasma density Np can be calculated after determining a value of Tp. The transport region shows no luminescence except in the case of argon gas to a point where a very small negative luminescence is observed. No measurable charge is created in this area. Under these conditions, temperature and pressure correspond to the temperature $T\ \infty$ and the pressure $P\ \infty$ of the test. The density $N_\infty = f\ (T_\infty, P_\infty)$ is known and calculable. The two regions of discharge and their conditions of temperature and pressure are presented in the figure 3. Discharge in argon gas, has been created in the range of pressure (P_∞) from 0.1 MPa to 10 MPa.

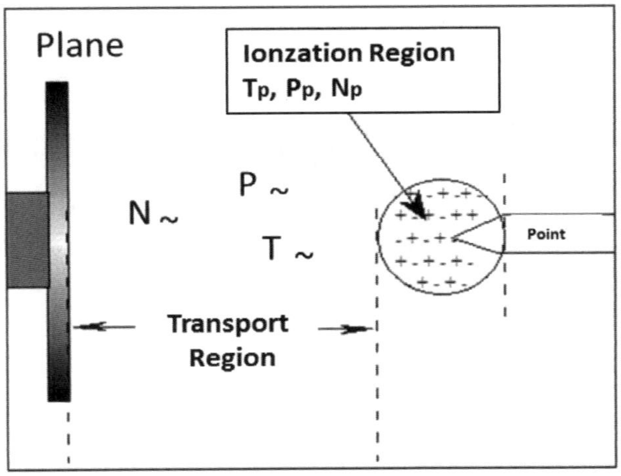

Figure 3. Sketch of region in the corona

3.1 Current-Voltage Characteristics

In all our experiments with purified gas well (= 20 successive passes through the purification system), a current is always observed beyond the threshold voltage Vs corresponding to the onset of corona discharge process. Indeed, for V = Vs, the current rises abruptly from a value not detectable ($<10^{-14}$ A), reaching about 10 μA. Ionic wind phenomenon has been cause of light emission from the recombination process between electron and ions. The light emission proper follow the electric field lines that occur between point and plane electrodes. To illustrate this phenomenon, we present patterns of events in figure 5 and the corresponding photographs in figures 4.

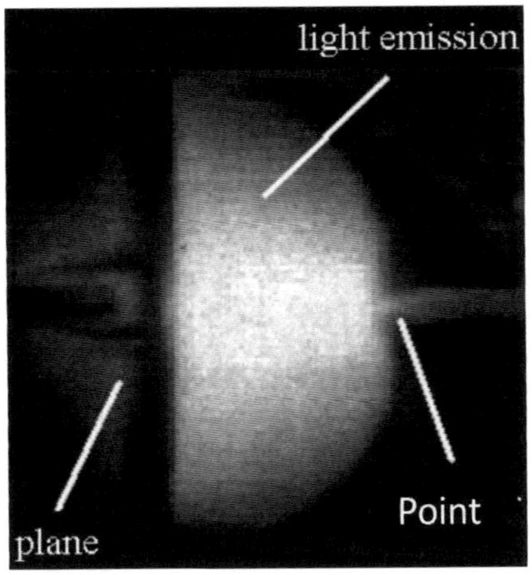

Figure 4. Photo of a negative corona discharge continues in pure argon. P = 1,2 MPa, T = 297 K, I = 110 uA and V = 2,5 kV. The black line is not visible in the photo.

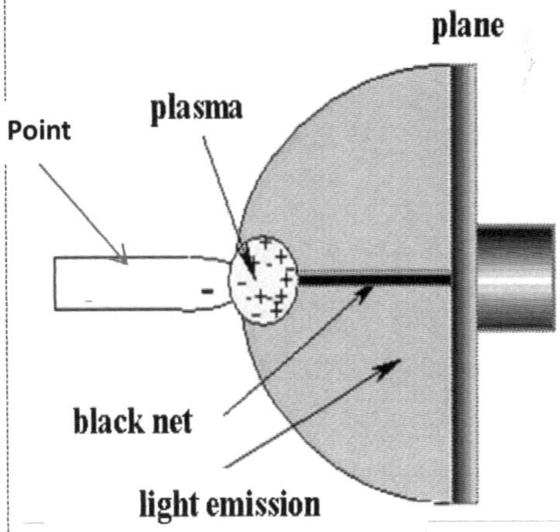

Figure 5. Detailed diagram of the corona of the photo in the figure 4

The typical shape of the current, depending on the voltage applied to tip, is shown in figure 6. The typical shape of the current-voltage curves in the argon gas is presented by scheth in the figure 7. In zone A, the measured current is always lower than the saturation current of Sigmond [10], so it is an unipolar current. After the transition in discharge region C, the current becomes bipolar and, in extreme cases, an arc current is directly detected. Since region A is that we can be analyzed according to the gas density, we will restrict the following presentation to the study of this part of the characteristics I(V). Firstly, the current increases steadily with V (area A) then becomes approximately constant (area B) and increas sharply (zone C). The value of current, there is the pseudo-saturation depends on the density of gas. In zone A, the very bright area is confined near the tip. When discharge condition with current in the area C, we found forms a bright discharge channel that completely crosses the space between the two electrodes (point and plane).

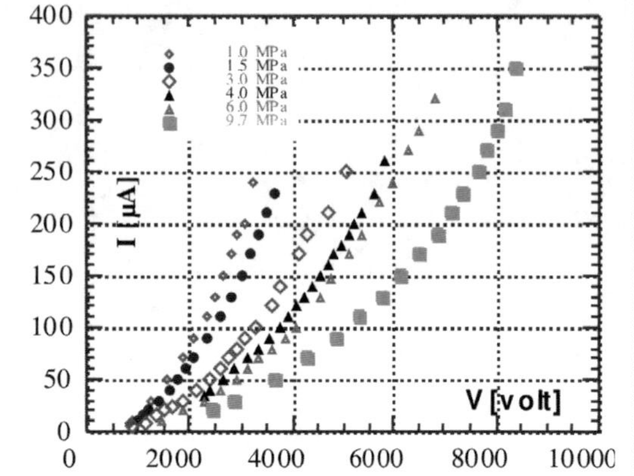

Figure 6. Characteristics I-V for argon gas at different pressures, limited to the region A. Negative polarity of the tip.

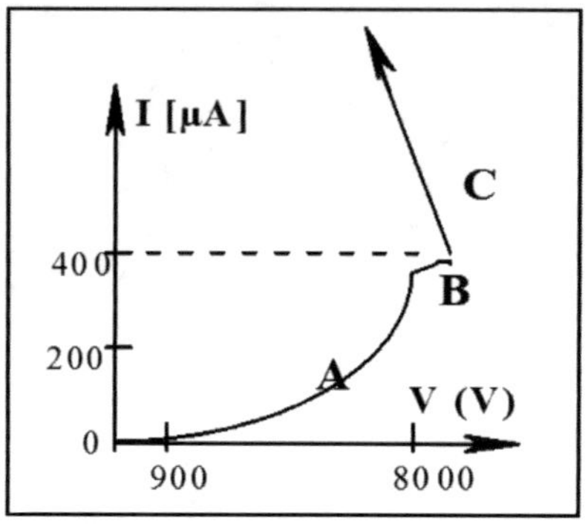

Figure 7. Typical shape of the current-voltage curves in the argon gas.

Another important aspect to consider is the role that impurities (N_2, O_2) in argon gaseous . Weislers observed a continuous glow discharge in contaminated by at least 0,3% nitrogen while it has obtained the same result with a content of 10% nitrogen in argon. In our case, we use a very pure gas whose nitrogen content is less than 0,3 ppm-volume before passing through the purification system. We have also used a titanium purifier that completely eliminates the residual nitrogen or desorbed from the walls . The glow discharge that we obtained in our experiments is therefore not the result of a residual nitrogen. In the presence of oxygen ($\geq$ 0,1%), Weislers got a pulse discharge rate Trichel type and also a very significant reduction of the current. We also obtained this result using a gas and its supply circuit unpurified. In figure 8, we show the variation of I/V versus V at different pressures. We can see that the voltage V (which is the value of V extrapolated to zero current) does not depend on the pressure. For all pressures studied, we note that Vo $\cong$ 400 volt. This value is low compared to the values of the corresponding threshold voltages. For example, a pressure of 1 MPa, $V_{threshold}$ $\cong$ 900 volt volt and at a pressure of 9,7 MPa, $V_{threshold}$ is equal to 2500 volt. In this case, the discharge rate is close to a current region limited by space charge [10].

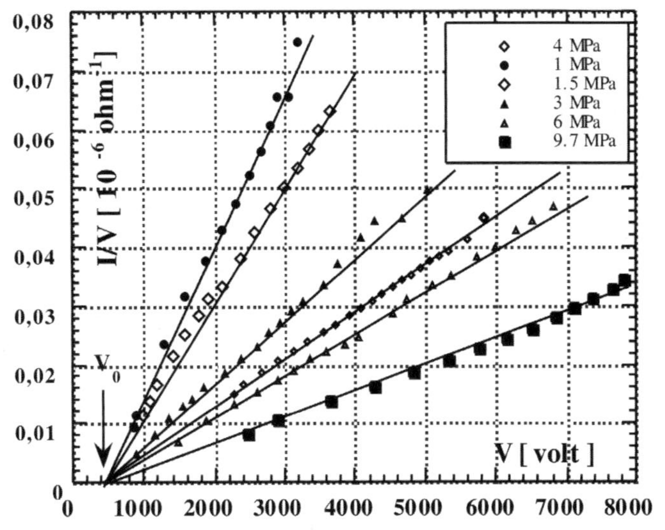

Figure 8. Curves I/V(V) to the argon gas in negative polarity at different pressures.

3.2 Nonthermal electronic Mobility

They are derived from tests where negative polarity of the voltage on the tip is applied (negative corona discharge). This corresponds to the creation and transport of electrons (or if the medium is impure to negative ion transport). Using spectral analysis, we measured the maximum length of the field glow discharge (ionization zone, excitement, creation of non-equilibrium plasma). It is at most 1,5 mm to the lowest pressures used. Under these conditions, the transport area and being close to the curves √I(V) are all straight , we can calculate the "apparent" mobility of charge carriers. They were calculated from the slope of the lines √I(V), using the model Sigmond. In Figure 9 we can observe the evolution of this "apparent" mobility with the gas density. Mobility decreases and approximately follows a variation of Langevin type until the density $N / Nc \leq 0,25$ (here Nc is the critical density of argon). Values mobility obviously can not be attributed to thermal electrons . Indeed, the reduced electric field (E/N) in our experiment was too high for it to strike a balance between the gas and electrons. For point-plane geometry, the actual field in the transport can be approximated by the value of the mean field V/d, and the reduced field can be written as E/N = V/ND. However, V may be varied between the threshold voltage (V_{th}) and the breakdown voltage (V_b) , and Vb and Vth as similarly depend on the density of the gas , wherein the field can vary the field is reduced very limited and it is practically independent of the density . For all ranges of density of our experiments, the reduced field varies only slightly between 0.5 x 10^{-17} and 3 x10^{-17} Vcm^2. We note in figure 9 , our results are in good agreement with those obtained by a direct method and Prew Allen [12] and Dutton [14] for non- thermal electrons in the same field reduced field. Therefore, in our case, we can say that our charge carriers are non- thermal electrons . We also present the results obtained in argon using a cell placed in a cryostat [11] and where it was possible variation of density from 2,5 10^{21} to 2 10^{22} cm^{-3} (liquid state near the triple point) . It is noted that these results are far from those achieved when Dutton the highest densities. One suspects that, in our case, the low mobilities observed are the result of argon purification still insufficient. The higher impurity content should be low to achieve the same result at lower density .

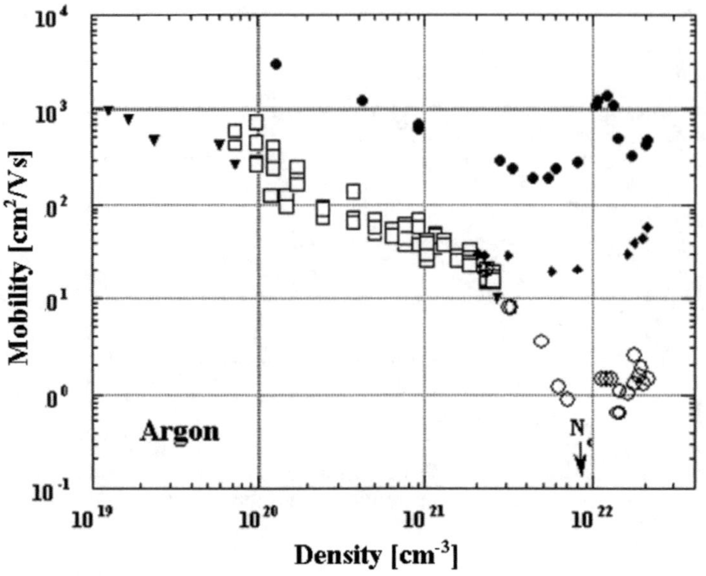

Figure 9. Mobilities depending on the density of negative charges.

¤ K_gas, our results.
○ K_app. gas and liquid [11]
▼ μ_e non-thermal electron in Ar gas [12].
• μ_e the thermal electron in Ar gas and ArL [13].
♦ μ_e non-thermal electron in Ar gas [14] and ArL [15].

3.3 Mobility of Ions

In the positive polarity, we studied positive corona discharge plasma which corresponds to a generation of positive ions. In argon, beyond the threshold voltage V_{th} a DC was recorded. The measured values of the current are very low compared with those obtained in negative polarity of the point (at least in the most pure gas). The threshold voltage increases with gas density. Characteristics I (V) being straight, the "apparent" mobility of the charge carriers was calculated from the measurement of their slope again using the model Sigmond [10]. In Figure 9, the behaviour of the apparent mobility with the gas density is presented. The apparent mobility inversely proportional to the density or follow the Langevin variation law [17]. On the same graph, we also present the results obtained by Freeman [13] and Hornbeck [15] . In both cases, these authors have performed a direct measurement (time of flight) of the speed of positive carriers between flat parallel electrodes.

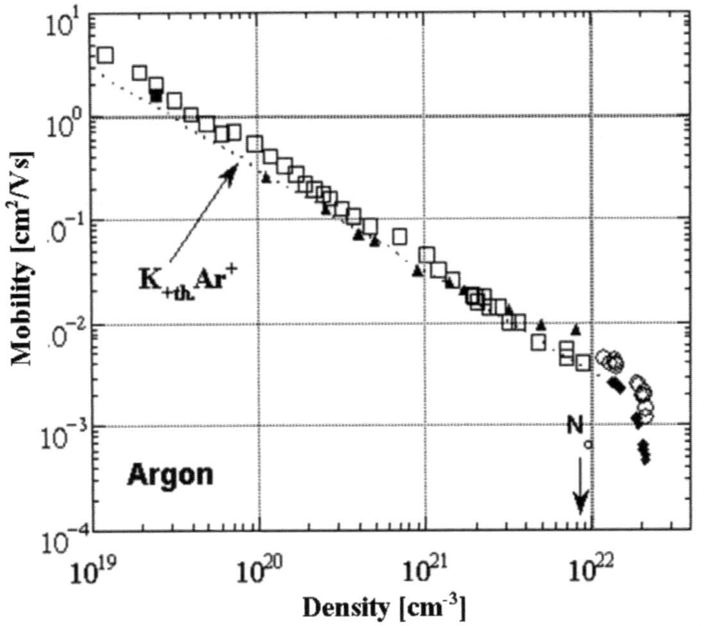

Figure 10. Mobility of positive ions depending on the density.
¤ K_{+app} gas, our results.
○ K_+ liquid [11]
♦ K_+ liquid [16].
▲ K_+ gas [13].
■ K_+ gas [15].
--- $K_{+th.Ar}^+$ theoretical curve according to the theory of Langevin corrected[17]

We can see that our values are in good agreement with measurements of others authors [13,15,16]. On the other hand if we consider that the charge carriers are ions Ar^+ and its evolved the Langevin model corrected [15,16], for the effect of charge transfer resonance is applied. We can see that the theoretical curve and the calculated mobility overlaps substantially all of our experimental values and those of Freeman and Hornbeck [13,15]. It seems that the charge carriers in these conditions are well Ar+ ions. The results obtained in the liquid argon are also shown in figure 10. These results show that when the density exceeds the critical density (Nc), the apparent mobility no longer follows the law of variation of Langevin. The same behavior is shown by the results of Davis [16].

4. Conclussion

In the very pure argon, corona discharge plasma can produce ionic wind phenomenon that has been shown by light emission from the recombination process. The light emission was observed visually and proper follow the electric field lines that occur between point and plane electrodes. Our results on mobility of charge carrier, we found that non-thermal electron, mobility decreases and approximately follows a variation of Langevin type until the density $\leq 0{,}25$ the critical density of argon. In the case of ions, we found that the behaviour of the apparent mobility inversely proportional to the density or follow the Langevin variation law. Determination of mobility of charge carrier of corona discherge plasma can be done with variation of density from 2,5 10^{21} to 2 10^{22} cm^{-3}.

Acknowledgment
The authors would like to express their thanks to Dr. Vlademir Atrazhev from Institute of High Temperature Physics, Russian Academy of Sciences, Moscow for his helpful suggestions.

References

[1] Jones, J.E. 2008 "On corona-induced gas motion and heating I:Field equations, modelling and vortex formation", *Journal of Electrostatics*, 66 84–93

[2] Nur, M. 1997 "Etude des décharges couronne dans l'argan et l'azote trés purs: transport des charges, spectroscopie et influence de la densité", *PhD. Thesis*, Joseph Fourier University, Grenoble, France.

[3] Coelho, R., Debeau, J. 1971,"Properties of the tip-plane configuration, *J. Phys. D: Appl. Phys.*, 4, p. 1266

[4] Timothy I.J. Goodenough P., Goodenough, W., and Goodenough, S. M. 2007, "The efficiency of "corona wind" drying and its application to the food industry", *Journal of Food Engineering*, Volume 80, Issue 4, , Pages 1233-1238

[5] Kim, C. Park, D.. Noh, K.C. and Hwang J. 2010 "Velocity and energy conversion efficiency characteristics of ionic wind generator in a multistage configuration", *Journal of Electrostatics* 68, 36–41

[6] Gelfan, P.C., 1975 "Corona Wind Generating Device", U.S. Patent No. 3,896,347 (22 July 1975)

[7] Robinson, M. 1961, "Movement of air in the electric "wind" of the corona discharge", *AIEE Trans.* 80, pp. 143–150

[8] Coelho R. and Debeau J., "Properties of the Tip-Plane Configuration", *J. Phys. D : Appl. Phys.,* Vol. 4, pp. 1266-1280, 1971.

[9] H. Kalman, E. Sher,1997 Enhancement of heat transfer by means of a corona wind created by a wire electrode and confined wings assembly, *Applied Thermal Engineering* 21 (2001) 265±282

[10] Sigmond, R.S. (1982),"Simple approximate treatment of unipolar spacecharge-dominated coronas: the Warburg law and the saturation current", *J. Appl. Phys.* 53 (1982) 891–898.

[11]Bonifaci N., Denat A., and Malraison B., "Transport phenomena in point-plane geometry in dense argon", J. Electrostatic 40&41 pp.91-96, 1997

[12]Allen N. L. and Prew B. A., 1970 "Some measurements of electron drift velocities in compresed gases ", *J. Phys. B : Atom. Molec. Phys* , Vol **3** , pp. 1113-1126

[13] Gee N. , Floriano M. A., Wada T., Huang S. S. S. and Freeman G.R., 1985"Ion and electron mobilities in cryogenic liquids: Argon, Nitrogen, Methane and Ethane", *J. Appl. Phys.* Vol. 57, pp. 1097-1101

[14]Dutton J.,1975 "A survey of electron swarm data", *J. Phys. Chem.* Ref. Data, Vol. **4**, pp 577-857

[15]Hornbeck J. A.,1951 "The drift velocities of molecular and atomic ions in helium, neon and argon", Phys. Rev. Vol. 84, pp. 615-620

[16] Davis H.T. , Stuart A.R.and Mayer L. , *J. Chem. Phys.*,37 (1962) 947

[15] McDaniel E. W. and Mason E. A., 1964 "Collision phenomena in ionized gases", John Wiley & Sons, New York , chapter 7 and 9

[16] Reininger R., Asaf U., Steinberger I. T., Basak S.,1983 "Relationship between the energy V_0 of the quasi-free-electron and its mobility in fluid argon, kripton, and xenon", *Phys. Rev. B,* Vol. **28** (8), pp. 4426-4432

[17] McDaniel E. W. and Mason E. A.1973, "The mobility and diffusion of ions in gases", John Wiley & Sons, New York , chapter 5

Improvement Text Compression Performance Using Combination of Burrows Wheeler Transform, Move to Front, and Huffman Coding Methods

Mohammada Aprilianto[1] and Maman Abdurohman[2]
Telkom University
Jl. Telekomunikasi, 1 Bandung, West Java, Indonesia
E-mail: mohammada.aprilianto@yahoo.com

Abstract. Text is a media that is often used to convey information in both wired and wireless-based network. One limitation of the wireless system is the network bandwidth. In this study we implemented a text compression application with lossless compression technique using combination of Burrows wheeler transform, move to front, and Huffman coding methods. With the addition of the compression of the text, it is expected to save network resources. This application provides information about compression ratio. From the testing process, it concludes that text compression with only Huffman coding method will be efficient when the number of text characters are above 400 characters, meanwhile text compression with burrows wheeler transform, move to front, and Huffman coding methods will be efficient when the number of text characters are above 531 characters. Combination of these methods are more efficient than just Huffman coding when the number of text characters are above 979 characters. The more characters that are compressed and the more patterns of the same symbol, the better the compression ratio.

1. Introduction

Compression is a technique to shrink the size of data. There are two kinds categories in compression based on the output, which is lossless and lossy compression. In lossless compression, the outcome data after decompression is exactly as before compression. While on lossy compression, there is the possibility of losing some data after decompression [4].

This compression can also be applied to SMS (Short Message Service). SMS that delivered by mobile phone to SMS Centre is sent in PDU (Protocol Data Unit) format. As well SMS that received by mobile phone from SMS Centre is saved on mobile phone in PDU format [11].

Burrows wheeler transform change the order of the characters so that there will be more redundancy characters, to take advantage of it, so it is forwarded by move to front method so that there will be many index zero as recurring characters will be placed in index zero by this method [6]. Due to the large frequency of zeros, Huffman coding will be used which will transform the character into a shorter codeword. The greater the frequency of characters, the shorter the codeword.

2. Research Method

In This research try to improve text compression performance by using burrows wheeler transform, move to front, and Huffman coding. After that, the author analysed the compression ratio. This research is done by making a text compression application using Net Beans 7. The Huffman Coding frequency table will be in the text. Here is the explanation about the methods and the system design.

2.1. Burrows Wheeler Transform

Burrows wheeler transform is a reversible transformation algorithm. This algorithm is used in lossless compression like bzip2 [8]. This method just change the order of characters, so when the data is reverted, the outcome will be just like before. In this algorithm the author will use text 'title' for the example. Perform shift one character to the right. Repeat until (n-1) times where n is the sum of characters on the input data. The result would be like in figure 1.

title
etitl
letit
tleti
itlet

Figure 1. Burrows Wheeler Transform First Step.

After that sort those words in lexicography and the result would be on figure 2.

1	etitl
2	itlet
3	letit
4	**title**
5	tleti

Figure 2. Burrows Wheeler Transform Second Step

The output of this transformation is the last character in every line from top to bottom and the line number where the original text is, so the output is 'lttei, 4'.

2.2. Move to Front

Move to front is a data transformation algorithm that is designed to improve the performance of entropy encoding compression technique. The idea of this algorithm is change every symbol with an index from a recent list. A symbol that shows often will be changed with index zero which will be make an advantage to entropy encoding compression technique like Huffman coding [5]. In this algorithm, first we need to make a list of all possible character that will be shown from the input data, for example of recent list is 'abcde'. The left character has index zero, move to right, and the index increase by one. For example, the first input data is character 'b', character 'b' is on index one, so write 1 on the output, then move 'b' on the recent list to the most left.

2.3. Huffman Coding

Huffman coding is a compression algorithm founded by David Huffman [9]. This algorithm works by building a tree based on data frequency, the more the frequency of a character, the shorter the codeword. The codeword is in binary form. First we need to calculate the frequency of every character that is shown from the input data. Then make a tree by take two characters with lowest frequency first. Mark the left leaf of tree with '0' and the right leaf with '1'. After that, make a codeword for a character by browsing where the leaf of that character from the root is. Finally, change every character on the input with codeword that has been made.

2.4 System Design.

Here is the flowchart of compression process.

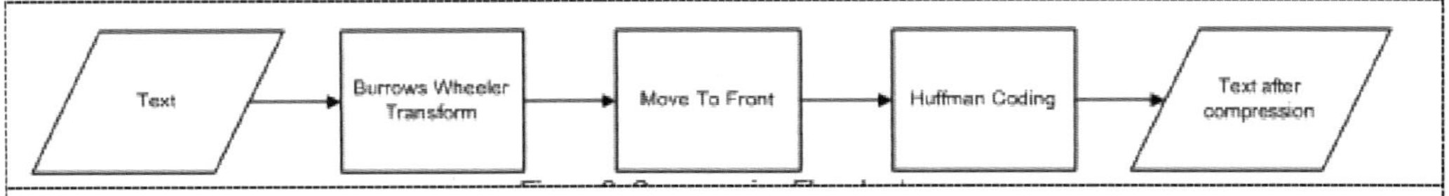

Figure 3. Compression Flowchart

In this application, the text is first transformed by burrows wheeler transform method so the arrangement of the text is changed with more redundancy characters. After that, the text is processed by move to front method. Each character will be changed to index number where recent character that showed will be change to index 0. In the end, Huffman coding will compress the text that has been changed to index number.

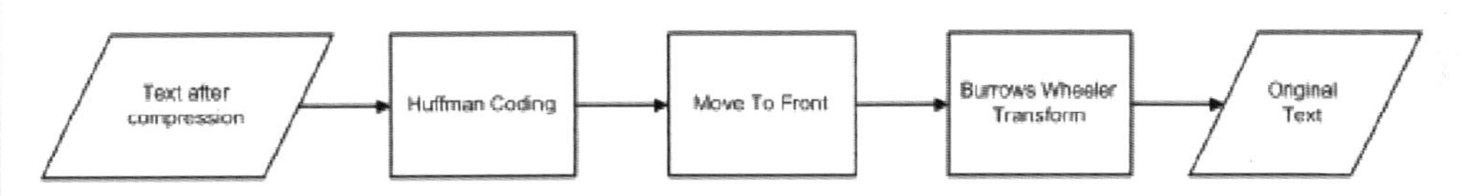

Figure 4. Decompression Flowchart

The decompression process is first change the text to index number with Huffman coding method, after that move to front will change back the index number back to text that has been reordered from the original text, and then burrows wheeler transform will change its text to original text before the compression.

3. Result and Analysis

This research analyse the compression ratio where compression ratio is ratio of the number of characters after compression by the number of characters before compression multiply 100%. This research used two scenarios, first scenario focused on number of characters. There were 80 number of texts that was tested. 40 texts were regular sentences, and the other 40 texts were sentences with more complicated characters. Second scenario focused on the pattern of characters used. Analysis based on worst case, regular text, and best case. Worst case has one frequency of every characters. Regular text has an increasing same pattern of characters. The best case has one kind character with 265 frequencies.

3.1. Analysis of Parameter Compression Ratio Testing

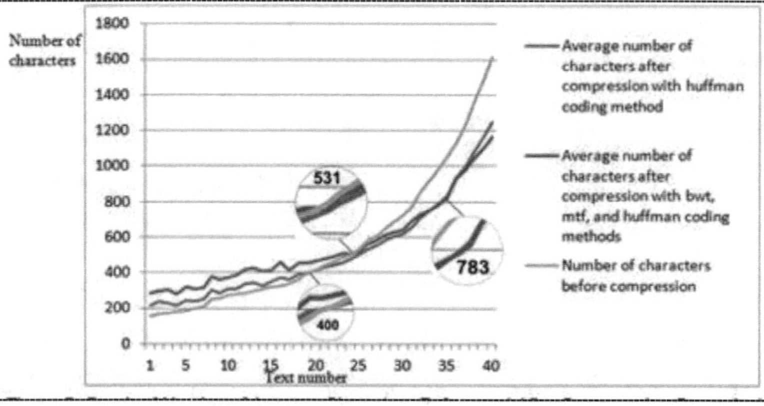

Figure 5. Graph of Number of Average Characters Before and After Compression Scenario 1

Figure 5 above is a graph about number of average characters from text with regular sentences and complicated sentences before and after compression scenario 1. The graph tells that the Huffman Coding compression method compress the text start from number of 400 characters, while the BWT, MTF, and Huffman Coding compression method compressed the text start from number of 531 characters. The BWT, MTF, and Huffman Coding compression method is better when the number of character are exceed 979 characters. At the number of 500 characters and more, the decompression time exceed 30 minutes.

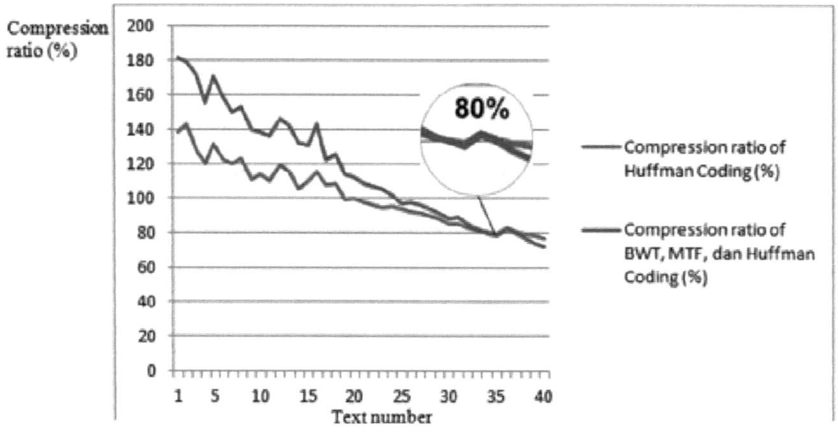

Figure 6. Graph of Average of Compression Ratio Scenario 1

Figure 6 above is a graph about average of compression ratio scenario 1. The graph tells that the more characters compressed, the better the compression ratio is, and the ratio of BWT, MTF, and Huffman Coding Compression Method is better than only Huffman Coding when it achieves 80%.

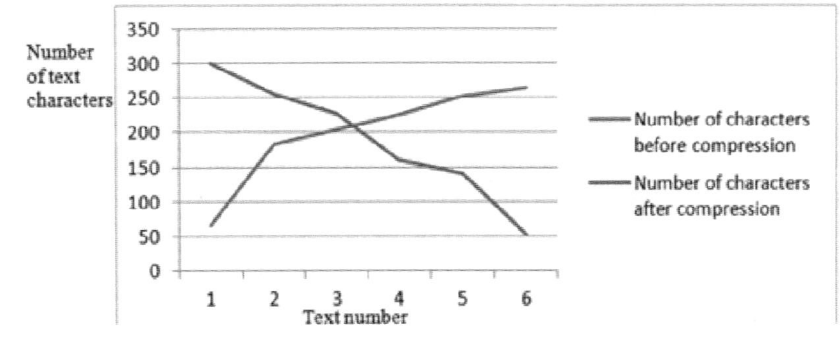

Figure 7. Graph Number of Character Before and After Compression Scenario 2

Figure 7 tells that number of characters after the compression on the worst case is increase than the original. Besides, the number of characters after the compression on the best case is greatly reduce than the original.

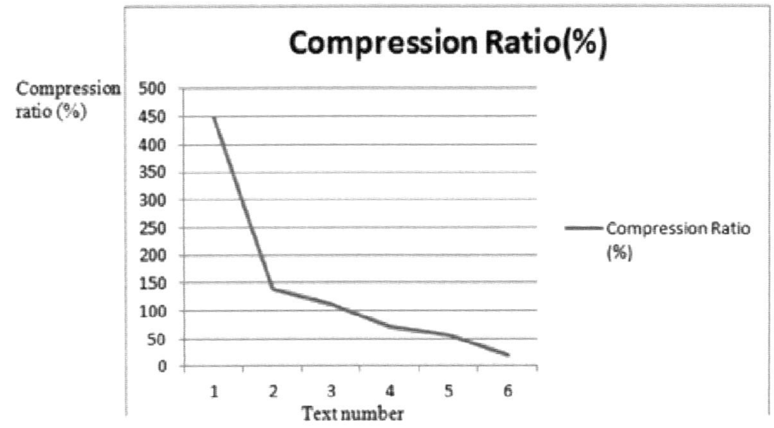

Figure 8. Graph of Compression Ratio Scenario 2

Figure 6 tells that the compression ratio is reducing, the more the number of same pattern of characters, the better the compression ratio is.

4. Conclusion

From the testing results it can be concluded that text compression with BWT, MTF, and Huffman Coding method will compress the text when the number of characters exceed 531 characters, while text compression with only Huffman Coding method will compress the text when the number of characters exceed 400. Combination of these methods are more efficient than just Huffman coding when the number of text characters are above 979 characters. The pattern of characters influence the compression ratio, the more the same pattern of characters, the better the compression ratio is.

For next research, this application can be developed with another method or using some method so the Huffman frequency table is not necessary to be included in text so the compression ratio will be better.

References

[1] Abdullah, M.M. Kompresi String Menggunakan Algoritma LZW dan Huffman. Bandung. 2008
[2] Amrullah, A., (et.al). Kompresi dan Enkripsi SMS dengan Metode Huffman Code dan Algoritma Enigma. Institut Teknologi Sepuluh Nopember.
[3] Ayuningtyas, N. Implementasi Kode Huffman dalam Aplikasi Kompresi Teks pada Layanan SMS. Bandung.
[4] Blelloch, G. E. Introduction to Data Compression. Camegie Mellon University. 2010.
[5] Burrows, M., Wheeler, D.J. A Block-sorting Lossless Data Compression Algorithm. California. 1994.
[6] Effros, M., Visweswariah, K., Kulkarni, S. R., Verdu, S. Universal Lossless Source Coding With the Burrows Wheeler Transform. 2002.
[7] Mahmoud, T. M., Abdel-latef, B. A., Ahmed, A. A., Mahfouz, A. M. Hybrid Compression Encryption Technique for Securing SMS. Minia University.
[8] Manzini, G. The Burrows-Wheeler Transform: Theory and Practice. Italy.
[9] Patra, N., Sankar, S.S. Data Reduction by Huffman Coding and Encryption by Insertion of Shuffled Cyclic Redundancy Code. Rourkela. 2007.
[10] Riswan. Mengenal SMS (Short Message Service). 2006.
[11] Achmad, B. (et.al). Sistem Alarm Mobil Menggunakan Mikrokontroler AT89S52 Berbasis SMS. TELKOMNIKA. April 2008; 6(1): 15-20.

Journal of Physics: Conference Series **495** (2014) 012043 doi:10.1088/1742-6596/495/1/012043

Implementation of the Intelligent Voice System for Kazakh

Zh Yessenbayev[1], N Saparkhojayev[2], T Tibeyev[3]
[1]Nazarbayev University Research and Innovation System, 53 Kabanbay Batyr Ave., Astana. 010000, Republic of Kazakhstan
[2]ISMB Research Institution, POLITO, Torino, 10129, Italy
[3]Suleyman Demirel University, 1/1 Abylaikhan St., Kaskelen, Almaty, 040900, Kazakhstan

E-mail: zhyessenbayev@nu.edu.kz

Abstract. Modern speech technologies are highly advanced and widely used in day-to-day applications. However, this is mostly concerned with the languages of well-developed countries such as English, German, Japan, Russian, etc. As for Kazakh, the situation is less prominent and research in this field is only starting to evolve. In this research and application-oriented project, we introduce an intelligent voice system for the fast deployment of call-centers and information desks supporting Kazakh speech. The demand on such a system is obvious if the country's large size and small population is considered. The landline and cell phones become the only means of communication for the distant villages and suburbs. The system features Kazakh speech recognition and synthesis modules as well as a web-GUI for efficient dialog management. For speech recognition we use CMU Sphinx engine and for speech synthesis- MaryTTS. The web-GUI is implemented in Java enabling operators to quickly create and manage the dialogs in user-friendly graphical environment. The call routines are handled by Asterisk PBX and JBoss Application Server. The system supports such technologies and protocols as VoIP, VoiceXML, FastAGI, Java SpeechAPI and J2EE. For the speech recognition experiments we compiled and used the first Kazakh speech corpus with the utterances from 169 native speakers. The performance of the speech recognizer is 4.1% WER on isolated word recognition and 6.9% WER on clean continuous speech recognition tasks. The speech synthesis experiments include the training of male and female voices.

1. Introduction

Analysis of several speech recognition and synthesis engines such as Sphinx4, HTK, Julius, FreeTTS, Festival, etc., showed that they require solid background knowledge in speech technologies to configure and get a functional system. However, this is a difficult task for most of the end-users such as governmental, educational or other industry entities that may not have adequate specialists. Moreover, to our best knowledge, there are no deployments of speech-based services in Kazakhstan to date. All the current call-centers use DTMF-based software. In view of this, we started a project of developing a prototype of an intelligent voice system (IVoS) to help fast and easy deployment of speech-enabled contact-centers and information desks for those who lack special knowledge in speech technologies.

2. System description

The intelligent voice system features Kazakh speech recognition and synthesis modules as well as a web-GUI for efficient dialog management. The high level architecture of the system can be outlined as in Figure 1. At the core of the system, there is an application server that manages the system's data and work flows as well as access to other modules. The server consists of a Control module that handles run-time work flow and an administration web-GUI that enables operators to quickly create, manage and browse the status of the dialogs in user-friendly graphical environment. The dialogs are compatible with VoiceXML format. For speech recognition we use CMU Sphinx engine and for speech synthesis – MaryTTS. However, the system defines several own interfaces to interact with the third-party speech engines based on JavaSpeech API, so that one can prefer other solutions. The call routines are handled by Asterisk PBX and forwarded using FastAGI protocol to our remote application server. All application related data are stored in MySQL database and accessed via Java Persistence.

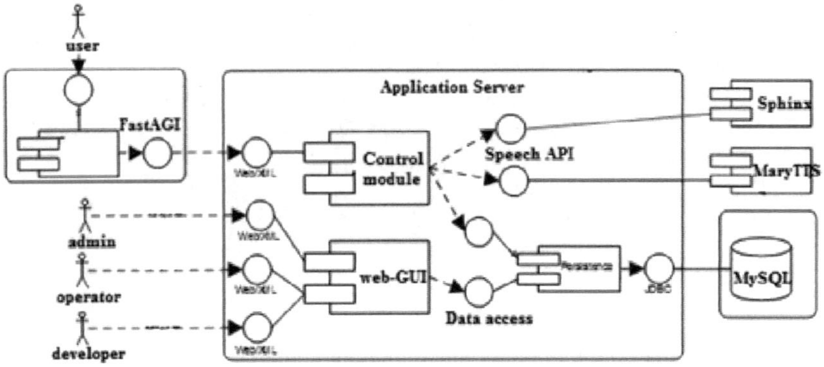

Figure 1 System architecture

3. Speech recognition and synthesis

3.1 Data

Most modern state-of-the-art speech recognition systems require audio data for training the acoustic models. Depending on the type of an application data needed varies from high quality microphone speech (WSJ0) to telephone speech (Switchboard or CALLHOME), from continuous speech (TIMIT) to connected (TIDIGITS) and isolated words (PhoneBook). In our current work, we compiled the first Kazakh speech corpus which was initiated as a part of the Kazakh Language Corpus presented in [1]. The speech corpus consists of three parts: 1) telephone speech of isolated and connected words (names, cities, dates, etc.) for the medium vocabulary (up to 2000 words) command-and-control applications; 2) high quality microphone speech of 169 native speakers for the large vocabulary continuous speech recognition; 3) long recordings of two professional actors for the speech synthesis experiments. Audio data were captured using professional vocal microphone as well as our call-collecting system using Asterisk PBX built on top of the existing university-wide infrastructure. All the recorded audio files were manually post-processed to have each utterance in a separate file. The total duration of the audio files is about 60 hours. The speakers that took part in the recordings are volunteers were balanced by region, age and gender. There are 15 region groups: 14 official regions ("oblast") of Kazakhstan and one group for those who lived outside of the country. The age groups are divided into four ranges between 18 and 65 years. We tried to keep the number of speakers of one gender per profile not more than 3. Additionally, two professional actors, one male and one female, were selected among these speakers to read excerpts from news articles and novels. The following Table 1 presents the distribution of the recorded speakers across the regions, gender and age groups.

Table 1. The distribution of the speakers

Age group	I		II		III		IV		
Region	F1	M1	F2	M2	F3	M3	F4	M4	Sum
1	3	3	2	1	2	1	2	1	15
2	2	3	2	1			2	1	11
3	1	1	2	3	2	1	1		11
4	3	2		1		1			7
5	2	2	2	1	2	2	2	1	14
6	2	2	2	2	2		1	2	13
7	2	2	1	2	2		2	1	12
8	2	1	1	2	1	1	2	1	11
9	3	2	2	1	3	1	1	1	14
10	1	1	2	2	1	1	2	1	11
11	2	1	2	1	1		2		9
12	2	2	2		2	1	2	1	12
13	2	2	2	1	1	1	1	1	11
14	2	1	1	1	1	2	1	2	11
15	1	3		1	2				7
Total	30	28	23	20	22	12	21	13	169
	34%		25%		20%		20%		

3.2. Experiments

Our first experiment was conducted for the isolated and connected words recognition. An acoustic model was trained using CMU Sphinx toolkits. In the experiments we aimed to build an adequate baseline acoustic model to be used in our intelligent voice system. Since most of the parameters in Sphinx are optimally pre-configured, there were basically two parameters that we were interested in to improve the recognition performance: LDA dimension and number of tied states (senones). Tables 2 and 3 show the results of the experiments conducted.

The front-end module was set to output default parameters such as 13 mel-frequency cepstral coefficients with their first and second derivatives. Additionally, speaker adaptation techniques such as CMN [2], LDA [3] and MLLT [4] are performed on feature vectors. We used a context-dependent tied-state continuous Hidden Markov Model with 8 Gaussian mixtures per state. The total number of test words is 5052. The dictionary is compiled from the text materials used for telephone speech and contains 1684 words with their spellings as a phonetic transcription, since the orthographic transcription of Kazakh roughly corresponds to a broad phonetic transcription.

It can be seen from the tables that the optimal size of reduced space was found to be 23 as opposed to the system's default value 29. As for the senones, the best result was obtained with 2000 senones.

In the next experiment we used our data for the large vocabulary continuous speech recognition task. Here we used almost the same parameters for the acoustic modeling as in the previous experiment. The values for the LDA dimension and the number of senones are chosen to be 23 and 2000, respectively. The audio data was separated into training and test sets. The test set is balanced based on gender and includes one representative from each region. The quantitative information about both sets is given in Table 4. For the language modelling we used our transcripts to build a trigram based model. The overall performance of recognition on test data is 6.9% WER.

Table 2. Performance for different LDA dimensions (with 2000 senones)

Dimension	WER, %
29	4,4
25	4,3
24	4,5
23	**4,1**

Table 3. Performance for different numbers of tied states

# senones	WER,%
3500	5.8
3000	4.9
2500	4.8
2200	4.4
2000	**4.1**
1500	4.8

Table 3. Distribution of data in training and test sets

	Train set	Test set
# of speakers	153	16
# of audio files	11367	1176

For the speech synthesis we used MaryTTS to train one male and one female voice. There are about 6 hours of audio data for each of the voices. The main approach is also based on Hidden Markov Models. Some NLP components such as grapheme-to-phoneme models, basic POS-tagging are built during the experiments. The work on speech synthesis is still in progress.

4. Conclusion and Future Work

In this work we implemented the intelligent voice system for the development of speech-enabled call-centres supporting Kazakh language. Also we conducted the experiments on Kazakh speech recognition and synthesis. Although the results for the speech recognition are encouraging, they are obtained on clean speech data. Therefore, our next work will concern collecting and processing the real data in office environment as well as improving the system functionality.

Acknowledgements

The work is supported by the Ministry of Education and Science of the Republic of Kazakhstan.

References

[1] Z. Yessenbayev, O. Makhambetov, and M. Karabalayeva, “Kazakh Text Corpus: Description, Tools and Statistics,” Int. scientific-theoretical conference “Modern Kazakh Linguistics: Actual Problems of Applied Linguistics” (2012), pp. 61-65.

[2] Liu, F.-h., Stern, R.M., Huang, X., Acero, R., “Efficient Cepstral Normalization for Robust Speech Recognition,” In Proceedings of the workshop on Human Language Technology (1993), pp. 69–74.

[3] Haeb-Umbach, R., Ney, H., “Linear discriminant analysis for improved large vocabulary continuous speech recognition,” IEEE Int. Conf. on Acoustics, Speech, and Signal Process (1992), vol. 1, pp. 13–16.

[4] Gopinath, R.A., "Maximum likelihood modelling with Gaussian distributions for classification," IEEE Int. Conf. on Acoustics, Speech and Signal Processing (1998), vol.2, pp. 661-664.

Low Power Shoe Integrated Intelligent Wireless Gait Measurement System

Y. Wahab[1], M. Mazalan, N. A. Bakar, A.F. Anuar, M.Z. Zainol, F. Hamzah
Advanced Multidisciplinary MEMS Based Integrated Electronics NCER Centre of Excellence (AMBIENCE),
School of Microelectronic Engineering,
Universiti Malaysia Perlis (UniMAP)

E-mail: yufridin@unimap.edu.my

Abstract. Gait analysis measurement is a method to assess and identify gait events and the measurements of dynamic, motion and pressure parameters involving the lowest part of the body. This significant analysis is widely used in sports, rehabilitation as well as other health diagnostic towards improving the quality of life. This paper presents a new system empowered by Inertia Measurement Unit (IMU), ultrasonic sensors, piezoceramic sensors array, XBee wireless modules and Arduino processing unit. This research focuses on the design and development of a low power ultra-portable shoe integrated wireless intelligent gait measurement using MEMS and recent microelectronic devices for foot clearance, orientation, error correction, gait events and pressure measurement system. It is developed to be cheap, low power, wireless, real time and suitable for real life in-door and out-door environment.

1. Introduction

This paper proposed a complete system called i-SMART ShoE II, a newer version to i-SMART Shoe. The i-SMART SHoE II focuses on both foot clearance measurement and orientation of the swing phase and also plantar pressure measurement during stance phase. The foot clearance measurement is the measurement of the distance between foot or shoe bottom and ground. This system is the combination of the ultrasonic sensor readings and the inertial measurement unit sensor. The ultrasonic sensor is used for direct clearance measurement and the inertial measurement unit sensor is used for orientation measurement. Based on the orientation measurement, the clearance measurement is corrected by an algorithm processing by microcontroller. The correction is required due to the orientation of foot during swing phase, landing phase, and toe off phase. In those situations, the positioning of ultrasonic on the back of the shoes may cause its orientation to be not perpendicular to the ground. So, an algorithm is required to recalculate and correct the data so as to represent the real and accurate data. Besides that, the characterization of a low cost piezoceramic device is performed so as to evaluate its capability in capturing foot plantar pressure as an insole embedded pressure sensor array. The insole is to be inserted into the shoes so that it can record plantar pressure when the subject is walking, running or standing. This capability makes it to be able to produce the foot pressure mapping for biometric application and early stage of diabetic detection.

[1] To whom any correspondence should be addressed.

2. Experimental setup and Hardware Description

Figure 1 illustrates a block diagram of the i-SMART SHoE II prototype system. The system design is a stand-alone system consisting of six parts which are Inertia Measurement Unit (IMU), ultrasonic sensor, ceramic based piezoelectric sensor (piezoceramic sensor), radio transmitters and receivers (XBee modules produced by Digi International Inc., USA), Arduino processing unit and personal computer software. The piezo pressure sensor array is used to obtain the pressure map under the sole. Ultrasonic and IMU sensor operated as orientation and distance measurement parts. The sensors are attached to the data acquisition system via Xbee transmitter and receiver while the data acquisition system at microcontroller will be connected to the PC via a USB interface.

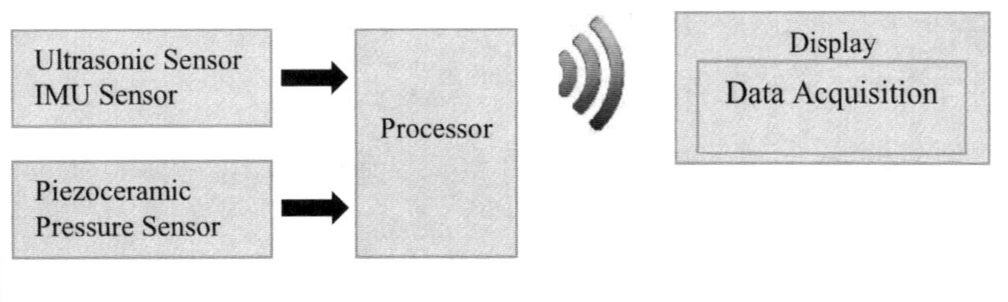

Figure 1. i-SMART SHoE Block Diagram

2.1 Foot Clearance and Orientation Measurement Requirement

Foot clearance is a critical parameter is identifying and prevention risk of falls among the elderly. It is also important for rehabilitation monitoring and sports training. The measurement should be done in real life activities to ensure the study of foot clearance is accurate and real. It is necessary to attach the sensor to shoe without interfering the subject's movement. The sensor should have the ability to measure clearance of both feet as explained in [1-4]. In order to have such ability, the sensor should be small and as light as possible with weight under 300 g as stated in [5]. Study in [6] suggests the closest maximum range for toe clearance of up to 20 cm is preferred. The study in [6] also suggests minimum foot clearance during the swing phase of walking to be within 3 cm above the walking surface. For sports application, this range will be significantly more [7]. The selections of sensors and hardware are depending on the requirement of the foot clearance and orientation measurement. The sensors that are used in this research are ultrasonic sensor and IMU for determination of the foot clearance as well as the orientation of foot position respectively. The monitoring system is not complete without other additional components that support the key functionality of the entire system. The hardware design consists of Arduino microcontroller, 2.4 GHz IEEE 802.15.4 XBee transceiver and power supply unit. Figure 2 shows the attachment unit for foot clearance and orientation measurement system.

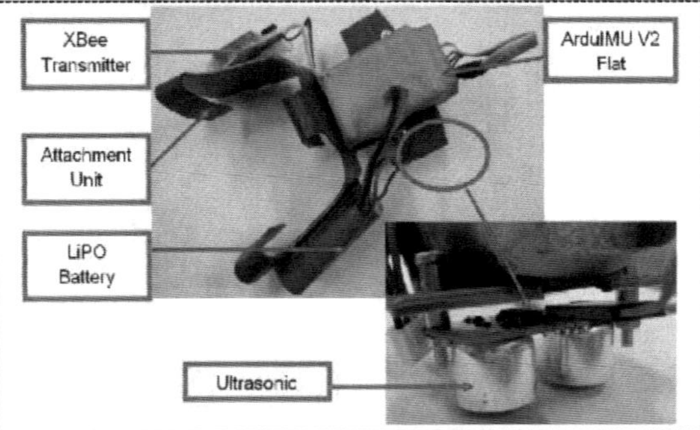

Figure 2. Shoe Attachment Unit

2.3 Insole Pressure Sensor

In our previous research as reported in [1], a MEMS pressure sensor for foot plantar pressure measurement was completed. However, in this work, we are trying a new cheaper and lower power device based on piezoelectric sensing. The piezoceramic pressure sensors as discussed below are originally the part of a cheap buzzer. It is taken out of the buzzer package carefully and repackaged with wires in an insole at three difference pressure points. These highly ruggedized sensors are being connected to XBee transmitter. The electronic board for RF reception is attached to a computer using USB connectivity and the computer placed nearby within certain range with the shoes integrated wireless acquisition board. In the computer, Arduino software has been used for the analysis of the wirelessly transmitted foot pressure and movement pattern readings.

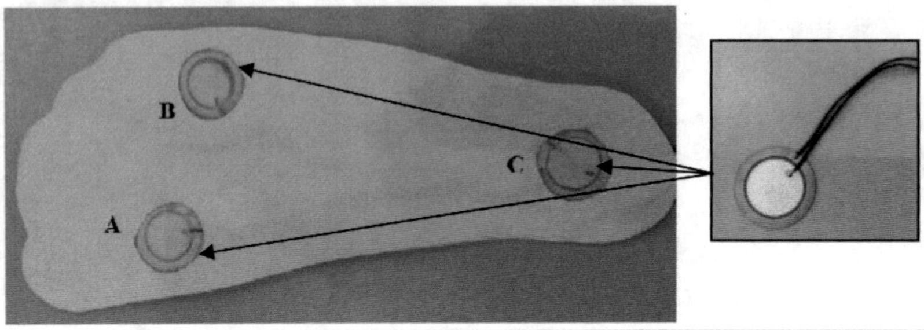

Figure 3. Top View of Sole Piezoceramic Pressure Sensors

Voltage to pressure converter forms the heart of this system. Voltage induced by the pressure of the foot plantar points is the key raw data. It is related to the mass (m) of its equivalent pressure unit either in N/m^2, Pa or Psi. The converter consists of a basic force formula with respect to the piezoceramic sensing area, *A*. The total force can be calculated from the output data using the equation (1) and (2). The equivalent value for pressure unit is shown in Table 1.

$$F = ma \quad (1)$$

$$P = \frac{F}{A} \quad (2)$$

Table 1. Equivalent Value for Pressure Unit

N/m^2	Parameter
1.0	1 Pa
1.0	0.0001450377 Psi

2.4 Power System

Based on the foot clearance, orientation and pressure measurement requirement, the real life measurement can be realized by wirelessly. For i-SMART SHoE II design, the system is powered up by a Lithium Polymer battery. It is a rechargeable battery and can be used up to 4 hours. Based on the limitation of battery lifetime, this paper comes out with the idea to reduce overal power consumption by turning on and off the different units beginning and ending at certain events.

2.5 Uniqueness of this work
The human gait consists of two phases namely stance phase, which is the period when the foot is on the ground, and the swing phase that corresponds to the period the foot is swung above the ground. By measuring the parameters during these two phases, a complete gait analysis measurement can be achieved. As explained in the subsections above, this work is unique as it integrates both foot plantar pressure measurements and foot clearance together in a new system, enabling both gait phases to be measured. Moreover, it innovatively uses very low cost piezoelectric sensors for pressure and event sensing. In addition, this work also enables intelligent system power control based on gait phase that saves electrical power usage of the system. For diabetic ulceration, pressure is the indicator. Therefore, by being able to detect the pressure beneath the foot especially at a number of points, possibility of ulceration can be detected and thus corrective measures can be implemented. This way the developed system may help the communities such as athletes, the elderly, patients undergoing rehabilitation, those diagnosed with diabetes and many more.

3. Experimental Result
The i-SMART SHoE II is a complete attachment unit with custom designed software for foot clearance measurement system. The total weight of attachment unit is about 115g. So, the weight of attachment unit fulfills the weight requirement of attachment unit which is less than 300g [5,7].

3.1 Ultrasonic and IMU System Evaluation
The testing is performed involving few subjects by attaching the attachment unit on their shoes and allow them to walk naturally in the outdoor. Figure 4 shows the gait pattern for walking subject using combination of ultrasonic sensor and IMU sensor. For this system, the data collected is more detail compared to the previous system. The system will analyzed the gait pattern based on foot clearance that classified to gender, feet, age range, toe off phase, maximum foot clearance phase and landing phase. The system is fixed on the attachment unit and attached to the subject shoes. The subjects need to walk as their daily activity. To ensure this system can be used in real life, this experiment is conducted out of laboratory. In this version, the error correction features is implemented where the error correction features is the new approach for foot clearance measurement which is being the novelty of this research.

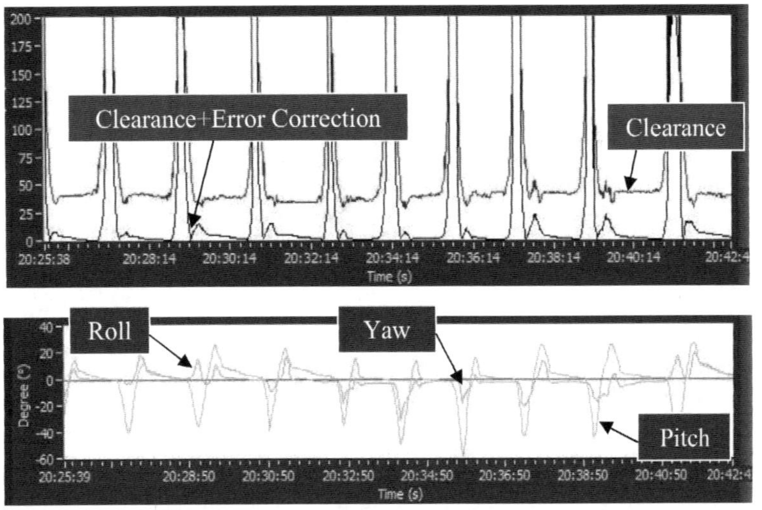

Figure 4. Movement Pattern by IMU and Ultrasonic Sensor

3.2 Pressure Test Bench

The piezoceramic sensors developed was tested in a custom made test bench with varying loads for calibration and to identify sensor linearity and response. The system is also evaluated during real walk and run activities to further characterize the system. Sensors are calibrated under linearly varying weight by the custom made test bench. The changes in output values with different loads applied and voltage generated is monitored. Voltage generated is found to vary proportionally with the applied load. Pressure is applied uniformly on the piezoceramic disk causing the piezoelectric effect to be generated. The experimental result is shown in Table 2.

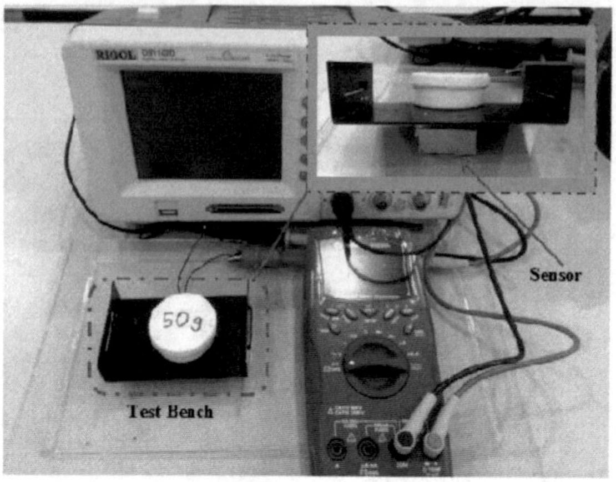

Figure 5. Calibration Procedure Undertaken by Custom Made Test Bench

Table 2. Voltage generated by applied force.

Mass (kg)	Voltage (V)
0.5	2.749
1.0	6.335
1.5	8.129
5.5	18.952

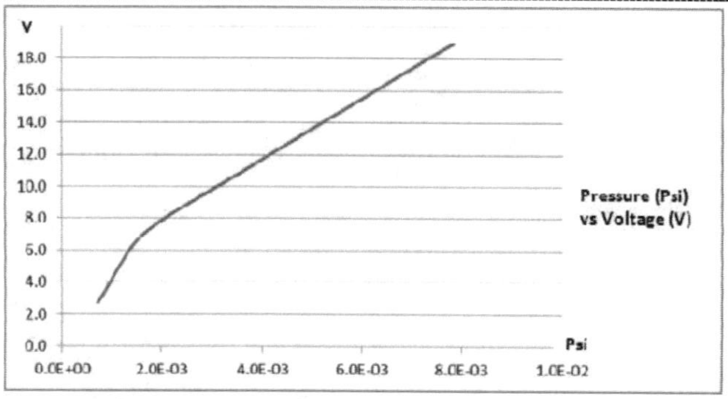

Figure 6. Comparison of Voltage (V) Generated and Equivalent Pressure Applied (Psi)

3.3 Characterization of Insole Pressure Sensor

To characterize the insole pressure system in real activities, the subject place the insole in a fitted shoe and the data is recorded. First the system is calibrated with no load pressure and measure the free condition. The pressure distribution curves were plotted in two difference activities; namely walk and run. Each subject first walked freely for 4 times. Following the calibrations, another 4 trials of "natural gait" were collected followed by ten other trials of various walking styles activities [8]. The same is done for running evaluation. The following figure shows the characterization of insole pressure sensor during walking and running.

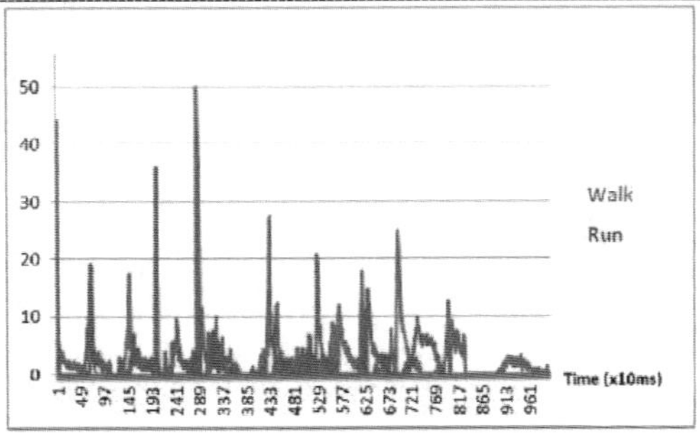

Figure 7. Movement Pattern for Insole Pressure Sensor

4. Conclusion

The development of the custom designed system which includes ultrasonic sensor, piezoceramic sensors, IMU, Arduino microcontroller and XBee modules is successfully performed and the performance of the system is analyzed in real life activities. The novelties of this research are wearable shoe attachment for wireless monitoring of foot clearance, orientation and sole pressure measurement during walking or while performing the daily routines in the house or outside of the house. The applications of this system are for monitoring the gait of the elderly, rehabilitation, injury prevention among athletes, biometric and early detection of diabetic ulceration. This system has high potential to be marketed especially for rehabilitation centers, sports centers, health centers, hospitals, research organizations and even stage performance training centre. We have presented a successful development and characterization of the system proved by the presented results. It is an evident to support the claim that this low cost low power consumption ultra-portable wireless i-SMART SHoE II system and its underlying concepts are realizable and practical.

5. References

[1] Wahab, Y., Zayegh, A. and Begg, R.K., 'Comparison and Silicon Realization of Custom Designed MEMS Biomedical Pressure Sensors', in Proc. 21st IEEE International Conference on Microelectronics, , Marrakech, Morocco. (2009), December 19-22, pp 418-421.

[2] Lai, D.T.H., Begg,R.K., Charry, E.,Palaniswami, M. and Hill, K., 'Measuring toe clearance using a wireless inertial sensing device', in Proceedings of International Conference on Intelligent Sensors, Sensor Networks and Information Processing, Sydney, Australia. (2008), December 15-18, pp. 375-380.

[3] Aminian, K. and Najafi, B., 'Capturing human motion using body-fixed sensors: Outdoor measurement and clinical applications', Computer Animation and Virtual Worlds, (2004), vol. 15, no. 2, pp. 79-94.

[4] Morris, S.J. Shoe-Integrated Sensor System for Wireless Gait Analysis. PhD Thesis, MIT, USA. (2004).

[5] Stacy J. Morris, Joseph A. Paradiso, "Shoe-Integrated Sensor System for Wireless Gait Analysis

and Real-Time Feedback," *IEEE Transactions on Information Technology in Biomedicine,* July (2008), vol. 4, pp. 413-423.

[6] Begg, R.K., Best, R.J., Taylor, S. and Dell'Oro, L., 2007. 'Minimum foot clearance during walking:Strategies for the minimization of trip-related falls',Gait & Posture,vol. 25, pp.191-198.

[7] Wahab, Y., Bakar, N.A., 'Gait analysis measurement for sport application based on ultrasonic system', in Proceedings of 2011 IEEE 15th International Symposium on Consumer Electronics (ISCE), Singapore, 2011, pp 20-24.

[8] S. J. Morris, "A shoe-integrated sensor system for wireless gait analysis and real-time therapeutic feedback," Sc.D. dissertation, Harvard/MIT Division Health Sci. Technol., Cambridge, MA, Jun. 2004.

Journal of Physics: Conference Series **495** (2014) 012045 doi:10.1088/1742-6596/495/1/012045

Design and Analysis of Various Microcantilever Shapes for MEMS Based Sensing

H. F. Hawari, Y. Wahab, M. T. Azmi, A. Y. Md. Shakaff, U. Hashim, S. Johari

Advanced Multidisciplinary MEMS-Based Integrated Electronic NCER Centre of Excellent (AMBIENCE), School of Microelectronic Engineering, Universiti Malaysia Perlis (UniMAP), 02000 Kuala Perlis, Perlis, MALAYSIA

E-mail: huzeinfahmi@gmail.com

Abstract. One the major requirements for sensing devices is its sensitivity. This paper presents the design, analysis and simulation of various types of MEMS based microcantilevers shapes and their influence on sensing sensitivity. A commercial finite element analysis software ANSYS is used to analyze the designs. Six different analytical model of the microcantilever were investigated for its Equivalent (von-Mises) stress, maximum deflection and fundamental resonant frequency. Finally, the best sensitive structure is then identified. It is found that the trapezoidal design microcantilever would have significant sensing advantages compared to other remaining microcantilever shape.

1. Introduction

In the mid-1990s, the interest of cantilever-based sensing has initiated due to the advent of atomic force microscopy (AFM) application [1]. However, recently it has taken wide surge interest since it is now made commercially available. Apart from that, microcantilever is the most MEMS versatile structure as it can be used as physical, chemical or biological sensor by detecting the changes in cantilever bending or vibration frequency. A general structure of microcantilever is shown in Figure 1 where a cantilever is actually a beam rigidly supported at one end, or in the middle, but not at both ends, which has forces applied along the free arm or at the free end [2]. Compared to conventional sensor, microcantilever based sensing has very high sensitivity, short response time and low power operation due its miniature in size.

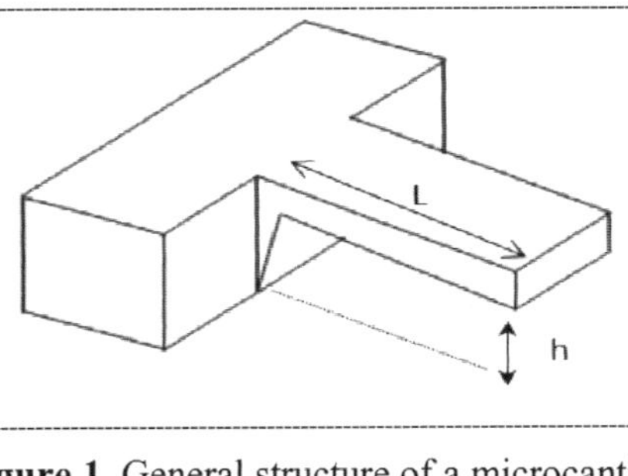

Figure 1. General structure of a microcantilever

One of the sensing methods done was to coat the microcantilever with a piezo material thin film which in this case is zinc oxide, to achieve the resistive deflection readout [3]. In the case of chemical-sensing applications, due to chemical reactions, such as the platinum-assisted catalytic conversion of hydrogen and oxygen into water, the deflection of the cantilever happened at very high sensitivity [4].

One of microcantilever most current research is as a platform for gas sensing application. Microcantilevers that are used to sense the presence of a certain particle or analyte are coated with a chemically sensitive material. This material will provide a high degree of specificity in detecting certain particles or analytes within a sample.This deflection of micro cantilever occurs when a specific mass of an analyte is specifically adsorbed on its surface. As a result, it would produce a deflection at the free end when a force is applied.

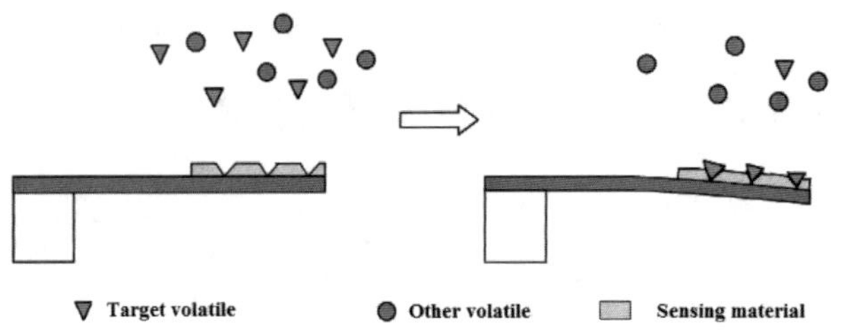

Figure 2. General structure of a microcantilever for gas sensing application

For sensing, microcantilever relies on their flexibility or elasticity to create some type of measurable change when exposed. The reaction of the cantilever is referred as mechanical stress and would result in a change either the cantilever mechanical or electrical properties. The most common properties used to measure this change the microcantilever's natural resonant frequency, angular deflection or resistivity. This technique can be used to directly measure surface stress changes and as a sensing mechanism by functionalizing one surface of the cantilever with a specific detector layer [5].

To obtain an application specific optimum design parameter and predict the cantilever performance, ANSYS workbench V12.0 software was used. The microcantilever size leads to significant advantages in the absolute device sensitivity. Precise measurement of the deflection at the end of the cantilever was achieved through this arrangement.

II. Design Parameter of Cantilever

A cantilever is a beam fixed firmly at only one end and left freely at other end. The beam carries the load to the support where it is resisted by moment and stress.

The structure is made up of silicon single crystal where the fracture strength of silicon is 7000 MPa [6] while other properties as shown in the table 1.

Table 1 Material Properties used

	Value	*Conditions*
Density	2330 kg/m^3 [7]	Solid Density
Poisson's ratio	0.22[8]	Silicon Substrate, isotropic & linear thermo elastic
Young's modulus	165GPa[8]	Silicon Substrate, isotropic & linear thermo elastic

Resonant frequency is the frequency of a system at which it oscillates at maximum amplitude. With little damping, this frequency is usually equal to the system's natural frequency. When a system reaches this resonant frequency, this state is resonance. As the mass of the system changes, so does the resonant frequency.

$$(\omega^{o}) = \sqrt{\frac{k}{m}} \tag{1}$$

Natural frequency

$$(k) = \frac{Et^{3}w}{4t^{3}}) \tag{2}$$

Spring constant

The natural frequency (ω^{o}) of a cantilever is related to its spring constant (k) and mass (m). This is also applies for both macro and microcantilever. For a rectangular cantilever beam, the spring constant (k) is a function of

E = Young's modulus of Elasticity (a property of the material)
w = Width of the beam
l = length of the beam
t = thickness of the beam

Young's modulus of Elasticity (E) is the measure of the stiffness or elasticity of a given material. The stiffer or less elastic a material is, the higher the E value. Young's modulus allow the behaviour of a material to be evaluated under load or stress.

The fundamental frequency of vibration is

$$f = \frac{1}{2\Pi} \bullet \sqrt{\frac{k}{m}}) \tag{3}$$

For a rectangular profile microcantilever, the differential surface stress (Δ□) and deflection (Δz) are related to Stony equation given as [3] :

Journal of Physics: Conference Series **495** (2014) 012045 doi:10.1088/1742-6596/495/1/012045

$$\Delta z = \frac{4(1-v)\,\Delta\sigma}{E}\left(\frac{l}{t}\right)^2) \quad (4)$$

where l and t are the length and the thickness of the cantilever, and E and v are the elastic modulus and Poisson ratio of the cantilever material.
The fundamental resonant frequency (f_o) for rectangular profile cantilever of mass density (m) is given as

$$f_o = \sqrt{\frac{E}{m}} \bullet \frac{1}{t^2}) \quad (5)$$

For a microcantilever of trapezoidal profile, $t(x) = t_l + (t_o - t_l)\, x/l$, the Stoney equation(4) can be given as

$$\Delta z = \frac{3(l-v)\Delta\sigma l^2}{E(t_o - t_1)^2}\left[\ln\left(\frac{t_o}{t_1}\right) + \frac{t_1}{t_o} - 1\right] \quad (6)$$

The geometric structure of the cantilever beam and also the properties of the material used to construct the shapes of cantilever determine the stiffness of the cantilever. Thus the sensitivity of the cantilever beam changes with the change is shape of the cantilever.

The design and dimension of the microcantilever used in the research is summarized in figure 3. Different microcantilever design based on trapezoid and rectangular structure with double and single legged structure were designed and simulated.

Microcantilever Dimension x= 110 µm , y =50 µm, z= 40 µm		*Microcantilever Design*
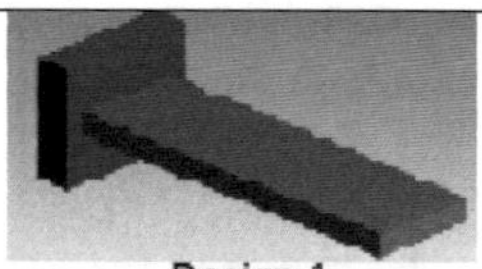 Design 1	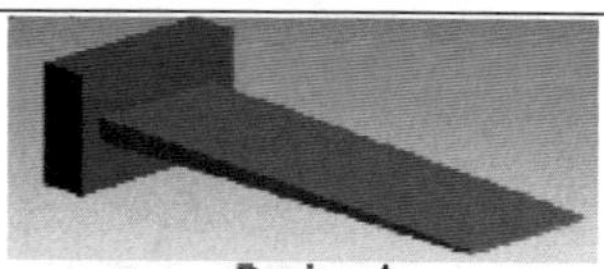 Design 4	*Rectangular/ Trapezoid*
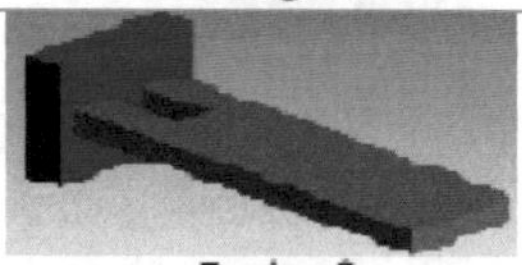 Design 2	 Design 5	*Double-legged Rectangular/ Trapezoid*
 Design 3	 Design 6	*One-legged Rectangular/ Trapezoid*

Figure 3. Schematic Design of Microcantilever

Journal of Physics: Conference Series **495** (2014) 012045 doi:10.1088/1742-6596/495/1/012045

III. Simulation results

Using static structural analysis in Mechanical (ANSYS Multiphysic) which uses the ANSYS solver, static structural analysis were performed. Structural analysis determines the Equivalent Elastic Strain, Equivalent Stress and total deformation in structure or components caused by loads that did not induce significant. In this static structural analysis, all the designs in figure 4 were applied with 50uN on the surface in the downward direction. It is assume that there were steady loading and response conditions for area.

For Equivalent (von-Mises) stress analysis, equation (7) are used for finding the strain value.

$$\sigma_e = \left[\frac{(\sigma_1 - \sigma_2)^2 + (\sigma_2 - \sigma_3)^2 + (\sigma_3 - \sigma_1)^2}{2} \right]) \qquad (7)$$

For Equivalent (von-Mises) stress comparison (figure 4), maximum strain for all six microcantilever design was simulated and compared. It was found that the trapezoidal design microcantilever would have a slight higher strain compared to rectangular design microcantilever.

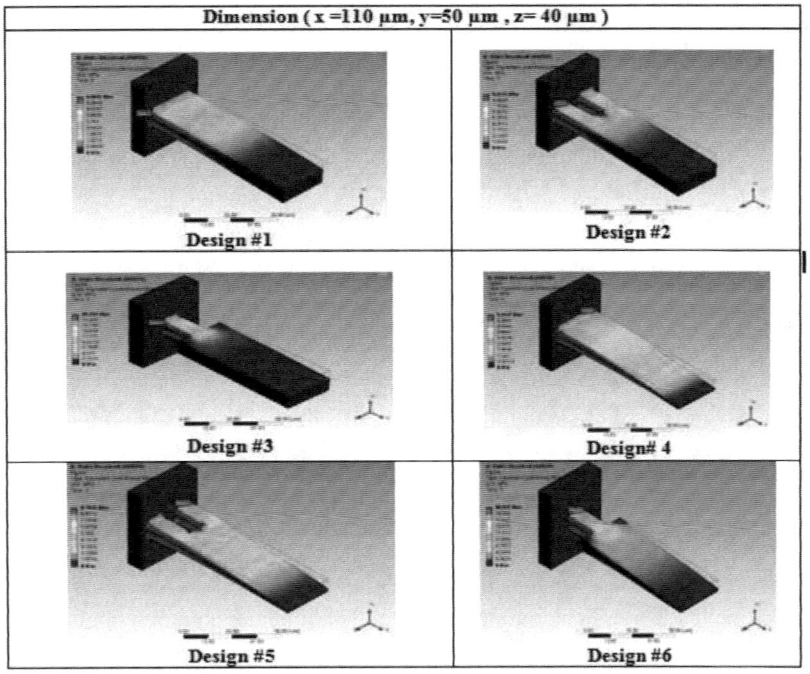

Figure 4. Simulated Equivalent (von-Mises) stress towards designed microcantilever

For example, in figure 5, there was 0.13% strain difference between rectangular microcantilever and trapezoid microcantilever. When double-legged design was simulated, strain difference between rectangular microcantilever and trapezoid microcantilever was found to increase by 59%. The highest Equivalent (von-Mises) stress was simulated on the one-legged trapezoid microcantilever where the maximum strain increased significantly to 20.361Mpa compared to the double-legged trapezoid microcantilever.

Journal of Physics: Conference Series **495** (2014) 012045 doi:10.1088/1742-6596/495/1/012045

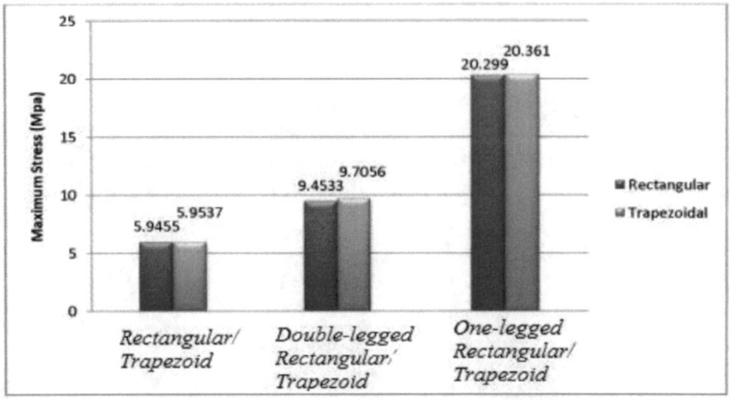

Figure 5. Comparison of Equivalent (von-Mises) stress result of the microcantilever design

For displacement comparison (figure 6), maximum displacement for all six microcantilever design was simulated and compared. For the displacement comparison, there is a significant displacement difference between the rectangular microcantilever and trapezoid microcantilever design.

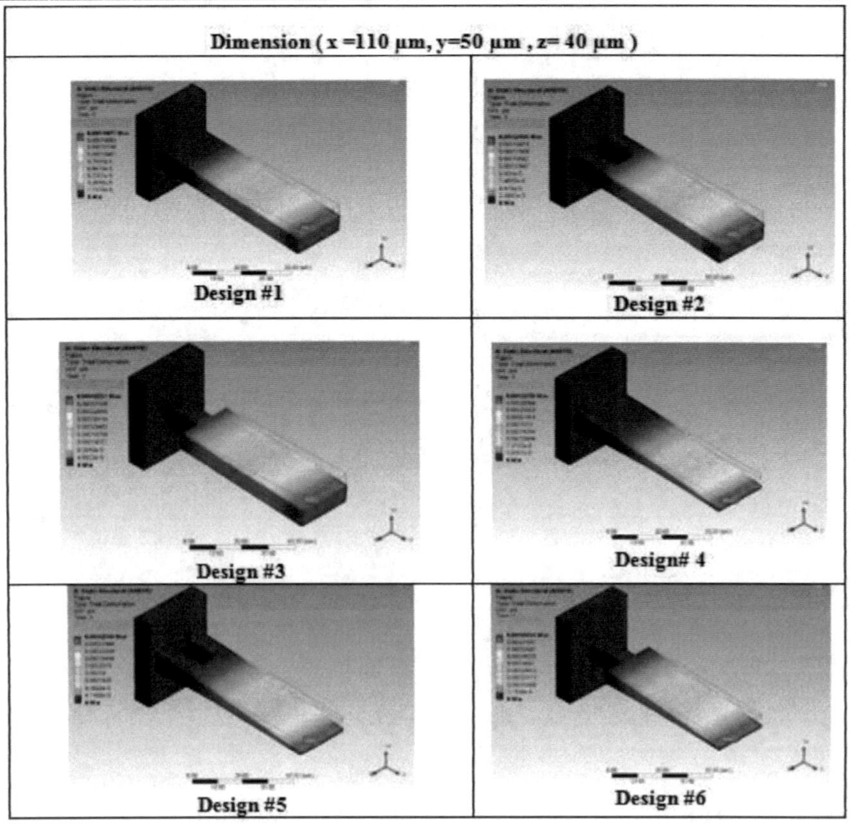

Figure 6. Simulated displacement result of designed microcantilever

As in figure 7, shows the comparison between rectangular and trapezoidal shape. The 50 μN load is given to the microcantilever surface. For rectangular microcantilever shape, it showed an increased of displacement from between design 1 to design 3. However, the displacement value for the trapezoidal microcantilever shape increased to about 2 times more for the rectangular microcantilever shape.

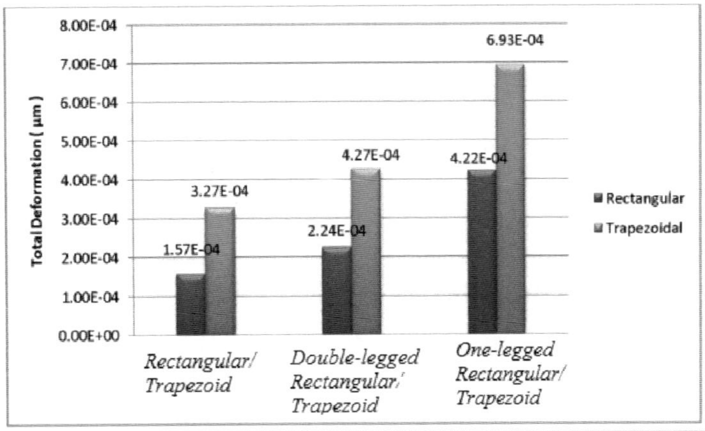

Figure 7. Comparison of displacement result of the microcantilever design

Next, using ANSYS solver, natural frequency for each mode using was compared as in figure 8. Natural frequencies and mode shapes are important parameters in the design of a structure for dynamic loading conditions. In this analysis, we determine 6 modes of natural frequency and mode shape for each design to determine the level of flexibility of each design.

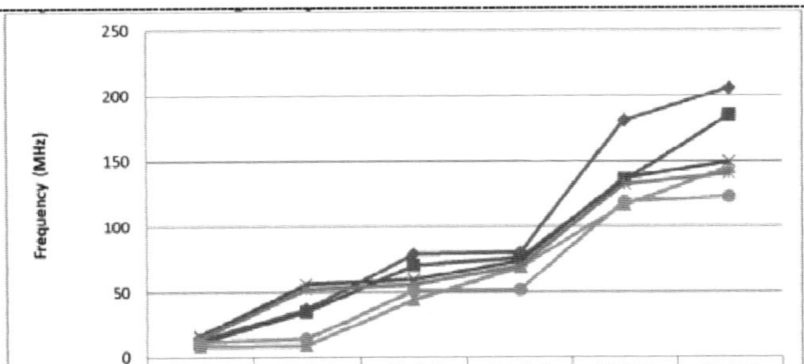

	Mode 1	Mode 2	Mode 3	Mode 4	Mode 5	Mode 6
Design 1	13.349	37.261	79.158	80.406	181.15	205.63
Design 2	11.507	35.289	70.509	76.279	135.5	184.9
Design 3	8.6436	9.1618	44.016	68.878	116.21	144.67
Design 4	16.428	56.065	59.513	73.797	136.84	149.63
Design 5	14.636	52.378	56.053	69.759	132.28	140.07
Design 6	11.58	14.143	51.446	51.98	118.67	122.45

Modes

Figure 8. Comparison of natural frequency of microcantilever design

In figure 8, we can see that microcantilever trapezoid design has a higher natural frequency than rectangular microcantilever design.

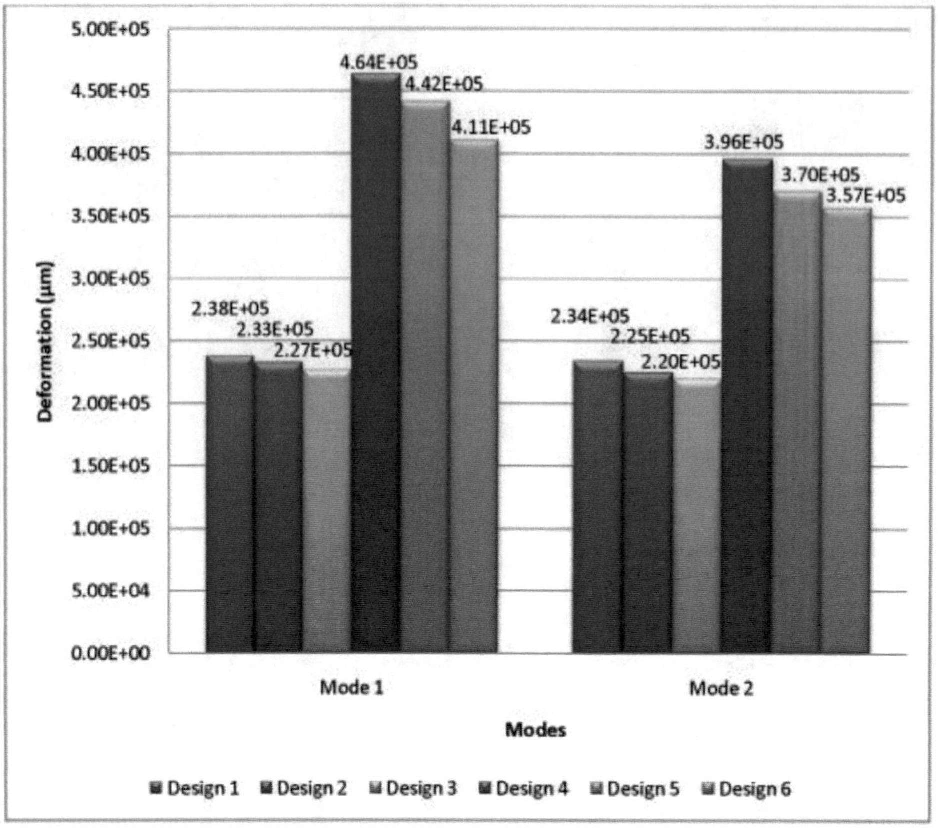

Figure 9. Simulated displacement result of designed microcantilever using mode 1 and mode 2

Next, using the mode 1 and mode 2 value, comparison of the displacement of the microcantilever design were analyzed. In figure 9, the sensitivity of the microcantilever were calculated by determing the value ΔZ* Freq.

ΔZ is the deformation value and freq is the frequency of mode 1. From the table 2, design 4 shows highest value of ΔZ* Freq followed by design 5 and design 4. This shows that the design 4 would be the most suitable design for dynamic mode operation.

Table 2 Sensitivity values of microcantilever design shape

	Deformation, ΔZ	*Frequency, MHZ*	ΔZ* *Freq*
Design 1	2.38×10^5 µm	13.349	3.18
Design 2	2.33×10^5 µm	11.507	2.68
Design 3	2.27×10^5 µm	8.6436	1.96
Design 4	4.64×10^5 µm	16.428	7.62

Journal of Physics: Conference Series **495** (2014) 012045 doi:10.1088/1742-6596/495/1/012045

Design 5	4.42×10^5 μm	14.636	6.47
Design 6	4.11×10^5 μm	11.58	4.76

IV. Conclusion

One of the most MEMS versatile structures available is microcantilever where it is used in many type sensing application. However, typical microcantilever shape available is in the form of uniform rectangular beam. In this research, microcantilevers of various shapes have been designed and their displacement, stresses, natural frequency and mode shape were analyzed when the same amount of force is applied. The highest Equivalent (von-Mises) stress was simulated from the one-legged trapezoid microcantilever shape where the maximum strain increased significantly to 20.361Mpa compared to the other microcantilever. However, since the fracture stress value for the material is much higher, this indicates that the cantilever is reliable enough for detection and will break under normal conditions. Another observation is on the microcantilever displacement analysis. The trapezoidal design microcantilever would have significant displacement compared to rectangular design microcantilever where the value for the trapezoidal microcantilever shape increased to about 2 times more than the rectangular microcantilever shape. This is important as with the same amount force applied, the trapezoidal design microcantilever was able to deflect more thus contributing to higher resistive value contribute by the piezoresistive material and ultimately, increasing the microcantilever sensing sensitivity. Design 4 showed the highest value of ΔZ* Freq than the other microcantilever design. For dynamic mode operating modes where the microcantilever is used especially to extract and sensing mass changes, design 4 would be the most suitable design due its sensitivity. Finally, based on the result of this investigation, we can conclude that the trapezoidal microcantilever shape has better overall sensitivity than the rectangular microcantilever shape due its better deflection and resonant frequency characteristics.

References

[1] Binnig G, Quate CF, Gerber C. Atomic force microscope. Phys Rev Lett. 1986;56:930–933.
[2] Anja Boisen ,Soren Dohn,Stephan Sylvest Keller,Silvan Schmid and Maria Tenje. "Cantilever like micromechanical sensor".IOP Publishing.(2011) February 28.pp:30
[3] Itoh T, Suga T. Force sensing microcantilevers using sputtered zinc-oxide thin-film. Appl Phys Lett. 1994;64:37–39
[4] Gimzewski JK, Gerber C, Meyer E, Schlittler RR. Observation of a chemical-reaction using a micromechanical sensor. Chem Phys Lett. 1994;217:589–594.
[5] Thundat T, Chen GY, Warmack RJ, Allison DP, Wachter EA. Vapor detection using resonating microcantilevers. Anal Chem. 1995;67:519–521.
[6] Petersen, K.E., Silicon as a mechanical material, *Proc. IEEE.*, Vol. 70, No. 5, 1982
[7] James F. Shackelford, William Alexander. "CRC Material Science and Engineering Handbook". CRC Press (2006), 3rd Edition, page 46
[8] David Roylance, Mechanics of Materials, John Wiley & Suns, Inc (1996) , p. 314

Journal of Physics: Conference Series **495** (2014) 012046 doi:10.1088/1742-6596/495/1/012046

Error Correction for Foot Clearance in Real-Time Measurement

Y. Wahab[1], **N. A. Bakar, M. Mazalan**
Advanced Multidisciplinary MEMS Based Integrated Electronic NCER Centre of Excellence (AMBIENCE),
School of Microelectronic Engineering,
Universiti Malaysia Perlis (UniMAP),
02000 Kuala Perlis, Perlis, Malaysia

E-mail: yufridin@unimap.edu.my

Abstract. Mobility performance level, fall related injuries, unrevealed disease and aging stage can be detected through examination of gait pattern. The gait pattern is normally directly related to the lower limb performance condition in addition to other significant factors. For that reason, the foot is the most important part for gait analysis in-situ measurement system and thus directly affects the gait pattern. This paper reviews the development of ultrasonic system with error correction using inertial measurement unit for gait analysis in real life measurement of foot clearance. This paper begins with the related literature where the necessity of measurement is introduced. Follow by the methodology section, problem and solution. Next, this paper explains the experimental setup for the error correction using the proposed instrumentation, results and discussion. Finally, this paper shares the planned future works.

1. Introduction

Due to the large incidence of fall among the elderly populations, the healthcare cost are increasing globally especially due to gait related injuries. As higher percentage of the world population, including Australia, is made up of the elderly, more incidents of elderly related fall injuries is expected each year. In Australia, the fall related injuries consume about $3 billion as reported in 1999 [1]. Foot clearance is an important gait parameter that directly influences the risk of fall among the elderly. The distance of shoe sole above the ground is represented by foot clearance which is the vertical spatial parameter of the foot during the swing phase of the gait cycle. In a recent analysis study of the tripping and falls risks among the elderly individuals during walking [1,2,3] , it is reported that the most critical event that can initiate the possibility of trip-related fall is the movement of the foot during mid-swing phase. This atmospherically critical parameter is called minimum foot clearance (MFC). The pattern of foot clearance during gait is depicted in Figure 1 where MFC of below 5 cm and foot trajectory of up to about 17 cm is shown [2].

To date, gait analysis study is still mostly performed in a specially set-up motion analysis laboratory using high-end motion imaging systems. In another measurement set-up, the analysis requires a

[1] To whom any correspondence should be addressed.

physician to observe visually the gait. These two approaches are undoubtedly expensive [4]. Despite the current practices, the reality proves that the demand for real life portable measurement and monitoring devices are raising fast. For that reason, a portable motion detector shoe is developed with ultrasonic and inertial measurement unit combined together to perform better in real life environment to allow comprehensive analysis and intensive monitoring. In situ measurement where the actual activities and measurement are performed such as reported in [5] reduces cost as the time requiring the presence of physician is reduced in addition to the cheaper device cost demanded. Potential of ultrasonic and its modelling for this specific application have been reported before [6,7,8].

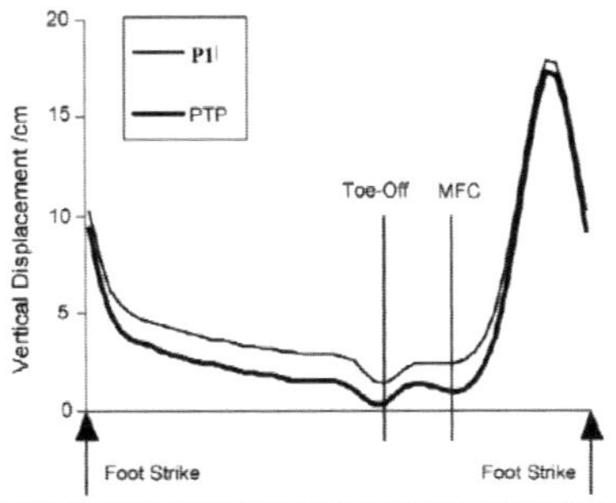

Figure 1. Foot trajectory during gait detailing the vertical displacement of foot for one gait cycle showing MFC during mid swing [1].

2. Methodology

2.1. Ultrasonic Distance Measurement

In order to measure the foot clearance measurement, the distance between foot and ground during walking is required. For this application, the ultrasonic system is used. The system consists of transceiver, controller board and wireless communication. The system is easily to attach to the shoes as wearable unit. This wearable unit is user friendly and low cost.

Figure 2 shows the block diagram of the measurement system. The ultrasonic signals with a frequency of 40 kHz are used with a target of maximum distance about 17cm. The maximum range was referred to the maximum of foot clearance as stated in [2]. The burst transceiver is recorded during sends by wireless communication processing consist of controller board and plotted as graph in custom design Gait Analysis Software. The foot clearance measurements directly plotted in the graph clearance versus time. Hence, the gait pattern captured. But, it is necessary to wait for the data to complete its transfer via wireless in order to get a stable data and optimum gait pattern. The power supply is the 7.4 V 1000 mAh Li-Po battery.

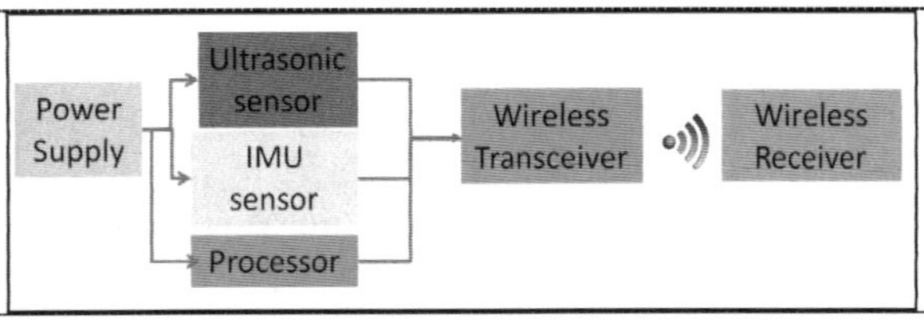

Figure 2. Foot Clearance Measurement Block Diagram.

2.2. Problem Statement

Ultrasound based sensing requires straight line arrangement as the signal travels that way [9]. However, when measuring the foot clearance during gait events and phases such as toe-off, landing, stance, and swing, the ultrasonic sensor is not parallel to the ground. It is resulting non actual distance measured for foot clearance. The situation is described in Figure 3 and Figure 4.

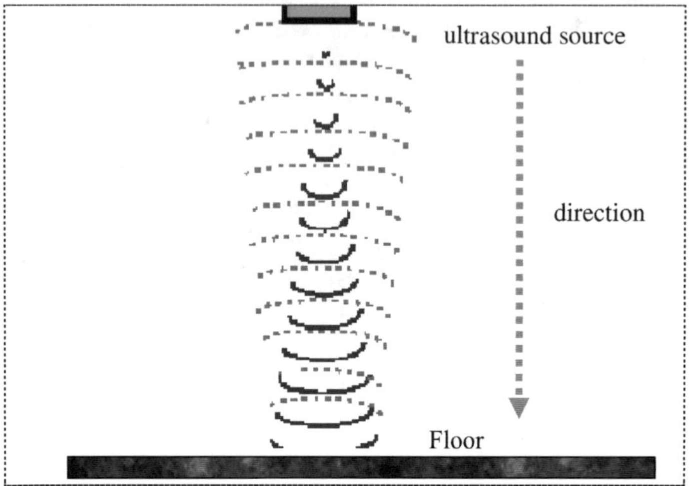

Figure 3. Direction of transmitted (blue) and returning (red) signals.

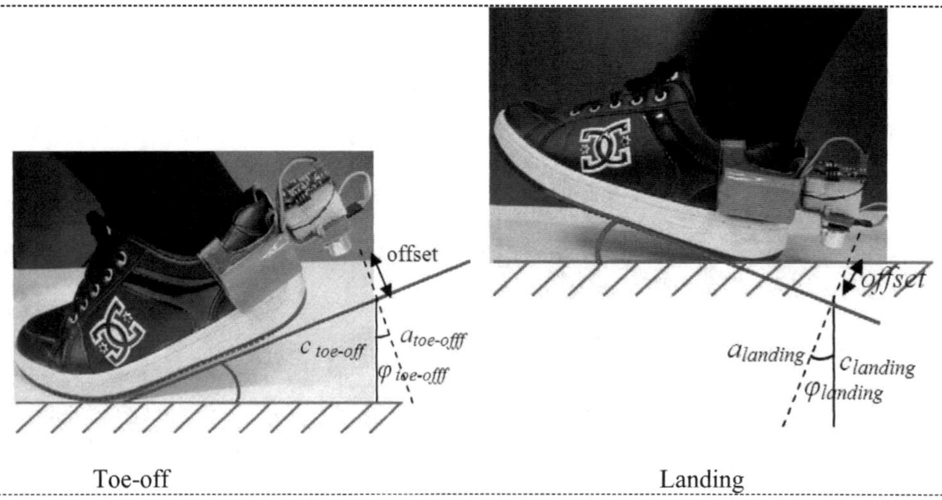

Figure 4. Foot angle ($\varphi_{toe\text{-}off}$) during toe-off phase where the distance measured is not the actual foot clearance. The same is for the landing part.

2.3. Proposed Solution : Error Correction

To solve the problem occur, the inertial measurement unit sensor (IMU) is used to detect the angle between foot and ground as shown in Figure 5. All parameters including yaw, roll and pitch are shown and can be detected by the IMU.

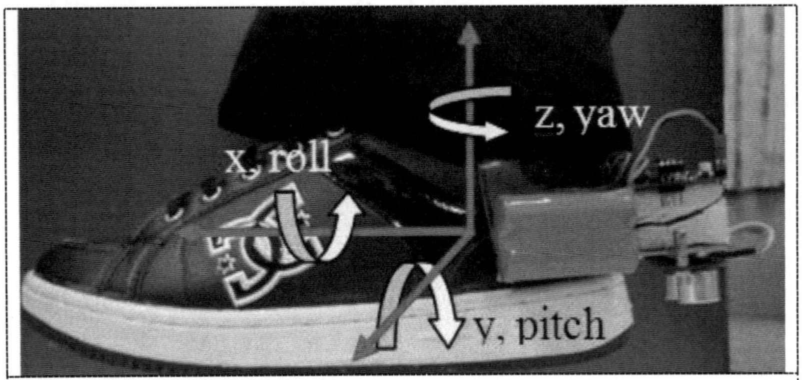

Figure 5. Yaw, Roll and Pitch

Then, the angle is used to correct the distance measured by ultrasonic sensor for foot clearance purpose. Besides that, the inertial measurement unit also gives the foot orientation data for roll and pitch data. Using trigonometry approach as shown in Figure 6, the angle identified is fed to the algorithm for real time foot clearance data correction. The equation (1) is valid to be applied during toe-off phase while equation (2) is valid for landing phase.

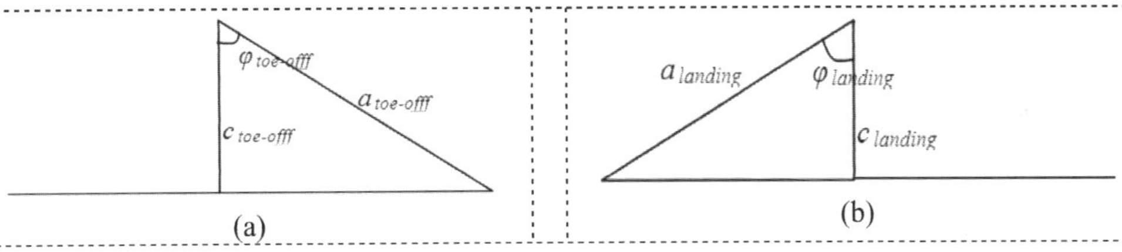

Figure 6. Trigonometry of (a) toe-off phase and (b) landing phase.

$$c_{toe-off} = a_{toe-off} \cos \varphi_{toe-off} \quad (1)$$

$$c_{landing} = a_{landing} \cos \varphi_{landing} \quad (2)$$

where c is the foot clearance in mm, φ is the angle between the foot and the ground and a is distance measured by ultrasonic sensor in mm.

This system is attached at the back of subject shoes using a special attachment unit. So the subjects wore their own shoes. While the controller board was attached on top of the attachment unit with covered by heat shrink as shown in Figure 6. By applying the wireless transmitter to the board, this system can be operated wirelessly. Each subject is required to walk freely. The data captured and recorded immediately in graph form and numeric value. The data collected from the healthy subjects is used as a reference pattern for the un-normal gait analysis.

2.4. Data Processing

Output from the sensor was processed by controller board into digital representations and will be sent for data collection and captured through wireless communication as discussed before. The wireless communication can support up to 1.5 km in line-of-sight and up to 100 m for non line-of-sight. A second-order zero phase forward and reverse digital Butterworth low pass filter with a 20 Hz cut-off frequency was used to get the smoothed signal.

2.5. Experimental Setup

To verify that our newly developed system with the correction capability is functioning as intended and physically acceptable for our target application, a number of tests are performed. It started with the weight test using digital weight scale. After that, a complete system operation is performed which includes application of the attachment unit on a healthy subject and allow him to walk for 30 minutes freely in an outdoor setting. The third and last test is performed involving a ruler with a stand to allow straight vertical measurement representing clearance. A shoe attached with the measurement unit at the back is held with certain constant pitch angle. The readings are taken at certain clearance values. Further explanation is given in the next section with the results.

3. Results

Figure 7 shows the attachment unit with ultrasonic system. The whole attachment unit (without the shoes) is placed on a weigh scale, so that only the weight of the attachment unit with the battery is measured. The total weight of this attachment unit with ultrasonic system is 82g. The low weight of this system is not affecting the normal gait as state in [10] the attached system must not more than 300 g. Friction is not included in the experiments as it is not affecting foot clearance readings.

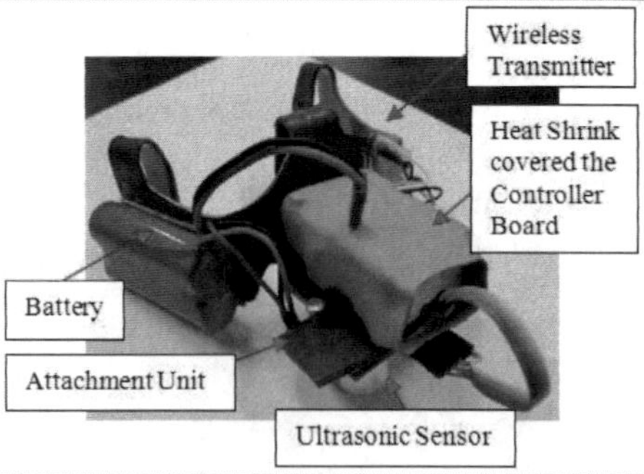

Figure 7. Attachment Unit with Ultrasonic System.

Figure 8 shows the Gait Analysis Software output screen. This Gait Analysis Software (GAS) has six tabs. There are main tab, right feet tab, left feet tab, raw data tab, com port configuration tab and data table tab. The clearance measurement and orientation graph are displayed in the GAS screen display.

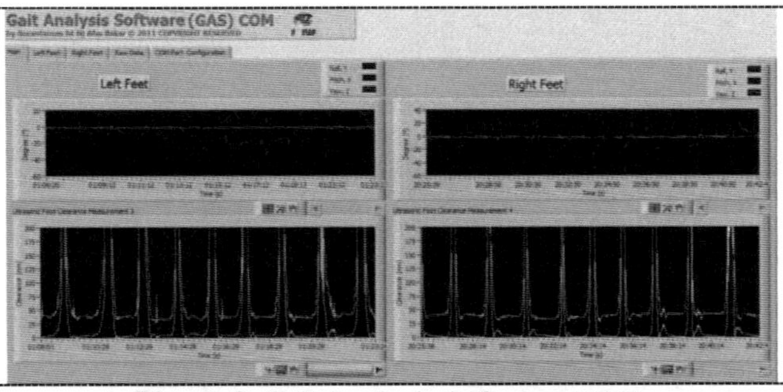

Figure 8. Gait Analysis Software (GAS).

In Figure 8 the corrected values and the ultrasound uncorrected values are shown. The correction can be seen more clearly in Figure 9. The correction is based on the algorithm using the trigonometry concept as represented by Equation (1) and Equation (2) as explained in the section 2.3. Figure 9 shows the technical view of clearance measurement for healthy subject during walking. Continuous clearance data collection for a period of about 30 minutes enables the system to be able to produce the gait pattern for the subject. The graph shows the gait pattern for right feet. The corrected data (clearance measurement) is represented by the white line. The numeric value for raw data from ultrasonic sensor and the clearance measurement are also displayed.

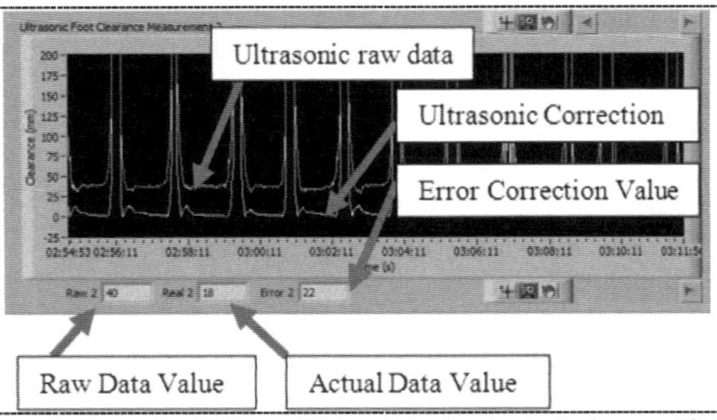

Figure 9. Technical View of The Gait Analysis Software (GAS).

Figure 10 shows the experimental setup for proving of error correction. In this experiment, the ruler is used for the reference data. The ruler is vertically so that it is parallel to the wall. The ultrasonic system and the shoe is raised gradually along the ruler and clearance is measured every 1cm. The distance measurement of the attachment is transferred wirelessly to the GAS with the error correction.

Journal of Physics: Conference Series **495** (2014) 012046 doi:10.1088/1742-6596/495/1/012046

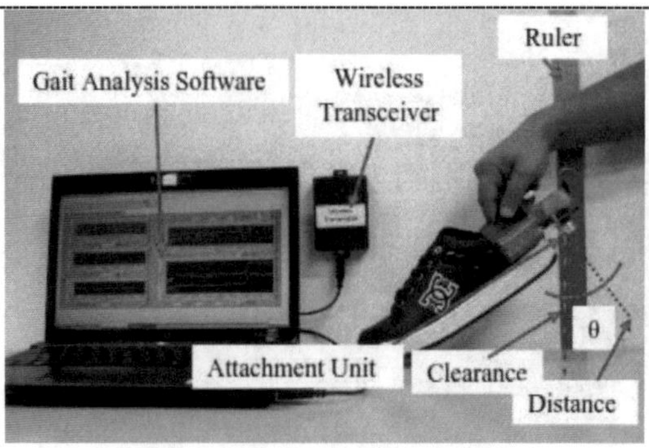

Figure 10. Experimental Setup.

Table 1 shows the experimental data of several distance values obtained from our ultrasonic system measurement, with and without error correction. From the table, the comparison between reference data and the corrected ultrasonic system reading proves that our ultrasonic system gives acceptable results. Therefore, our ultrasonic system can be used for foot clearance measurement.

Table 1. The Corrected and Uncorrected Readings from the test.

Reference Data (cm)	Uncorrected Reading (cm)	Ultrasonic System Reading (cm)
4	5.1	4.1
5	5.7	4.9
6	6.6	5.9
7	7.7	7.1
8	8.6	8.1
9	9.6	9.1
10	10.4	10.0

4. Discussion

Unfortunately, the current practice in measuring foot clearance mostly requires laboratory settings with the use of reflective or active markers, one or more video cameras, thread-mill or suitable floor and computer software running on suitable computers [1]. This type of foot clearance measurement may not be representative of real life activities of daily living, ADL based measurement in natural settings [11], such as at home or outdoor. Problems such as marker slippage may also occur even during laboratory measurement [1].

A more advanced technique is by the use the accelerometer, however, the required calculation that involves double integration of acceleration data yields erratic results due to the effect of drift and errors [11-12]. The sensing of MFC using accelerometer based measurement on surfaces that are uneven, bumpy or during stair descend or ascend is obviously problematic as it is not directly measuring clearance but rather calculate it using acceleration data.

As current state-of-the-art instruments are mostly requiring exclusive research, clinical or rehabilitation laboratories settings, plus the fact that they are limited in simulating the real world activities of an individual [1,11], an in-shoe approach is undoubtedly a better option of implementation.

With the weakness of the previous research, this research comes out more reliable for gait analysis. Where, our technology has more features enable to use in real time measurement. In addition, the features are self-corrected, simple set-up, user friendly, cost effective, in-situ measurement, convenient and affordable make our ultrasonic system is the best way for gait analysis based on the foot clearance and foot orientation measurement.

5. Future Work

Data collection for foot clearance measurement from huge sample group can be evaluated using statistical analysis using the custom designed newly proposed sensory board, ultrasonic system. Our intention is to design a fully integrated system on chip ultrasonic sensing system monolithically. We also plan to implement the smart Gait Analysis Software which is directly inform the user about the gait condition whether user in normal gait or abnormal gait using the classifier.

6. Conclusion

From the result and discussion, we can conclude that our technology is acceptable for gait analysis measurement especially for foot clearance measurement and foot orientation measurement. This is because of every requirement for gait measurement was considered and finally proven in this paper as fulfilled such as low system weight, real time measurement capability, a simple set-up, user friendly, cost effective, in-situ measurement, convenience and affordable as discussed, proves that our technology is acceptable. With real time capability, the system is able to measure, process, transmit to a remote computer, accessed or displayed on the computer screen at almost the same instant, which is especially very critical for example in rehabilitation or monitoring of the elderly.

Acknowledgments:

The authors wishes to thank Universiti Malaysia Perlis (UniMAP) and NCIA for the AMBIENCE laboratory and facilities provided. Also thanks to Ministry of Higher Education, FRGS for funding this research.

References:

[1] Best, R. and Begg, R.K., 2006. 'Overview of Measurement Analysis and Gait Features', in Computational Intelligence for Movement Sciences: Neural Networks and Other Emerging Techniques, eds Begg, R.K. and Palaniswami, M., Hershey, PA, USA, pp. 1-69

[2] Begg, R.K., Best, R.J., Taylor, S. and Dell'Oro, L., 2007. 'Minimum foot clearance during walking: Strategies for the minimization of trip-related falls', Gait & Posture, vol. 25, pp.191-198

[3] Winter, D.A., 1992. 'Foot trajectory in human gait: a precise and multi-factorial motor control task'. Physical Therapy, vol. 72, no 1 , pp 45-56

[4] Stacy J. Morris, Joseph A. Paradiso, 2008 "Shoe-Integrated Sensor System for Wireless Gait Analysis and Real-Time Feedback," *IEEE Transactions on Information Technology in*

Biomedicine, vol. 4, pp. 413-423

[5] K. Aminian and B. Najafi, 2004 "Capturing human motion using bodyfixed sensors: outdoor measurement and clinical applications" *Computer Animation And Virtual Worlds*, Vol. 15, pp.79 – 94

[6] Wahab, Y., A. Zayegh, R. Begg and R. Veljanovski, 2008 'A model for the measurement of foot-to-ground clearance and potential realization of micro-electro-mechanical systems', Modelling, Measurement and Control C, AMSE, vol. 69, no. 1, pp. 59-74

[7] Wahab, Y., A. Zayegh, R. Begg and R. Veljanovski, 2007. 'Analysis of foot to ground clearance measurement techniques for MEMS realization' Proceedings of the IEEE International Conference on Computer and Information Technology (ICCIT 2007), 27-29 December, Dhaka Bangladesh, pp. 413-417

[8] Norantanum Abu Bakar, Yufridin Wahab, Yasmani Awang, Safizan Shaari, 2013 'Modelling of Critical Slopes of Gait Patterns for the Realization of A Wireless Foot Clearance Measurement' in Proc. 2013 UKSim 15th International Conference on Computer Modelling and Simulation (UKSim), EEE, 2013.pp. 707-711

[9] S. Holm, 2009 "Hybrid ultrasound-RFID indoor positioning: Combining the best of both worlds," in IEEE Int. Conf. RFID, Orlando, FL, pp. 155–162

[10] Morris, S.J., 2004, Shoe-integrated sensor system for wireless gait analysis, PhD Thesis, MIT, USA

[11] Lai, D.T.H., Begg,R.K., Charry, E.,Palaniswami, M. and Hill, K., 2008. 'Measuring toe clearance using a wireless inertial sensing device', in Proceedings of International Conference on Intelligent Sensors, Sensor Networks and Information Processing, Sydney, Australia . December 15-18, pp. 375-380

[12] Aminian, K. and Najafi, B., 2004. 'Capturing human motion using body-fixed sensors: Outdoor measurement and clinical applications', Computer Animation and Virtual Worlds, vol. 15, no. 2, pp. 79-94

The Estimation Process in Bayesian Structural Equation Modeling Approach

Ferra Yanuar

Department of Mathematics, Faculty of Mathematics and Natural Sciences, Andalas University, Kampus Limau Manis, 25163, Padang, Indonesia.

Email: ferrayanuar@yahoo.co.id

Abstract. Structural equation modelling (SEM) is a multivariate method that incorporates ideas from regression, path-analysis and factor analysis. A Bayesian approach to SEM may enable models that reflect hypotheses based on complex theory. The development and application of Bayesian approaches to SEM has, however, been relatively slow but with modern technology and the Gibbs sampler, is now possible. The Gibbs sampler can be used to obtain samples of arbitrary size from the posterior distribution over the parameters of a structural equation model (SEM) given covariance data and a prior distribution over the parameters. Point estimates, standard deviations and interval estimates for the parameters can be computed from these samples. This study shows that the conditional distributions required in the Gibbs sampler are familiar distributions, hence the algorithm is very efficient. A goodness of fit statistic for assessing the proposed model is presented. An illustrative example with a real data is presented.
Key words: Structural equation modeling, Bayesian approach, Gibbs sampler, prior distribution.

1. Introduction

Structural equation modeling (SEM) is a collection of statistical techniques that provides a powerful set of tools for researchers in education, social, behavioral and other disciplines. SEM can suitably be used for the analysis of complex phenomena which involves hypothesized relationships between one or more independent latent variables and one or more dependent latent variables [1]. The general goal of SEM is to test the hypothesis that the observed variance matrix for a set of measured variables is equal to the covariance matrix implied by a hypothesized model. This relationship can be formally stated as:

$$\boldsymbol{\Sigma} = \boldsymbol{\Sigma}(\boldsymbol{\theta}) \qquad (1)$$

where $\boldsymbol{\Sigma}$ represents the population covariance matrix of a set of observed variables and $\boldsymbol{\Sigma}(\boldsymbol{\theta})$ represents the population covariance matrix implied by θ, a vector of model parameters. The vector $\boldsymbol{\theta}$ thus defines the form of a particular SEM through the specification of means and intercepts, variances and covariances, regression parameters, and factor loadings.

Certain assumptions need to be fulfilled before SEM can be applied to a particular data set. Under the traditional application of SEM, the data set are assumed drawn from a parent population which is assumed normally distributed; hence, the joint distribution of the variables which represent the data follows a multivariate normal distribution. If that condition is fulfilled, the standard estimation method in SEM which is recommended to be used to estimate parameters and standard errors is maximum likelihood (ML) method.

It is often found that the data gathered in a survey do not follow the assumption of multivariate normality. In the social science research, for example, social and behavioral attitudes are usually measured using ordered categories or dichotomous scaled. Clearly, such data are, by definition, not normally distributed

[2]. Often, the measured variables are continuous but their distributions depart dramatically from normality. We cannot use ML estimation method anymore for such conditions because model fit indices, parameter estimates and standard errors tend to be bias as non normality increase [3]. Hancock & Mueller [4] found that the effects of violating the assumption of non normality include large chi-square values (so too many models are rejected) and standard errors are too small (so significance testing of path coefficients will result an increase in Type I error rates).

The basic objective of this study is to demostrate the Bayesian approach for analyzing SEM, then the methodology is applied to a real data set. Different from classical SEM where computational algorithm is developed based on sampel covariance matrix, in Bayesian SEM, we focus on the use of the raw observations rather than the sampel covariance matrix [5] and apply some powerful tools in statistical computing. We treat the latent variables in the model and the latent measurements as missing data then we analyze the model on the basis of the complete data set. We use MCMC techniques to estimate the unknown parameters in the model. Gibbs sampler as a method in MCMC is applied to obtain a sequence of random samples for summarizing the posterior distribution of parameter model. Moreover, Bayesian technique has flexibility to use useful prior information for achieving better results.

2. Bayesian SEM Approach

In the basic SEM model, it consists of measurement and structural equation. Measurement equation in SEM approach is given by:

$$\boldsymbol{x}_i = \boldsymbol{\Lambda}\boldsymbol{\omega}_i + \boldsymbol{\varepsilon}_i, \quad i = 1, \dots, n \tag{2}$$

where $\boldsymbol{x}_i$ is an $p \times 1$ vector of indicators describing the $q \times 1$ random vector of latent variables $\boldsymbol{\omega}_i$, $\boldsymbol{\Lambda}$ is $p \times q$ matrices of the loading coefficients as obtained from the regressions of $\boldsymbol{x}_i$ on $\boldsymbol{\omega}_i$ and $\boldsymbol{\varepsilon}_i$ is $p \times 1$ random vectors of the measurement errors which follow $N(0, \boldsymbol{\psi}_\varepsilon)$. It is assumed that for $i = 1, \dots, n$, $\boldsymbol{\omega}_i$ is independent follows a normal distribution $N(0, \boldsymbol{\Phi})$ and uncorrelated with the random vector $\boldsymbol{\varepsilon}_i$.

Let the latent variable $\boldsymbol{\omega}_i$ be partitioned into $(\boldsymbol{\eta}_i, \boldsymbol{\xi}_i)$ where $\boldsymbol{\eta}_i$ and $\boldsymbol{\xi}_i$ are $m \times 1$ and $n \times 1$ vectors of latent variables respectively. The structural equation of SEM which explaining the interrelationship among the latent factors is expresessed by:

$$\boldsymbol{\eta}_i = \boldsymbol{B}\boldsymbol{\eta}_i + \boldsymbol{\Gamma}\boldsymbol{\xi}_i + \boldsymbol{\delta}_\mathrm{i}, i = 1, \dots, n. \tag{3}$$

where $\boldsymbol{B}$ is $m \times m$ matrix of structural parameters governing the relationship among the endogenous latent variables which is assumed to have zeros in the diagonal, $\boldsymbol{\Gamma}$ is $m \times n$ regression parameter matrix for relating the endogenous latent variables and exogenous latent variables, and $\boldsymbol{\delta}_\mathrm{i}$ is $m \times 1$ vector of disturbances which is assumed $N(0, \boldsymbol{\psi}_\delta)$ where $\boldsymbol{\psi}_\delta$ is a diagonal covariance matrix. It is also assumed that $\boldsymbol{\delta}_\mathrm{i}$ is uncorrelated with $\boldsymbol{\xi}_i$. Since only one endogenous latent variable involve in this study, or $\boldsymbol{B}\boldsymbol{\eta}_i = 0$, so equation (3) can be rewrite become $\boldsymbol{\eta}_i = \boldsymbol{\Gamma}\boldsymbol{\xi}_i + \boldsymbol{\delta}_\mathrm{i}$.

Under Bayesian approach in SEM, we consider $\boldsymbol{Y} = (\boldsymbol{y}_1, \boldsymbol{y}_2, \dots \boldsymbol{y}_n)$ and $\boldsymbol{X} = (\boldsymbol{x}_1, \boldsymbol{x}_2, \dots \boldsymbol{x}_n)$ be data matrices and let $\boldsymbol{\Omega} = (\boldsymbol{\omega}_1, \boldsymbol{\omega}_2, \dots, \boldsymbol{\omega}_n)$ be the matrix of latent variables and the structural parameter $\boldsymbol{\theta}$, a vector that includes all the unknown parameters in $\boldsymbol{\Phi}$, $\boldsymbol{\psi}_\delta$, $\boldsymbol{\psi}_\varepsilon$, $\boldsymbol{\Lambda}$ and $\boldsymbol{\Lambda}_\omega$ [5, 6]. We apply Markov Chain Monte Carlo (MCMC) methods to obtain the Bayesian estimates $\boldsymbol{\theta}$ and $\boldsymbol{\Omega}$. To achieve our goal, a sequence of random observations from the joint posterior distribution $[\boldsymbol{\theta}, \Omega|\boldsymbol{Y}]$ will be generated via the Gibbs sampler. The Gibbs sampler is a MCMC technique that generates a sequence of random observations from the full conditional posterior distribution of unknown model parameters. The user can create and implement the algorithm easily using WINBUGS [7]. The Gibbs sampler process starts with the setting of initial starting values $\left(\boldsymbol{\theta}^{(0)}, \boldsymbol{\Omega}^{(0)}, \boldsymbol{Y}^{(0)}\right)$, and then conduct the simulation for $\left(\boldsymbol{\theta}^{(1)}, \boldsymbol{\Omega}^{(1)}, \boldsymbol{Y}^{(1)}\right)$. At the *r*th iteration, by making use of the current values $\left(\boldsymbol{\theta}^{(r)}, \boldsymbol{\Omega}^{(r)}, \boldsymbol{Y}^{(r)}\right)$, the Gibbs sampler is carried out as follows:

a. Generate a random variate $\Omega^{(r+1)}$ from the conditional distribution $\left(\Omega|, \theta^{(r)}, Y^{(r)}\right)$
b. Generate a random variate $\theta^{(r+1)}$ from the conditional distribution $\left(\theta|\Omega^{(r+1)}, Y^{(r)}\right)$
c. Generate $\left(\boldsymbol{Y}^{(r+1)}\right)$ from $p\left(Y|\boldsymbol{\Omega}^{(r+1)}, \boldsymbol{\theta}^{(r+1)}\right)$

Under mild regularity conditions, the samples converge to the desired posterior distribution. The derivation of the conditional distribution that is required in the Gibbs sampler process above is discussed in Lee & Shi [8] or Lee [5]. In the process when determining the posterior distribution, the selection of prior distribution for $(\boldsymbol{\Lambda}, \boldsymbol{\psi}_{\varepsilon})$ and $\boldsymbol{\Phi}$ have to be made. In this study, we take the prior distribution for those three parameters via the following conjugate type distribution. Letting $\psi_{\varepsilon k}$ be the *k*th diagonal element of $\boldsymbol{\psi}_{\varepsilon}$ and $\boldsymbol{\Lambda}_k$ be the *k*th row of $\boldsymbol{\Lambda}$, we consider:

$$\psi_{\varepsilon k}^{-1} \sim Gamma(\alpha_{0\varepsilon k}, \beta_{0\varepsilon k}) \tag{4}$$

$$\left(\Lambda_k|\psi_{\varepsilon k}^{-1}\right) \sim N\left(\Lambda_{0k}, \psi_{\varepsilon k}\mathbf{H}_{0yk}\right) \tag{5}$$

$$\boldsymbol{\Phi}^{-1} \sim W_q(R_0, \rho_0) \tag{6}$$

where $Gamma(,)$ is the gamma distribution, $W_q(,)$ is an q dimensional Wishart distribution, parameters $\alpha_{0\varepsilon k}, \beta_{0\varepsilon k}, \Lambda_{0k}, \rho_0$, positive definite matrix $\mathrm{H}_{0\mathrm{yk}}$ and R_0 are hyperparameters which are assumed to be described by an uninformative prior distribution.

The next process in Bayesian SEM is convergence test of the model parameters. The convergence is assessed using a variety of diagnostics as detailed in the CODA package, plotting the time series to assess the quality of the individual parameters with different starting values graphically, and provide a diagnosis based on the trace plots [9, 10, 11]. We also use the Brooks Gelman-Rubin (BGR) convergence statistics [12]. This convergence statistics test compares the variation between and within multiple chains, denoted by $\boldsymbol{R}$. The estimated parameters converge if the value of $\boldsymbol{R}$ is close to 1. In addition, the accuracy of the posterior estimates are inspected by assuring that the Monte Carlo error (an estimate of the difference between the mean of the sampled values and the true posterior mean) for all the parameters to be less than 5% of the sample standard deviation [6].

For assessing the plausibility of our proposed model which includes the measurement equation and structural equation, we plot the residual estimates versus latent variable estimates to give information for the fit of the model. The residuals estimates for measurement equation $(\hat{\varepsilon}_i)$ can be obtained from

$$\hat{\varepsilon}_i = \boldsymbol{y}_i - \hat{\boldsymbol{\Lambda}}\hat{\boldsymbol{\xi}}_i, \quad i = 1, \dots, n \tag{7}$$

where $\hat{\boldsymbol{\Lambda}}$ and $\hat{\boldsymbol{\xi}}_i$ are Bayesian estimates obtained via the MCMC methods. The hypothesized models provide a good fit if the plots are centered at zero and lie within two parallel horizontal lines. The estimates of residuals in the structural equation $\left(\hat{\boldsymbol{\delta}}_i\right)$ can be obtained from following equation:

$$\hat{\boldsymbol{\delta}}_i = (\boldsymbol{I} - \hat{\boldsymbol{B}})\hat{\boldsymbol{\eta}}_i - \hat{\boldsymbol{\Gamma}}\hat{\boldsymbol{\xi}}_i, \quad i = 1,\dots,n \tag{8}$$

where $\hat{\boldsymbol{B}}$, $\hat{\boldsymbol{\eta}}_i$, $\hat{\boldsymbol{\Gamma}}$ and $\hat{\boldsymbol{\xi}}_i$ are Bayesian estimates that are obtained from the corresponding simulated observations through the MCMC. The proposed model fitted the data well or provided a reasonably good fit if the plots lie within two parallel horizontal lines that are centered at zero and no trends are detected [6].

3. Example: Modeling of Health Index

In this section we will illustrate the application of Bayesian SEM for constructing the model of health index. The data set used for the analysis is based on the Third National Morbidity Survey (NHMS III) that took place in Malaysia and conducted at the year 2006. NHMS III which organized by Institute for Public Health, Ministry of Health is a population based cross-sectional study using two stage stratified sampling design proportionate to population size throughout all states in Malaysia. The details of the methodology of survey have been reported previously [13]. The data which used to be analyzed are focused on the respondents who are living in Hulu Langat only, a district in Malaysia. There are 530 respondents involved in the analysis those who are 18 years old and older and had given a complete information required in this study. The information gathered in the survey includes information related to socio-demographic status, lifestyle, mental health condition and biomarkers of individuals. A questionnaire-form was used to obtain the information from the respondents. In addition, medical screening for measuring body height and weight, blood pressure, cholesterol level, HDL (high density lipoprotein) and blood glucose was done either during the house visit or during the consultation at the clinic. Respondents were also asked about the number of health problems experienced and perception about their health condition at the time of the interview.

3.1. Factors related to health

There are many factors which are related to health, some are observable and some are not, such as level of blood pressure, level of cholesterol, etc. Biomarkers can be considered as an example of observable measures which can be used to describe the level of health of an individual. The factors that are not directly observed such as lifestyle, socio-demography and mental health condition could be measured through the indicator variables. Following is a brief explanation for each unobservable variable considered in this study and its respective indicators.

Socio-demography status is an important determinant of health index. In several studies, such as Broadman [14], Cheadle et al. [15], Uitenbroek et al. [16] and Shi [17], it is found that people who are poorer, less well educated or having a lower job status tend to have a lower health index. The indicators of interest considered in this study for assessing the socio-demography factor are education level, employment status and age-group. It is quite well informed of the important role of lifestyle as one of a contributary factor in influencing the level of health of an individual. Unhealthy lifestyle, such as being a smoker or rarely having physical exercise, could have a negative influence to the health condition of an individual [15, 16, 17,18]. The indicators of lifestyle, based on the list of health related behaviors as suggested by Nakayama et al. [18] which has also been used by Broadman [14] that are considered in this study, are physical exercise, smoking habits, average working hours per day and average sleeping hours per day.

Apart from lifestyle, mental health is often recognized as one of the major health determinants. A study by Nakayama et al. [18] showed that health was influenced by many factors, including mental health. Hays et al. [19] found that there was a significant correlation between physical health and mental health. Hence, it is reasonable to assume that mental health as a contributory factor for determining the health index. Nakayama et al.[18] and Broadman [14] have suggested the use of stress level and the number of experiences on serious problem as the indicator for mental health. They found that people who have a high stress level as well as have experienced problems possess a low level of mental health.

In this study, it is hypothesized that socio-demography, lifestyle and mental health are latent factors that are related to the health index of an individual. The health index could also be measured directly based on certain indicators such as body mass index (BMI), blood pressure, cholesterol level, the number of common health problems experienced by the respondent. These hypothesis model is ilustrated in Figure 1.

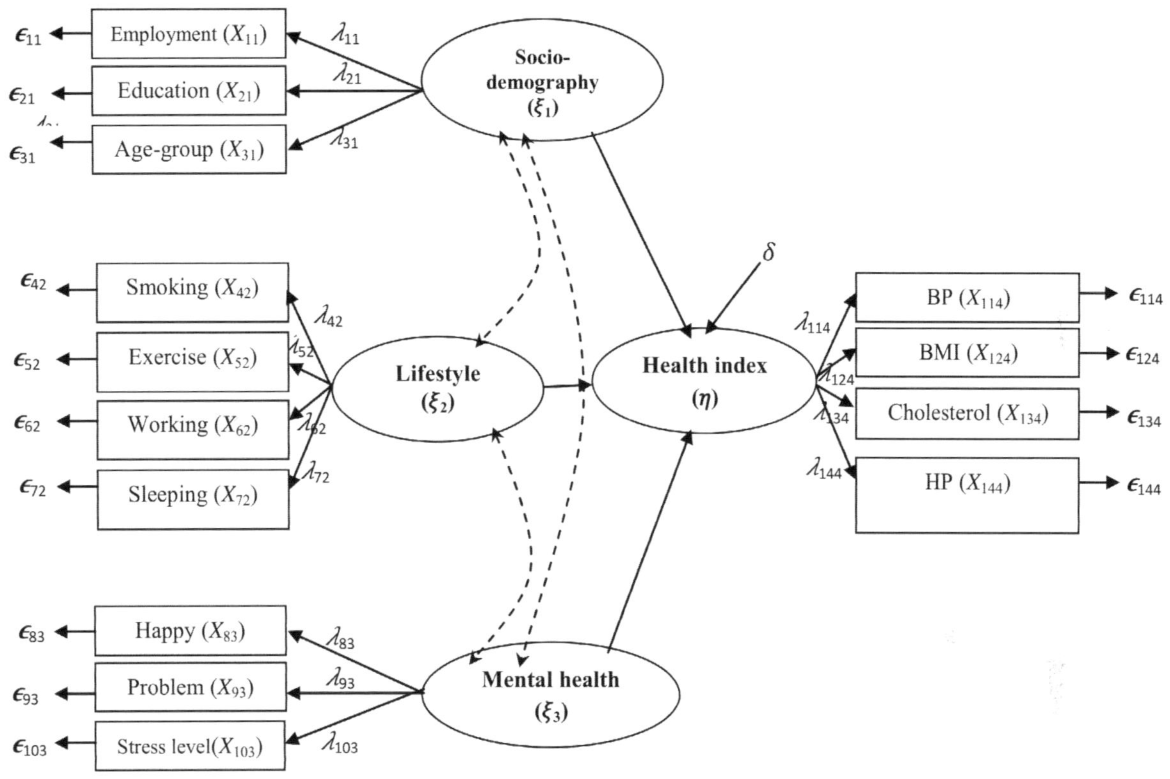

BP = Blood pressure
BMI = Body mass index
HP = Health problem

Figure 1. A Diagrammatic Illustration for Health Index Model.

3.2. Results

The first step we do test the sensitivity of the Bayesian anaysis by considering three types of prior inputs. We found that the statistics yielded based on Bayesian SEM is not sensitive to the different prior inputs or we could say that Bayesian SEM applied in this study is robust enough. Then, test of convergence statistics for all parameters of interest shows that the vaue of R are close to 1. Plots of sequences of observations corresponding to some parameters generated by two different initial values are also indicate converge. Then, test of accuracy for the posterior estimates prove that Monte Carlo error for all parameters are less than 5% of the sample standard deviation. Based on all tests we observe that Bayesian SEM estimates are close to the true values and that the standard error estimates are reasonable and we conclude here that the proposed measurement equation and structural equation are adequate. Figure 1 provides the fitted model for the structural model in Bayesian SEM which includes the standardized parameter estimates and their standard errors, covariance among latent variables and structural errors of the model. Table 1 shows the estimate values of loading factors which relate indicator variables and corresponding latent variables.

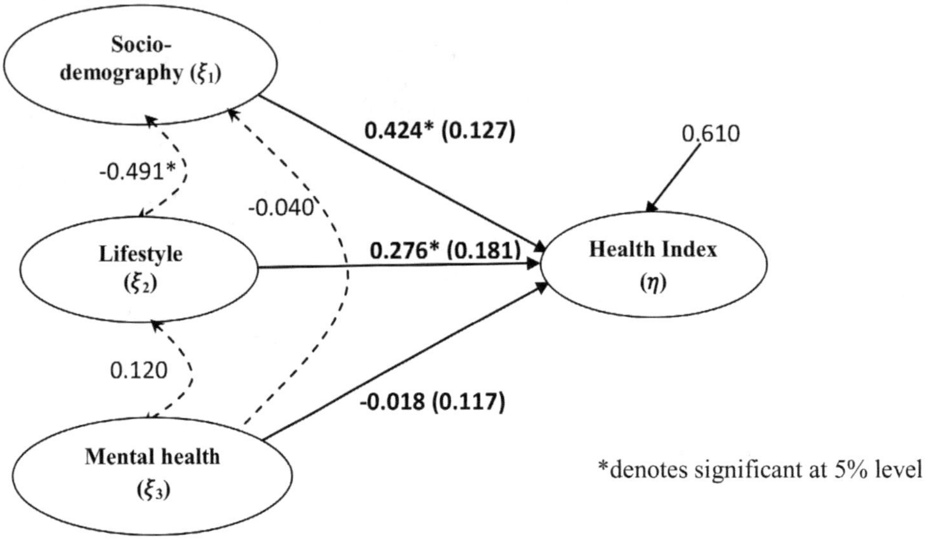

Figure 2. Parameter Estimates for Structural Model

Based on Figure 2, we could also present the structural model that address the relationship between socio-demography status, lifestyle, mental health with health index in the following equation :

$$\hat{\boldsymbol{\eta}}_{Bay\,SEM} = 0.424\xi_1 + 0.276\xi_2 - 0.018\xi_3 \qquad (13)$$

This equation indicate that the greatest effect to the health index is socio-demography then followed by lifestyle and mental health. Socio-demography and ifestyle give the significant effect to the health index meanwhile mental health does not. It is possible since there are other indicator variables for mental health should be incorporated into the model. This study aso inform that there is significant correlation between socio-demographic status and lifestyle.

Table 1 shows the value of coefficient of factor loading and the associate standard errors for each indicator variable in the measurement equations obtained based on Bayesian SEM approach.

Table 1. Factor Loading Estimates in Measurement Model

Construct	Estimate	
	Factor Loadings (SE)	95% CI
Socio–demography → Employment	1	-
Socio–demography → Education	0.424 (0.133)*	(0.650 , 0.987)
Socio–demography → Age group	-0.334 (0.123)*	(-0.582 , -0.103)
Lifestyle → Smoking	1	-
Lifestyle → Exercise	-0.071 (0.128)	(-0.430 , 0.306)
Lifestyle → Working hours	0.358 (0.107)*	(0.155 , 0.771)
Lifestyle → Sleeping hours	0.309 (0.195)	(-0.064 , 0.606)
Mental Health → Happy	1	-
Mental Health → Problem	0.884 (0.157)*	(0.590 , 0,921)
Mental Health →Stress leve	0.824 (0.167)*	(0.500 , 1.029)
Health Index → BP	1	-
Health Index → BMI	0.862 (0.103)*	(0.667 , 0.989)
Health Index → Cholesterol	0.401 (0.108)*	(0.189 , 0.601)
Health Index → Health Problem	0.662 (0.106)*	(0.456 , 0.869)

SE =standard error, CI=confidence interval, BP= blood pressure, BMI= body mass index,

*Significant at 5% level

4. Discussion

In classical SEM, computational algorithm is developed based on sampel covariance matrix and normal assumptions for the observations. But in many studies, It is often found that the data gathered in a survey do not follow the assumption of multivariate normality. Bayesian appproach in SEM is believed as a potential tools to overcome nonnormal assumption [8, 9, 11].

The basic objective of this present study is to demostrate the Bayesian approach for analyzing SEM. In contrast to maximum likelihood method, in Bayesian estimations, parameters are considered as random with prior distribution and a prior density function [5]. Once the data is collected, it is combined with prior distribution using Bayes theorem, next posterior distribution is calculated reflecting the prior knowledge and empirical data. Joint posterior distribution is summarized using MCMC simulation techniques in terms of lower dimensional summary statistics as posterior mean and posterior standard deviations.

The methodology of Bayesian SEM then applied to a real data set. We observe that the structural and measurement equation obtained from this study are adequate and in general we could accept the proposed model.

Journal of Physics: Conference Series **495** (2014) 012047 doi:10.1088/1742-6596/495/1/012047

Acknowledgments
The authors thank to the Institute for Public Health, Ministry of Health, Malaysia, who furnished us the health survey data used in this study. We also thank several anonymous referees for their constructive comments which have improved the final version of this paper.

References

[1] Ullman, J. B. 2006. Structural equation modeling: reviewing the basics and moving forward. *Journal of Personality Assessment*, 87, 35-50.

[2] Kaplan, D. 2000. *Structural equation modeling*. Thousand Oaks, CA: Sage.

[3] U.H. Olsson , T. Foss, S.V. Troye, and R.D. Howell. *The performance of ML, GLS, and WLS estimation in SEM under conditions of misspecification and non-normality.* Structural Equation Modeling 7 (2000), pp. 557–595.

[4] Hancock G.R., & Mueller R.O. 2006. *Structural equation modeling: A second course*. Information Age Publishing, Inc.

[5] Lee. 2007. *Structural equation modeling: A Bayesian approach*. John Wiley & Sons, Ltd, New York.

[6] Yanuar, F., Ibrahim, K. & Abdul, A.J. 2013. Bayesian structural equation modeling for the health index. *Journal of Applied Statistics*, 40(6): 1254-1269.

[7] Spiegelhalter, Thomas, Best, and Lunn. *WinBUGS User Manual, Version* 1.4, software available at http://www.mrc-bsu.cam.ac.uk/bugs, 2003.

[8] S.Y. Lee and J.Q. Shi. *Bayesian analysis of structural equation model with fixed covariates*. Structural Equation Modeling. 7 (2000), pp. 411–430.

[9] A. Ansari, K. Jedidi, and L. Dube L. *Heterogeneous factor analysis models: A Bayesian approach*. Psychometrika. 67 (2002), pp. 49–78.

[10] R.E. Kass and A.E. Raftery. *Bayes factors*. Journal of the American Statistical Association. 90 (1995), pp. 773–795.

[11] R. Scheines, H. Hoijtink, and A. Boomsma. *Bayesian estimation and testing of structural equation models*. Psychometrika 64 (1999), pp. 37–52.

[12] J. Palomo, D.B. Dunson, and K. Bollen. *Bayesian structural equation modeling*. Handbook of Computing and Statistics with Application. 1 (2007), pp. 163–188.

[13] Institute for Public Health, 2008, The Third National Health and Morbidity Survey (NHMS III) 2006, Vol 1. Ministry of Health, Malaysia.

[14] J.D. Boardman. *Stress and physical health: the role of neighborhoods as mediating and moderating mechanisms*. Social Science & Medicine. 58 (2004), pp. 2473–2483.

[15] A. Cheadle, D. Pearson, E. Wagner, B.M. Psaty, P. Diehr, and T. Koepsell. *Relationship between socioeconomic status, health status, and lifestyle practices of American Indians: Evidence from a plains reservation population*. Public Health Report. 109 (1994), pp. 405–413.

[16] D.G. Uitenbroek, A. Kerekovska, and N. Festchieva. 1996. Health lifestyle behaviour and socio-demography characteristics. A study of Varna, Glasgow and Edinburgh. *Social Science & Medicine*. 43, pp. 367–377.

[17] Shi. 1998. Socio-demography characteristics and individual health behaviors. *Socio demographic factors and health behaviors* 9, pp. 933–942.

[18] Nakayama K., Yamaguchi K., Maruyama S. & Morimoto K. 2001. The relationship of lifestyle factors, personal character, and mental health status of employees of a major Japanese electrical manufacturer. *Environmental Health and Preventive Medicine* , 5: 144-149.

[19] Hays, R.D., Revicki, D., & Coyne, K.S. 2005. Application of structural equation modeling to health outcomes research. *Eval Health Prof*, 28; 295-309.

Prediction and Validation of Heat Release Direct Injection Diesel Engine Using Multi-Zone Model

Bagus Anang Nugroho[1,2], Bambang Sugiarto[1], Prawoto[2] and Lukman Shalahuddin[3]

[1]Internal Combustion Engine Laboratory, Mechanical Engineering, University of Indonesia
[2]Agency for the Assessment and Application of Technology - BPPT- Indonesia
[3]Ministry of Research and Technology, Indonesia

E-mail: gusanang@rocketmail.com

Abstract. The objective of this study is to develop simulation model which capable to predict heat release of diesel combustion accurately in efficient computation time. A multi-zone packet model has been applied to solve the combustion phenomena inside diesel cylinder. The model formulations are presented first and then the numerical results are validated on a single cylinder direct injection diesel engine at various engine speed and timing injections. The model were found to be promising to fulfill the objective above.

Keywords: *diesel engine, heat release, multi-zone model, working fluids, step size.*

1. Introduction

A valuable information regarding with combustion study is heat release data. Heat release data expresses the rate of fuel mass burned during combustion process. Base on this data, various combustion phenomena can be explained. In the field of combustion modelling, the accurate prediction of heat release provides accurate performance prediction and as an initial stage for further prediction like emission and abnormal combustion.

Several model frameworks are used for the simulation of the diesel combustion, can be classified as zero, multi-zone and multi-dimensional models. While zero model also termed single-zone models cannot be used for heat release prediction and multi-dimensional models still need long computation time, therefore multi zone model also termed phenomenology models is promising framework to be used for heat release prediction. Multi-zone model is considered as an intermediate step between zero-dimensional and multi-dimensional models. It means that the multi-zone model combine efficient computation time almost like zero-models and good accuracy almost like multi-dimensional models.

Among of various multi-zone model, the model proposed by Hiroyasu and his coworkers [2] has been proven as the most comprehensive model and well documented up to date. The model dividing the spray into zones in the radial and penetrating directions and tracking the evolution of the zones over time. Correlations based on the constant volume vessel experiments to prescribe the spray cone angle, the spray break-up length and droplet size distribution, and the spray tip penetration has been implemented successfully. Hiroyashu model can also capture the effects of swirl and spray impingement on walls.

Jung [3] stated that within multi-zone models, combustion models that can simulate both the premixed and diffusion-controlled phases are relatively scarce. It because of the model assumed that the combustion rate was only related to the total amount of air entrainment during premixed combustion and was independent of the fuel-air mixing process. It must be noticed that the simple stoichiometric combustion concept is so sensitive to air entrainment that ad-hoc calibration coefficients are often applied to air entrainment rates in order to match the heat release rate with experimental data. The values of those empirical coefficients vary very widely among different references [3].

The model has been developed for similar engine by Rakopoulos et al [9] and give well results compare with experiment at engine speed 2500 rpm for three variation loads 40%, 60% and 80% of full load and two static injection timings of 15 and 20 deg CA BTDC. This paper adopts a similar principle but using several different empirical equations and validated at three various engine speed and timing injection.

2. Model Description

2.1. Hiroyashu Multi-Zone Model Concept

The spray injected into the combustion chamber from the injection nozzle is divided into many small packets of equal fuel mass as shown in figure 1. No intermixing among the packages is assumed. During the duration of injection, one new spray packets are generated (axial - *i* direction) is created for every time step, whereby individual packet content five ring layer (radial – *j* direction).

The fuel remains at liquid condition until time break up is reached. After the time break up, fuel is atomized into small drops. The small drop with same size will be arranged to fill the same packet. Parameter to represent the size of drop is called Sauter Mean Diameter or SMD. The entrainment of gasses from surrounding fresh air to the packets begins during droplet process. The amount of fresh gas fill the packet is assumed mixing stoichiometric., base on the assumption This process will continue during ignition delay period. After the end of ignition delay, the mixtures begin to burn.

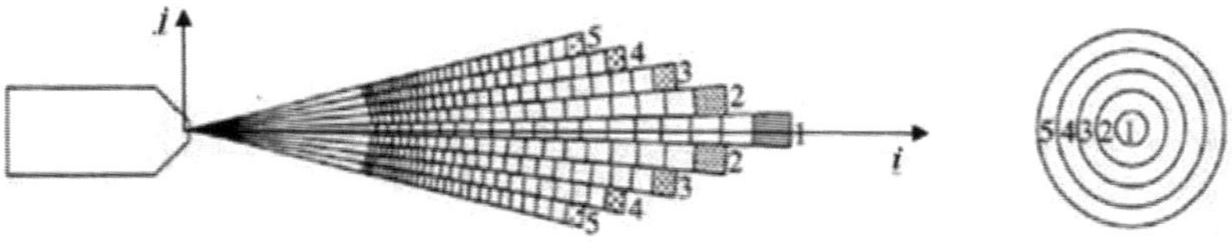

Figure 1. Packet Model Hiroyasu [4]

2.2. Model Formulation

Refer to the description of Hiroyasu model, the only effect was not counted at this model is impingement effect. The multi-zone model will solve the problem above base on continuity mass and energy. The complicated problem is solved by availability of empirical equations which it follows the correlation described by Kom [4] and Merker et. al [7]. Combination of the equations were simplified to the ordinary differential equation (ODE). Euler's method then can be applied to solve the ODE equations. At this sub section, several sub model are described more detail to distinguish the current approach with other approaches.

2.2.1 Nozzle Injection

Considering the injection pressure and nozzle open area is constant during the injection period, mass of the fuel injected, m_{inj}, for each time step is calculated using the relation as given at equation (1), where C_d is discharge coefficient of nozzle, ρ_{fl} (kg/m^3) is density of fuel at liquid condition, A_{noz} is area of nozzle hole (m^2) and Δp_{noz} (Pa) is the difference between injection and cylinder pressure at any time of analysis. At this model, the value of 0.6 has been used for C_d. Payri et. al [8] stated that injection profile is important to the heat release prediction, the application of Δp_{noz} was assumed adequate to represent injection profile.

$$\frac{dm_{inj}}{dt} = C_d A_{noz}\left(2\Delta p_{noz}\rho_{fl}\right)^{0.5} \quad (1)$$

2.2.2 Ignition Delay

Ignition delay time is the difference time between combustion starts and the time at which injection starts. In general, ignition delay is a complicated function of mixture temperature, pressure, equivalence ratio, and fuel properties. Refer to this definition, the correct data of start of injection is important . Refer to Seykens [10], the lack of injection which is caused by injection system must be involved to the model. While the only information of start of injection only from setting the injection pump, then it was considered as uncorrected start of injection. At this model, the correction value of 20 deg CA was used.

During injection time, the average mean cylinder pressure, $\bar{p}_{cyl}$,(Pa) and average mean cylinder temperature, $\bar{T}_{cyl}$, (K) has been calculated. Base on information of Cetane number of fuel (CN) which is 40 and average piston speed, $\bar{S}_p$, the ignition delay can be estimated by using classical Hardenberg and Hase model as given at equation (2). However this model cannot predict well compare with experiments as also described by Lakshminarayanan [6]. To match with experiment, the exponential fitting regression has been applied as shown at figure 2.

$$\tau_{id}(CA) = \left(0.36 + 0.22\bar{S}_p\right).exp\left[\left(\frac{618840}{CN+25}\right).\left(\frac{1}{\bar{R}\bar{T}_{cyl}} - \frac{1}{17190}\right).\left(\frac{21.2}{\bar{p}_{cyl}-12.4}\right)^{0.63}\right] \quad (2)$$

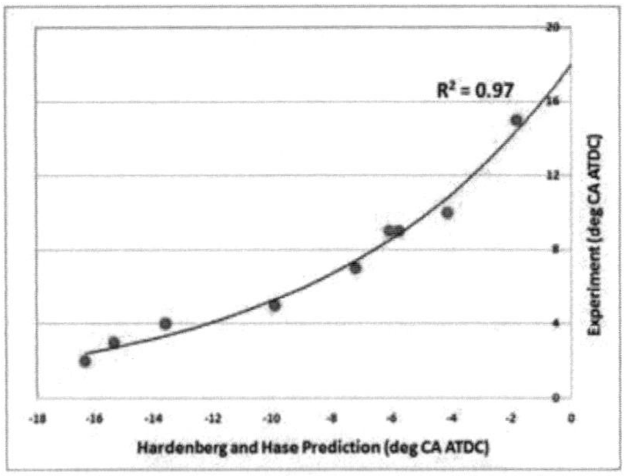

Figure 2. Start of Combustion

The next task is to find out the portion of packets was ready to burn. As described previously, during spray progression the fuel would interact with fresh air therefore the total equivalent ratio, Φ_{packet} can be calculated beside pressure, p_{cyl} and T_{packet}. The second criteria were applied to each packet zone as given at equation (3) and (4) [4]. The ignition of each packet occurs when the ignition delay integral reaches the value of 1.

$$\tau_{id} = 4.10^{-3}.\left(p_{cyl}.10^5\right)^{-2.5}.\left(\Phi_{packet}\right)^{-1.04}.exp\left(\frac{6000}{T_{packet}}\right) \quad (3)$$

$$\int_0^t \frac{1}{\tau_{id}}.dt \geq 1 \quad (4)$$

2.2.3 Heat Release

Following Kom [4] and Merker et. al [7], there are three aspects that limited rate of mass fuel burned, $dmfb/dt$, those are:

- If the mixture is stoichiometric or lean, it is assumed that all fuel available will be burned all.
- If the mixture is rich, the amount of fuel burned depend on the amount of fresh air is limited by the amount of air through into the packet.
- The third limitation is given by maximum rate of reaction of the fuel. By knowing the density of mixture, ρ_{mix}, the mass fraction of fuel (vapour), x_{fv}, and the fraction of oxygen, x_{O2}, the third limitation for each volume packet, V_{packet}, can be calculate by using equation (5).

$$\frac{dm_{fb}}{dt} \leq 5.10^{10}.\rho_{mix}{}^2.x_{fv}{}^1.x_{O_2}{}^5.exp\left[\frac{-12000}{T_{packet}}\right].V_{packet} \quad (5)$$

Additional empirical correction for combustion after half of total injection fuel burned was also included for the current model. The assumption behind this approach was the heat release tends to decline after half of injection fuel burned. The incline of heat release was estimated because of the ultra rich mixture. Therefore after half of total injection fuel burned and if the heat release higher than previous step condition, the rate of mass fuel burned needs to be corrected. The correlation as function of equivalent ratio, Φ, was originally used by Khrisnan et. al [5] for correcting ultra lean combustion mixture has been extended for correcting ultra rich mixture by using exponential factor as shown at equation (6). The equivalence ratio exponent C_Φ is set equal to 0.25 for the current study. As comparison, Jung et. al [3] proposed several model which differ between premixed combustion and controlled combustion

$$\frac{dm_{fb}}{dt} = \frac{dm_{fb}}{dt}\left\{exp\left(1 - \left|1 - \Phi_{packet}\right|^{C_\Phi}\right)\right\} \quad (6)$$

3. Model Validation

Reconstruction of multi zone model has been done by using software Matlab. Single step size crank-angle has been used to get stability [3]. Using minicomputer with 1.86 Ghz processor and 32 bits capacity, the model need less than five minutes to solve the problem with error tolerance 0.1% . It is considered to meet the objective of efficient computation time.

Experimental based on single cylinder DI diesel engine is used in this study for validation purpose. Specifications and operating conditions of the engine are shown in table 1 and table 2. A schematic diagram of system is shown in figure 3. Base on cylinder pressure and volume data each step time, the rate ,dQ_{hr}/dt, and total ,Q_{hr}, of net heat release can be calculated based on equation (7) and (8) [1], where ratio heat specific γ is equal to 1.3. The results for measurement and prediction are given at figure 4 to 6.

$$\frac{dQ_{hr}}{dt} = \frac{\gamma}{\gamma-1}p_{cyl}\frac{dV_{cyl}}{dt} + \frac{1}{\gamma-1}V_{cyl}\frac{dp_{cyl}}{dt} \quad (7)$$

$$Q_{hr} = \int_t^{\Delta t}\frac{dQ_{hr}}{dt}dt \quad (8)$$

Refer to figure 4 to 6, the graph shows that total heat release predictions give similar trend with experiments with slightly over estimate at the end of process. The error comes from two sources, the first error can be explained from graphic of heat release rate, where after half combustion process there are increasing of heat release rate at several points and it is appears more clearly at figure 6. The second error comes from the difference when combustion ends. Both errors can be solved easily by improvement of definition of when the equation (5) can be applied and the proper value of C_Φ.

However in general, the validation results from three different engine speed and three different timing injection for each engine speed shows acceptable accuracy of multi zone model predictions

Table 1. Engine specification

Subjects	Description
Engine type	DI diesel
Number of cylinder	1
Number of valve	2
Swept volume (ltrs)	0.45
Bore (mm) x Stroke (mm)	80.26 x 88.90
Compression ratio	20.3 : 1
Combustion chamber shape	Bowl in piston
Swirl ratio	3.57
Injector operation pressure (bar)	200
Nozzle Diameter (mm)	0.25
No of Nozzle Hole	4
Valve timing	IVO 8 BTDC, IVC 42 ATDC, EVO 60 BBDC, EVC 12 ATDC

Table 2. Engine Operating Set-Up

Case	Uncorrected Injection Timing	Diesel Fuel Mass Flow [g/h]	Air Mass Flow [kg/s]	Temperature of Diesel Fuel [deg C]	Inlet Manifold Pressure [mbar]	Air Inlet Temperature [deg C]	Engine Speed [rpm]
1	32.19	1844.33	0.01006	31.60	996.70	28.00	3000.00
2	28.10	1731.48	0.01035	31.60	996.00	28.00	3000.00
3	24.12	1917.49	0.01580	31.50	998.20	28.00	3000.00
4	38.01	2217.44	0.01127	32.30	992.70	27.00	3600.00
5	36.17	2239.40	0.01261	32.30	993.40	26.00	3601.00
6	28.78	2152.69	0.01146	32.20	992.10	27.00	3600.00
7	39.40	1328.49	0.01252	32.80	988.70	26.00	4200.00
8	30.28	1359.05	0.01229	32.60	988.80	27.00	4200.00
9	27.23	1379.56	0.01250	32.70	988.90	27.00	4200.00

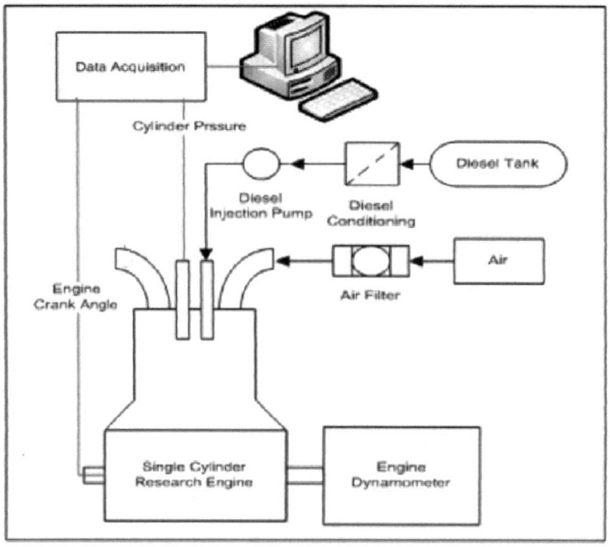

Figure 3. Schematic of experimental setup

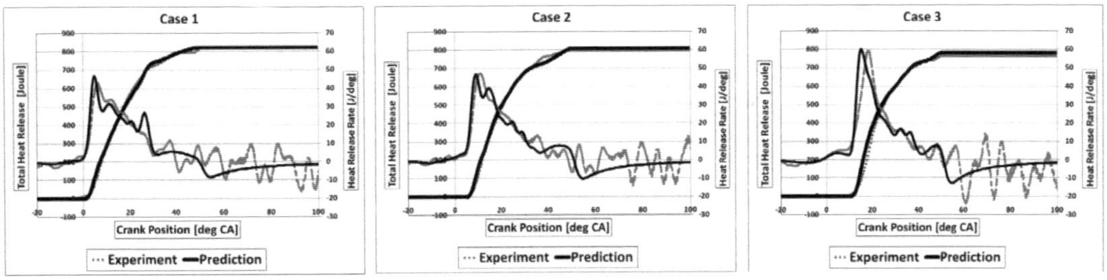

Figure 4. Heat Release at 3000 rpm

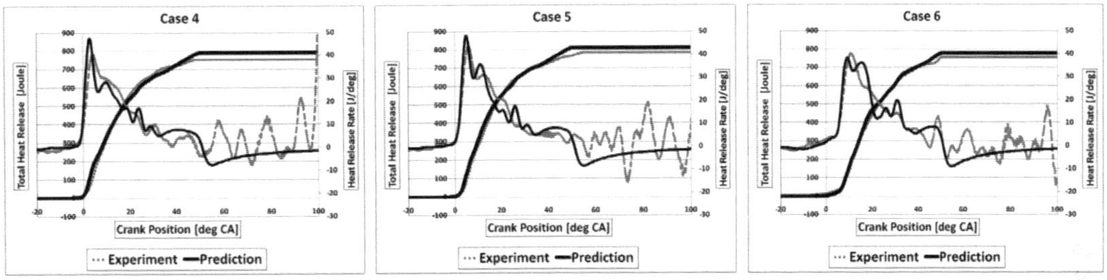

Figure 5. Heat Release at 3600 rpm

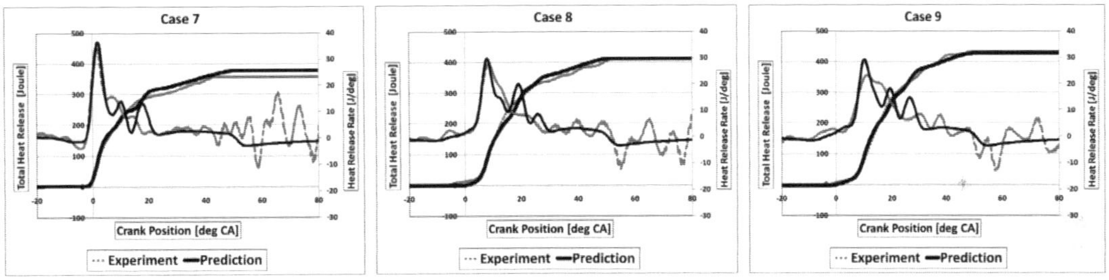

Figure 6. Heat Release at 4200 rpm

4. Conclutions

In the present study the multi-zone modeling of diesel combustion was implemented for heat release prediction. The documentation of several corrections and simplification approach was also presented. It is also reported that efficient computation time can be achieved by this model. The validation of the multi zone model by using heat release data from three various engine speed and three various injection timing for each engine speed shows acceptable accuracy compared to experimental data. However, further improvement still needed especially on start of combustion prediction, while at this study the model still depend on experimental data input. More extensive validation is also needed to find out the also the sensitivity of the current multi-zone model.

Acknowledgement

Authors would like to express thanks to BTMP-BPPT (Internal Combustion Engine Laboratory - Agency for the Assessment and Application of Technology) Indonesia for providing engine test data and Indonesia Ministry of Research and technology for financial support.

Journal of Physics: Conference Series **495** (2014) 012048 doi:10.1088/1742-6596/495/1/012048

References

[1] Heywood, John B., 1988, *Internal Combustion Engine Fundamentals*, Mc Graw – Hill Book, New York.

[2] Hiroyasu, H,. 1985, Diesel Engine and Its Modeling, Comodia

[3] Jung, D. and Assanis, D. N., 2001, Multi-Zone DI Diesel Spray Combustion Model for Cycle Simulation Studies of Engine Performance and Emissions, SAE paper 2001-01-1246

[4] Kom, M., 2004, Onderzoe'k naar verloop van de Seiliger parameters met behulp van een multizone model van het cylinderproces in een dieselmotor, TU Delft

[5] Krishnan, S. R., and Srinivasan, K. K., 2010, Multi-zone modelling of partially premixed low-temperature combustion in pilot-ignited natural-gas engines, Proc. IMechE Vol. 224 Part D: J. Automobile Engineering, JAUTO1472.

[6] Lakshminarayanan, P. A. and Aghav, Yogesh V., 2009, *Modelling Diesel Combustion*, Springer Book,

[7] Merker, Gunther P., Schwarz, C., Stiesch, G., Otto, F.,2005: *Simulating Combustion – Simulating of Combustion and Pollutan Formation for Engine Development*, Springer Book, Berlin.

[8] Payri, F., Olmeda, P., Martín, J., García, A., 2011, A complete 0D thermodynamic predictive model for direct injection diesel engines, Applied Energy 88, Elsevier

[9] Rakopoulos, C.D., Rakopoulos, D.C., Giakoumis, E.G., Kyritsis, D.C., 2004, Validation and sensitivity analysis of a two zone Diesel engine model for combustion and emissions prediction, Energy Conversion and Management 45, Elsevier

[10] Seykens, Xander Lambertus Jacobus, 2009, Development and validation of a phenomenological diesel engine combustion model, Dissertation, Eindhoven University of Technology

Journal of Physics: Conference Series **495** (2014) 012049 doi:10.1088/1742-6596/495/1/012049

A well-balanced numerical scheme for shallow water simulation on adaptive grids

HJ Zhang[1], JZ Zhou[1]*, S Bi[1], QQ Li[2], Y Fan[1]
[1]School of Hydropower and Information Engineering, Huazhong University of science and Technology, Wuhan 430074, P R Chia

Email: Zhanghuajie@hust.edu.cn (HJ Zhang); jz.zhou@hust.edu.cn (JZ Zhou)

[2]Changjiang River Scientific Research Institute, Wuhan, 430010, P R China

ABSTRACT The efficiency of solving two-dimensional shallow-water equations (SWEs) is vital for simulation of large-scale flood inundation. For flood flows over real topography, local high-resolution method, which uses adaptable grids, is required in order to prevent the loss of accuracy of the flow pattern while saving computational cost. This paper introduces an adaptive grid model, which uses an adaptive criterion calculated on the basis of the water lever. The grid adaption is performed by manipulating subdivision levels of the computation grids. As the flow feature varies during the shallow wave propagation, the local grid density changes adaptively and the stored information of neighbor relationship updates correspondingly, achieving a balance between the model accuracy and running efficiency. In this work, a well-balanced (WB) scheme for solving SWEs is introduced. In reconstructions of Riemann state, the definition of the unique bottom elevation on grid interfaces is modified, and the numerical scheme is pre-balanced automatically. By the validation against two idealist test cases, the proposed model is applied to simulate flood inundation due to a dam-break of Zhanghe Reservoir, Hubei province, China. The results show that the presented model is robust and well-balanced, has nice computational efficiency and numerical stability, and thus has bright application prospects.

1 Introduction

The two-dimensional Shallow-Water Equations (2D SWEs) are widely used in simulating long wave flow phenomenon, where the movement in vertical direction is so small that can be neglected, and the flow can be assumed to be horizontal. Godunov type finite volume scheme associated with Riemann solvers has a good property for capturing shocks that is essential to preserve discontinuous. However, the accuracy of numerical solution in space mainly depends on the scale of cell size, and in order to gain high resolution results in areas of interest (AOI), tiny grid must be used, leading to large cell numbers and high computational cost. In most problems, flow features vary sharply in small area and the solutions may be relatively smooth in the rest of the computation domain. Hence it is necessary to design a grid structure that involves fine grids in AOI and coarsens grids in rest areas so as to obtain the balance between the accuracy of numerical solutions and computational cost.

The adaptive mesh refinement (AMR) is widely used to seek high spatial resolutions by organizing the computational grid in tree-like structure. One coarse grid is subdivided into small grids when

*Corresponding author. Tel.: +86 02787543127.

confronting a sharp slope of water surface. The most well know structured adaptive refinement grid model was developed initially by Berger [1] and later improved by many researchers [2-4, 7]. And many Godunov-type numerical schemes are developed to suit for adaptive quad-tree grids, e.g. [5, 6].

Recently, considerable effort has been made on incorporate the balance of discrete design of flux and source terms. Such scheme are termed well-balanced (WB), and are carefully designed to prevent an unreasonable flow due to spatial-varying topography. Researchers [8-9] have developed different well-balanced schemes using additional source terms. These extra terms makes the schemes well-balanced, but increase computational demands and reduces the running efficacy.

This paper aims to introduce a simplified and fast strategy for solving 2D SWEs with a WB scheme. Coarse rectangular grids fill the computational domain initially, and fine grids are the child grids of coarse grids and are attached according to the flow gradients. The additional terms are no longer needed under the modification of Riemann states reconstruction, simplifying the well-balanced scheme.

2 Dynamically adaptive grid model

Dynamically adaptive structured grid model is used for capturing wet/dry fronts, and free surface discontinuities. Each grid is a leaf-grid, or has four child-grids when been divided small to gain a higher resolution in space. This grid model is designed hierarchical and quadric, can be generated fast and automatically, and can be changed easily whether a grid is enriched into four or four child grids gather as one.

2.1 Grid generation

The grids are created in a recursive spatial decomposition of root grids according to following procedure:

Step 1 Generate root grids, and set the values of subdivision level to zero.

Step 2 Input seeding point data to describe the boundaries of the flow domain and any areas of interests, usually where the bed gradients are large.

Step 3 Enrich a grid, which contains seeding points, until the subdivision level reach a maximum level prescribed.

The grids generated in Step 1 are coarse and usually contain undesired grids where water flow never reaches. Therefore, seeding points are added to describe the irregular domain boundaries and gain a higher resolution in areas of interests, such as regions where initial water surface or bottom elevations have steep slopes. A primary root grid is then divided into four equal quadrant grids recursively, when it contains more than one seeding points. Boundary identification cut off regions, which are far away from the route of water flow, and minimizes the total number of cells and hence reduces memory storage requirement and overall computational cost.

2.2 Adaption criterion

For leaf grids, the grid adaption variable is set to zero in dry cells, and calculated in wet cells based on the gradient of free surface elevation

$$\theta = \sqrt{\left(\frac{\partial \eta}{\partial x}\right)^2 + \left(\frac{\partial \eta}{\partial y}\right)^2} \tag{1}$$

and for grids containing with sub-grids, θ is set to the maximum of those of its child grids.

A grid is subdivided when θ is greater than θ_{max} and the subdivision level is less than L_{max}, in which θ_{max} and L_{max} are prescribe constants. Grid coarsening for wet grids is as follows: each grid with one-depth value is identified, and a Boolean status is set false if θ is less than a predefined value θ_{min}. The efficiency of an ARM algorithm is related to the reliability of the mesh adaption procedure, which is controlled by the threshold parameters. However these threshold values of θ_{max} and θ_{min} are usually problem dependent, and are determined by trial and error.

2.3 Variable reassignment

For a shallow-water simulation over complex topography, bottom elevations distribute spatially, and therefore, it's quite necessary to calculate bottom values of each child grids by interpolating from raw data of Digital Elevation Model (DEM) during grid enrichment. While the bed elevation changes, the flow values (especially h and η) of child grids should be assigned properly to maintain the conservation of mass and momentums. The bottoms at grid centers do not usually equal to the average of those in child grids, thus the new water lever changes to prevent the violation of maintain mass conservation. Giving the initial value as η_p, water surface elevation alters according (2). The iterative computing procedure ends within four loops, when the increment of water lever is zero.

$$\Delta\eta = (\eta_p - b_p) - \tfrac{1}{4}\sum\nolimits_i \max(\eta - b_i, 0)$$
$$\eta' = \eta + \Delta\eta \tag{2}$$

The coarsen procedure compute the flow variables of parent grid directly through conservation laws.

$$\eta = b_p + \tfrac{1}{4}\sum_i \max(h_i, 0) \tag{3}$$

3 Numerical model

3.1 Governing Equations

The 2D SWEs, which can be derived from the Navier-Stokes equations by average integrating over water depth, under the assumptions that a pressure distribution is hydrostatic and vertical accelerations can be ignored, are widely used for simulations of flood flows. These equations can be expressed in matrix form as follows.

$$\frac{\partial \mathbf{U}}{\partial t} + \frac{\partial \mathbf{f}}{\partial x} + \frac{\partial \mathbf{g}}{\partial y} = \mathbf{s}_f + \mathbf{s}_b \tag{4}$$

with

$$\mathbf{U} = \begin{bmatrix} h \\ uh \\ vh \end{bmatrix}, \mathbf{f} = \begin{bmatrix} uh \\ u^2h + \frac{1}{2}gh^2 \\ uvh \end{bmatrix}, \mathbf{g} = \begin{bmatrix} vh \\ uvh \\ v^2h + \frac{1}{2}gh^2 \end{bmatrix}, \mathbf{s}_f = \begin{bmatrix} 0 \\ -ghS_{fx} \\ -ghS_{fy} \end{bmatrix}, \mathbf{s}_b = \begin{bmatrix} 0 \\ ghS_{bx} \\ ghS_{by} \end{bmatrix} \tag{5}$$

in which, h is the water depth; u and v are the flow velocities in each Cartesian coordinate; g represents the acceleration due to gravity, and depends on latitude of computation grid , with a default value of 9.81m/s^2 as; S_b and S_f are bottom and frictional slops, respectively. Viscosity, wind effects and the Coriolis term can be included in the source term, but are neglected in this paper.

3.2 Finite volume method

Integrating (4) over a finite volume $\mathbf{\Omega}_i$, the integral form of the SWEs obtained by applying Green theorem is given by

$$\frac{\partial}{\partial t}\int_{\Omega_i} \mathbf{U}d\Omega + \oint_{\partial\Omega_i} \mathbf{F}\cdot\mathbf{n}dl = \int_{\Omega_i} \mathbf{s}d\Omega \tag{6}$$

The integrand $\mathbf{F}\cdot\mathbf{n}$ is the outward normal flux vector, in which $\mathbf{F}$= [$\mathbf{f}$, $\mathbf{g}$]. The finite volume method based on (6) ensures that the mass and momentum will be conserved across a discontinuity. Discretizing (6) in space and time, the equation becomes

$$\mathbf{U}_i^{n+1} = \mathbf{U}_i^n + \Delta t \cdot \left(-\frac{\mathbf{f}_E - \mathbf{f}_W}{\Delta x} - \frac{\mathbf{f}_N - \mathbf{f}_S}{\Delta y} + \mathbf{s}_i \right) \tag{7}$$

where the superscript n and n+1 represent the time level. The normal flux **f** can be estimated in different approximate Riemann technologies, and the HLLC approximate Riemann solver associated with wave speed estimated by two rarefactions assumption [7] is suggested in following work.

3.3 Well-balanced scheme

A new WB scheme is proposed in this section. The WB scheme is obtained by modifying bottom reconstruction values of Riemann states on cell interfaces. In reconstruction, flow variables on interfaces are estimated with a non-linear slope limitation for a spatial second-order accuracy. The minmod slop limiter [7] is used for restrain the flow gradients, given by $\mathbf{\Psi}_{x,i} = \text{minmod}(\mathbf{U}_e - \mathbf{U}_i, \mathbf{U}_i - \mathbf{U}_w)$, $\mathbf{\Psi}_{y,i} = \text{minmod}(\mathbf{U}_n - \mathbf{U}_i, \mathbf{U}_i - \mathbf{U}_s)$, and the interpolation is $\mathbf{U}_i(\Delta x, \Delta y) = \mathbf{U}_i + \mathbf{\Psi}_{x,i}\Delta x + \mathbf{\Psi}_{y,i}\Delta y$. Consider the computation of flux from cell i to its east cell e in one dimension, the interface values are given by

$$\eta_L = \eta_i + \tfrac{1}{2}\mathbf{\Psi}_{x,i}^{(\eta)}, h_L = h_i + \tfrac{1}{2}\mathbf{\Psi}_{x,i}^{(h)}, b_L = \eta_L - h_L \tag{8}$$

$$\eta_R = \eta_e - \tfrac{1}{2}\mathbf{\Psi}_{x,e}^{(\eta)}, h_R = h_e + \tfrac{1}{2}\mathbf{\Psi}_{x,e}^{(h)}, b_R = \eta_R - h_R \tag{9}$$

Based on the face values from (8) and (9), Riemann states are modified to ensure non-negative of water depth. A unique value of bottom elevation at the interface may be defined, and the water depth on either side are re-computed subsequently.

$$b_* = \min(\eta_L, \eta_R, \max(b_L, b_R)) \tag{10}$$

$$h_L^* = \eta_L - \max(b_*, b_L), h_R^* = \eta_R - \max(b_*, b_L) \tag{11}$$

Obviously, h_L^* and h_R^* given by the above procedure are non-negative. Correspondingly, the reconstruction values of water surface elevation and unit discharge is then given by

$$\eta_L^* = h_L^* + b_*, \eta_R^* = h_R^* + b_* \tag{12}$$

The single value of bottom elevation at interface b*, and the reconstruction values of η_L^*, η_R^* are recorded for later bed slop computation. The reconstruction procedure can be extended to other faces in a similar way. These Riemann states are then employed by the HLLC approximate Riemann solver to compute the interfaces fluxes $\mathbf{f}_E$, $\mathbf{f}_W$, $\mathbf{f}_N$, and $\mathbf{f}_S$.

3.4 Bed slope source term

The bed slope source term is given by $\mathbf{s}_b = [0, gh_x \, \partial b/\partial x, gh_y \, \partial b/\partial y]^T$, where η is the average of the reconstructed values of water surface elevation on the two interfaces, and slopes of bottom elevation are directly discretized.

$$gh_x \frac{\partial b}{\partial x} = g \cdot \frac{h_E^* + h_W^*}{2} \cdot \frac{b_E^* - b_W^*}{\Delta x} \tag{13}$$

The proof or well-balance property is straightforward. Assuming a steady state, the values of velocities u and v equal to 0 everywhere, It is easy to find the equivalence between convective flux terms and bottom source terms, by simply substituting (5) with (8)~(13). The scheme is pre-balanced, and no special treatments should be done on a wet/dry front.

3.5 Time marching

In the framework of Godunov-type finite volume scheme, the time marching formula is show as (7). However, it gives diffusive results with first-order accuracy in time. In this paper, predictor-corrector technique [5], which computes Riemann flux only once during one time-marching step, will be employed to achieve a second-order accuracy in time. In summary, the time marching process is as follows.

Step 1 For all leaf cells, refine the grids according to θ values when θ larger than θ_{max}.

Step 2 For the grids, whose children are all leaf grids, remove the child grids when θ is smaller than θ_{min}.

Step 3 Generate an array to store the information of all leaf cells (for later use in Step 4~8).

Step 4 Compute the gradients of flow variables.

Step 5 Evaluate the update rate of conserved variables, and predict them to an intermediate time step.

Step 6 Reconstruct Riemann states and estimate the fluxes for each interface.

Step 7 Calculate source terms.

Step 8 Correct flow variables of leaf grids and step to a new time step.

Step 9 Update non-leaf cells and step to next time step.

Step 9 Return to Step 1 until the process is terminated by $t>t_{end}$.

3.6 Stability criteria

The stability of the explicit numerical model is governed by the Courant-Friedrichs-Lewy (CFL) condition. For a two-dimension Cartesian grid, the CFL criterion for choosing an appropriate time step may be expressed as

$$\Delta t = C \cdot \min(\Delta t_x, \Delta t_y) \tag{14}$$

with

$$\Delta t_x = \min_i \left\{ \frac{M_i}{|u_i| + \sqrt{gh_i}} \right\}, \Delta t_y = \min_i \left\{ \frac{N_i}{|v_i| + \sqrt{gh_i}} \right\} \tag{15}$$

where C is the Courant number in the range $0<C\leq 1$. In the present scheme, an adaptive time step based upon (14) takes Courant number to 0.5. The overall time step is generally constrained by the smallest cells. Higher computational efficiency may be achieved by implementing a local time step [10-12].

4 Results and discussions

The numerical scheme is validated against several benchmark tests, and results are compared with analytical solutions. The test cases included preservation of still water, water sloshing over one-dimensional parabolic containers. In all cases, default values are specified as: g=9.81m/s^2, θ_{max}=0.08, θ_{min} =0.03.

4.1 Preservation of still water over a two-dimensional bump

The shallow flow solver associated with adaptive refinement mesh is validated in an initially still water case to prove that the present model preserves the flow steady with wet/dry front over non-uniform bottom topography. The bump is a cone located at the center of a 2m×2m square container. The boundaries are solid walls, and no friction is imposed by the smooth bottom, which is defined by

$$b(x, y) = 0.8\exp\left\{-25\times[(x-1)^2 + (y-1)^2]\right\} \text{ m} \tag{16}$$

The computational domain is approximated in 1600 rectangular cells (dx=0.05m), and no grid adaption is employed. Initially the water inside the container is at rest, and the bump is partially or absolutely submerged. The steady state should be maintained, and velocities in the cells must be zero anywhere and anytime. Two initial situations are tested for the validation, and both are performed for 100s. In Case 1a, bump is partially submerged as the water surface elevation is 0.5m. In Case 1b, the water surface is 1.0m and the bump is absolutely under water. Figure 1 profiles the numerical solutions and analytical values for water depth along y=0m in Case 1a and 1b, and shows the steady state is well preserved, verifying the *C*-property of the numeric model.

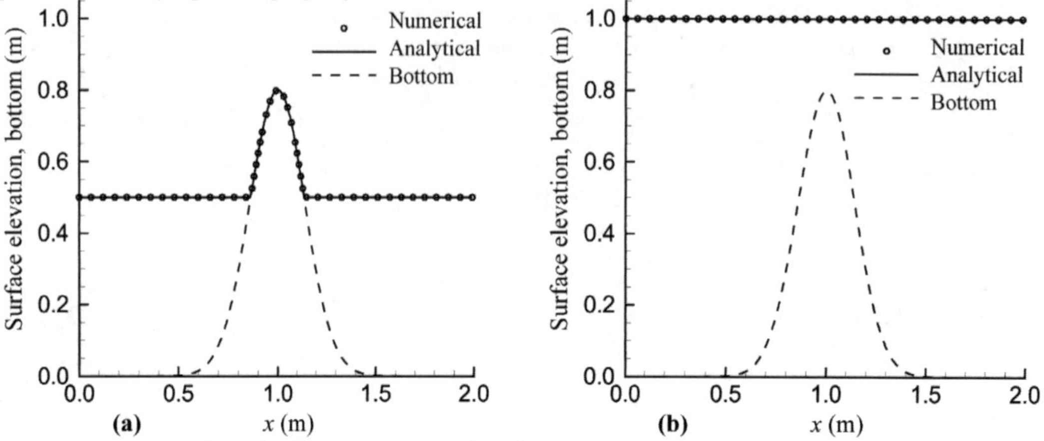

Figure 1. Preservation of still water: comparison between numerical and analytical results of water surface profiled along y=0m.

4.2 Moving shorelines over a frictional parabolic bottom topography

In 2006, Sampson derived analytical solution of the SWEs for a perturbed flow in a container with a one-dimensional parabolic bed bottom with friction. This benchmark test is adopted to validate the stability of the numerical model with moving wet/dry front, as well as provide a further check for source terms, including bed friction and bed slopes. The bed of the domain is defined by

$$b(x) = h_0 (x/a)^2 \tag{17}$$

in which, h_0 and a are constants. The bottom is uniform in *y*-direction. The analytical solution describes a bed resistance in parameter τ (relation with bed friction is given by $C_f = h\tau / \sqrt{u^2 + v^2}$, or $S_{fx} = u\tau$ in an equivalent form) and a hump parameter $p=(8gh_0/a^2)^{0.5}$. We consider a case where $\tau<p$, and the analytical solution of the water stage is

$$\eta(x,t) = h_0 - \tfrac{B^2 e^{-\tau t}}{4g} - \tfrac{e^{-\tau t}}{g}\left[Bs\cos(st) + \tfrac{\tau B}{2}\sin(st)\right]x + \tfrac{a^2 B^2 e^{-\tau t}}{8g^2 h_0}\left[-s\tau\sin(2st) + (\tau^2/4 - s^2)\cos(2st)\right] \tag{18}$$

where B is a constant and $s=0.5\times(p^2-\tau^2)^{0.5}$. As t tends to infinite, the water level along the channel converges to h_0, defining the still water elevation above the datum. And the perturb flow is damped by bed friction.

The computation domain is 10000m×500m. Slip conditions are invoked at the lateral boundaries. The coefficients are a=3km, h_0 =10m, τ=0.001s^{-1} and B=5m/s. Since the water surface is quite smooth in spatial, the grid adaption is not carried on, and the computation domain is filled with cells in uniform size of 100m×100m. The simulation lasts 6000s. Figure 2 shows a good agreement between the numerical and analytical solutions for water depth at observed points located at A(x=50, y=0), B(x= -2750, y=0), C(x=2750, y=0).

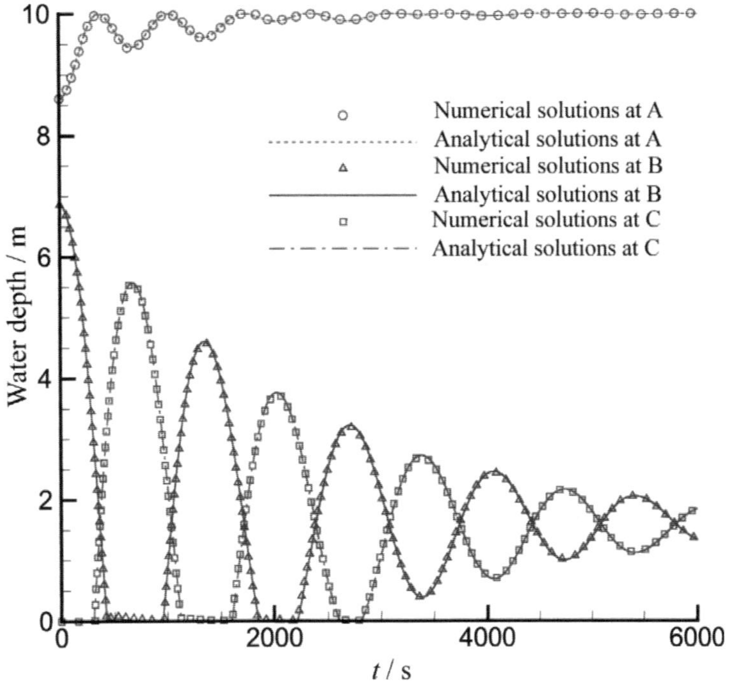

Figure 2. Hydrograph of observing points during dam-break wave propagation.

4.3 Application: flood wave propagation due to dam-break of Zhanghe Reservoir

Zhanghe River is a branch of Yangtze River. The river length is 202km, and the watershed area is about 400km^2. Zhanghe Reservoir was built maintaining huge water of nearly twenty billion tons. We will make a general analysis on flood propagation due to an imaginary dam-break.

The computational domain should contains both reservoir area and downstream river. However the dam-breach type is not defined uniquely, and the bathymetry of the reservoir is not available, therefore we only concern the river valley and floodplain on the downstream side of the reservoir. The computational domain is about 129km^2, and five survey points along the river.

The discharge at the dam site must be predefined, and a suggested inflow discharge plan is obtained by using empirical formulation. The flood water flushing from reservoir into the domain for 4.5h. The discharge of tributaries is relatively small and can be neglected during the flood inundation, thus the lateral boundaries are set to slip wall condition. And the downstream boundary is set to be free out.

Since the domain contains both narrow river valley and relatively wide floodplain, the domain is suggested to be discretized into coarse cells in floodplain and fine grids in river valley. The max cell size cannot be too large, otherwise river valley is unable for modelling. Considering that the average width of river is 300m, we suggest 100m as the max cell size and 25m as the smallest size.

The simulation runs in different cases, fixed fine mesh, and dynamic mesh. Table 1 shows the detail configurations and computational cost of time in each case. In Case 3a, the domain is discretization by the finest grids, no cell-density change is made, and the numerical results for water depth can be considered relatively rea, and thus be used as reference values in the subsequent comparison. In Case 3b, configuration of the 100m×100m coarse cells with 2 max subdivision level is employed, and the grid dynamically varies according the adaptive criterion in Section 2.2.

The number of leaf cells in Case 3b increases gradually from 13,689 to 44,919, and is much smaller than the fixed cells numbers in Case 3a, which is 206,903. The relative time t_r calculated by $t_{r,(i)}= t_{CPU,(i)}/t_{CPU,(a)}$, is shown in Table 1, indicating an improvement in computational efficiency of time by the local refined grids. The AMR algorithm saves 60% in time, gaining an excellent running efficiency.

Table 1. Grid configuration of dam-break wave propagation over Zhanghe floodplain

Case	AMR mode	cell size /m	L_{max}	cells number	t_{CPU} / s	t_r
3a	static	25	--	206,903	21,307	100.0%
3b	dynamic	100	2	time varying	8,319	39.0%

Figure 3 and Figure 4 compares numerical solutions for water depth and unit discharge at survey points respectively. The comparisons show a general agreement between the three cases. The information of flood arrival time and peak time is easily found from Figure 3. Flood waves arrive Point A and B soon after the dam-break, and a sharp rise in water depth is observed at these points. Due to a tributary flowing into Zhanghe River from the left bank, the area of debouchment can store amount of water during the flood propagation. The Point D and E, which locate at the downstream of the debouchment, has later arrival time and smaller peak values in unit width discharge, reflecting an attenuate peak discharge along the floodplain.

One phenomenon is observed at Point E, that as adaptive mesh refinement technology is employed, the flood arrives about 5 minutes earlier in the hydrograph line, yet the curve of unit width discharge arises nearly 8 minutes later, in contrast with fixed fine mesh. An explanation is that flood information transfers downstream faster in adaptive refined cells. However, water moves at a similar speed behind the wave front in different cases, herein the rise of unit width discharge lags. In order to simulate the flood propagation more accurately, the mesh adaption criterion is worth studying in future work.

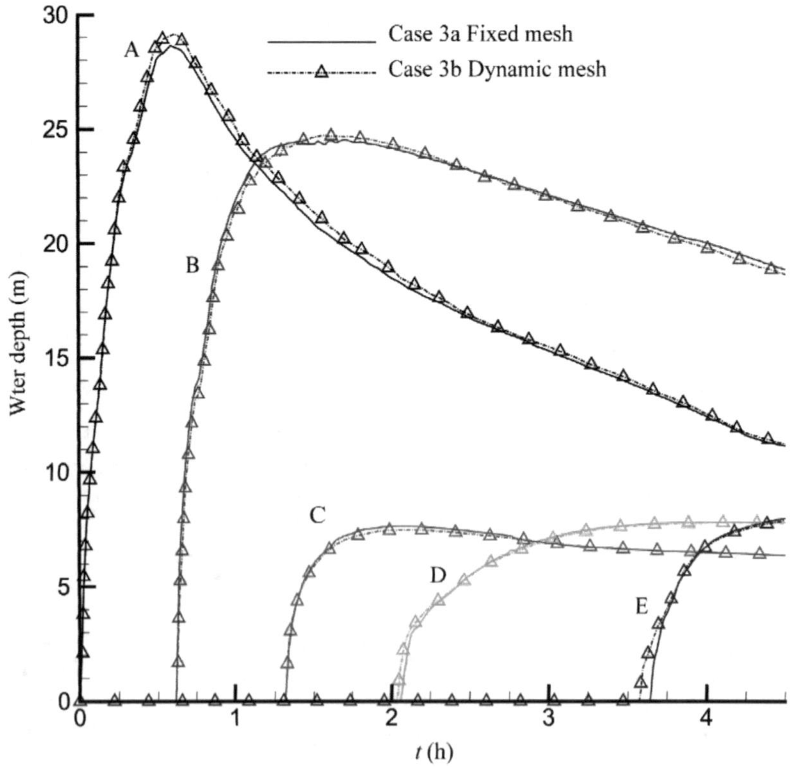

Figure 3. Hydrograph of observing points during dam-break wave propagation.

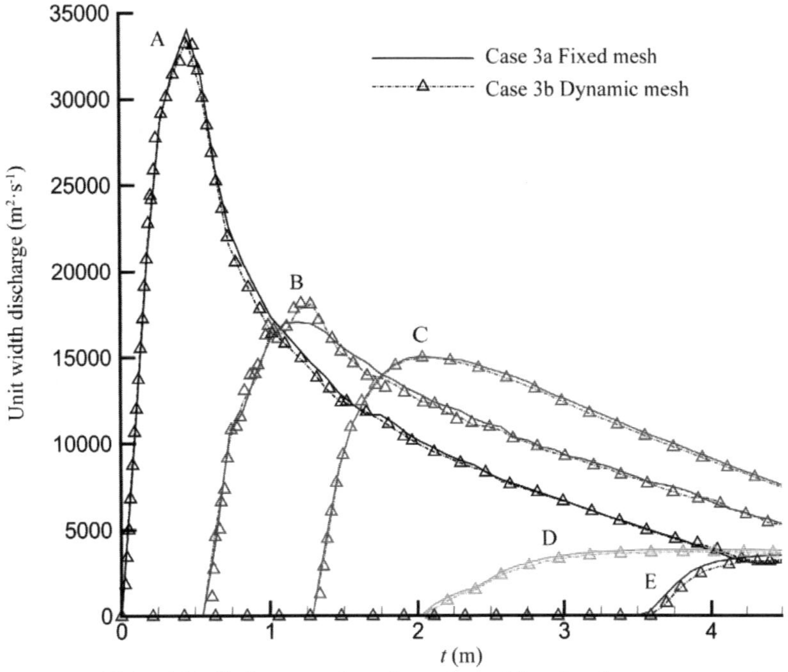

Figure 4. Unit width discharge at observing points of dam-break wave propagation.

5 Conclusions

A well-balanced scheme is developed in this paper. The presented scheme modifies the bed elevations at cell interfaces during the reconstructions of Riemann states, and need no special treatment on source terms, thus makes the numeric procedure easier to actualize. Adaptive grid model is introduced and well designed for seeking local high-resolution, reducing overall computational cost, and improving the running efficiency of large-scale shallow-water simulation over irregular topography.

The well-balanced property of proposed model is tested in two test cases of the preservation of still water with or without wet/dry fronts. Besides, the numerical model is validated against another two theoretical benchmark tests and a laboratory experiment. The simulated predictions agree excellently with analytical solutions or observed values in the validation tests. Finally, the model is applied to Zhanghe Reservoir, where a flood propagation event occurs due to an imaginary dam-break. Comparisons between different AMR strategies are made, and show that the AMR algorithm improves overall efficiency of numeric computations, and can account for flood propagation over highly irregular topography. Therefore the proposed model is suitable for problems with complex topography and thus has bright application prospects.

Acknowledgement

This work was financially supported by a grant from the Key Program of the National Natural Science Foundation of China (No. 51239004) ; the Research Fund for the Doctoral Program of Higher Education of China (No.20100142110012).

Reference

[1] Berger MJ and Oliger J 1984 Adaptive mesh refinement for hyperbolic partial differential equations. *J. Comput .Phys.* **53** 484-512

[2] Rogers B, Fujihara M and Borthwick AGL 2001 Adaptive Q-tree Godunov-type scheme for shallow water equations *Int. J. Numer. Methods Fluids* **35** 247-280

[3] Liang Q, Borthwick AGL and Stelling G 2004 Simulation of dam- and dyke-break hydrodynamics on dynamically adaptive quadtree grids *Int. J. Numer. Methods Fluids* **46** 127-162
[4] Baeza A and Mulet P 2006 Adaptive mesh refinement techniques for high-order shock capturing schemes for multi-dimensional hydrodynamic simulations *Int. J. Numer. Methods Fluids* **52** 455-471
[5] Liang Q 2012 A simplified adaptive Cartesian grid system for solving the 2D shallow water equations *Int. J. Numer. Methods Fluids* **69** 442-458
[6] Ji H, Lien F and Yee E 2010 A new adaptive mesh refinement data structure with an application to detonation *J. Comput. Phys.* **229** 8981-8993
[7] Song L, Zhou J, Guo J, Zou Q, and Liu Y 2011 A robust well-balanced finite volume model for shallow water flows with wetting and drying over irregular terrain *Adv. Water Resour.* **34** 915-932
[8] Liang Q and Marche F 2009 Numerical resolution of well-balanced shallow water equations with complex source terms *Adv. Water Resour.* **32** 873-884
[9] An H and Yu S 2012 Well-balanced shallow water flow simulation on quadtree cut cell grids *Adv. Water Resour.* **39** 60-70
[10] Crossley AJ and Wright NG 2005 Time accurate local time stepping for the unsteady shallow water equations *Int. J. Numer. Methods Fluids* **48** 775-799
[11] Krámer T and Józsa J 2007 Solution-adaptivity in modelling complex shallow flows *Comput. Fluids* **36** 562-577
[12] Sanders BF 2008 Integration of a shallow water model with a local time step *J. Hydraul. Res.* **46** 466-475

A compositional approach to building applications in a computational environment

V.V. Roslovtsev, L.D. Shumsky and V.E. Wolfengagen

Department of cybernetics, National Research Nuclear University "MEPhI", Kashirskoe sh. 31., Moscow, Russia

E-mail: vladiros@gmail.com, shumskyld@gmail.com, vew@jmsuice.msk.ru

Abstract. The paper presents an approach to creating an applicative computational environment to feature computational processes and data decomposition, and a compositional approach to application building. The approach in question is based on the notion of combinator – both in systems with variable binding (such as λ-calculi) and those allowing programming without variables (combinatory logic style). We present a computation decomposition technique based on objects' structural decomposition, with the focus on computation decomposition. The computational environment's architecture is based on a network with nodes playing several roles simultaneously.

1. Introduction

Today software engineers face a number of challenges: the evergrowing complexity of information processes and related computations in various subject areas, the exponential growth of processed data volumes coupled with growing meta-data volume rate and complexity; many of the problem domains has a high degree of mutability as their distinguishing feature, and that means one has to upgrade the model of the problem domain and its processes in a timely manner. Under such circumstances a key to information and computation systems efficiency is computational processes decomposition and reuse. That leads to two relatively independent problems:

- (computational) task formulation, preferrably, introducing to the model as few elements as possible, wich requires a high degree of task decomposition and reuse of previously defined constructs;
- definition of at least one solution to each identified task, and again, it is desirable to reuse existing code as much as possible to lessen costs and risks associated with additional testing and integration.

Solving the first of these tasks leads to conceptual integrity of the software and to minimality of the conceptual framework of the subject area, and its increased potential for analysis. Dealign with the second problem leads to various kinds of open architectures which, on one hand, ensure a relatively high degree of reuseability, and on the other, they go a long way towards systems' extensibility and adaptibility.

In the present paper we focus on a specific object (and in particular computational processes) decomposition methodology which leads to a particular design approach. We use this approach to design an environment suitable for an organization to host and execute a family of its (internal and otherwise) applications. The need for such environments occurs when

- a unified, domain-specific (organization-specific) framework for software development and support is required;
- hardware (e.g. high-performance CPUs and graphics cards, perhaps, sensors or other devices as well) and software (e.g. libraries for mathematical caclulations) resources are to be integrated, with a centralized and controlled access to them;
- there particular requirements for scalability, fault-tolerance, performance, flexibility (e.g. configuring a system to run as prototype for demostrations or as a high-performance fault-tolerant cluster).

Our methodology is based on applicative computational systems (ACSs, they include systems based on λ-calculus and combinatory logic, see [1]), because they implement what is called applicative computing: a way of performing computations that considers a computational block as a closed, relatively independent, entity with no free variables. The block are combined so as to create new, more complex computaions.

The rest of the paper is organized as follows. Section 2 contains a brief introduction to ACSs basics and outline the computation decomposition methodology. The section 3 deals with object interaction semantics, and section 4 outlines the design of the mentioned computational environment. In section 5 we present a brief related work review and a conclusion.

2. Computation decomposition

2.1. General considerations

In the present paper the term "applicative computational environment" denotes a set of objects (terms) defined by the corresponding applicative computational system (ACS). ACSs may be thought of as implementations of what is called the applicative computing, which suggest the "compositional way of constructing computational blocks out of previously constructed ones, the blocks being relatively closed and containing no free variables" [14]. An ACS is a formal system, generally based on combinatory logic or λ-calculus, with the following characteristic features:

- objects in ACSs are functional entities with unpredefined arity[1], that shows itself in course of computations;
- the main way of constructing new objects from existing ones is the application operation, where the object-function is applied to the object-argument;
- the function-argument roles are relative to the context in which an object appears, so that even self-applicability is allowed (in untyped systems; in typed systems this is often prohibited).

Various applied systems may be obtained by extending the set of initial objects. In some cases there may be more rules for producing complex objects, e.g. in categorical combinatory logic one would also add rules for pair $[\cdot,\cdot]$ and pairing $\langle\cdot,\cdot\rangle$ constructors. This suggests [11] that an environment's definition comprises two aspects: the set of initial objects and the set of ways of combining objects. A generalized scheme for environment definition may be suggested then [8, 9]:

(i) (*induction basis*)

[1] This applies mostly to complex objects in untyped systems; in practice, when typed system are used, objects arity is either part of objects definition, or may be derived from the object's structure.

   (a) there is an infinite set of variables and a set (possibly, empty) of atomic objects,
   (b) variables and atomic objects are atoms, and all atoms are objects;

(ii) (*induction step*)
   (a) there is a predefined set of term-forming entities pertaining to the meta-level (functions, as they are called), including, at least, the application operation;
   (b) if $A_1, \ldots, A_n$ are objects and σ is an n-ary term-forming function (TFF) then $\sigma A_1 \ldots A_n$ is an object.

Basically, given a (finite or infinite) set of initial (better still, primary) objects and a finite set of 'constructors' (ways of combinig existing objects into more complex ones), one consecutively construct new objects out of existing, simpler one, much like a building is constructed from bricks, panels and blocks, or a machine is assembled from parts, which, in turns, consist of other, smaller, parts. Combinatory logic and λ-calculus each embodies a distinc approach to object construction. That of combinatory logic (CL) follows the idea that any computable objects may constructed from constants exclusively, which ensures that:

a) there are no free variables in an object,
b) and hence, the object is closed in that it does refer to any part of the outside environment, eliminating any otherwise possible side effects.

Variables in CL are used for 'auxiliary' purposes, such as combinatory characteristics representation (i.e., speaking from the perspective of the classical logic, representing the rule of introduction-elimination of the symbol – ACSs are equational systems, and objects conversion is a bidirectional process). In λ-calculus, on the contrary, variables are used quite extensively, and one of the main things that λ-calculus focuses on is variable binding and substitution. Briefly, the $\lambda \cdot . \cdot$ is a binary operator that binds a variable (the first argument) in an object (the second argument); a variable in an object that is not bound is free in that object. λ-calculus, thus, implements explicit control over object's dependencies on the outside environment. On the conceptual level, with respect to objects construction, the difference between CL and λ-calculus is that CL focuses on modeling entities within a predefined framework, whreas λ-calculus makes it possible (e.g. via supercombinators, see [12]) to 'generate' a suitable basis on-the-fly.

2.2. Initial objects

Usually, part of a definition of a specific ACS (which is a formal system) is a specific class of *initial objects* of which all other objects are built using a specified set of constructors (or TFFs, as we call them). It may be so that some the initial objects may be decomposed and 'expressed' using other initial objects; some of the initial objects may be atomic in the sense they cannot be decomposed into smaller units without loss of the general logical integrity. These latter, atomic objects, we will also call *primary*. Initial objects that are not primary we call *secondary*, and all other objects are *derived* [14].

The question of choosing the set of atomic objects is a good one, and two points are worth noting here. The first is that, actually, objects atomicity is a matter of motivated assumption: further decomposition of such objects is either redundant or leads to inconsistencies [13]. Second, though one may choose different suitable sets of atomic objects for different purposes, those sets would share a common computing foundation – for which combinators K and S are good candidates, as argued in [14]. These objects have a sound interpretation on various contexts, and they offer a relatively easy way to constuct relatively complex entities, like lists.

2.3. Process constructors

Our approach, to an extent, was inspired by Dana Scott's flow diagrams [11]. A program may be seen as something, called 'flow diagram', through which information would *flow*, subjected to some transformations in the process. These diagrams are made of elementary, atomic transformation units connected via arcs that do not affect the information flowing through them and that transfer the information from one block's ouptut to another block's input. From the mathematical point of view these elementary transformation blocks are atomic combinators, and connectives represent the composition operation $(f \circ g) = f(gx)$. In ACSs this operation is actually an ordinary combinator (a detailed explanation may be found in section 2.2 in [12]), so that a program is basically a term in an ACS (a combinator, preferrably). In [11] it is also shown that some blocks may function as those 'controlling' (directing) the flow, conditional branching and parallel switch been the most obvious examples; they may be called 'diagram constructors' in that they join several smaller diagrams (branches) into a bigger one. A generalization of that idea was presented in [8], and later a somwhat refined version, concerning both computational processes and data objects, in [9].

Some of the above ideas are commonly used in (functional) programming languages, but hardly so when it comes to subject domain modeling and designing software architecture[2]. However, the idea of using the combinator model for computational processes, both at conceptual level and in implementation, has a number of obvious advatages:

- every process is closed, i.e. has no implicit dependencies or references to the outside environment[3];
- because of that, too, and because a process is, more or less, an ordinary object, an existing process, be it simple or complex, may be re-used to define other processes;
- a process may be passed to another process as an argument, thus allowing computation adjustment at run-time;
- very powerful, yet relatively simple, process algebras may be formulated.

With the idea of computation distribution in mind, in [9] it is suggested that data processing should proceed in three stages. First, the input object is decomposed onto several components and each of them, secondly, is sent to a corresponding specialized (sub)process (which may be either a structural part of the given process, or an independent entity). Finally, the partial results yielded by the 'subprocesses' are aggregated in some way, e.g. 'wrapped' with a TTF into a resulting object. One powerful approach in process construction is defining a set of 'standard' process 'schemes' to combine the already constructed processes into a more complex one.

Let $F : T_1 \to \ldots \to T_m \to T_{m+1}$ be a TFF. For every such TFF let us define a set of projecting objects $\pi^i_F : T_{m+1} \to T_i$ (with $i \in [1, m]$), such that:

$$\pi^i_F(Fa_1 \ldots a_m) = a_i,$$

– for every suitable set of objects $a_1 : T_1, \ldots, a_m : T_m$. Finally, consider the currying 'operator' $\mathbb{H}^n$ with the combinatory characteristics:

$$\mathbb{H}^n f[x_1, \ldots, x_n] = fx_1 \ldots x_n$$

Assume that $\langle f_1, \ldots, f_n\rangle x = [f_1x, \ldots, f_nx]$ holds for every suitable (e.g. type-compatible, according to a certain type system of choice) set of objects $f_1, \ldots, f_n$ and x. Then a process that takes an object $Fa_1 \ldots a_m$ and yields another object $Gb_1 \ldots b_l$ would in general look like this:

[2] We believe that the reasons for that are more or less the same as why functional programming, despite its growing popularity and everdeepening impact on programming culture, is still rejected by many developers as being 'excessively theoretical'.

[3] In real systems, there are many cases when side effects are necessary, e.g. logging; these issues are addressed through a more complicated type system which takes in account side affects.

$$\mathcal{F} = \mathbb{H}^l \mathcal{G} \circ \langle g_1, \ldots, g_l \rangle \circ \langle f_1 \circ \pi_F^1, \ldots, f_m \circ \pi_F^m \rangle \quad (1)$$

or

$$\mathcal{F} = \mathbb{H}^l \mathcal{G} \circ \langle g_1 \circ f_1, \ldots, g_l \circ f_l \rangle, \quad (2)$$

where all f_i are subprocesses that perform a certain pre-processing of the input (on a by-component basis in case of (1), and in case of (2) on a more general one). The responsibility of subprocesses $g_1, \ldots, g_l$ is to yield individual components $b_1, \ldots, b_l$ of the resulting object. Finally, $\mathbb{H}^l \mathcal{G}$ unpacks the tuple $[b_1, \ldots, b_l]$ and produces the final result. Of course, both in (1) and (2) one may add explicit post-processing handlers. An essential point is that every component $b_1, \ldots, b_l$ is evaluated independently of all others, which is rather handy in parallelized environments. If, on the other hand, some of these components depend on the others, an appropriate scheme may also be devised, but it is likely to be more complex and, obviously, yield a lower parallelization rate.

In (1) and (2) every $f_1, \ldots, f_m, g_1, \ldots, g_l$ may be either an internal part of $\mathcal{F}$ or an external process (and may be replaced independently), or it may stand for a 'process variable' $p_i (i = [1, k])$, or contain its (free) occurrence(s). Functional abstraction of $\mathcal{F}$ by such variables will produce a scheme:

$$\widehat{F} = \lambda p_1 \ldots p_k . F$$

Given already constructed processes $\mathbf{p}_1, \ldots, \mathbf{p}_k$, one may construct a new complex process $\widehat{F} \mathbf{p}_1 \ldots \mathbf{p}_k$. Note that, actually, the scheme $\widehat{F}$ may be replaced with another one, provided that all typing constraints are satisfied. Next, very naturally (most type systems would allow that) one can introduce variables ranging over schemes, thus leading to 'scheme constructors'.

Most of those schemes and scheme constructors are likely to be domain-specific and the purpose of their introduction is to facilitate (automate, if possible) the construction of processes in such environments where:

- the number of processes is exceedingly high and/or processes structure is very complex and requires extensive decomposition;
- processes makeup and/or structure is unfixed by the domain nature and adjustments are to be made in a timely fashion.

Futher discussion as to the TFFs' internal structure is to be found in [10].

3. π-calculus as process invocation semantics

Having a number of computational objects in an environment, it is then necessary to define how they are supposed to interact with each other. Convertional ACS give little or none attention to this question, presuming all objects being in the same 'space' and thus able to act 'directly' upon each other. This seems to work anyhow for both the lower level of the operational semantics, which is exatly what the numerous abstract machines are, and on the higher level of program architecture – so far as 'monolithic', (more or less) single-threaded programs of the old days are concerned. However, things get much more complicated for parallel and distribured systems. It is where the π-calculus seems a worthy candidate to help to formalize computational objects' interaction.

π-calculus is an extension of the standard calculus of communicating systems (CCS) allowing advanced process description: sending (and retrieving) named data blocks over named channels, the channels themselves being legal 'data blocks'. Every process in this theory is a sequence of subprocesses that either send or receive data blocks (purely computational processes are not considered). The main connectives to build complex processes out of simpler ones are: sequential execution, parallel execution and conditional branching. The main advantage of π-calculus is ability to describe processes with dynamic structure, i.e. process whose structure is determined in runtime depending on execution progress.

In our approach we consider computational processes which are represented by closed applicative objects (combinators), therefore we suggest using an extension of asynchronous polymorphic π-calculus as process execution semantics. Using semantic of π-calculus brings several benefits and we cover the most important of them in this part of paper. This calculus was designed for modelling interaction between systems [5, 6]. Its models better suits for information process, so it is required adaptation and a certain extension of the standard calculus model. The main aim of using π-calculus is to formalize not only data exchange but processes components execution as well. Also, such an extended model allows defining processes that involve data exchange between several systems. Assuming (1) as process' scheme and $\mathcal{X}$ standing for a certain input object of $\mathcal{F}$, here below are stated the main process execution rules:

(i) Every 'projection' mapping $\pi^i_{\mathcal{F}}$ is translated to $a_i(\mathcal{X}).([|\pi^1_{\mathcal{F}}\mathcal{X}|]\,1|\ldots|[|\pi^m_{\mathcal{F}}\mathcal{X}|]\,m)$, i.e. execution of the process $\mathcal{F}$ begins with distribution, over a number of channels a_i, data required to get input objects components. The statement $[|Kx_1\ldots x_n|]c$ means execution of the combinator K with a mount point c. A mount point is a channel through which a function would return its result. This way of execution functions in processes by modeling int π-calculus is known as "encoding" [4].

(ii) Every subprocess $f_1,\ldots,f_n$ must not contain any data exchange actions, i.e. it must be a purely computational, closed (at least, within $\mathcal{F}$) process, in which case the following holds: $f_i \to i(d).[|f_i d|]i$.

(iii) Every subprocess $g_1,\ldots,g_l$ receives, as its input, a data block containing all the processed components of $\mathcal{X}$ and yields a certain component of the resulting output object. Accordingly,

$$g_i \to t(1(y_1).\bar{t}y_1|\ldots|m(y_m).\bar{t}y_m|t(z_1\ldots z_m).[|g_i z_1\ldots z_m|]o)$$

Note that such an execution scheme is possible in an asynchronous environment only [3].

These rules outline the use of π-calculus as a mechanism underlying execution of our constructive processes. If all computations are performed by a single actor – executing the provided process is identical to ordinary reduction of a correspondent combinator. Thisngs are different in heterogeneous environments. For example, if projection functions or subprocesses are executed in different systems.

The first benefit from implementing π-calculus semantics is implementation of complex dynamic routing of computational flows in a computational network. We face this problem of routing requests over the network when various nodes are able to execute several different types of projection functions or subprocesses. In that case the network has to be able to find the right (sub)set of nodes to execute th computational task in a corrent and, and, preferredly, optimal (from the resource consumption) way. Process execution model must deal with two main problems: selection of the target nodes to participate in the process execution and load-balancing. The second benefit from using π-calculus is a runtime type checking for computations, hosted by different systems. These checks can be implemented by using typed π-calculus. This version of calculus assigns to every name of term some type constant and forbids interaction if types of names involved into interaction don't suit each other.

We suggest the following changes in the π-calculus for implementing routing for different systems. Let $a_i(\mathcal{X}).([|\pi^1_{\mathcal{F}}\mathcal{X}|]\,1|\ldots|[|\pi^m_{\mathcal{F}}\mathcal{X}|]\,m)$ be a process for projecting input data. This is a correct process when all projecting functions are hosted by single computation node, described by channel a. Computation process doesn't need routing in that case – all requests may be safely sent via the channel a. In case of a distributed system (meaning that various parts of the system are separated, isolated from each other, not necessarily be means of a physical network) the previous scheme tranforms into the following:

$$\left(a_{i_1 \ldots i_n}(x_1).(\left[\left|\pi_{\mathcal{F}}^{i_1} x_1)\right|\right] i_1 | \ldots | \left[\left|\pi_{\mathcal{F}}^{i_n} x_1)\right|\right] i_n\right) | \ldots | \left(a_{i_m \ldots i_p}(x_1).(\left[\left|\pi_{\mathcal{F}}^{i_m} x_m)\right|\right] i_m | \ldots | \left[\left|\pi_{\mathcal{F}}^{i_p} x_m)\right|\right] i_p\right)$$

In order to manage this 'zoo' of channels a_i, we will introduce a centralized routing channel whose responsibility is to select valid targets for every particular message (which, obviously, involves some run-time type-checking, as was mentioned above). We define two rule for target (channel) selection. First, we use any meta-information supplied for the input data (basically, its type; note that nearly any kind of metadata may be fit into a data type, given the type system is strong enaugh, like that with dependent types; but this is far too complex a question to discuss here). We introduce a combinator $Tx = [i_n, \ldots, i_k]$ that, given an obect (and its type along it), produces a list of required 'projections'. There is also another required combinator is L which does the actual consistency checking, so that a given set of channels is able indeed to process the given set of subcomponents of a given input object. This model is implemented as such:

$$r(x).ve\,([|Tx|]e|e(y_1, \ldots, y_n)\,[|Ly_1 \ldots y_n a_{i_1} \ldots a_{i_n} a_{i_m} \ldots a_{i_k}|]\,a_{i_1} \ldots a_{i_n} a_{i_m} \ldots a_{i_k})$$

The second rule uses external knowledge about input objects and selects projection channels by the name of the passed input. This rule is applied when a computation can take inputs from a finite set of objects. It doesn't require the computation performed by the router, but fails on objects not from the expected set. Implementation of the router uses construction of process on the case-by-case basis.

4. Computational environemnt design

The usage of various kinds and variations of service-oriented atchitectures (SOA) is beneficial in that it helps in addressing the problems we stated in the introductory section – and besides, improves software scalability, naturally supporting distributed software configurations. Note that in this paper we stick to a wider understanding of SOA which is not limited to "classical" web-services (usually hosted by web-servers and relying on SOAP and the likes of it for interaction) but includes other options as well. Services interaction may be carried out in variuos ways: from a "local" bus like DBus to standard technologies like WCF[4] or RMI[5] to custom TCP/IP-based protocols implementations.

We assume that applications are build, in a sense, in a 'task-oriented' manner, meaning that every application has a definite list of (formally written) tasks that it should perform. Such an application contains a data and meta-data object model (basically, a list of entities found in the subject area, their attributes and relationships) and computational processes model. An advantage of ACSs is that they provide a uniform framework suitable to represent both kind of models. We see a task analogous to a request to a database written in relational calculus: one defines what is it exactly that he or she wants to get, leaving to the system to figure out the best way to do it. In case of a relational database, such a request is translated to a computational process consisting of elementary relational operations (and, perhaps, additional operations as well).

[4] Windows Communication Foundations is .NET technology to allow interaction between objects located in different, including remote, processes. Its main feature is that it offers developers high-level interfaces that abstract away many details of working with low-level network protocols.

[5] Remote Method Invocation is, roughly speaking, the Java analogue for WCF.

Consequently, the environment in question must primarily contain a number of elementary domain-specific processes, each equipped with some meta-information as to the kind of transformation the process does. At the very least, it should be a list of 'key words' – references to concepts from the subject domain model. As an option, a more interesting one though, seemingly, yet poorly explored, this meta-information is to be included in the process type (using Curry-Howard isomorphism). When a user invokes a function in an application, a 'request', which is basically a term in λ-calculus (or CL) either assembled on-the-fly or defined and stored in advance, is sent to the environment for execution. This term is translated into another one that uses only those elementary computational processes that are defined in the environment. The environment then acts as an abstract machine and reduses this term, not unlike how the conventional abstract machines reduce terms representing programs in functional languages.

The environment is a set of nodes forming a connected graph (each pair of connected nodes may be connected either with undirected edges in case of a symmetric protocol, like TCP, or with two directed arcs in case of use of assymetric protocols, as may be the case with nodes hosted via web services). The basic functionality of all nodes is only that they can establish connection, interchange with messages of various kinds and break the connection. On top of this, two mechanisms are to be implemented: task execution and event propagation. Task execution consists in selecting a set node of nodes capable of executing a specific task, chosing one (usually) or several (e.g. for service tasks) nodes, sending the task to these nodes and then ensuring that the result is produced and stored for further use. Event propagation is a mechanism to inform an intended set of nodes that a particular situation has occured (e.g. a specific node has gone offline, which might mean that a particular application or part of its functionality is no longer available). Above that, a node must run one or more loadable modules (plugins, of sorts) and according to this play one or more roles in the network, of which the main are:

- transport: routing messages throughout the network, ensuring that a message may be delivered to any given node, including plotting optimal paths;
- decomposer/composer: a task is an object constructed from atoms using TFFs and the same applies to the result of computations, so executing a task means, roughly speaking, invoking the initial computing objects with some parameters in some order and combining the partial results;
- computational: the node actually hosts a number of initial computing processes;

The set of transport node should form a connected graph and is called the transport system of the network. On top of this functionality implemented are: task and result sending, Besides these three roles, a number of other roles may be identified, important as they may be in practice, they are secondary to the environment's functioning, such as:

- load-balancing between node;
- gateway to enable interaction with external systems;
- caching;
- support to allow remote control over nodes, self-diagnostics, and other usefull features concerning internal management.

Another important notion is that of a resource. Resource is an object with a 'mobility policy' assigned to it that determines whether this object may or may not be copied, or moved to, or accessed from another node. There are two basic applications to this. First, we may or may not transfer data between two nodes, be it due security reasons or simply availability of storage space to accomodate the data on the target node. The other application is that a computing object, which is basically a piece of (compiled) executable code, may or may not be transferred to another node – again, security considerations and code dependencies availability may be concerned. The basic framework for formal reasoning about resources is to be found in [7], but the actual formal system is a matter of future work.

5. Conclusion

There is a number of approaches to model business/information/computational process. Many are based on network, state-diagram (e.g. Petri Coloured Nets), event-oriented models, and often either lack proper formal basis or this basis is simply ignored, the being used more for visualizing purposes. These models may be useful for subject areas analysts, but they are of littly help when it comes to software design, complicating architecture development and increasing the probability of design errors (which are most dangerous to software development). To counter this problem, a number of process algebras [2] were developed, covering a range of aspects of process semantics execution issue, including temporality.

Using ACSs offers two major advantages: it is a very natural tool for objects structure studying and represntation, and it offers a conceptual basis that is much easier to translate to software architecture and then to program code, thus being a very promissing option for filling the two big gaps: the one between sibject area model and architecture, and the one between the architecture and the source code.

In this paper we described an approach to processes structure and, to an extent, execution semantics modeling. For process structure representation we suggested using an approach based on an applicative computational systems extension that explicitly deals with objects constructors – TFFs. This approach has an additional advantage that, using it, an 'environment algebra' may be developed [10], which may be used to address various problems of semantic interoperability between different applications. To counter the problem of (possibly, remote) process interaction we suggested to use the π-calculus and showed how ACSs terms are represented and executed in it.

We suggested an approach and an architecture outline for an environment in which applications may be continuously constructed, debugged and executed. The environment is scalable and may equipped with many useful features, uncluding fault-tolerance. The environment is (distributed) heterogeneous computational network, nodes playing one or more roles; the network's functioning consists in solving 'tasks' wich are decomposed into atomic components, distributed over the nodes suitable for each atomic task, and the partial results are then aggregated into the 'answer' of the orignal task. Compared to the previous version, described in [9], the architecture has been slightly changed to increase overall consistency.

An experimental prototype is presently under development, and we hope in some near future to be able to publish experimental results.

References

[1] Barendregt, H., 2012. *The Lambda Calculus, its Syntax and Semantics.* Studies in Logic, Mathematical Logic and Foundations, vol. 40.

[2] Bergstra J.A., Ponse A., Smolka S.A., 2001. *Handbook of Process Algebra.* (Elsevier Science)

[3] Boudol G, 1992. *Asynchrony and the pi-calculus.* Rapport de Recherche RR-1702, INRIA Sophia Antipolis.

[4] Milner R., 1992. Functions as processes. *Mathematical structures in computer science*, vol. 2, no. 02, pp. 119141

[5] Milner R, 1993. *The polyadic π-calculus: a tutorial.* (Springer)

[6] Milner R., Parrow J., Walker D, 1992. A Calculus of Mobile Processes. *Information and computation*, vol. 100, no. 1, pp. 140

[7] Murphy T, Crary K, Harper R, Pfenning F, 2004. *A Symmetric Modal Lambda Calculus for Distributed Computing.* (Carnegie Mellon University, Computer Science Department. Paper 465.
Url: `http://repository.cmu.edu/compsci/465`)

[8] Roslovtsev, V.V., Luchin, A.E., 2009. Concept of Higher-Order Applicative Computational Environment. *Proc. of the 11th Int. Workshop on Computer Science and Information Technologies CSIT'2009*, pp 48-53.

[9] Roslovtsev V, Shumsky L, Wolfengagen V, Bohulenkov A, Sakhatskiy A, 2013. Applicative Approach to Information Processes Modeling. Towards a Constructive Information Theory. *Proc. of the 15th Int. Conf. on Enterprise Information Systems ICEIS2013* (Anger, France) pp 221-227.

[10] Roslovtsev, V.V., 2013. Applicative Computational Environments Construction. *SWorld: Conf. Series, Issue 2, Volume 6* (also avaialble at `http://sworld.com.ua/index.php/ru/technical-sciences-213/informatics-computer-science-and-automation-213/18159-213-745`)

[11] Scott D., 1971. The Lattice of Flow Diagrams. *Lecture Notes in Mathematics*, Vol. 188. (Springer) pp 311-366

[12] Wolfengagen V.E., 2003. Combinatory logic in programming. *Computations with objects through examples and exercises.* 2nd ed. (Moscow, Russia: "Center JurInfoR" Ltd)

[13] Wolfengagen W.E., 2004. *Methods and Means for Computations with Objects. Applicative Computational Systems.* (Moscow, Russia: "Center JurInfoR") [In Russian]

[14] Wolfengagen V.E., 2010. *Applicative computing. Its quarks, atoms and molecules.* Ed. Dr. L.Yu. Ismailova. (Moscow, Russia: "Center JurInfoR")

Using multipliers analysis in order to get another perspective related to the role of ICT sectors in national economy of Indonesia: 1990-2005

Ubaidillah Zuhdi

Department of Industrial Administration, Tokyo University of Science, Noda 278-8510, Japan
School of Business and Management, Institut Teknologi Bandung, Bandung 40132, Indonesia

ubaidillah.zuhdi@gmail.com

Abstract. The purpose of this study is to get another perspective related to the role of Information and Communication Technology (ICT) sectors in national economy of Indonesia. The period of analysis of this study is 1990-2005. This study employs Input-Output (IO) analysis as a tool of analysis. More specifically, this study uses simple output multipliers method in order to achieve the purpose. Comparison with previous study is conducted in order to get the objective of this study. Previous study, using Structural Decomposition Analysis (SDA), showed that ICT sectors did not have an important role in Indonesian national economy in above period. The similar results also appear in this study. In other words, from this study, another perspective related to the role of these sectors in Indonesian national economy in analysis period is not found.

1. Introduction

In recent times, technology has become important thing in society. As evidence, every aspect of human life has a relation with technology. For example, mining industries use heavy technologies in doing their activities. Other example could be shown on agricultural sectors. These sectors employ technologies in conducting cultivation activities. Besides, technology changes have the potential to improve the service delivery efficiency within the market of career support, to increase existing services, and to generate career support's new paradigms [1]. These facts show that the role of technology in improving human life cannot be underestimated in this age [2].

One kind of technology that available in society is an Information and Communication Technology (ICT). ICTs have fundamental roles in modern economic growth and development. These are 1) to reduce the costs of transaction and thereby to increase productivity, 2) to offer a quick connectivity, 3) to replace for other such as physical travel, 4) to add choice in the marketplace and to provide access to goods and services which are unavailable, 5) to extend the scope of area of potential markets, and 6) to distribute all kinds knowledge and information [3]. [4] argued that economic growth theories forecast that investment in ICT drives the growth of economy. [5] explained that economists recognize that ICT and the Internet-related topics are significant parts of activity of economy. These arguments show that ICT has an important role in this age and cannot be separated from human life.

There are many previous studies discussed about ICT. For example, [6] focused on developing a hierarchy of applications and definitions of ICT. [7] examined the relation of ICT and farming community in Uttarakhand state, Indian Himalayan region and explored the ways of integrating ICT with the existing system of public extension. [8] analyzed the impacts of ICT and Foreign Direct Investment (FDI) on growth of productivity of observed countries. Their analysis was based on panel data of 42 developed and developing countries on the period between 1993 and 2001. On the other hand, [9] explored the ways in encouraging Indonesian ICT sectors using Input-Output (IO) analysis.

However, the study compares the results of previous study, which the topic of discussion is the role of ICT sectors in national economy of specific country, is still lack. The kind of study, actually, is important since it can describe more detail about this role. Besides, the kind of study will give another point of view related to this role. [10] has tried to conduct the kind of study which the object was Japan, one of developed country. The study focuses on this topic, which the object is a specific developing country, actually, is also needed. This study is conducted in order to fulfill this gap.

This study focuses on Indonesia, one of developing country in the world. The purpose of this study is to get another perspective related to the role of ICT sectors in national economy of Indonesia. The period of analysis of this study is 1990-2005. This study employs IO analysis as a tool of analysis. More specifically, this study uses simple output multipliers method in order to achieve the objective. This paper is arranged as follows. Chapter 2 describes the methodology of this study while the results of calculation and analysis of these results are explained in chapter 3. Chapter 4 explored the conclusions of and suggested further researches from this study.

2. Methodology

In this study, IO tables of Indonesia for 1990, 1995, and 2005 are used as data. Following explanation describes the methodology of this study. First, before conducting the calculation, the process of aggregation for above tables is done. The aim of this process is to get the compatibility among different periods. 159 industrial sectors of Indonesia are obtained from this process. ICT sectors of Indonesia used in this study are described from outcome of aggregation process afterwards. Table 1 shows these sectors. The process of adjustment, actually, in order to get the proper data, is addressed to 2005 IO table of Indonesia before conducting the aggregation process.

The previous study have similar topic with this study, actually, is needed in achieving the objective of this study. Therefore, as a second step, we explore the study conducted by [11] as a compared previous study. This previous study analyzed the role of ICT sectors in national economy, refers to the national economic structural changes, of Indonesia on the period between 1990 and 2005 using Structural Decomposition Analysis (SDA), one of analysis tool in IO analysis. In other words, the results of previous study were based on the calculation using this analysis. The data used in their study were same with the data of this study, namely IO tables of Indonesia for 1990, 1995, and 2005. The aggregate and adjustment processes conducted by them are also same with this study, as well as the ICT sectors used.

As described in previous explanation, simple output multipliers method is employed as a tool of analysis. This method is chosen since it can analyze the total production value of all sectors in economy that is needed in satisfying a monetary unit's worth of final demand for output of particular sector [12]. In other words, this method can analyze the effect of one unit increasing in particular sectors' final demand on national economy, the total value of production made by all industrial sectors in economy in order to fulfill the escalation. Therefore, the third step in this study is to conduct the calculation using above multipliers method.

As mentioned in previous explanation, method of simple output multipliers is a part of IO analysis. Therefore, the concepts used in this analysis are also applied in this method. The detail of the concepts is described by [12]. According to them, the following equation is used in the simple output multipliers calculation:

$$m(o)_j = \sum_{i=1}^{n} l_{ij} \quad (1)$$

where $m(o)_j$, n, and l_{ij} are simple output multiplier for sector j, the numbers of sectors, and sector-to-sector multipliers matrix, respectively. Conclusions of this study and suggestions related to the further researches are described on final step.

Table 1. ICT sectors of Indonesia used in this study.

No.	Sector Number	Sector Name
1	135	Construction and installation on electricity, gas, water supply, and communication
2	146	Communication services

(Source: [11])

3. Results and analysis

Tables 2 and 3 describe the results of previous study, namely the top five sectors influencing structural changes in Indonesian national economy, refers to the national economic structural changes, in 1990-1995 and 1995-2005, respectively. ICT sectors, actually, do not exist in these tables. Therefore, according to [11], ICT sectors did not have important role in national economy of Indonesia on the period between 1990 and 2005.

On the other hand, Tables 4, 5, and 6 show the top five industrial sectors of Indonesia viewed from values of simple output multipliers on 1990, 1995, and 2005, respectively. ICT sectors, once again, do not include in the tables. This fact describes that, though using simple output multipliers method, ICT sectors did not have important role in national economy, refers to the total value of production made by all industrial sectors in economy in order to fulfill the increase of one unit in particular sectors' final demand, of Indonesia on the period between 1990 and 2005.

Sector of machinery and apparatus, industrial sector of Indonesia which the number is 112, appears in Tables 4, 5, and 6. The values of simple output multiplier of this sector on these tables are 2.94, 2.78, and 3.00, respectively. This phenomenon indicates that, from the point of view of simple output multipliers analysis, this sector has an important role in national economy of Indonesia from 1990 through 2005.

Table 2. Top five sectors influencing structural changes in the Indonesian national economy, 1990-1995.

No.	Sector Number	Sector Name	Value (100 Million Rupiah)
1	137	Trade	487268.52
2	132	Residential and non-residential buildings	306269.37
3	150	Business services	238193.25
4	138	Restaurant	212883.49
5	147	Banking and other financial intermediaries	202307.68

(Source: [11])

Table 3. Top five sectors influencing structural changes in the Indonesian national economy, 1995-2005.

No.	Sector Number	Sector Name	Value (100 Million Rupiah)
1	137	Trade	4289107.11
2	132	Residential and non-residential buildings	2369657.07
3	138	Restaurant	1599208.04
4	134	Public work on road, bridge, and harbor	1437422.16
5	31	Crude oil	1396709.27

(Source: [11])

Table 4. Top five industrial sectors of Indonesia viewed from values of simple output multipliers, 1990.

No.	Sector Number	Sector Name	Value of Simple Output Multiplier
1	97	Plastic products	2.97
2	112	Machinery and apparatus	2.94
3	66	Made up textile goods except wearing apparel	2.78
4	68	Wearing apparel	2.69
5	111	Prime movers engine	2.68

Table 5. Top five industrial sectors of Indonesia viewed from values of simple output multipliers, 1995.

No.	Sector Number	Sector Name	Value of Simple Output Multiplier
1	124	Aircraft and its repair	2.79
2	112	Machinery and apparatus	2.78
3	66	Made up textile goods except wearing apparel	2.74
4	68	Wearing apparel	2.69
5	111	Prime movers engine	2.62

Table 6. Top five industrial sectors of Indonesia viewed from values of simple output multipliers, 2005.

No.	Sector Number	Sector Name	Value of Simple Output Multiplier
1	112	Machinery and apparatus	3.00
2	124	Aircraft and its repair	2.99
3	111	Prime movers engine	2.81
4	127	Musicals instruments	2.61
5	66	Made up textile goods except wearing apparel	2.59

The similar phenomenon also appears on other sectors. These are made up textile goods except wearing apparel and prime movers engine sectors. The values of simple output multiplier of made up textile goods except wearing apparel sector for above tables are 2.78, 2.74, and 2.59, respectively. On the other hand, simple output multiplier values of prime movers engine sector are 2.68, 2.62, and 2.81.

Generally, the composition of sectors in Tables 4, 5, and 6 is similar. This fact indicates that, from the point of view of simple output multipliers, few sectors, in 1990-2005, have dominant positions in Indonesian national economy. More specifically, the national economy of Indonesia on that period did not drastically change.

There are some possibilities in describing above phenomenon, the domination of few sectors in national economy of Indonesia from 1990 through 2005 viewed from simple output multipliers approach. First, these sectors were very attractive on this period. In other words, in 1990-2005, the movements of these sectors in market activities of Indonesia were very mobile so they could control the market. Consequently, in this period, other sectors contributed many inputs to and depended on them when demands came from market.

The second possibility is because of trend of market of Indonesia on the period between 1990 and 2005. Indonesian market in this period might focused on textile products. This trend, actually, would increase the demands for the sectors produce these products. Besides, the other sectors contribute indirectly to this trend, such as the sectors provide the tools and machines for producing textile products, also gain the advantage. Consequently, both sectors have dominant positions in market.

The third possibility is national economy of Indonesia in above period tended to stagnant. The stagnation in national economy of one country, actually, causes its market situation is immovable. Consequently, in 1990-2005, the demands from and the composition of main players in Indonesian market did not drastically change.

Previous explanations show that, using simple output multipliers method, ICT sectors did not have important role in national economy of Indonesia from 1990 through 2005. In other words, in this study, comparing with previous study, the other perspective related to the role of these sectors in Indonesian national economy in above period does not appear. This phenomenon is interesting to be discussed.

Both methods, SDA and simple output multipliers methods, actually, have different characteristics. As illustration, former method focuses on the change of gross output of industrial sectors between two periods while latter one focuses on specific period. Besides, the values coming from SDA depend on values of decomposition factors while the values of simple output multipliers method relied on final demands of sectors.

The decomposition factors influence the results of SDA are divided into four parts. These are 1) domestic final demand expansion, 2) export expansion, 3) import substitution, and 4) technical coefficient changes. Moreover, expansion of domestic final demand could be elaborated into the four sub-parts, namely 1) household consumption expenditure, 2) consumption expenditure of private, 3) government consumption expenditure, and 4) capital formation and changes in inventory [13]. In other words, seven decomposition factors were employed in previous study. On the other hand, the method of simple output multipliers only has one exogenous variable, namely final demand. Therefore, previous study considers more factors than current study.

Although the characteristics of above both methods are different, the results of calculation using these methods are similar. This phenomenon underlines that ICT sectors did not have an important role in national economy of Indonesia on the period between 1990 and 2005. The growth of these sectors in this period, actually, could describe above situation.

Based on the results of previous calculations, the development of Indonesian ICT sectors in 1990-2005 was not rapid. This condition, actually, caused the movements of these sectors in market were relatively slow. Consequently, the demand for and the dependency of other sectors on these sectors were low. This domino effect, actually, explains why ICT sectors did not have an important role in national economy of Indonesia in this period.

Figures 1, 2, and 3 describe the values of simple output multipliers of all Indonesian industrial sectors on 1990, 1995, and 2005, respectively. Based on information in these figures, the patterns of these values on these years are similar. This phenomenon, actually, indicates that the economic activities of Indonesia in 1990-2005 did not drastically change. This phenomenon, once again, underlines that ICT sectors did not have an important role in national economy of Indonesia in this period.

Table 7 shows the simple output multipliers of ICT sectors of Indonesia from 1990 through 2005. Based on the information in this table, two patterns of simple output multiplier appear on these sectors. These are decreasing-increasing and increasing-decreasing patterns.

The sector has first pattern, decreasing-increasing, is construction and installation on electricity, gas, water supply, and communication. On the other hand, the second pattern, increasing-decreasing, is owned by communication services sector. Therefore, in 1990-2005, ICT sectors of Indonesia had different pattern of simple output multiplier. This difference indicates that the activities conducted by both sectors in market in this period were different. Moreover, this difference also indicates that the dependency level of other sectors on each ICT sector was not same in analysis period. Figure 4 describes more detail about these patterns.

Table 7. Simple output multipliers of Indonesian ICT sectors.

No.	Sector Number	Sector Name	1990	1995	2005
1	135	Construction and installation on electricity, gas, water supply, and communication	2.48	2.34	2.49
2	146	Communication services	1.52	1.53	1.41

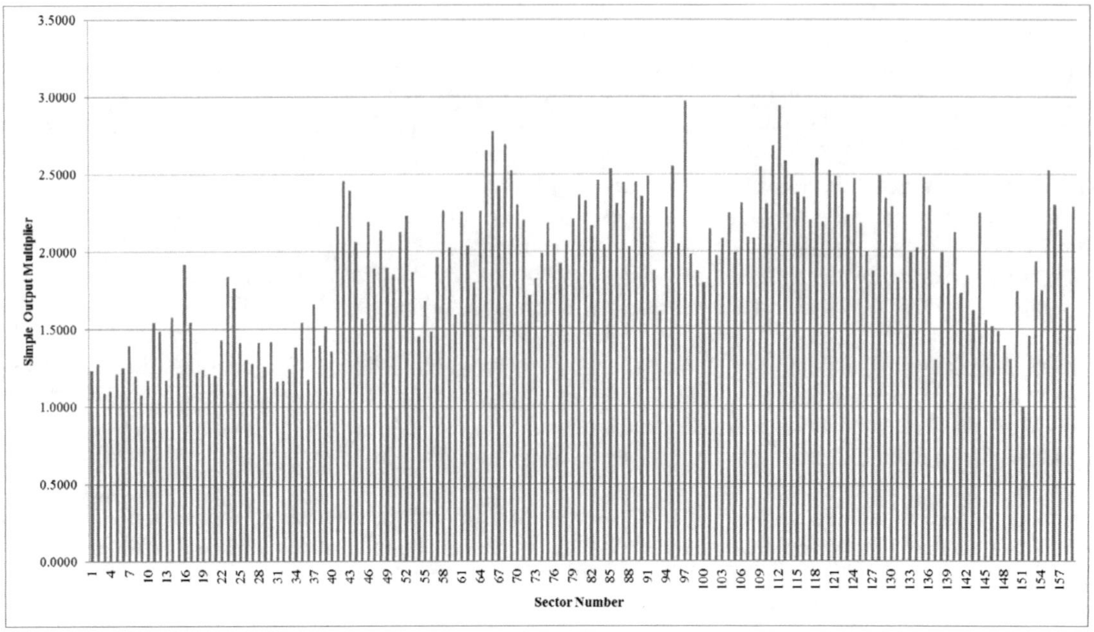

Figure 1. The values of simple output multipliers of all industrial sectors of Indonesia, 1990.

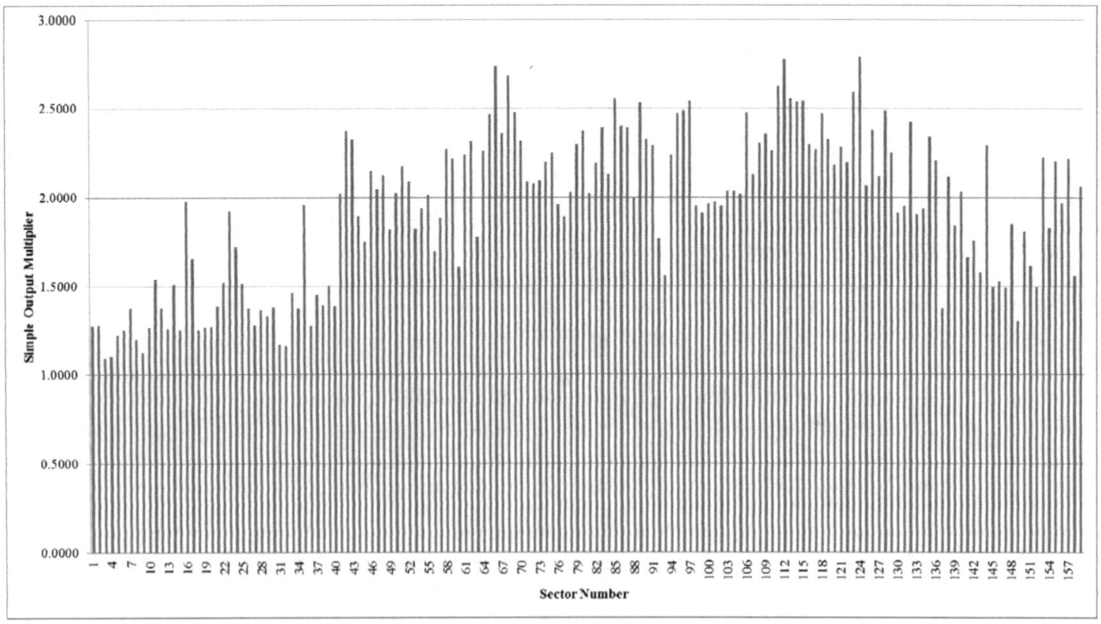

Figure 2. The values of simple output multipliers of all industrial sectors of Indonesia, 1995.

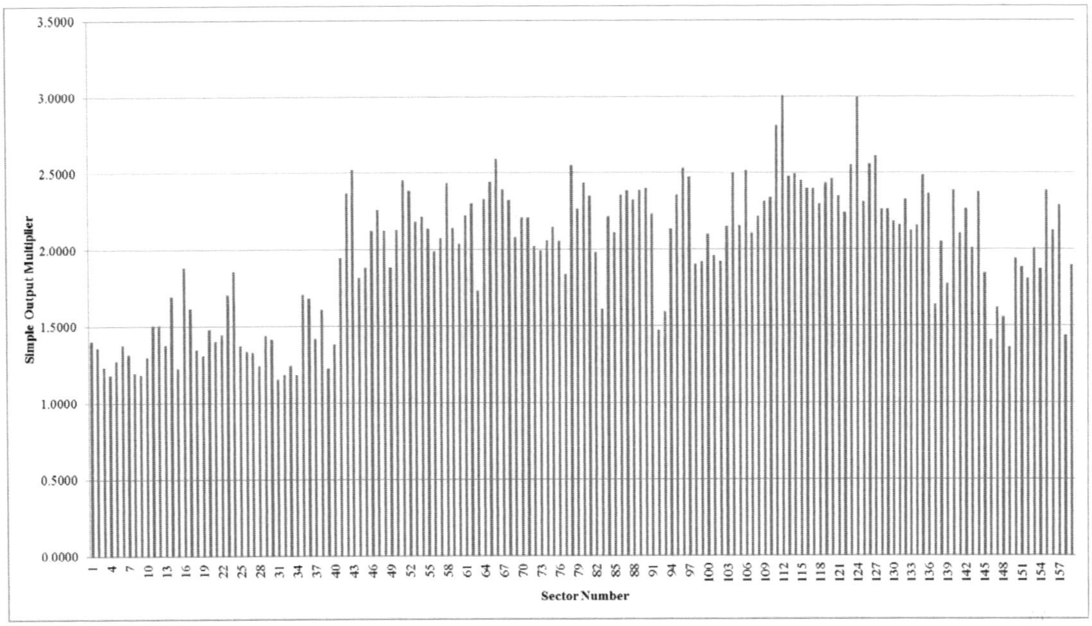

Figure 3. The values of simple output multipliers of all industrial sectors of Indonesia, 2005.

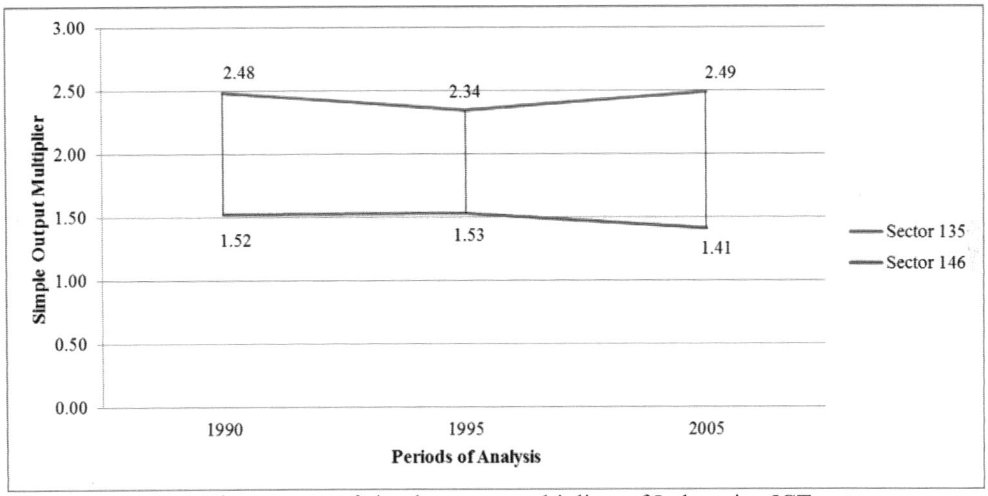

Figure 4. The patterns of simple output multipliers of Indonesian ICT sectors.

4. Conclusions and further researches

This study tries to get another point of view related to the role of ICT sectors in national economy of Indonesia on the period between 1990 and 2005 using simple output multipliers method. Comparison with previous study is conducted in order to get this perspective. Previous study, using SDA, showed that ICT sectors did not have an important role in Indonesian national economy in this period. The similar results also appear in this study. In other words, from this study, another perspective related to the role of these sectors in Indonesian national economy in analysis period is not found.

The other methods, such as income / employment multipliers, value-added multipliers, and statistical analysis methods, however, are not applied in this study. Using these methods in this topic, actually, will generate deeper analysis related to the characteristics of ICT sectors of Indonesia. Consequently, the role of these sectors in national economy of Indonesia can be explored more detail. Therefore, using other methods in this topic is a suggested further research from this study.

The comparison between two or more countries, such as developed and developing countries, related to the role of ICT sectors in national economy, viewed from simple output multipliers method, is also not conducted in this study. This comparison, actually, will show the differences and similarities of characteristics of these sectors in compared countries. This comparison will also describe the sectors have an important role in national economy of each compared country. Based on these reasons, this study also suggests this comparison as further research.

References

[1] Hooley T, Hutchinson J and Watts A G 2010 Enhancing choice? The role of technology in the career support market *http://www.ukces.org.uk/assets/ukces/docs/publications/enhancing-choice-the-role-of-technology-in-the-career-support-market.pdf* Accessed 6 January 2014

[2] Langmia K 2005 The role of ICT in the economic development of Africa: The case of South Africa *International Journal of education and Development using Information and Communication Technology* **2** 144156

[3] Kramer W J, Jenkins B and Katz R S 2007 The role of the information and communications technology sector in expanding economic opportunity *http://www.hks.harvard.edu/m-rcbg/CSRI/publications/report_22_EO%20ICT%20Final.pdf* Accessed 6 January 2014

[4] Nasab E H and Aghaei M 2009 The effect of ICT on economic growth: Further evidence *International Bulletin of Business Administration* **5** 4656

[5] Mohan R 2007 Determinants of ICT expenditure using logit transformation for proportion data analysis *Journal of Information Technology Impact* **7** 145158

[6] Zuppo C M 2011 Defining ICT in a boundaryless world: The development of a working hierarchy *International Journal of Managing Information Technology* **4** 1322

[7] Kameswari V L V, Kishore D and Gupta V 2011 ICTs for agricultural extension: A study in the Indian Himalaya region *The Electronic Journal of Information Systems in Developing Countries* **48** 0112

[8] Dimelis S P and Papaioannou S K 2009 FDI and ICT effects on productivity growth: A comparative analysis of developing and developed countries *European Journal of Development Research* **22** 7996

[9] Zuhdi U, Putranto N A R and Prasetyo A D 2014 Encouraging information and communication technology sectors using input-output approach: The case of Indonesia *Advanced Science Letters* **20** 199202

[10] Zuhdi U 2014 The other perspective related to the role of information and communication technologies sectors in national economy: The case of Japan *Advanced Science Letters* **20** 483486

[11] Zuhdi, U, Mori S and Kamegai K 2012 Analyzing the role of ICT sector to the national economic structural changes by decomposition analysis: The case of Indonesia and Japan *Procedia-Social and Behavioral Sciences* **65** 749754

[12] Miller R E and Blair P D 2009 *Input-Output Analysis: Foundations and Extensions* (Cambridge: University Press)

[13] Zuhdi U 2012 Analyzing the influence of creative industry sector to the national economic structural changes by decomposition analysis: The case of Indonesia *Procedia-Social and Behavioral Sciences* **65** 980985

AUTHOR INDEX

AUTHOR INDEX